P9-CQM-014

Gallium Arsenide Technology

Volume 2

Gallium Arsenide Technology

Volume 2

David K. Ferry
Editor-in-Chief

HOWARD W. SAMS & COMPANY
A Division of Macmillan, Inc.
11711 North College, Carmel, Indiana 46032 USA

FIRST EDITION
FIRST PRINTING — 1989

International Standard Book Number: 0-672-22555-7
Library of Congress Catalog Card Number: 85-50442

Acquisitions Editor: Greg Michael
Development Editor: Jennifer Ackley
Manuscript Editor: J. L. Davis
Indexer: Northwind Editorial Services
Compositor: Beacon Graphics

Printed in the United States of America

Contents

Preface

Since the first book of this series, we have seen a dramatic growth in the fabrication of actual integrated circuits in gallium arsenide. The Defense Advanced Research Projects Agency (DARPA) has driven this growth by the establishment of a series of "foundries" for various technologies utilizing this material and heterojunctions of this material and its alloys. In addition, a sizable program has begun to push the development of monolithic microwave integrated circuits. This growth has been based upon the realization that we are now in a position from which GaAs ICs are now feasible. Yields have become quite acceptable in various foundries, and the variety of chips available has grown. Indeed, we should soon see the emergence of GaAs microprocessor chip sets operating at clock rates of more than 100 MHz.

In this volume, we have tried to bring together a series of chapters dealing with the materials characterization as well as the integration of the technology. In addition, we include chapters on various surface/interface effects such as oxides, contacts, and Schottky barriers. Several approaches to modeling the various devices are also included. These chapters provide an accumulated knowledge of the current state of the art in GaAs processing, characterization, integration, and understanding. Once again, the authors are drawn from industry, academia, and government laboratories and each is an expert in his or her field. The work in the book is of course due to these individual authors and to them goes the credit (and my thanks) for the resulting quality of the book.

David K. Ferry

Contributors

Stephen G. Bishop received the PhD in physics from Brown University, Providence, Rhode Island, in 1965. After one year as a postdoctoral research associate at Brown University and two years as a postdoctoral research associate at the Naval Research Laboratory (NRL), Washington, D.C., he joined the Naval Research Laboratory as a research physicist. During his tenure at NRL he also served as a visiting scientist and physicist in the United Kingdom and West Germany. In 1980 Dr. Bishop became head of the Semiconductors Branch, Electronics Science and Technology Division, NRL. He was also an adjunct professor in the physics departments of the State University of New York at Buffalo and the University of Utah, Salt Lake City. In 1989 he joined the faculty of the University of Illinois at Urbana-Champaign as a professor in the Department of Electrical and Computer Engineering and as the director of the Center for Compound Semiconductor Microelectronics. His research interests encompass many aspects of the optical and electronic properties of semiconductors. He has coauthored 120 research publications and made an equal number of presentations at APS meetings and international conferences.

Dr. Bishop is a fellow of the American Physical Society and a member of Phi Beta Kappa, the Sigma Xi Society, the IEEE, the Materials Research Society, and the AAAS.

David M. Bloom was born on October 10, 1948 in Brooklyn, New York. He received the BS degree in electrical engineering from the University of California, Santa Barbara, in 1970 and the MS and the PhD degrees in electrical engineering from Stanford University in 1972 and 1975, respectively.

From 1975 to 1977 he was employed by Stanford University as a research associate. During this period he was awarded the IBM Postdoctoral Fellowship. From 1977 to 1979 he was employed by Bell Telephone Laboratories, Holmdel, N.J., where he conducted research on optical phase conjugation, ultrafast optical pulse propagation in fibers, and tunable color-center lasers. From 1979 to 1983 he served on the staff and later as a project manager at Hewlett-Packard Laboratories, Palo Alto, California. While there he conducted and managed research on fiber-optical devices, high-speed photodetectors, and picosecond electronic measurement techniques. In late 1983 he joined the Edward L. Ginzton Laboratory, W. W. Hansen Laboratories of Physics, Stanford University, where he is currently an associate professor of electrical engineering. His current research interests are ultrafast optics and electronics.

He was awarded the 1980 Adolph Lomb Medal of the Optical Society of America for his pioneering work on the use of nonlinear optical processes to achieve real time conjugate wavefront generation. In 1981 he was elected a fellow of the Optical Society of America in recognition of his distinguished service in the advancement of optics. He was the 1985 IEEE LEOS traveling lecturer. In 1986 he was elected a fellow of the Institute of Electrical and Electronics Engineers for contributions to nonlinear optics and ultrafast optoelectronics.

David K. Ferry is a Regent's Professor and Chairman of the Department of Electrical Engineering at Arizona State University. He heads an active research group studying the physics/electronics of ultrasmall microelectronic devices, both theoretically and experimentally with electron-beam lithography. He is the author or coauthor of some 270 scientific works.

Dr. Ferry received his BS and MS degrees from Texas Tech University and a PhD degree from the University of Texas, Austin. His career includes fellowships at the University of Vienna and the Boltzmann Institute for Solid State Physics in Vienna. He has held faculty positions previously at Texas Tech University and Colorado State University, as well as having worked at the Office of Naval Research. From 1983 to 1989 he directed the Center for Solid State Electronics at ASU.

Dr. Ferry has been active in organizing several NATO Advanced Study Groups and other specialist workshops. Since 1982, he has been a member of DARPA's Materials Research Council. He is a fellow of both the American Physical Society and the Institute of Electrical and Electronic Engineers.

John Michael Golio received the BSEE degree from the University of Illinois, Urbana, in 1976. After graduation he worked at Watkins-Johnson company, Palo Alto, California, in the Microwave Tunable Devices Section. His work there was primarily concerned with the design of YIG-tuned oscillators and of automated test equipment. He received the MS and PhD degrees in electrical engineering from North Carolina State University in 1980 and 1983, respectively. His graduate work plus one-year postdoctoral experience at NCSU focused on solid-state device issues related to microwave devices and circuits—especially ion-implanted MESFETs for MMIC applications. After spending two years as an assistant professor of electrical engineering at Arizona State University, he joined Motorola Government Electronics Group, Chandler, Arizona, with the RF Subsystems Section. There he is responsible for the development of device models, circuit simulation techniques, parameter extraction techniques, and design procedures for large-signal circuit applications.

Dr. Golio is a member of Sigma Xi, Tau Beta Pi and Pi Mu Epsilon.

Karl Hess was born in Trumau, Austria, on June 20, 1945. After studying physics and mathematics he received his PhD degree (Dr. Phil) at the University of Vienna, Austria, in 1970. He is currently a professor of electrical engineering and a research professor in the Coordinated Science Laboratory, College of Engineering, University of Illinois at Urbana-Champaign. In 1982 he was named a Beckman Associate in the Center for Advanced Study, the University of Illinois. The research of Dr. Hess focuses on effects which are basic to the theory of semiconductor devices, such as impact ionization, hot electrons, and high field transient electronic transport. At present he is interested in electronic properties of III-V compound quantum-well heterostructures and superlattices, and in 1979 he conceived the real space transfer effect. His most recent research concentrates on the use of large computational resources (supercomputers) for the simulation of semiconductor transport.

Dr. Hess is a fellow of the Institute of Electrical and Electronics Engineers.

Evelyn L. Hu is a professor of electrical and computer engineering at the University of California at Santa Barbara. In addition, she serves as Associate Director for Microelectronics, at the Center for Robotic Systems in Microelectronics, an engineering research center. Her current interests center about high-resolution fabrication of III-V based electronic devices and heterostructures. She received her BA in physics (summa cum laude) from Barnard College and her MA and PhD in physics from Columbia University.

From 1975 to 1981 Dr. Hu was a member of the technical staff at Bell Laboratories at Holmdel, N.J. Her research was concerned with superconducting and semiconducting devices, encompassing high-resolution fabrication techniques such as e-beam lithography (at dimensions of several hundred angstroms) and reactive ion etching to form low-dimensioned structures for transport studies. From 1981 to 1984 she served as a supervisor for VLSI patterning processes at Bell Laboratories at Murray Hill, New Jersey. The emphasis of the work there was on the development of reactive ion etching and optical lithographic processes commensurate with 1- to 1.25-μm NMOS and CMOS technologies. In 1984, she joined the ECE Department at U.C. Santa Barbara.

Dr. Hu is a member of the APS, IEEE, AVS and Sigma Xi.

Sridhar V. Iyer was born on August 6, 1961, in Sivagangai, India. Upon completing his undergraduate education in electrical engineering at Madras University, Madras, India, in 1983, he joined the graduate program in electrical and computer engineering at the University of Illinois at Urbana-Champaign. Between 1985 and 1987, he worked with the hardware accelerator based design verification group at IBM. Since mid-1987, he has been back at the University of Illinois at Urbana-Champaign working towards his doctoral degree. His current research focuses on the epitaxial growth of high-quality gallium arsenide and other III-VI compound semiconductors on silicon and the monolithic integration of optical and electronic devices. His other research interests include expert systems and optical interconnects for parallel processing systems.

George N. Maracas is an associate professor of electrical and computer engineering at Arizona State University in Tempe, Arizona. He is currently performing research on III-V semiconductor heterojunction materials and devices for integrated optoelectronics. Whenever possible he hikes the Grand Canyon.

Dr. Maracas received his BA in physics in 1977 from New York University in New York City. He received his MEng in engineering physics (1978) and PhD in electrical engineering (1982) from Cornell University in Ithaca, New York. He then became an assistant professor at North Carolina State University in Raleigh, where he worked with Research Triangle Institute and the Microelectronics Center of North Carolina (MCNC) to establish III-V semiconductor microwave and high-speed optical device fabrication and characterization processes and facilities. He has continued these efforts at ASU and is also working on III-V materials growth by molecular-beam epitaxy (MBE).

He is a member of the IEEE, SPIE, and a recipient of the National Science Foundation Presidential Young Investigator Award.

Winfried Mönch is a professor of physics at the Universität-Gesamthochschule-Duisburg (Federal Republic of Germany). Dr. Mönch studied at the Universität Göttingen, where he received his diploma degree in physics and his Dr. rer. nat. After this, he joined first the AEG Forschungsinstitut in Frankfurt am Main and then, in 1965, the physics department of the Technische Hochschule Aachen. It was at the latter that he received the venia legendi in 1968 and was appointed as an associate professor in 1970. In 1975, Dr. Mönch joined the recently founded Universität Duisburg. Since 1965, his research interests have been surface and interface science of semiconductors.

In 1981, Dr. Mönch was appointed as the first Walter Schottky Visiting Professor at Stanford University, and he received the 1984 E.W. Müller Award from the University of Wisconsin at Milwaukee.

Hadis Morkoç was born in Turkey in 1947. After receiving BSEE and MSEE degrees from the Instanbul Technical University in 1968 and 1969, respectively, he began work on his PhD in electrical Engineering at Michigan State University and later transferred to Cornell, where he received his PhD in 1975. Following a postdoctoral fellowship at Cornell, he held positions at Varian Associates in Palo Alto and Bell Laboratories in Murray Hill. In 1979 he joined the University of Illinois. His research covered many aspects of heterojunction and thin-film materials and devices including superlattices. He has authored and coauthored some 500 book chapters, reviews, and journal articles. Dr. Morkoç received the *Electronics Letters* Best Paper Award in 1978 for his work in InGaAsP and InGaAs field-effect transistors. His research on high-speed devices and GaAs on Si have been covered widely in technical and popular journals. In 1987 he spent a sabbatical year at Caltech and the Jet Propulsion Laboratory.

Dr. Morkoç is a life member of Sigma-Xi, member of the American Physical Society, fellow of IEEE, and a fellow of the American Association for the Advancement of Science.

Johannes K. Notthoff received the BS in electrical engineering from the Chamber of Industry and Crafts in Würzburg, Germany, in 1948. Following graduation he worked as an electronics engineer and designed electronic hardware, digital equipment, and automatic film processing equipment. He has also designed a radiation-hardened linear amplifier and ICs for GaAs digital devices, and he was the principal designer of a low-noise cyrogenic preamplifier. Before his retirement Mr. Notthoff was responsible for all GaAs circuit design mask layout, device testing, and radiation hardening at the McDonnell Douglas Microelectronics Center. He performed design and testing of radiation-hardened digital circuits, including a variety of GaAs JFET SSI and MSI circuits, 256-bit, 1K, 4K, and 16K random-access memories, and linear amplifiers, and he originated the contemporary E-JFET gate array circuits. He holds three patents and is the author of numerous technical papers in advanced GaAs electronics.

Gordon C. Osbourn received his BS and MS degrees (1974, 1975) from the University of Missouri at Kansas City and his PhD degree (1979) from the California Institute of Technology. He has been at Sandia National Laboratories since 1979, and is currently a division supervisor. His research interests include the physics and device applications of III-V semiconductors and strained-layer superlattices. His most recent studies have involved the band structure properties and IR detector applications of InAsSb strained-layer superlattices. Dr. Osbourn has published 70 technical articles and has received three patents on strained-layer superlattice devices.

He received the 1985 E. O. Lawrence award for his theoretical work on the electronic properties of strained-layer superlattices. Dr. Osbourn is a fellow of the American Physical Society.

Umberto Ravaioli obtained the Laurea Dr. in electronics engineering (1980) and the Laurea Dr. in physics (1982) from University of Bologna, Italy, and the PhD in electrical engineering (1986) from Arizona State University, where his dissertation work was on Monte Carlo simulation of the HEMT. He is currently an assistant professor in the Department of Electrical and Computer Engineering, and a research assistant professor in the Coordinated Science Laboratory, at the University of Illinois at Urbana-Champaign. His main research interest is in the simulation of semiconductor devices, with emphasis on Monte Carlo methods and supercomputing applications. Dr. Ravaioli is a coauthor of the book *Measurements of Optical Fibers and Devices* (Artech House, 1984) and is the author and coauthor of a number of technical papers and conference presentations. He is a member of the IEEE.

Mark Rodwell was born on January 18, 1960, in Altrincham, England. He received the BS degree in electrical engineering from the University of Tennessee, Knoxville, in 1980, and the MS degree from Stanford University in 1982. From 1982 through 1984 he worked at AT&T Bell Laboratories, designing fiber-optic digital transmission systems. In 1984 he returned to Stanford University. There he pursued research in electro-optic sampling of GaAs integrated circuits, developed nonlinear transmission lines, monolithic GaAs devices for picosecond electrical pulse generation, and supervised the development of 130-GHz-bandwidth monolithic sampling circuits based on nonlinear transmission lines. He received the PhD in electrical engineering in January 1988, and remained at Stanford as a research associate until September 1988. Currently, he is an assistant professor in the Department of Electrical and Computer Engineering, University of California, Santa Barbara. His research interests are picosecond electronic and optoelectronic devices and circuits, millimeter-wave integrated circuits, and picosecond instrumentation.

Ronald J. Roedel was born in Brooklyn, New York, on January 22, 1950. He received his BS in electrical engineering at Princeton in 1971 and his PhD in electrical engineering at UCLA in 1976.

In 1976 he joined Bell Telephone Laboratories in Murray Hill, New Jersey, where he was involved in research in GaAs light-emitting diodes, including their design, fabrication, and characterization. In 1981 he joined the faculty of the Department of Electrical and Computer Engineering at Arizona State University in Tempe, Arizona, as an associate professor. He was promoted to professor in 1987. His work at ASU is equally divided between his teaching and research activities. As a researcher, his activities include compound semiconductor processing (diffusion, sputtering, metallization), devices (optoelectronics, solar cells, heterojunction transistors), and characterization (electron microscopy, luminescence, defect imaging). As a teacher, his work includes teaching an equal number of graduate and undergraduate courses. Dr. Roedel received both the Tau Beta Pi and the College of Engineering Outstanding Teaching awards in 1988.

Gary L. Troeger received the BS in physics from UCLA in 1968, the MS in physics and the PhD in solid-state physics in 1970 and 1975, respectively, from the University of Colorado. He was also a postdoctoral research associate at the University of Colorado, Boulder, where his technical contributions included determination of the roles of substitutional nickel and manganese in a I-III-IV_2 semiconductor. Currently he is Manager of Electronics, GaAs Pilot Line, at the McDonnell Douglas Microelectronics Center. He is responsible for IC fabrication technology on the GaAs pilot line. He developed fully ion-implanted 1-μm design rule complementary JFET process utilizing enhancement-mode n- and p-channel devices. His specific areas of effort are ion implantation and annealing of GaAs, ohmic contact improvement, metal-over-metal crossovers, dielectric and metal deposition, plasma etching, ion milling, and yield analysis. Dr. Troeger has published many papers on GaAs technology. He is a member of the Electrochemical Society, the IEEE, and Sigma Pi Sigma.

Hilmi Ünlü received the BS degree in physics from the University of Istanbul, Turkey, in 1978. He earned MSEE and PhD degrees in electrical engineering from the University of Minnesota, Minneapolis, in 1984 and 1986, respectively. His graduate studies were supported by the Scientific and Technical Research Council of Turkey with a NATO scholarship. After the completion of his PhD degree he joined the University of Illinois at Urbana-Champaign as a postdoctoral research associate in Prof. H. Morkoç's group. Dr. Ünlü's current interests include the fundamental understanding of heterojunctions and metal-semiconductor interfaces as well as their applications for high-speed transistors. He has made contributions to the modeling of equilibrium and nonequilibrium properties of heterojunction devices. He is a coauthor of journal articles on these subjects.

Kurt J. Weingarten was born on January 30, 1961, in St. Petersburg, Florida. He received the BS degree in electrical engineering from the Georgia Institute of Technology in 1983, and his MS in 1985 and the PhD in January 1988 in electrical engineering from Stanford University, where he worked as a research assistant in the Edward L. Ginzton Laboratory, developing an electro-optic sampling system and applying this to the measurement of GaAs ICs. His thesis was "Gallium Arsenide Integrated Circuit Testing using Electrooptic Sampling." He has authored more than a dozen papers on this topic and coauthored several papers on related topics such as the timing stabilization of mode-locked lasers. In 1985 he received an IBM Predoctoral Fellowship and in 1986 and 1987 the Newport Research

Award. He currently works for Lightwave Electronics in Mountain View, California, developing a commercial version of an optical tester for GaAs ICs. His research interests are electro-optic sampling and charge-sensing of GaAs ICs, high-speed electronic testing, ultrafast optical pulse generation, and picosecond synchronization of mode-locked lasers to microwave signal generators.

He is a member of the Institute of Electrical and Electronics Engineers and the Optical Society of America. During his leisure hours, he may be found windsurfing during the warmer times of the year or snow skiing during the colder months.

Carl Wilmsen received the PhD in electrical engineering from the University of Texas in 1967. He has investigated the oxidation of III-V compounds since 1973 and has authored over 60 journal papers in this area along with several review articles and book chapters. He is a professor of electrical engineering at Colorado State University and serves as the associate director of the NSF Center for Optical Computer Systems. He has spent sabbatical leaves at the IBM Research Center in Yorktown Heights and the Naval Ocean Systems Center in San Diego, where he investigated the passivation of GaAs and InP surfaces using oxidation and chemical treatments. He is currently studying the correlation between the chemical, structural and electrical properties of insulator-semiconductor interfaces with special emphasis on unpinning the GaAs surface and inversion layer transport in InP.

Rainer Zuleeg studied physics and mathematics at the Hochschule Bamberg and University of Munich, and received the PhD degree in semiconductor physics from Tohoku University, Japan, in 1972. He has published over 100 papers in professional journals and he is the holder of 22 patents. He is now Staff Director and MDC Senior Fellow at the McDonnell Douglas Microelectronics Center, which was established in 1984. He directs and supervises research and development of GaAs semiconductor devices and integrated circuits. He has been a lecturer in semiconductor device physics and electronics since 1972 at the University of California in Irvine. In 1975 and 1981 he was a visiting professor at the Royal Melborne Institute of Technology (RMIT), Australia, performing research in ion implantation. In 1984, he was invited by the DFG (German Research Society) to be a guest professor at the University of Duisburg, West Germany, and engage in InGaAs/InP heterojunction device research.

Dr. Zuleeg is a member of the APS, the ECS, the IEEE, and the AIAA.

Acknowledgments

Chapter 1

Warren Seely, David Warren, Greg Gorrie, and Gary McGoff have contributed significantly to much of the work presented here. Their help is sincerely appreciated. The author also wishes to thank Jerry Brand and Don Holcomb for their support and encouragement.

J.M. Golio

Chapter 2

Partial support by the Naval Research Laboratory is gratefully acknowledged by the authors.

K. Hess and U. Ravaioli

Chapter 3

The authors are with the McDonnell Douglas Microelectronics Center (MDMC) in Huntington Beach, California, and are very grateful to the GaAs team for numerous and valuable contributions to the GaAs JFET technology. In the process development they acknowledge Mrs. Judy Bolen, Maria Hodgman, and Dr. T.S. Rao-Sahib, in the design and simulation, Mr. Carl Vogelsang, Mrs. Toni Nicalek, and Dr. Choong Hyun. Thanks go to the McDonnell Douglas Corporation management, Mr. Anatole Browde, and Mr. Arnie Maddow and Mr. Bill Geideman from McDonnell Douglas Astronautics Company, for support of this program. Mr. Sven Roosild of DARPA/DSO and Dr. Sherman Karp and Dr. John Egan of DARPA/ORD are sponsors and are advocating this technology for space and military applications.

R. Zuleeg, J. K. Notthoff, and G. L. Troeger

Chapter 5

The authors thank Joe Jensen of Hughes Research Laboratories (Malibu), Ross LaRue, and Majid Riazait of the Varian Research Center, J. A. Valdmanis of AT&T Bell Laboratories, K. Reed Gleason of Cascade Microtech, Inc., and TriQuint Semiconductor, Inc., and Stephan Swierkowski of Lawrence Livermore National Labs for their assistance with the GaAs circuits. Thanks to Tom Baer and Jim Kafka of Spectra-Physics, Inc., for their assistance with the optical pulse generator. We acknowledge the generous equipment donations of Cascade Microtech, Inc., Tektronix, Inc., and Hewlett-Packard Co. K. J. Weingarten acknowledges a Newport Research Award, and M. J. W. Rodwell acknowledges an IBM predoctoral fellowship. Our research was sponsored by the Air Force Office of Scientific Research under grant number F49620-84-K-0139, by the Joint Services Electronic Program under grant number N00014-84-K-0327, and by the Wright-Patterson Air Force Base Avionics Laboratory under contract number F33615-86-C-1126.

K. J. Weingarten, M. J. W. Rodwell, and D. M. Bloom

Chapter 6

This work was funded by the Air Force Office of Scientific Research. The authors would like to acknowledge the assistance of their colleagues at the University of Illinois, and thank R. Fischer, D. A. Neumann, H. Zabel, C. Choi, N. Otsuka, L. Davis, M. Longerbone, L. P. Erikson, and P. L. Gourley for their contributions and S. Norwood and P. Carlson for manuscript preparation.

H. Ünlü, H. Morkoç and S. Iyer

Chapter 8

The author wishes to thank Mr. Kent Geib for his help in preparing this chapter. The financial support of the National Science Foundation, Grant No. ECS-8610098, is also acknowledged.

C. W. Wilmsen

Chapter 11

The author wishes to thank all of the coworkers and colleagues, too numerous to list, whose published work forms the basis of this review. Their contributions are evident in the 35 figures and 261 references presented here. In addition, the author is grateful to J. S. Blakemore, H. G. Grimmeis, T. A. Kennedy, P. B. Klein, W. J. Moore, R. C. Newman, L. Sammuelson, B. V. Shanabrook, M. S. Skolnick, G. E. Stillman, and M. Tajima, for valuable discussions during the preparation of the manuscript. This work was partially supported by the Office of Naval Research.

S. G. Bishop

Chapter 1

Large-Signal Analog Circuit Simulation

J. M. Golio

1.1 Introduction

Microwave devices and components all exhibit nonlinear properties to some degree. For some applications the nonlinearities of a system may be neglected during analysis without loss of accuracy in the performance predictions. However, this only occurs if the signals of interest are small in amplitude. As signal levels increase, departure from ideal linear performance becomes evident and analysis techniques must account for these nonlinearities in order to continue to obtain acceptable performance predictions. Analog circuit applications that may be categorized as large-signal, and require nonlinear analysis techniques, include power amplifiers, oscillators, and mixers or frequency converters.

The distinction between small-signal and large-signal operation is illustrated in Fig. 1–1. The transconductance of the device in the simple circuit considered is the slope of the I_{ds} versus V_{gs} curve. When the gate-source voltage undergoes a small perturbation as a function of time (case *A* in the figure), the drain-source current, I_{ds}, varies almost linearly with voltage (i.e. the slope is a good approximation to the $I_{ds}(V_{gs})$ curve). When the perturbation becomes large, however (as shown in case *B*), the line representing the slope of the $I_{ds}(V_{gs})$ curve is no longer a good approximation to the $I_{ds}(V_{gs})$ characteristic, and the $I_{ds}(t)$ plot is clearly a highly distorted form of the input signal. Figure 1–1 also indirectly illustrates the relationship between large-signal operation and frequency conversion. Large-signal circuit operation involves the conversion of input signal frequencies to different output signal frequencies. In the case of oscillators, DC input (0.0 Hz) and noise is converted to a desired signal frequency plus some unwanted signals in the spectrum. Power amplifiers convert a desired input signal frequency to an amplified form of the signal plus some undesirable harmonics. Finally, mixers will convert two input signals (f_{RF} and

f_{LO}) to a desired intermediate frequency (IF) plus a number of undesirable signals. The illustration of Fig 1–1 is representative of amplifier operation. Case *A* represents small-signal amplifier operation, where a single-frequency low-level sinusoidal signal is input into the device, and an amplified form of the sinusoid with minimal distortion is present at the output. When the larger-amplitude signal of case *B* is applied to the device, however, the output is no longer a single frequency, but clearly contains significant harmonic distortion.

Large-signal analysis procedures cannot make use of many of the most valuable techniques that are exploited for the analysis of small-signal circuitry. Small-signal circuit analysis, for example, is greatly simplified by transformation of system properties into the frequency domain. This transformation is possible because of the linear nature of the problem. Passive circuit elements are described by fixed impedance, admittance, "*ABCD*" parameters, or equivalent parameter matrix. Likewise, active devices are typically described by measured scattering parameters (called *s*-parameters) at the bias level and frequencies of interest. Small-signal characteristics of devices can also be well described by equivalent circuit models which uti-

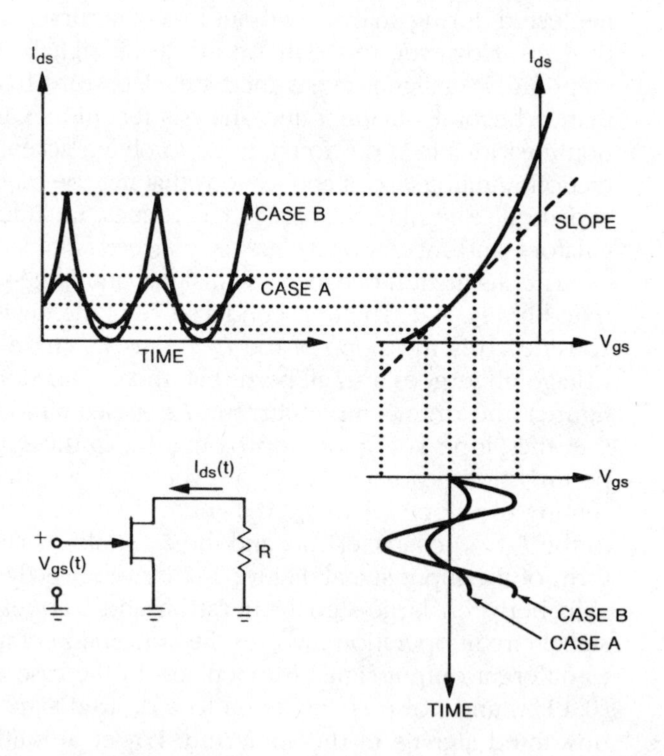

Fig. 1–1. Illustration of large-signal versus small-signal operation.

lize fixed-value circuit elements. Once such descriptions have been established, small-signal circuit analysis is accomplished through straightforward matrix operations. This approach has been used with astonishing success in a number of computer-aided design (CAD) packages, such as SUPERCOMPACT and TOUCHSTONE. The technique has proved to be both accurate and efficient.

The development of small-signal, frequency-domain CAD programs has had a tremendous effect on the achievements of the designer of linear microwave circuits. Part of the credit for the success of these designers must also be assigned to the development of the microwave network analyzer. This equipment has allowed the circuit designer to obtain accurate characterization data, describing circuit components and devices, in a relatively simple manner. The data obtained using this equipment is also directly usable in the small-signal circuit simulators and in the determination of the parameters which describe device models.

Large-signal circuit analysis has not yet advanced to the level attained by small-signal programs. The problem is fundamentally more complex, so that the capabilities of large-signal circuit analysis will probably always lag behind those of its small-signal counterpart. Before discussing the current state of large-signal modeling and which issues still have not been adequately addressed, it may be helpful to examine the issues which relate to the success of small-signal simulators. The discussion which was given above has identified three critical factors that are related to the success of small-signal analog circuit design. These three factors can be summarized as follows:

1. An accurate electrical description of the device exists. Measured *s*-parameters themselves model device behavior and this data can be incorporated into an equivalent circuit model.
2. A measurement capability exists which can easily produce the parameters required to describe the device characteristics (i.e. characterization and parameter extraction processes are straightforward).
3. Efficient numerical methods for analyzing overall circuit performance exist. Transformation into the frequency domain allows the system performance to be described by a set of linear algebraic equations instead of by differential equations.

Each of these three issues has been addressed to some extent by a number of researchers in nonlinear circuit techniques. Many questions remain unanswered, however, and no single technique for addressing the issues has yet emerged superior for all applications.

The first point listed above relates to device modeling. There appears to be no shortage of device models which can be used in large-signal simulation applications. Typically large-signal equivalent circuit models of microwave devices have topologies very similar to, or identical with, small-signal models of these devices. The difference between the two is the voltage dependence of the element values which is included in large-signal models. Both physically based[1–5] and empirical[6–13] models have been used in large-signal simulations with some success, and comparisons of predictions from different models have been examined.[14,15] In at least one study,[16] a two-dimensional device simulation was used in conjunction with a circuit simulator to predict large-signal performance. Despite the abundance of work in this area, a number of important issues, related to device modeling, still have not been adequately addressed. In addition, the relationship of the device model used to the device characterization is significant. Although most of the device models just referenced have been shown to be quite accurate for at least some applications, many of the advantages of this accuracy are lost if the characterization and parameter extraction process is tedious, time-consuming, and expensive. Such a process also needs to be generally applicable to a wide variety of devices if the model is to be useful.

Large-signal characterization of actual active devices has been done in a variety of ways. Load-pull measurements[17–18] and large-signal *s*-parameter measurements[19–20] have been used successfully to design nonlinear microwave circuitry for some applications. These techniques, however, do not lend themselves easily to incorporation into general large-signal circuit simulators. Characterization of devices at DC has been used to determine model parameters in some cases[10,11,12] and small-signal measurements made at multiple bias levels have also been used.[4,7,13] In theory, the physically based models[1–5] enjoy a certain advantage over empirical models in terms of required characterization, since performance predictions are derived from device dimensions and material properties in such models. It should be pointed out, however, that device dimensions and material properties of the semiconductor are often not part of the knowledge held by the circuit designer. In addition, physically based models often incorporate some physical mechanisms occurring in the device through the use of empirical parameters. Regardless of the kind of device characterization process used in the large-signal design process, extracting relevant device parameters from characterization data is not a trivial process for any of the models considered.

The final point cited above as contributing to the success of small-signal simulation packages is the existence of efficient circuit-simulation techniques. Large-signal simulation techniques fall into three major categories: time-domain techniques, harmonic-balance and related techniques, and quasi–large-signal techniques. All of these methods have been used with some success for certain large-signal applications. Each has certain limitations and certain advantages in comparison with the others. Time-domain simulators have been around for some time but have been used primarily to analyze digital switching circuits. The circuit simulator SPICE[21] is an ex-

Book Mark

HOWARD W. SAMS & COMPANY

Excellence In Publishing

DEAR VALUED CUSTOMER:

Howard W. Sams & Company is dedicated to bringing you timely and authoritative books for your professional library. Our goal is to provide you with excellent technical books written by the most qualified authors. You can assist us in this endeavor by listing subjects for which you need more information.

We appreciate your comments and will use the information to provide you with a more comprehensive selection of professional reference titles.

Thank you,

Vice President, Book Publishing
Howard W. Sams & Company

SUBJECT AREAS:

____

____

____

____

Name ______________________________

Title ______________________________

Company ______________________________

Address ______________________________

City ______________________________

State/Zip ______________________________

Daytime Telephone No. ______________________________

A Division of Macmillan, Inc.
4300 West 62nd Street
Indianapolis, Indiana 46268 USA

22555

Bookmark

HOWARD W. SAMS & COMPANY

ample of such a package. Although time-domain simulation has been used successfully to simulate several different types of nonlinear microwave circuits,[3,4,8,16,22] problems related to long–time-constant transient response within the circuit, and to the representation of transmission lines, are severe.

Harmonic-balance and related techniques are receiving a great deal of attention as possible alternatives to time-domain simulations.[9,13,15,23–26] Although the harmonic-balance methods do address many of the difficulties found in the use of time-domain simulations, not all of the questions related to the implementation of the techniques have been answered.

The third category of large-signal circuit simulation technique, quasi–large-signal methods, is included for completeness. In reality, many of the large-signal circuits being designed today are making use of small-signal techniques, which have been modified or interpreted slightly differently in order to obtain the extra design information required for a large-signal application. The techniques being applied typically are valid for only a particular circuit category, but often produce results which are acceptable. Oscillators, for example, have been designed using small-signal *s*-parameter data.[27–29] This design process, along with information concerning the large-signal power saturation characteristics of the device, can be used to predict both output power[30,31] and frequency of oscillation. Likewise, power amplifiers have been analyzed using small-signal methods in conjunction with a limited amount of nonlinear data.[10,14,32,33] The design of MESFET mixers is often aided by the use of the conversion matrix technique,[34,35] which is an extension of the small-signal analysis that includes some of the effects of the device nonlinearities.

1.2 Large-Signal Device Modeling

Any device model that operates within a circuit-simulation routine must ultimately describe the relationship between voltages at the device terminals and currents into or out of these terminals. A device model which is used in DC simulations is required to predict the magnitude of the terminal currents for any value of voltage applied to the device. Small-signal device models must predict the rate of change of the terminal currents with respect to voltage at one particular bias level. In contrast to these, large-signal models must be capable of determining both the magnitude of the current and the rate of change of the current with respect to voltage at all possible bias and signal conditions. Figure 1–2 represents the various modeling problems for a simple two-terminal device, and the illustration is summarized in Table 1–1.

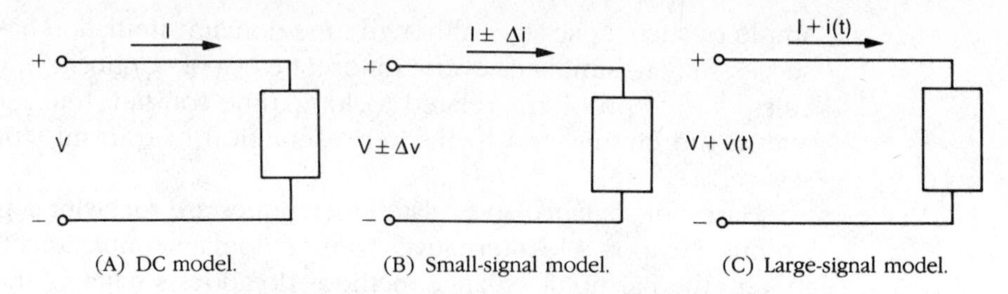

(A) DC model. (B) Small-signal model. (C) Large-signal model.

Fig. 1–2. Representation of a two-terminal device as it is modeled for DC, small-signal, and large-signal applications.

Because most of the work done recently in the area of large-signal modeling has been aimed at microwave metal-gate field-effect transistor (MESFET) applications, the discussion here will focus on nonlinear models describing MESFET characteristics. The GaAs MESFET has become the workhorse of the microwave industry, and the trend toward the exploitation of monolithic circuitry will tend to make its use even more widespread. Many of the issues discussed in relation to MESFET modeling will apply directly to the modeling of microwave HEMTs or bipolar transistors. It should also be noted that Schottky diodes are still an important element in many nonlinear microwave systems (primarily mixers). Large-signal simulations of diodes in nonlinear applications have indicated, however, that relatively simple models of device characteristics are sufficient to describe circuit performance[36] extremely accurately.

Table 1–1. Requirements for DC, Small-Signal, and Large-Signal Models

Type of Simulation	Independent Variable	Dependent Variable
DC	V	I
Small-signal	Δv	Δi
Large-signal	V, $v(t)$	I, $i(t)$

There are a number of ways to approach the large-signal modeling problem as described above. Perhaps the most direct method mathematically was taken by Madjar and Rosenbaum,[3] who describe the total terminal currents using explicit functions of the gate-to-source voltage, V_{gs}, the drain-to-source voltage, V_{ds}, and their derivatives. The model thus predicts instantaneous terminal currents for each instantaneous voltage pair from equations of the form

$$I_d = I_{\text{con}} + D_{VSG}\frac{dV_{gs}}{dt} + D_{VDS}\frac{dV_{ds}}{dt}, \qquad (1\text{–}1)$$

and

$$I_g = G_{VSG}\frac{dV_{gs}}{dt} + G_{VDS}\frac{dV_{ds}}{dt}, \tag{1-2}$$

where I_{con}, D_{VSG}, D_{VDS}, G_{VSG}, and G_{VDS} are all functions of the terminal voltages V_{gs} and V_{ds}. A more common approach to the modeling problem is to describe the device characteristics using an equivalent circuit. The circuit is comprised of common circuit elements, some of which are nonlinear. Within the circuit simulation, the equivalent circuit is analyzed and appropriate voltage and current levels are computed. The various models differ from each other in the assumed equivalent circuit topology, in the circuit elements which will be considered nonlinear, and in the functional form of these nonlinearities.

It should be pointed out that analysis of an equivalent circuit model can produce the information obtained from Eqs. 1–1 and 1–2. Such a formulation of the problem is required of many large-signal circuit simulations. Likewise, the data obtained from direct evaluation of Eqs. 1–1 and 1–2 can be approximated by an equivalent circuit model.

1.2.1 Equivalent Circuit Parameters

Figure 1–3 shows an equivalent circuit for the field-effect transistor (FET). The circuit topology illustrated is fairly generic. Some device models utilize more elements, some less, and in some instances node connections of the elements are altered slightly. The only fact that distinguishes such an equivalent circuit as being large-signal versus small-signal is whether or not any of the circuit element values are dependent on the instantaneous

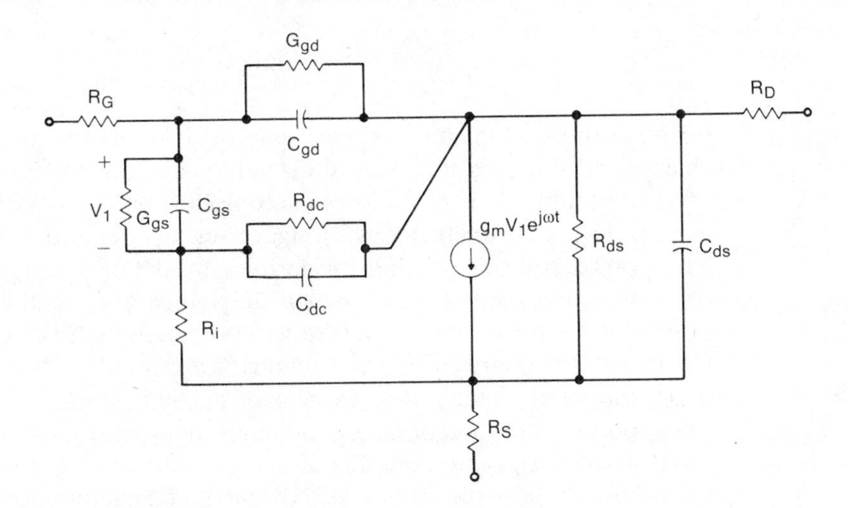

Fig. 1–3. An equivalent circuit model of the GaAs MESFET.

terminal voltages. Most of the large-signal models reported in the literature have assumed that at least four of the model's element values are nonlinear. At least one study has obtained good performance predictions using only two nonlinear elements,[37] and as many as seven nonlinear circuit elements have been utilized.[13]

At this point, a discussion of the physical significance of the various equivalent circuit elements as they relate to the MESFET is useful.

1.2.1.1 Parasitic Resistances R_S, R_D, R_G

These values represent the sum of all resistance contributions between terminal pads and the active device. Although measurements indicate that there is some voltage dependence to these values, it is normally neglected in large-signal models, so that the values are assumed to be fixed. Empirically derived expressions, which estimate the value of R_S and R_D, have been shown to be quite accurate[38] for many cases, and a similar expression for R_G can be used to estimate this value.[39] These values can also be calculated from DC forward-conduction measurements.[40]

1.2.1.2 Capacitances C_{gs}, C_{gd}, C_{ds}

The gate-source capacitance, C_{gs}, affects device characteristics and performance limits in a fundamental manner. Therefore, describing the gate-source capacitance with considerable accuracy is important. The gate-drain capacitance, C_{gd}, can also be important in many applications, while the drain-source capacitance, C_{ds}, is typically much smaller than the other two and less critical to overall device characteristics. The gate-source and gate-drain capacitances have been modeled in a number of different ways.[1,4,6,12,41] Both of these capacitance values result from changes in the depletion-region charge as a function of terminal voltages

$$\frac{dQ_D}{dV_{gs}} \quad \text{and} \quad \frac{dQ_D}{dV_{gd}}.$$

Because these capacitances are related, through the depletion-region charge and the gate terminal, their values are not independent of each other. Although many of the proposed models have successfully predicted overall device performance while ignoring this dependence, Statz *et al.*[12] have considered this relationship in their model of C_{gs} and C_{gd}. The value of C_{ds} is typically much smaller than the values of C_{gs} and C_{gd} and is considered to be fixed or is neglected for most models. This drain-source capacitance is primarily due to geometric capacitance effects between the drain and source electrodes. Expressions based on this assumption[1] have been found to be in generally good agreement with measurement.

It should be pointed out that direct measurements of these capacitance values are not possible. This is also true to some extent of other model element values. To determine these values, small-signal *s*-parameter measurements are usually made and the model element values are altered until the *s*-parameters predicted by the model agree with those measured for the device.

1.2.1.3 Gate-Drain and Gate-Source Conductances G_{gd}, G_{gs}

The gate terminal of the GaAs MESFET is a Schottky rectifying contact. It behaves electrically like a diode. If the gate-source or gate-drain potential becomes positive, forward-conduction gate current will flow through the diode. In addition, for large negative values of gate-drain voltage, reverse breakdown of this junction may occur. This phenomenon can be especially important for the case of power amplifier circuits.

From the preceding discussion, it appears obvious that models of these conduction mechanisms can be accomplished using diode models. This has indeed been the most widely used method[6, 9–11, 13] and has been shown to be successful. Because these mechanisms only contribute to device characteristics under conditions of very large signal levels, many models have also ignored these effects for certain applications.

1.2.1.4 Transconductance g_m

The transconductance of the FET is defined as the differential change in output current with respect to input voltage:

$$g_m = \left. \frac{dI_{ds}}{dV_{gs}} \right|_{V_{ds} = \text{constant}} . \tag{1–3}$$

The definition of transconductance given by Eq. 1–3 suggests that a simple DC measurement of the drain-source current, while varying gate-source voltage, will produce transconductance values. It should be noted, however, that the transconductance measured in this manner is an external transconductance, $g_{m\text{ext}}$, which differs from the model transconductance because of the voltage drop across the parasitic source resistance, R_S. The external transconductance can be corrected for this effect using the formula

$$g_m = \frac{g_{m\text{ext}}}{1 - R_S g_{m\text{ext}}} . \tag{1–4}$$

Most large-signal models utilize mathematical expressions for drain-source current as a function of the two terminal voltages. The transconductance is then calculated as the derivative of that function as expressed by Eq. 1–3.

1.2.1.5 Output Resistance R_{ds}

The output resistance of the device can also be defined in terms of a derivative of an *I-V* relationship as

$$R_{ds} = \left. \left| \frac{dI_{ds}}{dV_{ds}} \right|^{-1} \right|_{V_{gs} = \text{constant}} \tag{1–5}$$

Just as for the case of transconductance above, this definition suggests that a simple DC measurement of the drain-source current while varying the drain-source voltage will produce accurate output resistance values. As in the case with transconductance, most models compute the output resistance as the derivative of the drain-source current function as specified by Eq. 1–5. There is some indication that this approach may not be appropriate,[14,42] and this question will be examined in Sec. 1.3 below.

1.2.1.6 Charging Resistance R_i

A finite amount of time is required to charge the gate-source capacitance. A time constant is established in the equivalent circuit by including the charging resistance R_i. Empirical expressions which define this quantity as the ratio of the electron transit time, τ, to the gate-source capacitance, C_{gs}, are often employed:

$$R_i \propto \tau / C_{gs} .$$

Another technique employed to describe this resistance is to use measured *s*-parameter data to determine what resistance value best describes the device performance. No analytical expressions exist for this element and some modelers have chosen to neglect the effect.

1.2.1.7 Drain-to-Channel Resistance and Capacitance R_{dc}, C_{dc}

These elements are used in some models[7,11] to account for the effects of Gunn domains, which are hypothesized to form in the active channel at the drain end of the gate. The elements are useful in obtaining good agreement between measured and model-predicted performance for many cases.

1.2.2 Physical versus Empirical Descriptions

Device-modeling approaches can be classified along a continuum between those physically derived and those determined in a completely empirical fashion. A purely empirical approach is one in which we use a lookup table model. With this technique, values of the various circuit elements are measured, calculated, or estimated (usually indirectly), and stored for several different bias levels. Appropriate values are then looked up by the model when required. For voltage levels not included in the table of measured data, interpolation can be used. One limitation with such an approach is that errors in the data measured to characterize the device become part of the model. Another difficulty with such an approach is that a large amount of characterization is required to obtain a device description.

A more common approach to the modeling problem is to first examine the measured data and then look for a mathematical function which behaves in the same manner. This mathematical function will include adjustable parameters which, when assigned proper values, will cause the function to closely approximate the measured data. One of the best illustrative examples of this approach is the model described by Curtice.[6] Examination and simple physical modeling of the I_{ds} versus V_{gs} characteristics of a MESFET show an approximately quadratic response. Likewise, the I_{ds}-V_{ds} characteristics resemble a hyperbolic tangent function. Finally, when devices operate in current saturation, they exhibit a finite output resistance. Combining these facts, Curtice describes the *I-V* characteristics of the MESFET using the expression

$$I_{ds}(V_{ds}, V_{gs}) = b(V_{gs} - V_{po})^2(1 + LV_{ds}) \tanh(aV_{ds}), \tag{1–6}$$

where V_{po} is the pinch-off potential and a, b, and L are empirical parameters. In Eq. 1–6, the quadratic term describes observed pinch-off phenomena, the hyperbolic tangent term describes general I_{ds}-V_{ds} behavior, and the term containing the parameter L produces finite output resistance effects. Models derived from this expression or in a similar manner have been used by a number of researchers with considerable success.[9,10,12,43]

Equation 1–6 does not constitute a complete large-signal model, since capacitance expressions have not been addressed by this equation. To the first order, the development of capacitance-voltage relationships is independent of the development of current-voltage relationships. Thus, models which use identical expressions to evaluate terminal currents may use different capacitance expressions. As mentioned previously, the drain-source capacitance may often be assumed to be constant or even neglected. The gate-drain capacitance is also assumed constant[6,9,10,14] for some applications. One of the common empirical expressions used for gate-source capacitance is given by[6,9]

$$C_{gs} = C_{g0}\left(1 - \frac{V_{gs}}{V_{bi}}\right)^{-1/2}, \tag{1–7}$$

where V_{bi} is the built-in potential of the Schottky gate and C_{g0} is the zero-bias gate-source capacitance. The expression may be derived from first-order semiconductor junction theory applied to a two-terminal Schottky diode structure. It does not include drain-source voltage dependence, which may be important for some applications. More elaborate expressions developed specifically for FET structures provide more accurate predictions of device capacitance[4,12] and include drain-source voltage dependence. An expression which is capable of accurately describing gate-source capacitance is given by

$$C_{gs} = WC_{GS0} + L_{GS1}(\epsilon WL/a)\left(\frac{V_{gs} - V_{po}}{V_{bi} - V_{po}}\right)^m f_c(V_{ds}), \tag{1–8}$$

where

$$f_c = 1 + L_{GS2}\left(\frac{V_{bi} - V_{po}}{V_{gs} - V_{po}}\right)^{1/m}(V_{ds} - V_{d(\text{sat})}), \quad \text{for } V_{ds} > V_{d(\text{sat})},$$

and

$$f_c = (L_{GS2}V_{d(\text{sat})} + B_{gs} - 1)\left(\frac{V_{ds}}{V_{d(\text{sat})}}\right)^2 + (2 - L_{GS2}V_{d(\text{sat})} - 2B_{gs})\left(\frac{V_{ds}}{V_{d(\text{sat})}}\right) + B_{gs}, \quad \text{for } V_{ds} \leq V_{d(\text{sat})}.$$

In the above expressions, W, L, and a are the gate width, gate length, and epi-thickness of the device. The dielectric constant of the material is represented by ϵ. These expressions require that five empirical parameters be specified: C_{GS0}, L_{GS1}, L_{GS2}, m, and B_{gs}. When capacitance-voltage data is available, these parameters are easily determined from appropriately plotted characteristics.[4] Note also that these expressions do include the appropriate physical parameter dependence. The inclusion of physical relationships in such equations often simplifies the parameter extraction process.

Figure 1–4 shows the gate-source capacitance as a function of drain-source voltage, as calculated by the empirical expression above. The comparison made in Fig. 1–4 is with predictions of a two-dimensional ion-implanted device simulation.[44] Similar excellent agreement has also been obtained with capacitance data which was determined from actual device measurements on a wide variety of device structures.

Gate-drain capacitance may also be accurately expressed using such an empirical expression. An equation which has been found to produce excellent agreement with measurements is

$$C_{gd} = WC_{GDI}\left[1 - \left(\frac{L_{GD1}V_{gs} - V_{ds}}{L_{GD1}V_{bi}}\right)^r\right]^{-1}(1 - L_{GD2}V_{gs}) + WC_{GD0}. \tag{1–9}$$

Evaluation of the above expression requires that the five empirical parameters C_{GDI}, L_{GD1}, L_{GD2}, r, and C_{GD0} be determined. This is again easily accomplished when capacitance-voltage information is available.

Figure 1–5 compares the values computed from the above expression to those predicted by two-dimensional simulations. (In this figure V_{GS} is the DC gate-source voltage.) Similar outstanding agreement has been obtained for measured values of a number of different devices.

Willing, Rauscher, and deSantis[7] have also used an empirical approach similar to those referenced above. A key distinction between the Willing approach and those above is that the data which is used to determine model parameters is not DC current-voltage data, but measured micro-

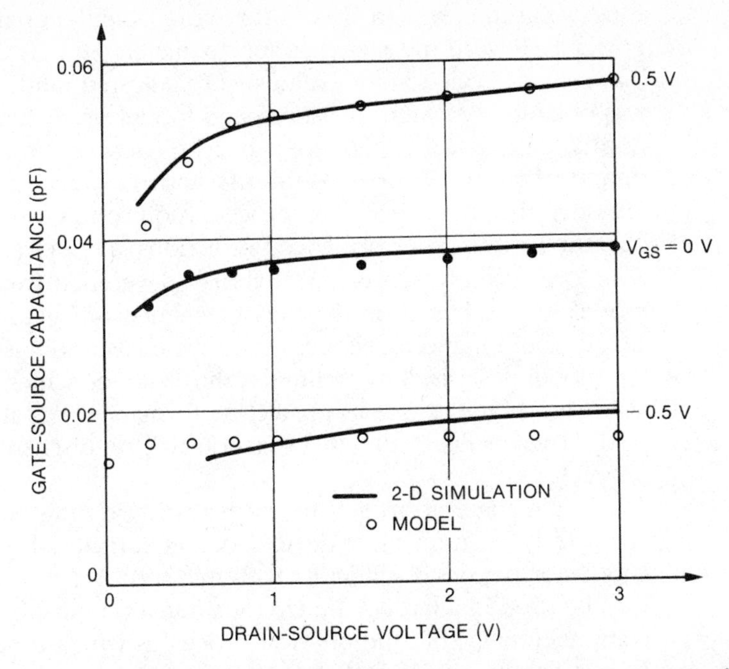

Fig. 1–4. Gate-source capacitance as a function of device terminal voltages. Empirical expressions are compared to values determined from two-dimensional device simulation.

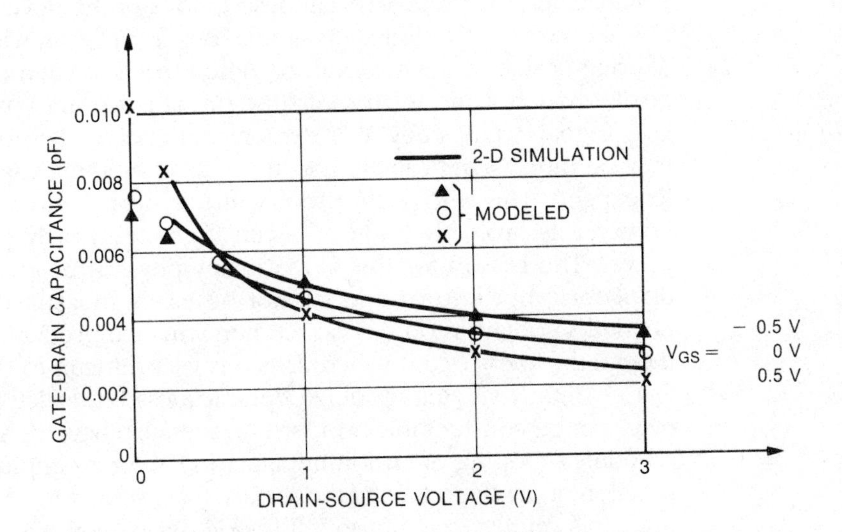

Fig. 1–5. Gate-drain capacitance as a function of device terminal voltages. Empirical expressions are compared to values determined from two-dimensional device simulation.

wave s-parameters. In this latter work, small-signal s-parameters of a power FET were measured at approximately 50 bias conditions. For each bias setting, device model values were adjusted until optimum agreement was obtained between measured and model predicted s-parameters. Polynomial expressions were then used to describe the functional dependence of the nonlinear elements with respect to terminal voltages. Although the characterization process required to use this model is laborious, the resulting performance predictions are very accurate.

More recently, Weiss and Pavlidis[14] developed two different empirical models—one based on DC measurements similar to those of Refs. 9, 10, 12, 14, and 43, and a second, based on measured s-parameters similar to the model described by Willing. A third model, which used physical characteristics of the device to predict performance, was also included in their study. The results of this investigation will be discussed in more detail in Sec. 1.3.

An alternate approach to empirical modeling is to describe device characteristics in terms of device geometry, material properties, and semiconductor physics. A difficulty with this approach is that any model which is to be used in a large-scale circuit simulation must be extremely efficient computationally, but the physical processes which describe device performance are complex. Definite compromises between accuracy and efficiency must be made for this application.

Two-dimensional device simulations which iteratively solve Poisson's equation and current transport equations on a large grid of points can produce very accurate physical descriptions of the device. Long computer times, however, prohibit such simulations from being widely used as engineering design tools. It should be noted that Snowden *et al.*[16] have incorporated such a model into a time-domain simulation for at least one large-signal circuit study. This demonstration of the feasibility of such a task may be more significant in the future as computing capabilities continue to expand. The role such a tool will play in the circuit design process, however, is probably limited—even if granted nearly infinite computing power. The reason for this is that many physical properties of a finished device are not known and cannot be easily measured to the accuracy needed to obtain required device performance predictions. For example, the epi-thickness of a finished device is very difficult to determine. The effect of slightly varying modeled epi-thickness on device predictions, however, can be considerable. Modern devices also have very complex doping schedules. Doping of an ion-implanted FET, for example, varies both as a function of depth and along the surface from source to drain. These doping gradients can have significant effects on device performance, yet are quite difficult to measure precisely. Other examples of difficult-to-determine physical parameters include deep-level densities and carrier capture properties at the semiconductor surface and in the active layer. These deep levels have been shown both experimentally[45] and theoretically[46] to have significant effects on device performance. To the circuit designer the method of determining these physical properties of the device becomes an empirical process or involves such an elaborate measurement process as not to be useful. An ultimate solution to these problems may come through the development of more accurate process simulation schemes.

The above discussion is not meant to imply that two-dimensional device modeling is not important to the circuit designer. Simulations of this type are very important in the process of determining which physical mechanisms are having significant effects on device performance. These effects can then be included into simpler analytical device models within large-scale circuit simulators.

Until computing and process simulation capabilities have matured considerably, physically based models used in large-scale circuit simulation will be primarily one-dimensional analytical models.

Shockley[47] developed the first analytic FET model in 1952. His model assumed constant mobility—independent of electric field—but predicted general device behavior. For accurate GaAs MESFET predictions, the incorporation of velocity saturation of electrons into the device model is critical. There have been two basic approaches to this problem. The first is to use an analytical expression to describe the field-dependent mobility. The most common velocity expression used for this purpose is given by

$$v(E) = \frac{\mu E}{1 - \mu E/v_{\text{sat}}}, \tag{1–10}$$

where μ is the low-field mobility and v_{sat} is the saturation velocity. Using this expression, along with standard assumptions used in one-dimensional modeling problems, an analytical model can be derived which requires no iterative computations. Expression 1–8 underestimates the velocity somewhat at low fields, but has been used effectively to model devices in conjunction with large-signal circuit simulations.[4,48] Current-voltage expressions derived from this approach are considerably more involved than the relationship expressed by Eq. 1–6. The resulting device description is a set of explicitly calculated algebraic expressions. The expressions are given in terms of physical dimensions of the device, material properties of the active layer, and the terminal voltages.

Yamaguchi and Kodera[49] have used Eq. 1–8 to develop a fairly unique device model. In addition to accounting for the velocity saturation of charge carriers in the semiconductor, they also remove the assumption of the abrupt change in the carrier concentration at the edge of the depletion region. In removing this approximation they also introduce the need for iteration to obtain solution. Madjar and Rosenbaum,[3] however, have developed a technique to reduce the iterations to one step. Using this technique they have incorporated the resulting model into a time-domain simulation and predicted power amplifier performance with considerable success.

A second approach to the field-dependent mobility problem is to use a two-piece velocity field approximation.[1] The velocity of electrons in the GaAs is then given by

$$\begin{aligned} v(E) &= \mu E \quad \text{for } E < E_0 \\ &= v_{sat} \quad \text{for } E > E_0 . \end{aligned} \qquad (1\text{–}11)$$

Use of Eq. 1–11, along with standard approximations, results in a device model which requires an iterative technique to obtain solutions. The resulting model has been used to obtain very accurate small-signal results.[1,50] One attractive feature of the model is that it can be modified in a fairly straightforward manner to include effects due to doping density changes in the epi-thickness.[5,46] The iteration required of these models detracts from their appeal as candidates for large-signal simulators. Higgins[5] and Weiss,[14] however, have both used such models in simulation techniques to obtain large-signal information. It should be pointed out that neither of these studies was accomplished using a full large-signal simulation program, but rather estimated the large-signal effects using approximations.

Both empirical and physically based modeling techniques are associated with certain advantages and disadvantages. For example, the use of physically based expressions for each of the elements is very appealing for monolithic microwave integrated-circuit (MMIC) applications. Using such an approach it is possible to obtain expressions for each of the nonlinear elements purely from physical data concerning the device (i.e. the geometry of the device and the GaAs material parameters). No electrical characterization of individual devices is required. In theory, this allows the circuit designer to optimize device and circuit simultaneously. Unfortunately, purely physical models are seldom as accurate as required for circuit design without some alteration of input data. The inaccuracies arise from the approximations required to perform the device analysis. For example, MMIC devices are typically fabricated by ion implantation. Most physically based models, however, are derived from assumptions of an active layer with a fixed doping concentration. The physical parameters defining doping density and epi-thickness required of the model become empirical parameters that must be determined. Another problem with physical models is that many times the circuit designer does not have access to information concerning the physical properties of the device. At the other extreme, table lookup techniques are capable of extreme accuracy. The difficulty with these models is that an extremely large amount of tedious device characterization is required before the model can be implemented. In addition, minor changes in device geometry require that complete recharacterization be performed. The functional empirical models typically require less characterization than table lookup models, but a unique model is still required for each minor processing change and simultaneous device/circuit optimization is not possible.

The ultimate large-signal analog circuit simulation package will need to incorporate both physical and empirical modeling capabilities. This can be accomplished in several different ways. More than one model can be available in the simulator. This is the technique that has been used with MOSFET models in SPICE, for example, where several different device descriptions are available. This allows the circuit designer to make use of the model which best fits his or her current application. A second technique is to use a physically based model which incorporates certain device mechanisms through empirically determined parameters. The physical parameters of the model, such as gate length or mobility, may also be treated as empirical parameters. This technique can be used for any physically based model but may not be an optimal solution. Ideally, empirical parameters may be determined or estimated directly from some measured characteristics. The parameters of the physical model, however, may be difficult to relate directly to measured data. Finally, an empirical model can be implemented with optional expressions which relate the parameters required of the model to device physics. The process of determining the relationships between model parameters and device physics, however, is not straightforward. The fabrication and characterization of a number of devices with widely varying geometry and material properties will be required to determine such relationships. The process could be a very expensive undertaking and the results would apply to only one particular fabrication process.

1.3 Characterization and Parameter Extraction

There is an important relationship between device model, parameter extraction technique, and device characterization. The accuracy of any device model is limited ultimately by how accurately the model parameters can be determined. In order to obtain device model parameters that adequately describe performance behavior, it is almost always necessary to perform some device characterization. Actual measured device characteristics are then compared to model predictions, and model parameters are adjusted until acceptable agreement between the two is obtained. A device model that involves only a small number of device parameters which are determined from a limited number of simple-to-perform measurements has tremendous advantages over device models which require that a large number of parameters be determined from several tedious measurements.

Parameter extraction from measured characteristics has been used with considerable success to obtain equivalent circuit element values for small-signal device models. In the case of small-signal applications the measured data which is used in this process typically consists of *s*-parameters for the device biased at the appropriate bias level and covering the frequency range of interest. This *s*-parameter data completely describes the small-signal performance of the device at the specified bias level.

It is less obvious what set of measured data should be used in this process when large-signal applications are of interest. An argument can be given for attempting to match large-signal model predictions to any or all of the following: DC current-voltage characteristics, large-signal *s*-parameter data, small-signal *s*-parameter data or load-pull contours. The amount of data represented by this list is not small. The current-voltage characteristics that could be examined include not only the typical I_{ds} versus V_{gs} and I_{ds} versus V_{ds} curves but also data related to forward-conduction gate current and to avalanche breakdown properties of the gate-drain junction. Large-signal *s*-parameter data must be measured over the entire frequency range of interest and also at all power levels to be considered. Small-signal measured *s*-parameters must also be obtained over the entire frequency range of interest, and, in addition, must be obtained at several bias levels to be useful in large-signal applications. Finally, to obtain load-pull contours is a very tedious process which requires a great deal of microwave equipment. Data must be obtained for a number of termination impedances at several power levels.

Clearly, it is desirable to minimize the number of measurements required to perform accurate parameter extraction. In order to intelligently choose a minimal set of measurements, which is sufficient for parameter determination purposes, it is helpful to understand which model parameters most affect performance predictions of the simulation. For example, if it is determined that, for a particular application, avalanche breakdown between the gate and drain does not occur, then measurements designed to characterize that mechanism obviously could be avoided.

Determining performance prediction sensitivity to model parameters can also be important in choosing how heavily to weigh various measured characteristics in the parameter optimization process. An ideal model will accurately match all important measured characteristics using a minimal set of measurements. Typically, however, obtaining an ideal match between model predictions and measurement of one particular parameter involves compromise in terms of another characteristic. Information which specifies the sensitivity of performance predictions to model parameter values determines what compromises are most appropriate—and ultimately what modifications should be made to obtain a better model.

1.3.1 Device Characterization

One of the earliest measurement techniques designed to characterize large-signal device properties is the load-pull measurement. Load-pull characterization requires that the device being measured is embedded in circuitry which can be impedance tuned and which is capable of simultaneously monitoring performance characteristics.[51] Tuning is done until some specified performance is achieved. The tuned impedance associated with that performance is then measured. By determining a number of different load impedance values associated with several performance specifications, a map of performance as a function of circuit impedance can be produced. It should be noted that a characterization performed in this manner applies only for the RF power levels used in the measurement. To make the characterization process more complete, measurements must be made at a number of different incident power levels.

Cusack *et al.*[17] succeeded in developing automated load-pull measurement apparatus in 1974. Their computer controlled equipment configuration was capable of mapping constant power and efficiency contours onto a Smith chart for dynamic matching of both input and output circuits. The technique has been used in various slightly modified forms to successfully design power amplifiers.[18] As normally realized, load-pull measurements are not capable of describing harmonic content of the measured device. Poulin,[51] however, describes a system capable of measuring harmonic content of the RF output signal as a function of device load. Such a system could, in principle, be used to completely determine large-signal device characteristics. The amount of equipment required to perform such a general characterization, however, is not insignificant. Two network analyzers, a spectrum analyzer, a traveling-wave–tube amplifier, and various other pieces of microwave equipment would be required. In addition to requiring a lot of expensive microwave equipment, the measurement also involves multiple de-embedding techniques, so it is best achieved using fairly powerful computer capabilities to control the equipment and perform the calculations. Because of these problems, the technique is not easily utilized for the extraction of parameters required to define large-signal device models.

Another technique which has been used for large-signal device characterization is to perform measurements similar to *s*-parameter measurements, but under conditions of high RF signal levels. The resulting data are referred to as large-signal *s*-parameters. As in the case of load-pull measurements, the data obtained in this manner is applicable at only the power level used to perform the experiment. Complete characterization requires that measurements be performed at multiple signal levels. The technique has been used with some success in the design process for power amplifiers[19] and oscillators.[20] Large-signal *s*-parameters have been shown to be closely related to load-pull contours[52] in terms of the information contained concerning the device. Thus, limitations and capabilities of the two techniques are similar. One particular limitation of the large-signal *s*-parameter technique is that the harmonic content of the measured signal is not considered in the measurement. This limitation, which also exists for standard load-pull measurements, makes it difficult to use the technique for the general device characterization required of large-signal circuit simulators.

Probably the most easily performed measurement, which can be used to extract model parameters, is the measurement of DC *I-V* behavior. This is an attractive approach because of the simplicity of performing such measurements. Once the drain current is described as a function of drain-source and gate-source voltages, both transconductance and output resistance can be computed from Eqs. 1–3 through 1–5. This sort of characterization has been utilized in some investigations,[10–12] but provides limited accuracy for many cases.[14,42] Such measurements also fail to specify parameters related to capacitance-voltage characteristics. Parameter extraction techniques, which make use of DC measurements, must also involve either some RF characterization or physical-device descriptions to specify the gate-source and gate-drain capacitance properties. DC measurements can certainly be considered in the specification of some device parameters, but cannot be used as the only method of characterization if accurate models are to be developed.

As mentioned previously, parameter extraction for small-signal models has made use of measured *s*-parameters with great success. The technique can be made to apply to the large-signal problem by performing small-signal–model parameter extraction at multiple bias levels throughout the operating range of the device. This results in an equivalent circuit with different element values for each bias level considered. The large-signal model must then describe the behavior of each element value as a function of terminal voltage condition. One advantage this technique has over others is that the measurement equipment and software developed for small-signal applications are directly applicable to the large-signal problem. The technique can be used to produce extremely accurate predictions,[4,7,13] but is fairly tedious. Previous studies have used *s*-parameters measured at as many as 50 bias levels to define the large-signal model.[7,14]

1.3.2 DC versus RF Characterization

To illustrate the effects of using DC or RF characterization in the parameter extraction process, a power amplifier application will first be considered.

The device and circuit of this investigation have been characterized extensively by Willing *et al.*[7] Two different device models are used in the simulations. Both models are derived from the physically based model of Lehovec and Zuleeg,[2] and both use the empirical expressions given by Eqs. 1–8 and 1–9 to describe capacitance-voltage relationships. The equivalent circuit topology which describes the two models is illustrated in Fig. 1–6. The distinction between the models lies in whether the equivalent circuit element representing output resistance is determined from DC or RF characteristics. A time-domain simulation (modified Berkeley SPICE[21]) is used to obtain performance predictions of the amplifier.

Initially, only DC characteristics are considered in determining the output resistance of the device. Figure 1–7 shows the comparison between measured and modeled DC current-voltage characteristics. Because the velocity-field characteristics assumed in performing the calculations are described by Eq. 1–10, the model cannot predict negative differential resistance, which is exhibited in the measured current-voltage characteristics. The limitation that this property of the model imposes on *I-V* predictions is clearly evident for gate-source voltage values between 0.0 and −2.0 V. The worst-case error between measured and modeled drain-source current is approximately 20 percent. For gate-source voltage values of −2.0 V or less, however, the model is seen to be quite accurate.

Using this DC current-voltage relationship, the output resistance of the device can be modeled by computing the derivative specified by Eq. 1–6. Thus, the accuracy of the DC output resistance predictions should correspond approximately to the accuracy of the current-voltage predictions. From the data of Fig. 1–7 it is seen that DC output resistance predictions should be most accurate for gate-source voltage values from about −2.0 V to pinch-off.

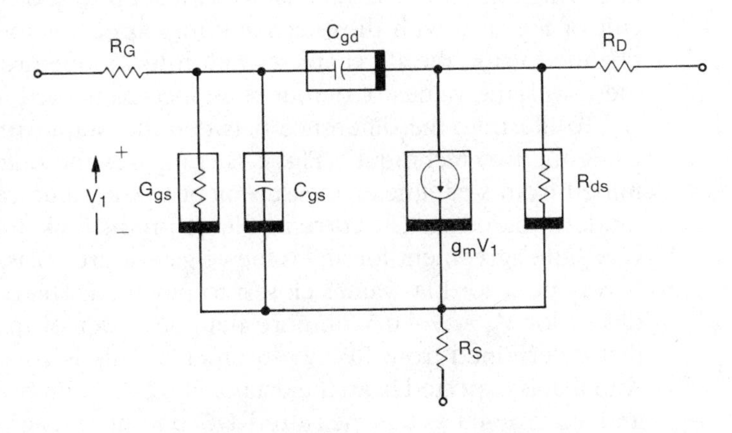

Fig. 1–6. A large-signal equivalent circuit model for the MESFET incorporating five nonlinear elements.

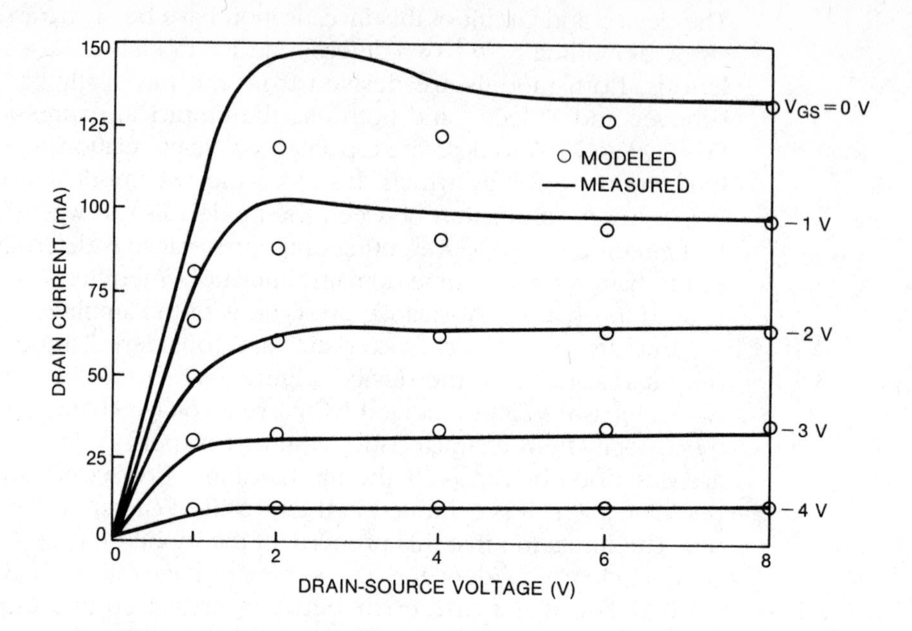

Fig. 1–7. Measured and modeled DC current-voltage characteristics of the power FET investigated by Willing. *(After Willing et al.*[7]*)*

The output resistance described for this first model depends only on the DC characteristics of the device. An alternate method for specifying the output resistance of the device is to use measured *s*-parameters to determine equivalent circuit element values. Using this approach, DC characteristics are ignored for the calculation of output resistance. The element values of the model are determined instead by using an optimization routine which matches the RF characteristics predicted by the equivalent circuit of Fig. 1–6 with those actually measured for the device. The values obtained using the RF characterization technique are not in good agreement with the values of output resistance computed from the DC curves.

To illustrate the difference between the output resistance determined using the two techniques, Fig. 1–8 compares the output resistance determined from *s*-parameter measurement to the value calculated by the first model—based on DC current-voltage predictions. It is seen that there is very little agreement for any value of gate-source bias, and that agreement is very poor for bias values closest to pinch-off. The output resistance predicted for $V_{gs} = -4.0$ V is more than an order of magnitude larger than that determined from RF measurements. This is contrary to the relationship that is expected from the data of Fig. 1–7. For a bias level of $V_{ds} = 6.0$ V and $V_{gs} = -4.0$ V, the measured DC output resistance is approximately 4000 Ω. The output resistance computed by the DC model is in good

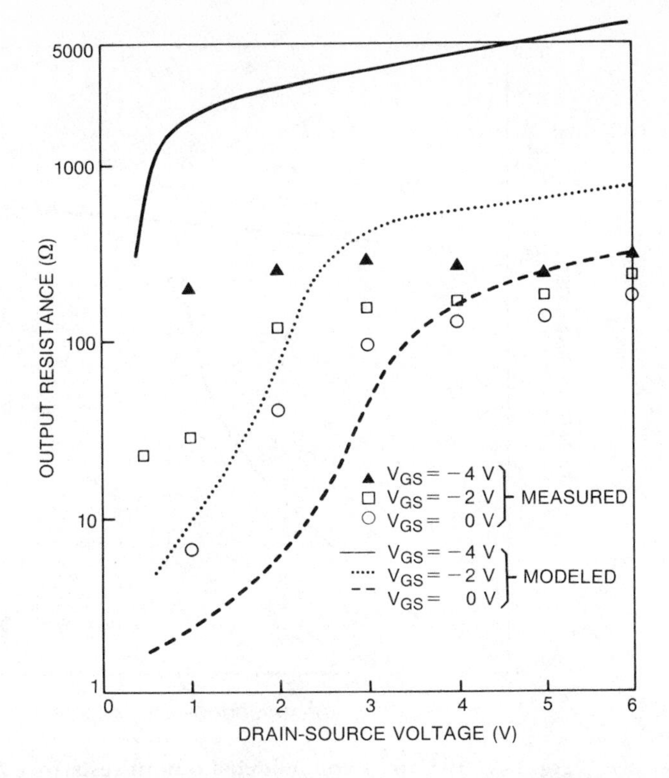

Fig. 1–8. Measured and modeled output resistance for a power FET. Measured characteristics are determined from high-frequency *s*-parameter data. Modeled characteristics are determined from DC current-voltage characteristics.

agreement with this value. Using the RF characterization technique, however, the output resistance is found to be only 230 Ω. Note also that negative resistance is computed for the device described by the measured curves of Fig. 1–7 when DC data is used. This is not the case for any bias level when RF characterization is considered. Clearly, output resistance computed from DC characteristics is significantly different from that computed from RF measurements.

The second model to be used in the simulations modifies the output resistance expression empirically to more closely approximate the output resistance determined from the measured *s*-parameters. A plot of the resulting predictions is compared to measured values in Fig. 1–9.

Using each model, characteristics of the device under large-signal conditions were simulated. The circuit simulated corresponds to the one fabricated and tested by Willing *et al.*[7] Fundamental, second-, and third-harmonic output powers, as a function of the input RF power level, were evaluated and compared to measurements. Both models were capable of predicting

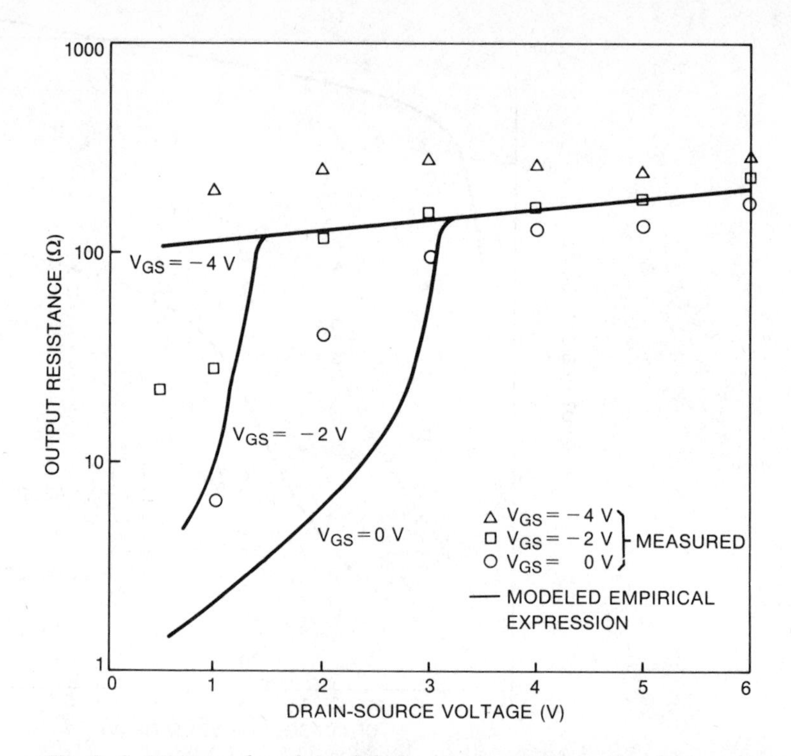

Fig. 1–9. Measured and modeled output resistance for a power FET. The resistance is modeled using an empirical expression to improve the agreement of the model with *s*-parameter measurements.

fundamental saturation characteristics and third-harmonic content of the signal in excellent agreement with measurement. Figures 1–10A and 1–10C show the agreement obtained. The agreement between measured and modeled characteristics of the second-harmonic output level was not as good for the DC characterized model. Significant improvement, however, was shown using the RF characterized model. A plot showing predictions from both models along with measured results is presented in Fig. 1–10B.

In the investigation just presented, the effect of using DC or RF characterization of only one element—output resistance—was considered. Weiss and Pavlidis,[14] however, have compared three different models: a model whose parameters are determined primarily from DC or low-frequency measurements, a model whose parameters are determined only from high-frequency *s*-parameter measurements, and a model which uses physical properties of the device for input. Their study compares predictions of gain-saturation characteristics of the three models with those measured for an actual circuit using the device. In agreement with the above investigation, they find that the DC characterized model produces predictions far less accurate than the RF characterized model. The physical model used in their work is based on the model of Pucel *et al.*[1] and also shows good agreement with measured characteristics.

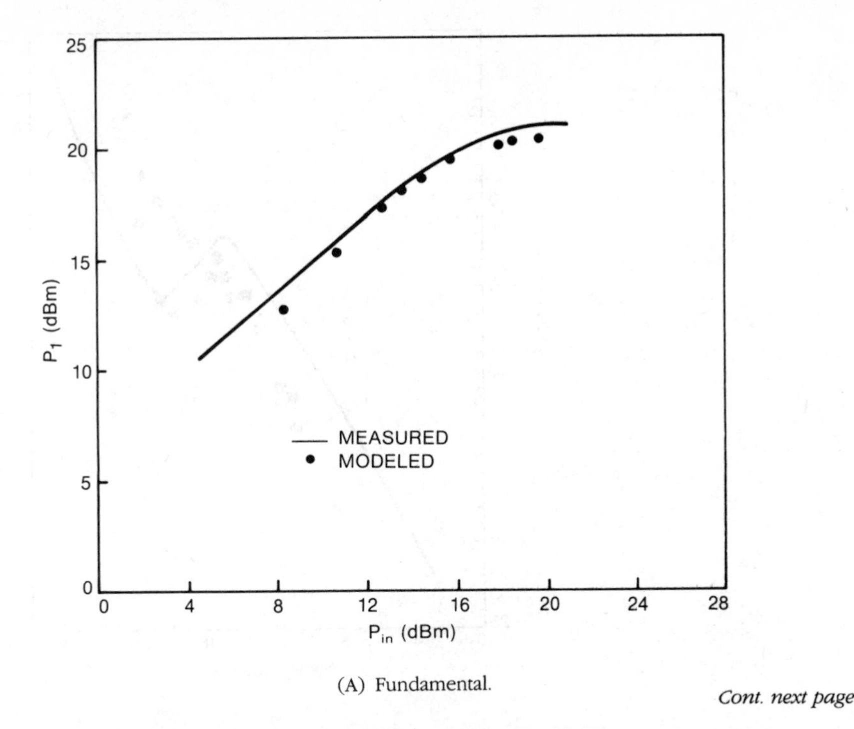

(A) Fundamental.

Cont. next page

Fig. 1–10. Measured and modeled output power as a function of input power level for a power FET in a 50-Ω system. Circles represent predictions of the model using output resistance as illustrated in Fig. 1–8. Squares represent predictions of the model using resistance as illustrated in Fig. 1–9.

A question arises concerning why models based on DC current-voltage characteristics, and evaluation of Eqs. 1–3 and 1–5 for transconductance and output resistance, are found to be less accurate than RF characterized models in some applications. Smith *et al.*[42] have performed experiments which provide some insight into this question. Their investigation utilized the equipment configuration shown in Fig. 1–11. For this measurement, a 1-MHz signal is amplified and used to drive the gate of an FET device under test. Part of the same signal is split from the source, phase shifted and applied to the drain of the same FET device. The resulting effect on the device is that a range of instantaneous gate-source and drain-source terminal voltage levels is cycled through at a 1-MHz rate. By altering the phase shift in the drain branch of the circuitry, a wide range of V_{ds}-V_{gs} combinations can be applied to the device. Using a digitizing oscilloscope and computer, the instantaneous terminal voltages and corresponding drain-source current can be monitored. The data obtained in this manner is then sorted to produce current-voltage characteristics of the device. It should be pointed out that a frequency of 1 MHz is low enough that capacitive effects in the transistor equivalent circuit model are negligible.

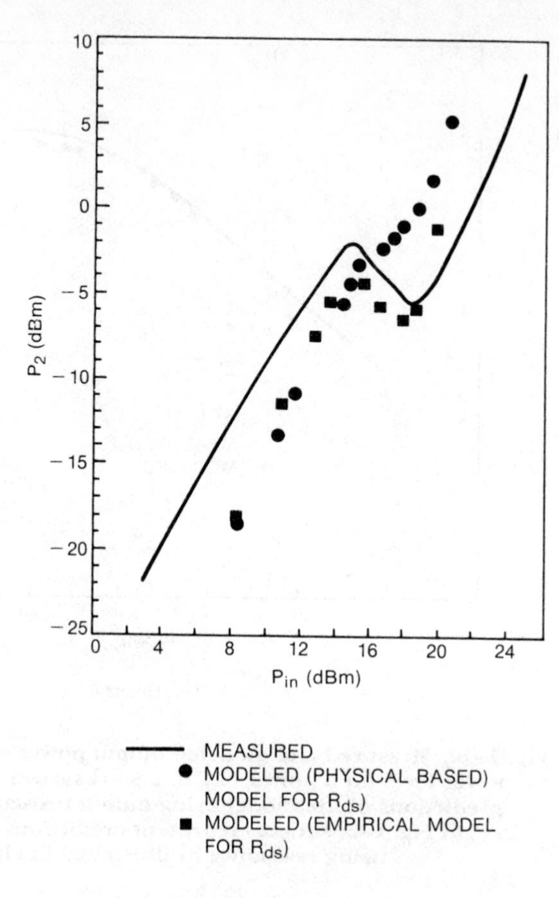

(B) Second harmonic.

Fig. 1–10. *continued*

The key finding in the study just cited is that the DC current-voltage characteristics differed significantly from the 1-MHz characteristics as illustrated in Fig. 1–11. Thus, the output resistance characteristics of the device are also a function of frequency below 1 MHz. Figure 1–11 also illustrates this phenomena. A slight variation of this measurement on another FET device shows qualitative agreement with these results.[53] The output resistance was found to obtain a maximum value at DC, to decrease dramatically at a measurement frequency near 100 Hz, to continue to decrease for frequencies between 100 Hz and 100 kHz, and then to remain fairly constant. The results of this study are again in agreement with the two investigations previously described and indicate that parameter extraction techniques which depend only on DC characteristics may not be valid.

There is similar evidence that the DC characterization used to specify gate-drain breakdown characteristics of the device is inadequate. In a study by Wemple *et al.*[54] it was determined that DC avalanche currents are

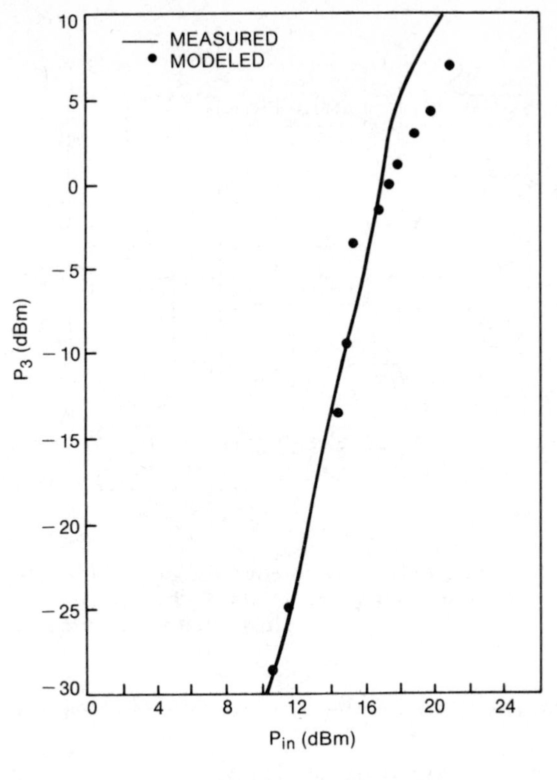

(C) Third harmonic.

Fig. 1–10. *continued*

generally much smaller than pulse values and are not directly applicable to RF performance questions. An option to measuring DC avalanche characteristics as proposed in this investigation is the measurement of avalanche processes using pulse excitation on the drain with DC applied to the gate.

It is still unclear as to what physical mechanism is responsible for the discrepancy between DC and low-frequency RF characteristics. The mechanism has a very long time constant typical of trapping states or thermal relaxation. Some studies seem to indicate that deep levels at the surface of the device and at the epi/substrate interface are involved in this frequency-dependent output resistance.[53] If further investigations can provide an explanation for the observed behavior, which can easily be modeled, or if fabrication techniques can eliminate it, then significant simplification of the characterization process can be realized. Such an explanation would allow measured DC characteristics to be converted to the RF characteristics required of accurate models, while the elimination of the effect would cause the two measurements to coincide. This capability would significantly reduce the amount of characterization required for parameter extraction.

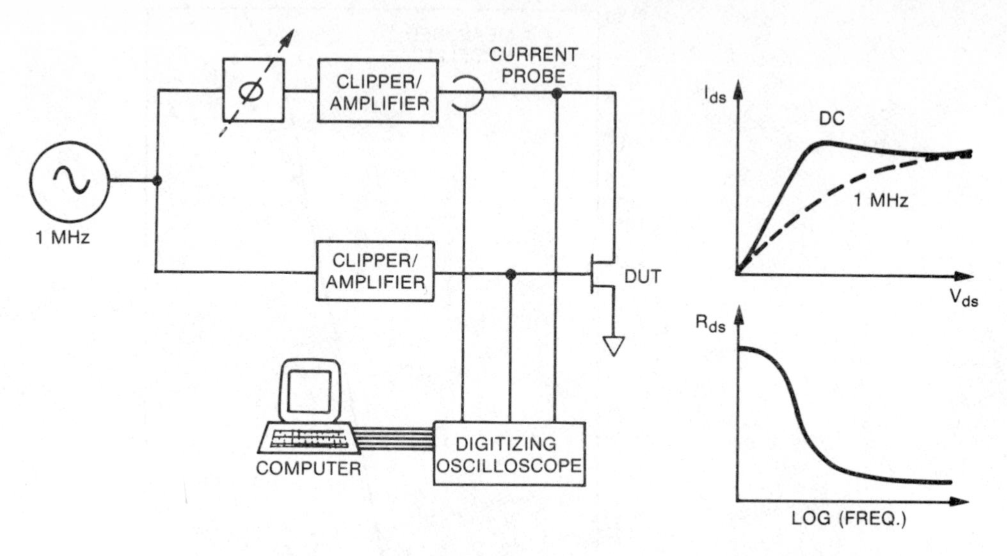

Fig. 1–11. Equipment configuration for the measurement of low-frequency current-voltage characteristics of a device. Results are illustrated schematically.

1.3.3 Sensitivity of Predictions to Model Parameters

Besides the need to determine which measurable device characteristics apply to the large-signal problem (as described in the previous section), it is also important to rank the relative significance of different measured characteristics. To some extent this will be dependent on the application under consideration and the particular device being used.

In this section, both a microwave mixer and an oscillator are considered. Both circuits make use of a 0.8 μm gate length device fabricated by ion implantation. The gate width of the device is 400 μm. Device *s*-parameters were measured at several bias levels and a variety of DC measurements were performed on the device.

Using the measured *s*-parameters, equivalent circuit element values for the circuit in Fig. 1–6 are determined at each bias level. Several different device descriptions were used in this work. The most complete model is described in the paragraphs to follow. Modifications to this model were then made to determine effects that various parameters have on performance predictions. Table 1–2 presents the resulting optimized element values for the most complete model at three bias levels. The parasitic resistance values, R_S, R_D, and R_G, were measured using the DC forward bias characteristics of the gate[40] electrode. Values were then assumed to be constant as a function of bias conditions. These forward-bias measurements were also used to determine the parameters required to describe

Table 1–2. Equivalent Circuit Element Values for 0.8-μm Device Used in the Investigations at Three Bias Levels

Bias Level (V)	V_{ds} = 3.0 V, V_{gs} = −2.0 V	V_{ds} = 5.0 V, V_{gs} = −0.8 V	V_{ds} = 1.0 V, V_{gs} = −0.3 V
R_G (Ω)	13.9	13.9	13.9
R_S (Ω)	3.60	3.60	3.60
R_D (Ω)	4.55	4.55	4.55
R_{ds} (Ω)	436	403	14.4
g_m (S)	22.3	46.7	67.0
C_{gs} (pF)	0.207	0.487	0.572
C_{gd} (pF)	0.126	0.103	0.150

the forward-conduction element G_{gs}. This element is modeled using a diode with ideal characteristics in series with a 12.5-Ω resistor. The diode characteristics are given by the expression

$$I = I_S[\exp(qV/NkT) - 1.0], \qquad (1\text{–}12)$$

where I is the current through the diode, V is the voltage across the diode, q is electronic charge, kT is the thermal energy, I_S is the reverse saturation current, and N is the ideality factor. For the devices examined here, it was found that

$$I_S = 6.8 \text{ pA} \quad \text{and} \quad N = 1.15$$

provided a good fit to the data.

Empirical equations 1–8 and 1–9 were used to model both the gate-source and gate-drain capacitance. Each of these expressions has been found to match measured capacitance-voltage characteristics of a wide variety of devices with extreme accuracy. For the present study, agreement between measured and modeled values for C_{gs} and C_{gd} is within 10% for all bias levels considered.

Although the device model used in these studies is physically based,[2,4] information concerning doping concentrations, electron mobility, saturation velocity, and epi-thickness was not available. These parameters were therefore treated as empirical parameters to match model predictions to measured device characteristics. During this process, both DC and RF characterization data were considered. Figure 1–12 presents a comparison between the resulting model predictions and measured DC characteristics. Although better agreement of the model predictions with these DC curves could be obtained, this agreement is gained at the expense of agreement with measured RF output resistance.

The comparison between measured and modeled output resistance is illustrated in Fig. 1–13.

Transconductance measured from DC characteristics is found to be very close to the values determined from RF measurements so that only the DC

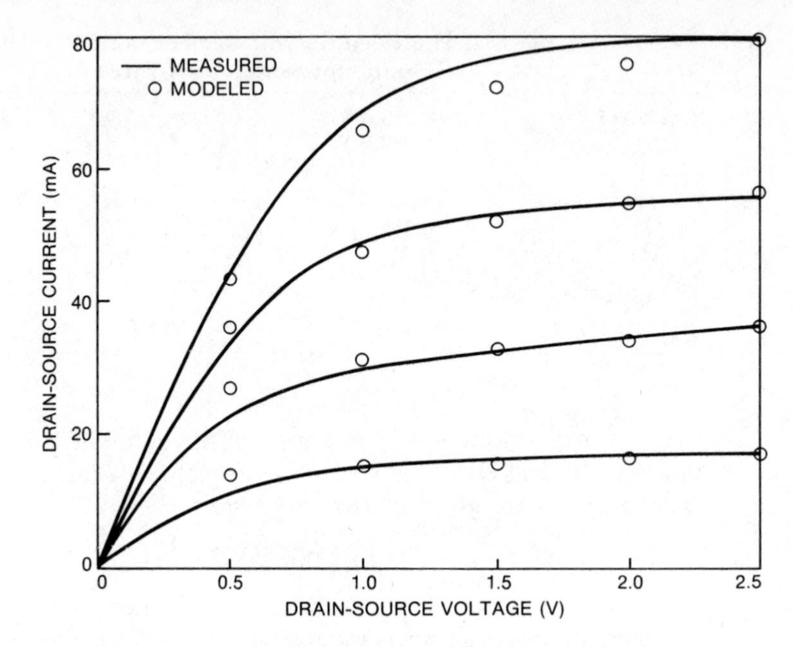

(A) As a function of drain-source voltage.

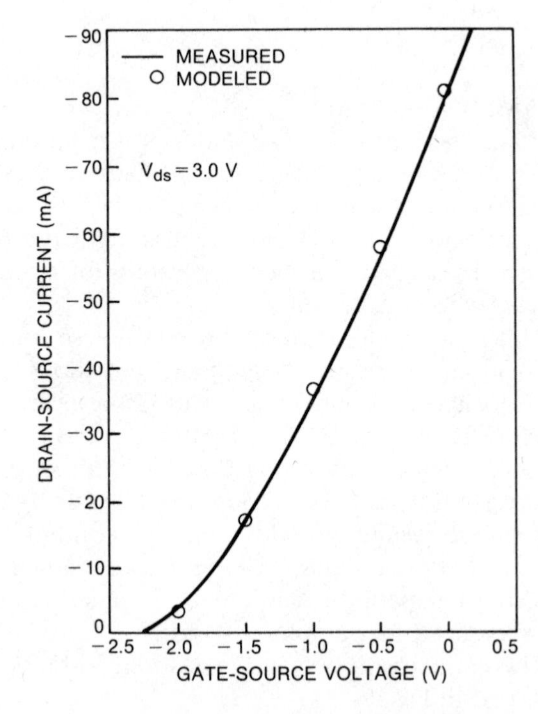

(B) As a function of gate-source voltage.

Fig. 1–12. Measured and modeled drain-source current for a 0.8 × 400 μm gate length device.

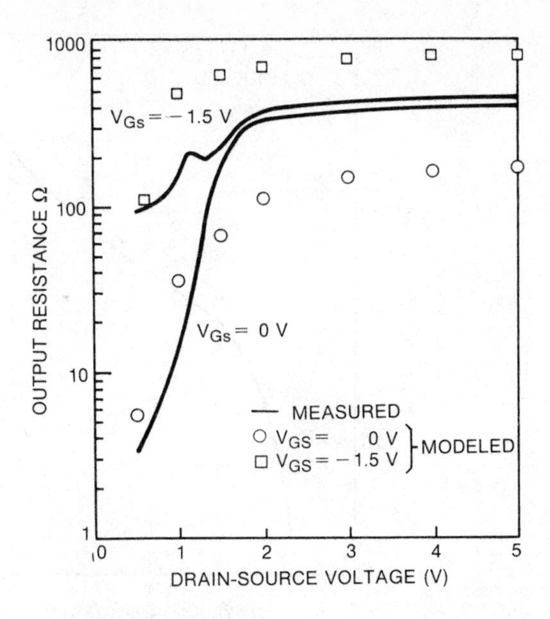

Fig. 1–13. Measured and modeled output resistance as a function of drain-source voltage for a 0.8 × 400 μm gate length device. Measured output resistance is determined from high-frequency s-parameter measurements.

data was considered here. Figure 1–14 shows the transconductance predicted by the model compared to that measured from DC characteristics for a drain-source voltage of 3.0 V. Agreement is seen to be excellent.

Using the device model as described above, the mixer circuit of Fig. 1–15 was simulated with ALLSPICE,[55] a modified version of SPICE which is run on personal computers. The local-oscillator (LO) frequency used in the simulations is 6 GHz while the RF frequency is 8 GHz. The circuit elements C_I and L_I form a simple matching circuit to improve the power transfer properties of the input circuit at the RF and LO frequencies. Inductor L_I also serves to short out any IF signal present at the input and is used to bias the FET. The output circuit is composed of a low-pass filter formed with the elements L_0 and C_0. This network tends to short-out RF, LO, and higher frequencies while passing the 2-GHz IF frequency. The conversion gain of such a network is defined as

$$A_{GV} = \frac{P_{\text{out}}(f_{\text{IF}})}{P_{\text{in}}(f_{\text{RF}})}, \tag{1–13}$$

where $P_{\text{out}}(f_{\text{IF}})$ is the output-power level of the IF signal and $P_{\text{in}}(f_{\text{RF}})$ is the incident-power level of the RF signal. Conversion gain is often expressed in dB as

$$A_{GV(\text{dB})} = 10 \log A_{GV}. \tag{1–14}$$

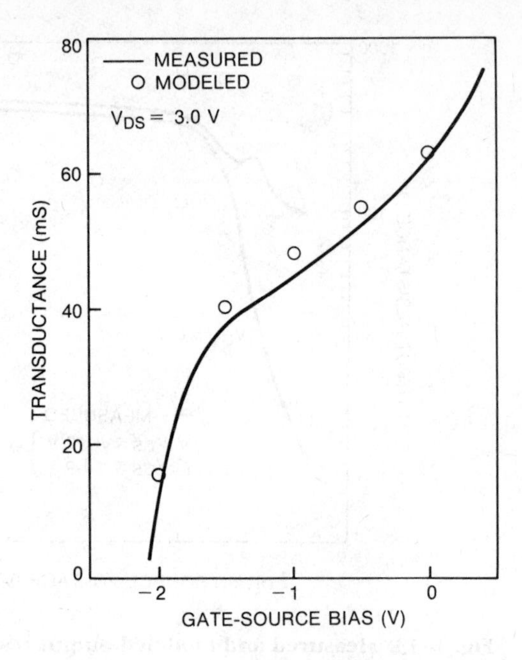

Fig. 1–14. Measured and modeled transconductance as a function of gate-source voltage for a 0.8 × 400 μm gate length device.

Conversion gain properties of the device, when embedded in these circuits, were analyzed as a function of input LO power. A number of model parameters were varied and the effect these variations had on performance predictions was noted. In the simulations, RF input power was kept at a level approximately 10 dB below the LO input power.

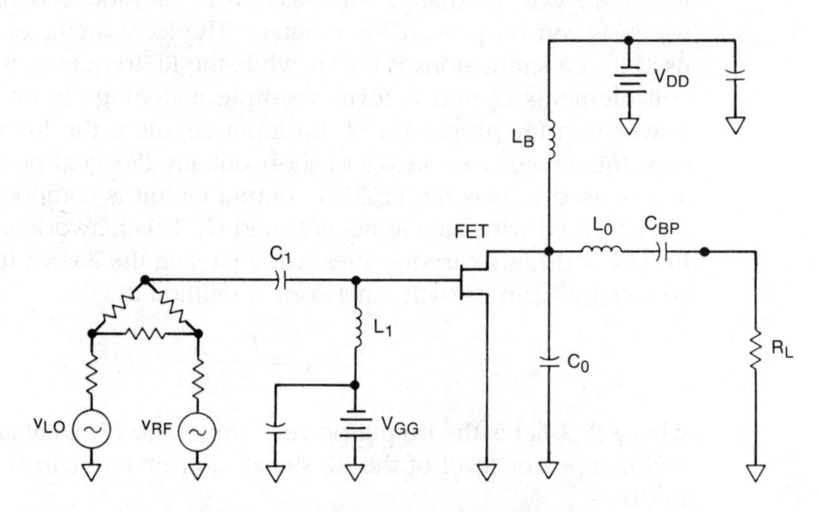

Fig. 1–15. Topology for the simulated mixer circuit.

Six different device descriptions were used in the simulations. The results are presented in Fig. 1–16. The model referred to as FULL MODEL is as described above. This is the only simulation that included the effect of forward conduction through the gate.

The absence of a forward-conduction mechanism is the only feature that distinguishes this model from that labeled NO FORWARD CONDUCTION. As is shown, there is little difference between the predictions using this model and the FULL MODEL except at input power levels above about 12 dBm. Notice that the conversion gain predictions of both models have begun to decrease for input LO powers greater than 13 dBm.

The third device description, CGS, CGD CONSTANT, neglects the voltage dependence of the gate-source and gate-drain capacitance. For this simulation, the capacitance values were fixed at their measured small-signal values. The bias used for these simulations is defined by V_{ds} = 3.0 V and V_{gs} = −2.0 V so that from Table 1–2, it is seen that C_{gs} = 0.207 pF and C_{gd} = 0.126 pF. The agreement with the FULL MODEL is still quite good, displaying a maximum 1.5-dB difference in predicted conversion gain. Prediction of the LO input power required to obtain maximum conversion gain is also in good agreement with the FULL MODEL.

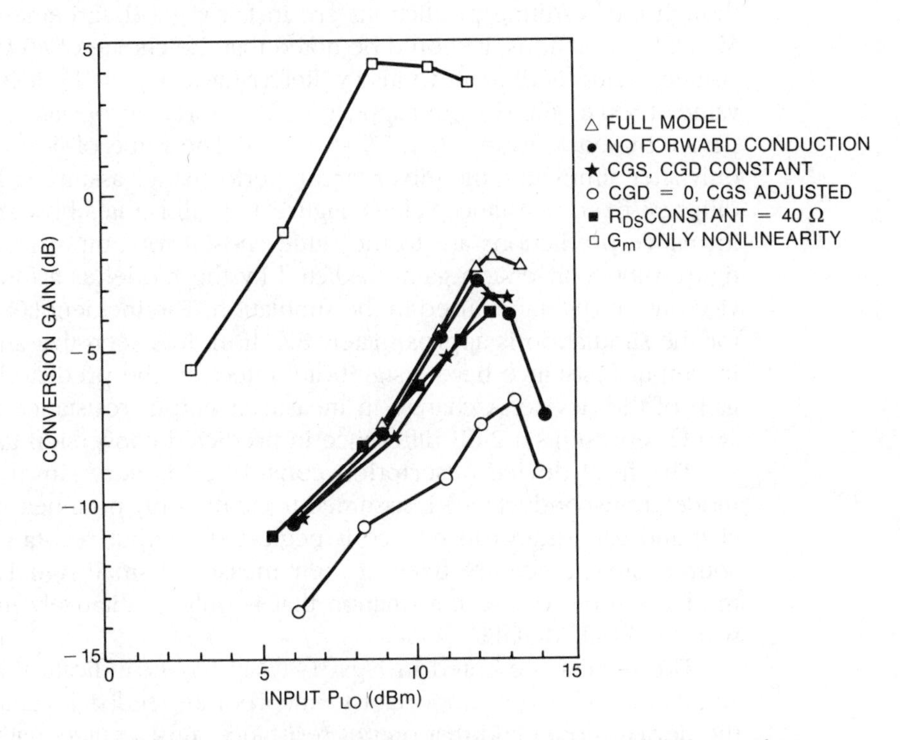

Fig. 1–16. Predicted conversion gain using several different device models. The modeled devices are assumed to be embedded in the circuit of Fig. 1–15.

The relatively minor effect this constant capacitance assumption has on performance predictions suggests that further simplifications in the device model might be possible. The model labeled CGD = 0, CGS ADJUSTED tests this hypothesis by neglecting the gate-drain capacitance. The gate-source capacitance is adjusted to attempt to account for the omission of a gate-drain capacitance. The adjustment was accomplished by reoptimizing the equivalent circuit element values to match measured *s*-parameters at all bias levels while requiring C_{gd} to be zero. Despite the adjustment in gate-source capacitance, conversion gain predictions for the simplified model were significantly different than for the FULL MODEL. This result is in agreement with the findings of Begemann and Jacob.[56] Although in qualitative agreement with the FULL MODEL predictions, the predicted conversion gain of this simplified model is less than the FULL MODEL values by as much as 5 dB.

The model labeled RDS CONSTANT = 40 Ω places a fixed 40 Ω of resistance in parallel with the output branch of the device model. The effect of this resistance is to set the maximum output resistance value of the device at 40 Ω. To keep this resistance from affecting DC characteristics of the device, a large capacitance is placed in series with this resistance. Although the resulting predictions are in fairly good agreement with FULL MODEL predictions, it should be noted that the choice of 40 Ω for this resistance value is difficult to justify. Referring to Fig. 1–13, it is seen that a value of 300 to 400 Ω would appear to be a more appropriate choice for the device biased at V_{ds} = 3.0 V, V_{gs} = −2.0 V. The value of 40 Ω was chosen only after simulating the mixer circuit performance assuming a number of different fixed resistance values. Figure 1–17 illustrates how sensitive performance predictions are to the value chosen for output resistance. The figure plots conversion gain predicted by the model as a function of device output resistance used in the simulation. The incident LO power level for the simulation is approximately 8.4 dBm. It is seen that small changes in output resistance have a significant effect on the predicted conversion gain of the device. A change in measured output resistance from 100 to 160 Ω represents a 2-dB difference in predicted conversion gain.

The final device description considered is very simplistic. In this model, transconductance is assumed to be the only nonlinearity of the device and gate-drain capacitance is neglected. Output resistance and gate-source capacitance are fixed at their measured small-signal values. The model is seen to give information that is only qualitatively in agreement with the FULL MODEL.

The results presented in Figs. 1–16 and 1–17 indicate that for mixer simulations it is very important to incorporate feedback capacitance into the device model and that output resistance must be adequately modeled if accurate predictions are to be obtained. Capacitance nolinearities and

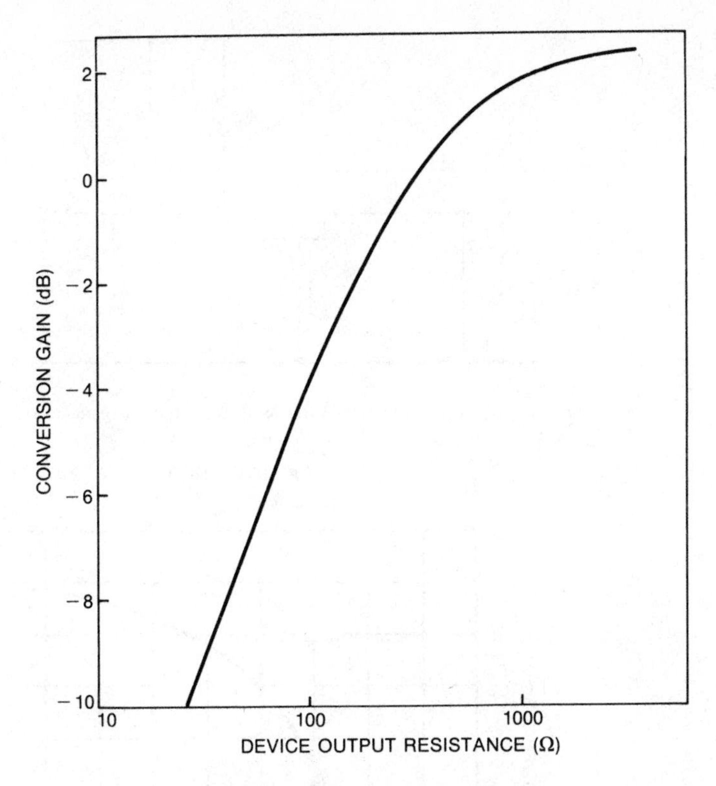

Fig. 1–17. Predicted conversion gain as a function of modeled output resistance. Input LO power is 8.4 dBm and the device is assumed embedded in the circuit of Fig. 1–15.

forward conduction are second-order properties, but do have an effect on model results for some signal levels.

The conclusions drawn from the data presented in Fig. 1–16 may not be applicable to simulations of oscillators or power amplifiers. To examine the sensitivity of oscillator performance predictions to model descriptions, many of the same models described in the above mixer study were used to simulate a fixed-frequency 2-GHz MMIC oscillator. The circuit design for the oscillator is shown in Fig. 1–18A. The circuit was fabricated using MMIC technology and the MMIC inductors were simulated using a modified Greenhouse model.[57] Measured oscillator performance as a function of applied bias is compared to modeled predictions in Fig. 1–18B. The device model used in these simulations is identical with the FULL MODEL referred to in the mixer study. Agreement between measurement and theory is excellent.

For a bias level of 5 V, simulations of the oscillator circuit were also performed using alternative device descriptions. Results of these studies are

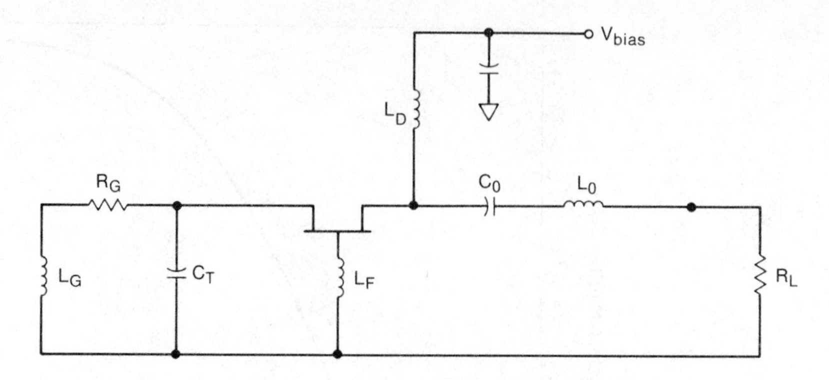

(A) Topology of the oscillator circuit investigated.

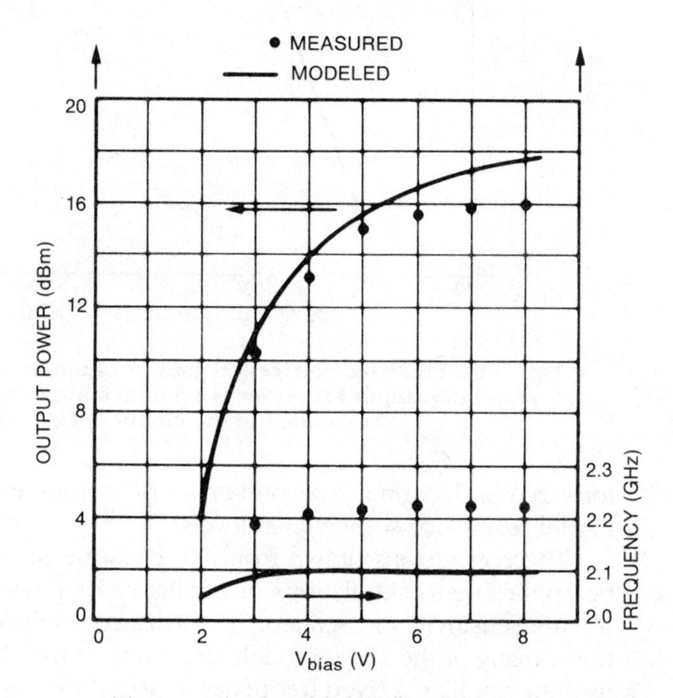

(B) Measured and modeled oscillator performance.

Fig. 1–18. Simulation of 2-GHz MMIC oscillator.

presented in Table 1–3. The device descriptions labeled as FULL MODEL, NO FORWARD CONDUCTION, and CGS, CGD CONSTANT are identical with those described for the mixer study with the same labels. The FULL MODEL is seen to give best overall predictions of frequency and power as measured for the actual circuit. The model labeled NO FORWARD CONDUCTION does predict an output frequency approximately 50 MHz closer

Table 1–3. Predicted and Measured Output Power and Frequency for the MMIC Oscillator and Several Different Device Models

Model Used or Actual Measurement	f_{out} (GHz)	P_{out} (dBm)
Actual measurement	2.24	10.0
FULL MODEL	2.17	9.4
NO FORWARD CONDUCTION	2.22	8.5
CGS, CGD CONSTANT	2.58	>13
RDS CONSTANT	2.22	>13

to the actual measured frequency than the FULL MODEL. The output power predictions of this model, however, are not as accurate. A model assuming fixed device capacitance was also investigated. These simulations predicted an output frequency significantly higher than the measured value. The simulation of this device did not reach steady state as quickly as the previous simulations. When the simulation was stopped, the predicted output power was approximately 13 dBm. Although the frequency had stabilized, oscillations were still growing in amplitude. For the final device description, labeled RDS CONSTANT, output resistance of the device was fixed. The value of the output resistor was kept at the value determined from small-signal measurements of the device at the appropriate bias. This corresponds to a resistance of approximately 400 Ω. The simulation again had not reached steady state when the program was stopped. Output power predicted by the simulation was greater than 13 dBm.

The results of the oscillator study are in significant contrast to the mixer results as they relate to the nonlinear capacitances of the device. The inclusion of a forward-conduction mechanism would also appear to be more important for the oscillator simulations. All of the data reported in this section suggests the importance of adequately describing output resistance of the device. The study described in Sec. 1.3.2 also indicates the importance of accurate output resistance descriptions for power amplifier simulations. Similar model element sensitivity studies performed by Brazil *et al.*[58] indicate that nonlinear gate-source capacitance and gate-drain avalanche breakdown can be important for power amplifier design.

Notice that transconductance properties of the device were not altered in the simulations reported. The transconductance is the essence of what the active device is. Clearly, the transconductance must also be described accurately to obtain reasonable results.

1.3.4 Parameter Extraction

The accuracy of a device model is dependent ultimately on how accurately device-model parameters can be determined. Without reliable parameter extraction techniques, performance predictions of a device model are at best of the right functional form. Despite this fact, much less attention has been focused on parameter extraction techniques than on device modeling efforts.

Parameter extraction methods can be classified into two distinct categories: graphical/analytical methods and numerical methods. Graphical or analytical methods depend on carefully devised experiments which isolate the relationships between the parameters in question and the device-modeling equations. In the case of graphical methods, data from specific experiments is plotted on appropriate linear or logarithmic scales. By extrapolating data to certain intersection points or computing slopes, parameters are determined. Analytical techniques involve the solution of a set of formulas derived from the device-model equations and applied to measured data. Both techniques have been used with a certain degree of success for the determination of some modeling parameters.[38] These graphical/analytical methods can be very powerful and very efficient when techniques can be developed for the applications under consideration. Typically, however, these methods can be applied at most to a few of the required model parameters or only to very simple device models which lack required accuracy. In addition, such techniques are specific to the one set of modeling equations for which they are derived. If modifications are made to the model or another model is to be used, the techniques must often be redetermined.

Numerical methods of parameter extraction treat the process of determining optimum parameters as a least squares error minimization problem. An error criterion is established which quantifies the disagreement between modeled and measured characteristics of the device. Conventional minimization techniques, such as steepest descent or other methods, are then employed[59–61] to determine what parameter set produces a minimum value for the established criterion. These methods are quite generally applicable to any known explicit algebraic function which can be expressed in terms of independent variables.

The material discussed in the previous sections does not directly identify a particular parameter extraction technique as superior. There are still significant unanswered questions. The information presented, however, does identify certain important characteristics that any parameter extraction routine should have.

Parasitic resistance values should be determined before other parameters are investigated because these values can be found independently using graphical/analytical techniques. Simple DC forward-bias measurements can be performed[40] to determine these values. Alternately, expressions requiring physical data[38,39] can be evaluated if necessary. Forward-conduction parameters can also be determined accurately and independently from forward-bias measurements.

Capacitance values can be determined either from *s*-parameter measurements and model element value determination, from physical expressions,[1,12] or from two-dimensional device simulation.[4] The sensitivity data examined here indicates that this capacitance information must be obtained as a function of bias when oscillator or power amplifier applications are being considered. Once capacitance versus voltage is determined, the parameters to be used for the modeled capacitance can be optimized to match the data using standard numerical parameter-extraction techniques. When simplified expressions such as that given by Eq. 1–7 are used, only the capacitance at a single bias level is required to determine modeling parameters. More general expressions similar to Eqs. 1–8 and 1–9 require that capacitance values be determined at a number of bias levels.

Optimization of parameters describing DC characteristics and differential output resistance must be treated with some care. The data discussed in Secs. 1.3.2 and 1.3.3 indicates that RF output resistance must be accurately modeled for all analog applications. The DC current-voltage relationships have a secondary effect on the performance predictions. This indicates that a useful parameter extraction algorithm will consider both DC and RF information, but weight the RF output resistance data more heavily. In addition, transconductance data should be included in the optimization process. An error criterion which satisfies these requirements can be expressed as

$$E^2 = E_I^2 + E_R^2 + E_G^2, \tag{1–15}$$

where

$$E_G^2 = \frac{1}{N}\sum_{i=1}^{N} \frac{c_i}{g_{mi}^2}(\overline{g}_{mi} - g_{mi})^2,$$

$$E_R^2 = \frac{1}{N}\sum_{i=1}^{N} \frac{b_i}{R_{dsi}^2}(\overline{R}_{dsi} - R_{dsi})^2,$$

and

$$E_I^2 = \frac{1}{N}\sum_{i=1}^{N} \frac{a_i}{I_{dsi}^2}(\overline{I}_{dsi} - I_{dsi})^2,$$

with N the number of data points considered; I_{dsi}, R_{dsi}, and g_{mi} the measured values of drain-source current, output resistance, and transconductance; and $\overline{I}_{dsi}$, $\overline{R}_{dsi}$, and $\overline{g}_{mi}$ the corresponding model calculated values for current, output resistance, and transconductance. The a_i, b_i, and c_i are appropriate weighting factors. A numerical parameter-extraction technique which minimizes the criterion expressed by Eq. 1–15 can be applied to any device-model application.

Finally, gate-drain avalanche mechanisms should be included in device models to be used for power amplifier studies. It is important that these characteristics be measured under dynamic conditions. Parameters to describe this breakdown which are determined from DC data may cause considerable inaccuracies in model predictions.[54]

1.4 Circuit Simulation Techniques

The simulation of circuits and devices which exhibit nonlinear properties can be accomplished by a variety of methods. The most commonly exploited general simulation techniques are time-domain methods and harmonic-balance methods. A number of variations on the standard harmonic-balance method have been investigated, so that the harmonic-balance classification defines a broader range of approaches to the problem than does the term time-domain.

In addition to these general approaches to the nonlinear analysis problem, several different techniques have been developed to estimate large-signal behavior of specific circuits. These techniques typically use a combination of small-signal and large-signal methods to quickly determine nonlinear circuit behavior to the first order. Such techniques are usually applicable to a fairly limited class of circuits and device configurations and are discussed here under an umbrella classification termed "quasi–large-signal" techniques.

1.4.1 Time-Domain Methods

Time-domain analysis is conceptually straightforward because output parameters can be related to input parameters in a causal fashion. A number of general circuit design packages have been developed which depend on time-domain solutions of the circuit problem. The most commonly used large-scale circuit simulation of this type is probably SPICE.[21] Although such packages have been generally available since the mid-1970s, they have been used primarily for digital circuit applications.

Time-domain analysis techniques involve the determination of state equations for each node of the circuit being considered. In general, these equations are nonlinear differential equations which are solved using numerical integration. In employing such an approach, circuit elements are described by their current-voltage properties in the form

$$i(t) = F\left[v(t), \frac{dv(t)}{dt}\right], \qquad (1\text{–}16)$$

where F is an appropriately chosen function for the circuit element being described. Nonlinear elements are linearized for individual time steps in the simulation. Thus, to obtain valid solutions, the time steps must be kept small.

Although the time-domain approach is logical from the point of view of circuit-analysis theory, such an approach fails to take advantage of many efficient frequency-domain analysis methods. Circuit elements, like inductors and capacitors, are more efficiently modeled using phasor notation in the frequency domain than they are in the time domain. Likewise, transmission lines are represented in the frequency domain simply by a phasor rotation in the complex plane. Long integration times can be required to simulate the same lines using a time-domain approach. Especially when short transmission line segments are to be simulated, this limitation of time-domain analysis can be severe. Another disadvantage of the time-domain analysis is that adding circuit components to the circuit being analyzed also adds matrix elements to the system description. This increased matrix size is associated with increased difficulty in solving the problem.

To avoid the difficulty of analyzing transmission lines, these elements can be simulated quite accurately using a lumped LC circuit model of the line. Repeated sections of series inductor and shunt capacitor elements with appropriately chosen values can model transmission lines of arbitrary electrical length and characteristic impedance. A simple method for converting ideal transmission lines of given characteristic impedance and electrical length is described below.

Initially a value for the cutoff frequency, f_c, of the artificial transmission line model is chosen. This frequency is chosen to be larger than the highest frequency of interest in the analysis. The number of series-L, shunt-C segments to be used in the artificial line is then determined from

$$N > \frac{\theta_0}{f_0} f_c, \tag{1–17}$$

where θ_0 is the electrical length of the line in radians specified at the frequency f_0. Once the right-hand side of Eq. 1–17 is evaluated, the value of N is rounded to the next highest integer. The value of the series inductance of each section of the transmission line model is next computed from

$$L_{ser} = \theta_0 Z_0 / 2\pi N f_0, \tag{1–18}$$

where Z_0 is the characteristic impedance of the transmission line segment being modeled. Finally, the shunt capacitance value of the line segments is given by

$$C_{sh} = L_{ser}/Z_0^2. \tag{1–19}$$

The resulting N-section series-L, shunt-C equivalent circuit has been used in time-domain simulations with negligible loss of accuracy. Brazil *et al.*[58] used a similar technique and compared the results to those obtained using the built-in transmission line models of SPICE. No noticeable differences were observed between the simulation results and of the power amplifier circuit which used the transmission lines and the circuit which used an *LC* equivalent circuit.

Another advantage of using lumped-element representations of transmission lines is that attenuation characteristics of lossy lines are easily incorporated into the model. This is accomplished by including series and/or shunt resistance into the equivalent circuit sections.

Typically, for analog circuit applications only the steady-state solution to the circuit problem is of interest. When a sine wave or other periodic signal is applied to a circuit containing energy storage elements, however, steady-state behavior is not reached immediately. Instead, transient effects will be apparent in the resulting waveforms for a period of time. When time-domain analysis is performed, these transient effects are part of the simulated solution. Thus, the circuit being analyzed must be simulated until all transient effects vanish for the analysis to be valid at steady-state.

Figures 1–20 through 1–22 illustrate the transient behavior in the solutions of an amplifier, mixer and oscillator circuit simulation. In Fig. 1–20, results of a simple amplifier circuit are plotted. The simulated circuit is shown in Fig. 1–19. Transmission line segments, which are a quarter wavelength long at the fundamental frequency, are located at the input and output of the device and used for biasing. These line segments will effectively short circuit the device at the second-harmonic frequency. Under steady-state conditions, therefore, no Fourier component of the voltage at the

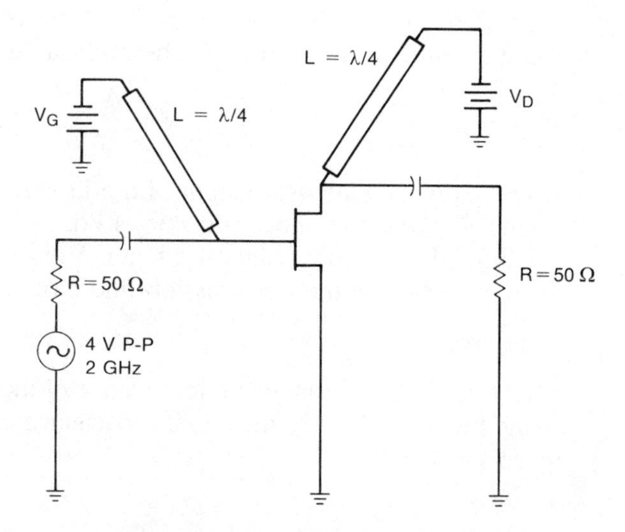

Fig. 1–19. Circuit simulated to determine approximate effects of transients in power amplifier simulations.

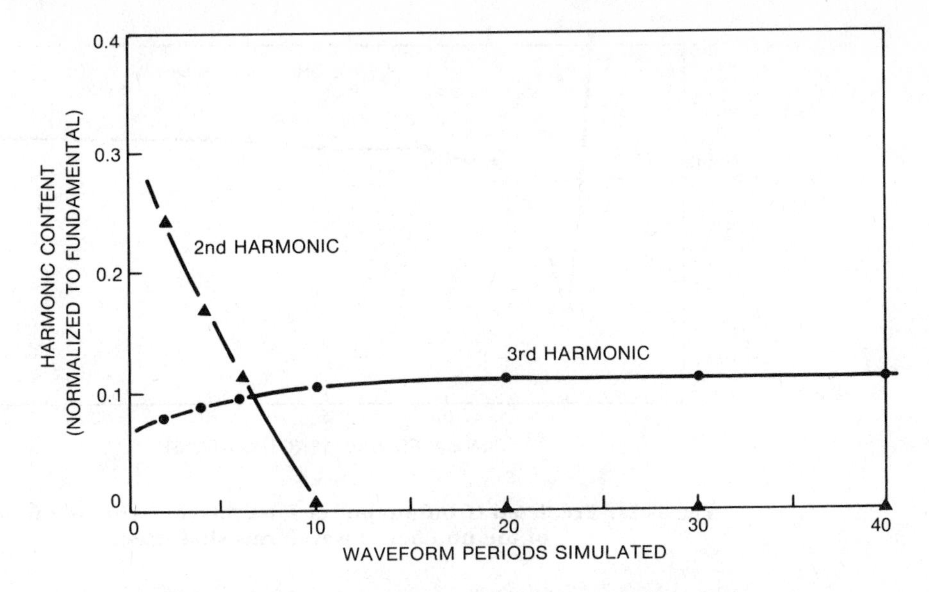

Fig. 1–20. Predicted fundamental, second-, and third-harmonic power amplifier output as a function of the number of waveforms simulated.

second harmonic frequency should be present across the output resistor. As is seen from Fig. 1–20, this condition does not exist until approximately 10 waveforms at the fundamental frequency have been simulated. Also plotted in the figure is the third-harmonic component determined from the simulation.

Results of a similar analysis for the mixer circuit of Fig. 1–15 are illustrated in Fig. 1–21. In the figure, predicted IF output power is plotted as a function of the number of waveforms simulated. It is apparent that significant transient effects are present for the first several wave cycles simulated. Second-order transients continue for up to 20 simulated wave periods.

Finally, Fig. 1–22 illustrates the effect transient phenomena have on the simulation of the oscillator of Fig. 1–18A. To simulate oscillator circuits using time-domain analysis, some voltage perturbation must initially be supplied to the circuit. In this example a 0.05-V pulse of 5-ps duration was applied through the gate bias. Figure 1–22 plots the voltage across the output resistor as a function of time. The number of cycles required for the oscillations to reach steady state will be a function of the Q of the circuit being simulated.

Despite the limitations of the time-domain approach, this technique has been used to produce accurate performance predictions for a wide variety of different circuits. All of the simulation results presented in Secs. 1.2 and 1.3, for example, were obtained using time-domain simulations. These

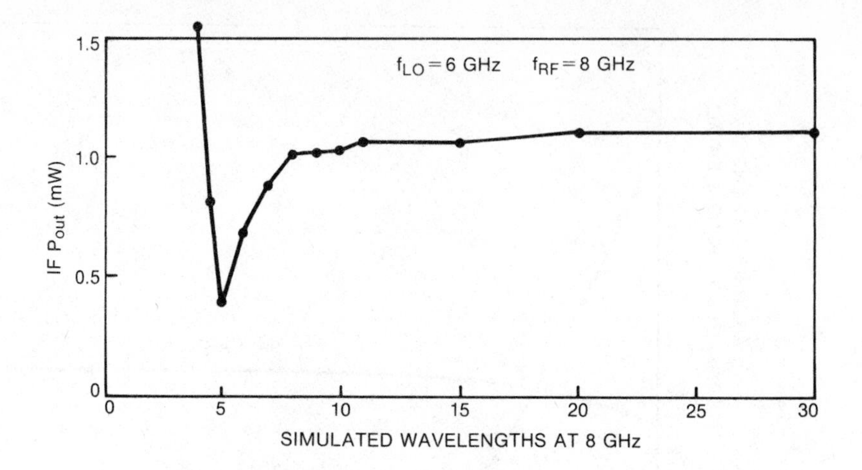

Fig. 1–21. Predicted IF output power for a mixer circuit as a function of the number of waveforms simulated.

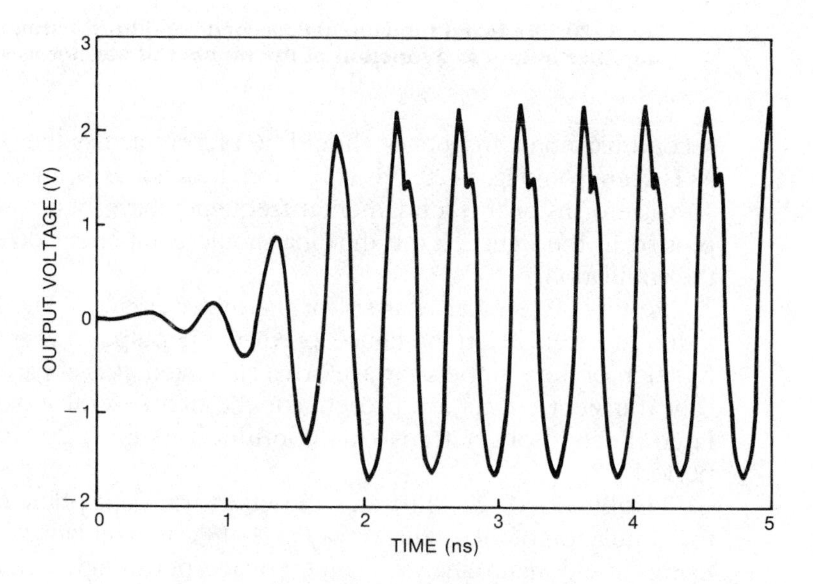

Fig. 1–22. Predicted output voltage as a function of time for a simulated oscillator circuit.

results demonstrate the use of the technique for amplifier, mixer, and oscillator applications.

Time-domain simulations of amplifier circuit properties have been examined by a number of researchers.[3,4,7–9] Prediction of output power saturation properties as well as harmonic content of the output waveforms have been found to be in good agreement with experimentally determined values. Oscillator design problems have also been investigated using time-domain circuit simulation techniques.[16,21] As illustrated by the results of Fig. 1–18B, excellent agreement with measured values is possible.

The time-domain simulation of frequency conversion circuitry is less frequently reported in the literature, but demonstration of such analysis has been accomplished.[58] General mixer analysis using the time-domain approach is associated with limitations related to discrete Fourier transform (DFT) algorithms. These limitations also apply to the prediction of intermodulation distortion which occurs as the result of two signals which are closely spaced in the frequency domain. Provided that the lowest frequency of interest in a simulation (the IF frequency in a mixer simulation) is harmonically related to the other frequencies being examined (RF, LO, image, and second harmonics typically) analysis of the waveforms is straightforward. When the frequencies of interest are not harmonically related, however, application of the DFT can become awkward. For example, if an LO signal at 11 GHz were combined with an RF signal at 11.7 GHz, the resulting IF signal is at 700 MHz. The lowest common denominator of these three signal frequencies is 100 MHz. To apply a DFT algorithm to the time-domain solution requires that 117 harmonic components of the signal be evaluated using a fundamental frequency of 100 MHz. If second-harmonic RF power is of interest, the number of Fourier components which must be evaluated increases to 234. In addition, one full 100-MHz wave cycle beyond the time when transient effects vanish must be simulated. In the case of an 11-GHz signal, transient effects will vanish in approximately 10 periods or 900 ps. One full 100-MHz wave cycle beyond this increases the required time to be simulated by 10 ns. This means that the required simulation time is increased by more than an order of magnitude over that required to observe single-frequency performance. Such requirements are clearly prohibitive.

It should be noted that these same restrictions will apply to standard harmonic-balance techniques. There has been a significant effort to overcome this problem as it applies to the harmonic-balance approach, and much of the work done in relation to that problem could be applied to the time-domain problem as well.

1.4.2 Harmonic-Balance and Related Methods

Harmonic-balance methods are iterative techniques which attempt to match frequency components of currents or voltages flowing between two subcircuits. The standard harmonic-balance approach[62,63] is illustrated in Fig. 1–23 as it applies to a three-terminal device. The circuit being analyzed is divided into two subcircuits. All nonlinear components involved in circuit performance are included in one subcircuit, while only linear elements are included in the second. By dividing the circuit as illustrated,

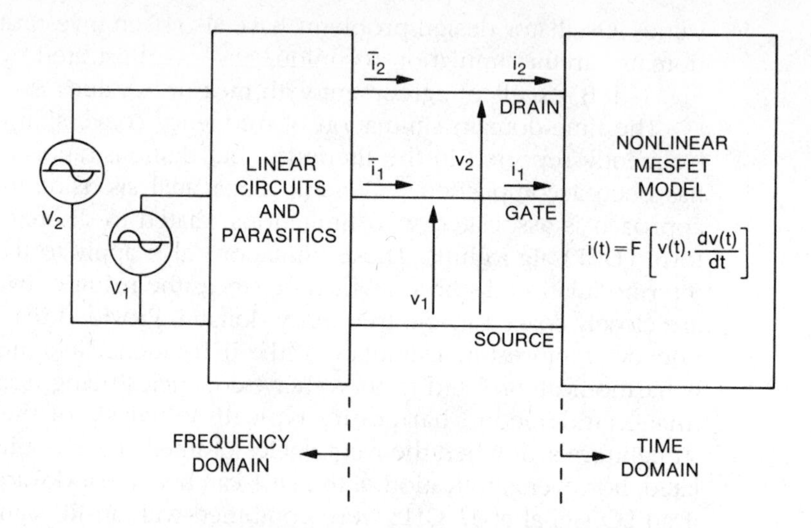

Fig. 1–23. Schematic representation of the harmonic-balance formulation for a three-terminal nonlinear device.

the full advantages of frequency-domain analysis can be exploited to analyze the linear subcircuit. The subcircuit containing nonlinearities, however, must still be solved using time-domain or other nonlinear analysis techniques.

Nodes are identified at branches which connect to two subcircuits. The current which flows from one subcircuit to an identified node must be equivalent to the current which flows from the node into the other subcircuit. This statement of current continuity is used to determine when the solution to the problem has been obtained. In Fig. 1–23, signals $V_1(t)$ and $V_2(t)$ are applied to the circuit. When the currents i_1 and $\bar{i}_1$ as well as the currents i_2 and $\bar{i}_2$ are consistent, then convergence has been achieved. In the standard harmonic-balance approach, as well as many of the derivative methods,[15,24,64] time-domain techniques are employed to analyze the nonlinear subcircuit. The relationship between the currents into and out of the identified nodes is then established using a Fourier transform.

When a time-domain method is used in the analysis of the nonlinear subcircuit, some of the limitations which apply to general time-domain analysis will also apply to the harmonic-balance problem. The nonlinear subcircuit will still exhibit transient behavior, for example. For most applications, these effects will be less significant for the subcircuit of a harmonic-balance simulation than they will be for the entire circuit considered for a time-domain simulation. Such effects must be considered, however, so that the simulation of multiple wave cycles will be required to obtain the steady-state solution for the nonlinear subcircuit. Restrictions which are encountered in time-domain analysis because of the properties of discrete Fourier transform algorithms will also apply to this problem.

Another limitation of standard harmonic-balance methods is that an initial guess of the resulting waveforms at identified nodes is required. If the guess is not appropriate, it can inhibit convergence. Convergence problems will be rare for some circuits, but difficult to overcome for others.[15]

Many of the problems with standard harmonic-balance algorithms have been addressed through modifications of this method. Gilmore[24] has applied bandpass sampling to overcome sampling problems related to the DFT. The approach gives rise to an aliasing problem, but this too is accounted for using Gilmore's modified harmonic-balance method. The devices which can be analyzed effectively using this approach are limited to low-Q devices. It should be noted that bandpass sampling does not eliminate the use of the time-domain analysis. Similar procedures could also be applied to time-domain simulations.

Rhyne and Steer[13] have chosen to avoid time-domain limitations by describing the nonlinear circuit in terms of power-series-like descriptions. The descriptions are related to Volterra series analysis with complex coefficients and time delays.[65,66] Employing such an approach allows the nonlinear problem to be solved using algebraic formulas as opposed to numerical integration and time step. The resulting method is a true frequency-domain technique for the analysis of nonlinear analog circuits. Such an approach has a strong intuitive appeal. Since the performance information desired is described in the frequency domain, it is attractive to work the problem entirely without transforming into and out of the time domain.

Other modifications to the standard harmonic-balance approach have addressed the problem of requiring an initial estimate to the solution. Kerr[36] developed a multiple reflection technique for diode analysis which eliminates this requirement. This technique seldom fails to converge and has been generalized by Maas for multiterminal devices.[39] One limitation of this modified process is that convergence is sometimes slow. Hwang and Itoh[16,67] have made further modifications to the multiple reflection method which have significantly enhanced the convergence properties. For one circuit examined in their study, the modified multiple-reflection method converged in 7 iterations while the original multiple reflection method required more than 50.

Harmonic-balance simulations have been used to investigate circuit performance of power amplifiers[9,13,15,24,26,58] with considerable success. Gain-saturation characteristics, harmonic content and intermodulation distortion have all been examined. Frequency doublers[23,24,64] and mixers[58] have also been the subject of harmonic-balance simulations. The predicted output power spectrum has been shown to be in excellent agreement with measurement. Simulation of oscillator circuitry has also been demonstrated[25] using harmonic-balance simulation. Both fundamental and harmonic output power predictions agree well with measured values.

Using a harmonic-balance approach for the simulation of oscillators does present special problems. By necessity, harmonic-balance simulations examine current-voltage characteristics at only a finite number of frequencies. For amplifier and frequency conversion circuitry, frequencies that must be considered in the analysis are easily determined. Applied signal frequencies, harmonics of these frequencies, and mixing products of these frequencies are candidates for consideration. In the case of oscillators, however, no external RF signal is applied. It is not known in advance at what frequency the circuit will oscillate. Rizzoli *et al.*[25] have applied harmonic-balance techniques to the analysis of a cavity-tuned oscillator. Their simulation technique allowed circuit elements to be altered in order to minimize harmonic-balance error. The error value was very high when circuit component values were such that oscillations would not occur at the design frequency. When circuit components were tuned to achieve the proper oscillation frequency, however, the value of the harmonic-balance error decreased dramatically.

Note that the method described above addresses the issue of oscillator design, but is not directly applicable to the analysis of an existing circuit. This could have consequences in the design process of circuits other than oscillators. For example, if a circuit being designed to perform as an amplifier is unstable, it may oscillate at a frequency other than the applied signal frequency and its harmonics. This sort of problem may be difficult to identify using a harmonic-balance approach.

Harmonic-balance and time-domain methods are often discussed as if they are competing techniques for large-signal analog circuit simulators. Such a viewpoint ignores the fact that the two methods have different strengths which can be exploited for various circuit applications. The two methods complement each other and both can be used to advantage in different applications and at different points in the design process. Two studies which have compared the techniques are worth discussion here. In the work of Brazil *et al.*,[58] two different time-domain simulation packages and a harmonic-balance routine were used to obtain performance predictions for identical circuits. Both a power amplifier and a frequency doubler were considered. The results obtained from the three simulations were essentially identical.

Kundert and Sangiovanni-Vincentelli[26] compared the performance of SPICE with a harmonic-balance routine. Both methods were applied to the simulation of a traveling-wave amplifier (TWA) and a noninverting amplifier. The traveling-wave amplifier contained six transmission line segments and four active devices, while the noninverting amplifier circuit contained sixteen transistors and no transmission line segments. Simulation time required to obtain a solution was monitored for each method. The results are produced in Table 1–4. The simulation time required for harmonic-balance routines is dependent on how many frequency components are considered. Thus, Table 1–4 presents harmonic-balance run times for various numbers of harmonics considered.

Table 1–4. Simulation Time in Seconds for Circuit Simulations. The Routines Were Run on a VAX 11/785 Running UNIX 4.3BSD[26]

		Harmonic Balance		
		Number of Harmonics Considered		
Circuit	**SPICE2**	**8**	**16**	**32**
Traveling-wave amplifier $V_{out} = 1$ V	62.5	7	22	56
Noninverting amplifier $V_{out} = 10$ V	14	No conversion	365	575

As is shown in the table, harmonic-balance techniques exhibit significant speed advantages over time-domain techniques when transmission line segments are a major part of the circuit design. This advantage is lost, however, for situations such as the noninverting amplifier which uses no transmission line segments.

This result has significant implications as it applies to hybrid and MMIC technologies. Hybrid microwave circuit designs typically make significant use of transmission line segments for matching and tuning. Harmonic-balance approaches should exhibit significant speed advantages over time-domain approaches for such circuits. In MMIC microwave circuits, lumped-element tuning is more common and transmission line tuning is rare. Speed advantages of the harmonic-balance approach may be lost for such applications.

1.4.3 Quasi–Large-Signal Methods

Harmonic-balance and time-domain techniques can produce very accurate performance predictions for large-signal circuit applications. The approaches are very general and capable of predicting the existence of many large-signal phenomena. For some applications, however, less detailed simulations may be useful and more efficient.

Power amplifier design is often simplified by referring to load-pull contours for the device being used. Such contours can be obtained by measuring them directly as described in Sec. 1.3.1 or using large-signal simulation techniques.[9] Both of these methods, however, can be very time consuming. An alternative method[32] for estimating load-pull contours has been described by Cripps. The method involves the use of a single measurement of the gain compression characteristics of the device in question. The method has been modified by Foust[33] to develop a design technique capable of predicting large-signal matching characteristics for GaAs MESFETs.

Tajima *et al.*[10] developed a fairly general technique which utilizes a large-signal device model and small-signal circuit simulation techniques.

Considering only the fundamental frequency and ignoring the effects of higher harmonics, the voltages present at the device electrodes are written as

$$V_{gs}(t) = V_{GS} + v_{gs} \cos(\omega t + \phi), \quad (1\text{–}20)$$

and

$$V_{ds}(t) = V_{DS} + v_{ds} \cos \omega t, \quad (1\text{–}21)$$

where V_{GS} and V_{DS} are the DC bias voltages, v_{gs} and v_{ds} are amplitudes of the signal-frequency components, and ϕ the phase difference between gate and drain voltage. The values of v_{ds} and ϕ are not known in advance. They are determined, rather, by iteration for this method.

Using the definitions of transconductance and output resistance as expressed by Eqs. 1–3 and 1–5, the instantaneous drain-source current can be written as

$$I_{ds}(t) = I_{DS} + g_m v_{gs} \cos(\omega t + \phi) + \frac{1}{R_{ds}} v_{ds} \cos \omega t. \quad (1\text{–}22)$$

Use of a large-signal device model and assumed values for voltage amplitudes and phase provides a direct evaluation of $I_{ds}(t)$.

A quasi–large-signal transconductance value can now be obtained by rearranging Eq. 1–22 and integrating over a complete period:

$$g_{m(\text{large})} = -\frac{\omega/\pi}{v_{gs} \sin \phi} \int_0^{2\pi/\omega} I_{ds} \sin \omega t \, dt. \quad (1\text{–}23)$$

Similar expressions for output resistance, gate-source capacitance, and other nonlinear equivalent circuit elements are obtained following this method.[10] The element values arrived at are used with standard small-signal analysis techniques to determine updated drain-source voltage amplitude and the phase difference between gate and drain. When this process converges, the equivalent circuit is obtained for a voltage drive level of the assumed v_{gs}. This drive level is easily converted to a corresponding power level if terminating circuit impedances are known.

The process described above is performed for several different assumed input power levels. The result of this process is a set of equivalent circuit models—each of which applies for a given input power level and which can be used in small-signal simulation routines. Although the technique is not capable of predicting such phenomena as harmonic content or mixing products of the signals, it can be applied to both oscillator[10] and power amplifier[10,14] design. Predictions of gain compression using the method show excellent agreement with measured characteristics. Oscillator performance predictions are also remarkably accurate.

Kurokawa[68] has discussed the behavior of oscillator circuits in significant detail. The criterion for the existence of stable oscillations in terms of circuit and device impedances is well established. Analysis is performed by splitting the circuit arbitrarily into two subcircuits, and evaluating the impedance seen looking into both sections. The resulting impedance values are then compared to the stable oscillation criterion. Strict interpretation of this analysis requires that the large-signal impedance of the device be used in testing the criterion for oscillation. It is common practice, however, to use small-signal parameters as a first-order estimate in the analysis. This approximation has been used to analyze and design oscillator circuits[27–29,69] with considerable success. The technique predicts output frequency in good agreement with measurements, but is not capable of output power predictions.

Small-signal impedance information was used to examine the oscillator of Fig. 1–18. The output frequency predicted was found to be dependent on where the circuit was split for evaluation purposes. For the same conditions as those of Table 1–3, the small-signal predicted output frequency was found to be between 2.0 and 2.17 GHz. The output frequency actually measured from the circuit was 2.24 GHz.

Both the technique described by Tajima[10] and the small-signal parameter method for oscillator frequency prediction are powerful analysis tools. A key to their strength is that they allow the circuit designer to make use of the considerable analysis capability of linear CAD tools. Efficient simulation and optimization capabilities are more readily available with small-signal simulation packages than with large-signal packages.

Mixer performance can also be estimated using a quasi–large-signal analysis approach. Pucel *et al.*[34] described FET mixing properties in terms of a single nonlinearity in the device, the transconductance. A sinusoidal LO signal is assumed at the input terminals of the device and only four frequency components are considered at the output. Transconductance as a function of time is then computed and Fourier transformed. This information is used with the device equivalent-circuit model and circuit impedances to predict conversion gain for the device. The original analysis was limited to gate-driven common-source mixers. Comparison with experiment shows that the technique produces conversion gain predictions in close agreement with actual circuits in some cases.[34]

A simplified device model was used to simulate the mixer circuit of Fig. 1–15 using the Pucel analysis and considering only four frequency components: the LO frequency, RF frequency, IF frequency, and image frequency. Results of the simulation are plotted in Fig. 1–24. Also shown are the results obtained by simulating the same circuit and device using a time-domain approach. The time-domain approach will consider all frequency components present. As is seen from the figure, agreement in this case is qualitatively good. More accuracy can be obtained by considering more frequency components.

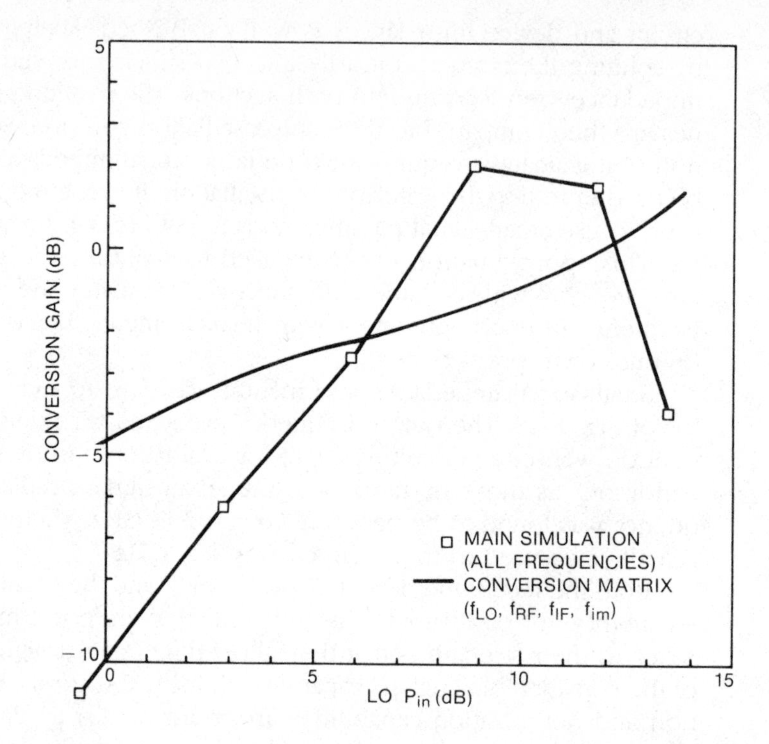

Fig. 1–24. Conversion gain predicted by a conversion matrix approach using only four frequency components versus that predicted by a time-domain simulation.

Begemann and Jacob[56] took the analysis of Pucel a step further by considering both the nonlinearity of the output resistance and the effect of gate-drain capacitance in the calculations. This allowed them to consider drain-driven as well as gate-driven mixers. This work has been further extended by Maas[35] using the techniques developed for diode mixers by Egami.[70] This conversion matrix approach is a generalization of the work of Pucel and of Begemann. It allows any equivalent circuit model with any number of nonlinear elements to be included in the analysis. Also, extension to include other frequencies is straightforward. Maas has used this conversion matrix approach to analyze fabricated mixers.[35] Excellent agreement of conversion gain predictions to measured values is seen over a wide range of input power levels.

The quasi–large-signal methods described in this section serve as excellent starting points in the design process of large-signal circuitry. Because these methods are computationally much more efficient than time-domain or harmonic-balance large-signal simulation techniques, more approaches to the design problem can be explored early in the design process. Initial optimization of the circuit design can be performed much faster using these methods. Because the quasi–large-signal methods do not constitute a complete large-signal description of the device, final optimization and performance tests should be accomplished using time-domain and harmonic-balance techniques.

References

1. R. A. Pucel, H. Statz, and H. A. Haus, "Signal and noise properties of gallium arsenide microwave field-effect transistors," *Adv. Electron. Phys.*, New York: Academic Press, 38:195–265 (1975).
2. K. Lehovec and R. Zuleeg, "Voltage-current characteristics of GaAs JFETs in the hot electron range," *Solid State Electron.*, 13(10): 1415–1426 (October 1970).
3. A. Madjar and F. J. Rosenbaum, "A large-signal model for the GaAs MESFET," *IEEE Trans. Microwave Theory Technol.*, MTT-29 (8): 781–788 (August 1981).
4. J. M. Golio, J. R. Hauser, and P. A. Blakey, "A large-signal GaAs MESFET model implemented on SPICE," *IEEE Cir. Dev. Mag.*, 1:21–30 (September 1985).
5. J. A. Higgins, "Intermodulation distortion in GaAs FETs," *IEEE Microwave Theory Technol. Symp. Dig.*, 138–141 (1978).
6. W. R. Curtice, "A MESFET model for use in the design of GaAs integrated circuits," *IEEE Trans. Microwave Theory Technol.*, MTT-28(5): 448–457 (May 1980).
7. H. A. Willing, C. Rauscher, and P. deSantis, "A technique for predicting large-signal performance of a GaAs MESFET," *IEEE Trans. Microwave Theory Technol.*, MTT-26(12): 1017–1023 (December 1978).
8. R. Soares, M. Goudelis, and B. Loriou, "Nonlinear equivalent circuit for broadband GaAs MESFET power amplifier design," *IEEE Microwave Theory Technol. Symp. Dig.*, 63–65 (1982).
9. A. Materka and T. Kacprzak, "Computer calculation of large-signal GaAs FET amplifier characteristics," *IEEE Trans. Microwave Theory Technol.*, MTT-33(2): 129–135 (February 1985).
10. Y. Tajima, B. Wrona, and K. Mishima, "GaAs FET large-signal model and its application to circuit designs," *IEEE Trans. Electron Dev.*, ED-28(2): 171–175 (February 1981).
11. L. O. Chua and Y. W. Sing, "Nonlinear lumped circuit model of GaAs MESFET," *IEEE Trans. Electron Dev.*, ED-30(7): 825–833 (July 1983).

12. H. Statz, P. Newman, I.W. Smith, R.A. Pucel, and H.A. Haus, "GaAs FET device and circuit simulation in SPICE," *IEEE Trans. Electron Dev.*, ED-34(2): 160–169 (February 1987).
13. G.W. Rhyne and M.B. Steer, "A new frequency domain approach to the analysis of nonlinear microwave circuits," *IEEE Microwave Theory Technol. Symp. Dig.*, 401–404 (1985).
14. M. Weiss and D. Pavlidis, "Power optimization of GaAs implanted FET's based on large-signal modeling," *IEEE Trans. Microwave Theory Technol.*, MTT-35(2): 175–188 (February 1987).
15. V.D. Hwang and T. Itoh, "Large signal modeling and analysis of GaAs MESFETs," *Proc. 16th Eur. Microwave Conf.*, 188–194 (September 1986).
16. C.M. Snowden, M.J. Howes, and D.V. Morgan, "Large signal modeling of GaAs MESFET operation," *IEEE Trans. Electron Dev.*, ED-30(12): 1817–1824 (December 1983).
17. J.M. Cusack, S.M. Perlow, and B.S. Perlman, "Automatic load contour mapping for microwave power transistors," *IEEE Trans. Microwave Theory Technol.*, MTT-22(12): 1146–1152 (December 1974).
18. H. Abe and Y. Aono, "11-GHz GaAs power MESFET load-pull measurements utilizing a new method of determining tuner Y parameters," *IEEE Trans. Microwave Theory Technol.*, MTT-27(5): 394–399 (May 1979).
19. R.A. Soares, "Novel large signal s-parameter measurement technique aids GaAs power amplifier design," *Proc. 7th Eur. Microwave Conf.*, 113–117 (September 1977).
20. Y. Mitsui, M. Nakatani, and S. Mitsui, "Design of GaAs MESFET oscillator using large-signal s-parameters," *IEEE Trans. Microwave Theory Technol.*, MTT-25(12): 981–984 (December 1977).
21. L.W. Nagel, "SPICE2: a computer program to simulate semiconductor circuits," Memo ERL-M520, Electronics Research Laboratory, Univ. of California, Berkeley (1975).
22. J.F. Sautereau, J. Graffeuil, K. Tautrarongraj, and P. Rossel, "High efficiency GaAs Schottky-barrier gate FET oscillator," *Electron. Lett.*, 16(13): 490–491 (June 1980).
23. C. Camacho-Penalosa, "Numerical steady-state analysis of nonlinear microwave circuits with periodic excitation," *IEEE Trans. Microwave Theory Technol.*, MTT-31(9): 724–730 (September 1983).
24. R. Gilmore, "Design of a novel FET frequency doubler using a harmonic balance algorithm," *IEEE Microwave Theory Technol. Symp. Dig.*, 585–588 (1986).
25. V. Rizzoli, A. Lipparini, and E. Marazzi, "A general-purpose program for nonlinear microwave circuit design," *IEEE Trans. Microwave Theory Technol.*, MTT-31(9): 762–770 (September 1983).

26. K. S. Kundert and A. Sangiovanni-Vincentelli, "Simulation of nonlinear circuits in the frequency domain," *IEEE Trans. Computer-Aided Des.*, CAD-5(4): 521–534 (October 1986).
27. R.J. Trew, "Design theory for broad-band YIG-tuned FET oscillators," *IEEE Trans. Microwave Theory Technol.*, MTT-27(1): 9–14 (January 1979).
28. P. M. Ollivier, "Microwave YIG-tuned transistor oscillator amplifier design: application to C band," *IEEE J. Solid-State Cir.*, SC-7(1): 54–60 (February 1972).
29. W. El-Kamali, J. Grimm, R. Meierer, and C. Tsironis, "New design approach for wide-band FET voltage-controlled oscillators," *IEEE Trans. Microwave Theory Technol.*, MTT-34(10): 1059–1063 (October 1986).
30. R. A. Pucel, R. Bera, and D. Masse, "Experiments on integrated gallium-arsenide FET oscillators at X-band," *Electron. Lett.*, 11(10): 219–220 (May 1975).
31. K. M. Johnson, "Large signal GaAs MESFET oscillator design," *IEEE Trans. Microwave Theory Technol.*, MTT-27(3): 217–227 (March 1979).
32. S. C. Cripps, "A theory for the prediction of GaAs FET load-pull power contours," *IEEE Microwave Theory Technol. Symp. Dig.*, 221–223 (1983).
33. R. Foust, "Using GaAs FET small signal models and DC load line characteristics to predict optimum power match in GaAs FETs," *Microwave J.*, 29(9): 151–156 (September 1986).
34. R. A. Pucel, D. Masse, and R. Bera, "Performance of GaAs MESFET mixers at X band," *IEEE Trans. Microwave Theory Technol.*, MTT-24(6): 351–360 (June 1976).
35. S. A. Maas, "Theory and analysis of GaAs MESFET mixers," *IEEE Trans. Microwave Theory Technol.*, MTT-32(10): 1402–1406 (October 1984).
36. A. R. Kerr, "A technique for determining the local oscillator waveforms in a microwave mixer," *IEEE Trans. Microwave Theory Technol.*, MTT-23(10): 828–831 (October 1975).
37. C. Rauscher and H. A. Willing, "Design of broad-band GaAs FET power amplifiers," *IEEE Trans. Microwave Theory Technol.*, MTT-28(10): 1054–1059 (October 1980).
38. H. Fukui, "Determination of the basic device parameters of a GaAs MESFET," *Bell Syst. Tech. J.*, 58(3): 771–797 (March 1979).
39. P. Wolf, "Microwave properties of Schottky-barrier field-effect transistors," *IBM J. Res. Develop.*, 14(3): 125–141 (March 1970).
40. S. A. Maas, *Microwave Mixers,* Dedham, Mass.: Artech House (1986).

41. M. Ino, M. Hirayama, K. Kurumada, and M. Ohmori, "Estimation of GaAs static RAM performance," *IEEE Trans. Electron Dev.*, ED-29(7): 1130–1134 (July 1982).

42. M. A. Smith, T. S. Howard, K. J. Anderson, and A. M. Pavio, "RF nonlinear device characterization yields improved modeling accuracy," *IEEE Microwave Theory Technol. Symp. Dig.*, 381–385 (1986).

43. S. E. Sussman-Fort, S. Narasimhan, and K. Mayaram, "A complete GaAs MESFET computer model for SPICE," *IEEE Trans. Microwave Theory Technol.*, MTT-32(4): 471–473 (April 1984).

44. M. F. Abusaid and J. R. Hauser, "Calculations of high-speed performance for submicrometer ion-implanted GaAs MESFET devices," *IEEE Trans. Electron Dev.*, ED-33(7): 913–918 (July 1986).

45. M. G. Adlerstein, "Electrical traps in GaAs microwave FET's," *Electron. Lett.*, 12(13): 45–53 (June 1977).

46. J. M. Golio and R. J. Trew, "Profile studies of ion-implanted MESFETs," *IEEE Microwave and Millimeter Wave Monolithic Cir. Symp. Dig.*, 22–26 (1983).

47. W. Shockley, "A unipolar 'field-effect' transistor," *Proc. IRE,* 40(11): 1365–1376 (November 1952).

48. J. M. Golio, P. A. Blakey, and R. O. Grondin, "A general CAD tool for large-signal GaAs MESFET circuit design," *IEEE Microwave Theory Technol. Symp. Dig.*, 417–420 (1985).

49. K. Yamaguchi and H. Kodera, "Drain conductance of junction gate FET's in the hot electron range," *IEEE Trans. Electron Dev.*, ED-23(6): 545–553 (June 1976).

50. J. M. Golio and R. J. Trew, "Optimum semiconductors for high-frequency and low-noise MESFET applications," *IEEE Trans. Electron Dev.*, ED-30(10): 1411–1413 (October 1983).

51. D. Poulin, "Load-pull measurements help you meet your match," *Microwaves,* 11(11): 61–65 (November 1980).

52. R. Tucker, "RF characterization of microwave power FET's," *IEEE Trans. Microwave Theory Technol.*, MTT-29(8): 776–781 (August 1981).

53. P. Canfield, J. Medinger, and L. Forbes, "Buried-channel GaAs MESFET's with frequency-independent output conductance," *IEEE Electron Dev. Lett.*, EDL-8(3): 88–89 (March 1987).

54. S. H. Wemple, M. L. Steinberger, and W. O. Sclosser, "Relationship between power added efficiency and gate-drain avalanche in GaAs MESFETs," *Electron. Lett.*, 16(12): 459–460 (June 1980).

55. *ALLSPICE User Manual,* Menlo Park, California: Acotech, 1986.

56. G. Begemann and A. Jacob, "Conversion gain of MESFET drain mixers," *Electron. Lett.*, 15(18): 369–370 (August 1979).

57. H. M. Greenhouse, "Design of planar rectangular microelectronic inductors," *IEEE Trans. on Parts, Hybrids and Packaging,* PHP-10(2): 101–109 (June 1974).

58. T. Brazil, E. Choo, S. El-Rabaie, J. A. C. Stewart, V. Fusco, and S. McKeown, "Analysis and optimization of the harmonic output of an FET amplifier," *Proc. 16th Eur. Microwave Conf.*, 195–200 (September 1986).

59. S. Wang, J. Lee, and C. Chang, "An efficient and reliable approach for semiconductor device parameter extraction," *IEEE Trans. Computer-Aided Des.*, CAD-5(1): 170–179 (January 1986).

60. D. W. Marquardt, "An algorithm for least-squares estimation of nonlinear parameters," *J. Soc. Indust. Appl. Math.*, 11(2): 431–441 (June 1963).

61. M. Powel, "On the maximum errors of polynomial approximation defined by interpolation and by least squares criteria," *The Computer J.*, 9(2): 404–407 (June 1967).

62. M. S. Nakhla and J. Vlach, "A piecewise harmonic balance technique for determination of periodic response of nonlinear systems," *IEEE Trans. Cir. Sys.*, CAS-23(2): 85–91 (February 1976).

63. R. G. Hicks and P. J. Khan, "Numerical technique for determining pumped nonlinear device waveforms," *Electron. Lett.*, 16(10): 375–376 (May 1980).

64. J. Dreifuss, A. Madjar, and A. Bar-Lev, "A full large signal analysis of active microwave mixers," *Proc. 16th Eur. Microwave Conf.*, 687–691 (September 1986).

65. M. B. Steer and P. J. Khan, "An algebraic formula for the output of a system with large-signal, multifrequency excitation," *Proc. IEEE,* 71(1): 177–179 (January 1983).

66. M. B. Steer and P. J. Khan, "Large signal analysis of nonlinear microwave systems," *IEEE Microwave Theory Technol. Symp. Dig.*, 402–403 (1984).

67. V. D. Hwang and T. Itoh, "Large signal modeling and analysis of the GaAs MESFET," Technical Report No. 86-P-2, Contract No. DAAG29-84-K-0076, U.S. Army Research Office (June 1986).

68. K. Kurokawa, "Some basic characteristics of broadband negative resistance oscillator circuits," *Bell Syst. Tech. J.*, 48(4): 1937–1955 (July 1969).

69. J. M. Golio and C. M. Krowne, "New approach for FET oscillator design," *Microwave J.*, 21(10): 59–61 (October 1978).

70. S. Egami, "Nonlinear, linear analysis and computer-aided design of resistive mixers," *IEEE Trans. Microwave Theory Technol.*, MTT-22(3): 270–275 (March 1973).

Chapter 2

Advanced Modeling of the High Electron Mobility Transistor

U. Ravaioli and K. Hess

2.1 Introduction

The high electron mobility transistor (HEMT) represents the most recent variation of the field-effect transistor, and exhibits features common to the MESFET and the MOSFET.[1] The HEMT is based on the *modulation doped structure,* which achieves spatial separation between the conduction electrons and the donor ions at a heterojunction, most commonly $Al_xGa_{1-x}As$ and GaAs.[2] The $Al_xGa_{1-x}As$, which has the larger bandgap, is heavily doped, while the GaAs is lightly doped or nominally undoped. Electrons diffuse from the $Al_xGa_{1-x}As$ to the GaAs, where they are confined in a narrow channel at the interface. Because of size quantization effects, motion of the electrons confined in this channel is restricted to the two dimensions parallel to the interface, and the energy levels are grouped into subbands (Fig. 2–1). Band I is GaAs with the density of acceptors $N_A = 10^{15}$ cm^{-3}, band II is $Al_{0.3}Ga_{0.7}As$ with $N_A = 10^{14}$ cm^{-3}, and band III is $Al_{0.3}Ga_{0.7}As$ with $N_A = 10^{14}$ cm^{-13} and density of donors $N_D = 5 \times 10^{17}$ cm^{-3}.

The electrons then constitute a quasi–two-dimensional electron gas (Q2DEG), which exhibits high mobility due to the reduced interaction of the carriers with the impurities, and to the extremely good quality of the interface, usually achieved using molecular beam epitaxy (MBE) or metalorganic chemical vapor deposition (MOCVD).

As mentioned, the HEMT is a field-effect transistor based on the modulation-doped structure,[1,3,4] as shown in Fig. 2–2. A Schottky barrier gate is used, as in conventional MESFET structures. The presence of the heterojunction, and of quantum effects, creates a completely new set of problems to be faced in the simulation of the device. Since electrons in general have a higher mobility than holes, we consider only n-channel devices in the following. However, analogous considerations may be used for p-channel devices as well.

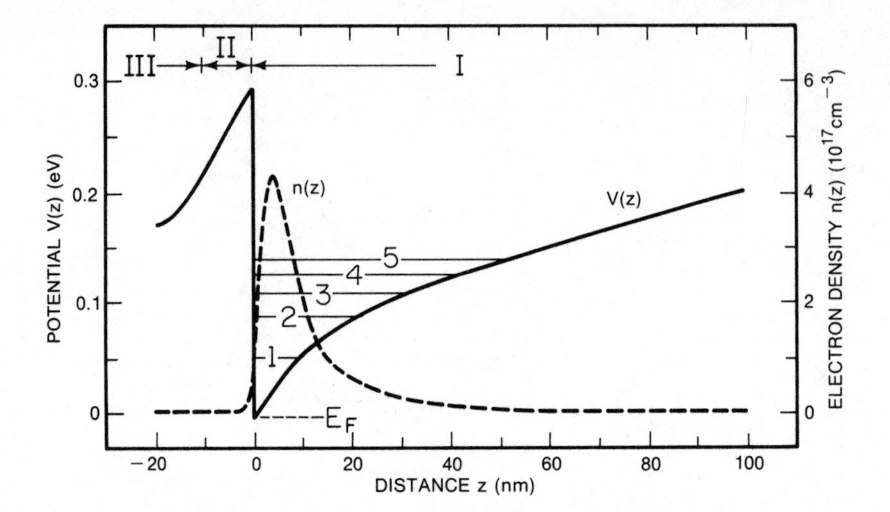

Fig. 2–1. Band structure and quantum levels at the heterojunction interface.

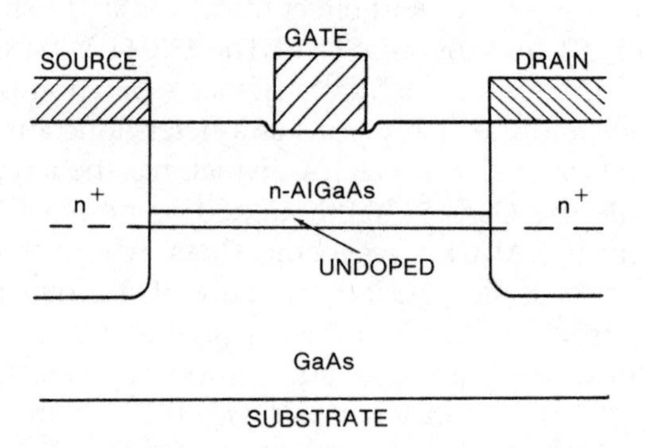

Fig. 2–2. Structure of the HEMT.

The typical structure of a HEMT includes a semi-insulating substrate upon which a layer of undoped GaAs is grown to a thickness of about 1 μm. Typically 1 to 10 nm of undoped $Al_xGa_{1-x}As$ is then grown, and then a 50- to 100-nm heavily doped layer of $Al_xGa_{1-x}As$ follows. The presence of the nominally undoped $Al_xGa_{1-x}As$ layer further separates the ionized impurities from the electrons in the channel. Source and drain contacts are typically obtained by metallization of AuGe/Ni/Au, which is alloyed after liftoff for a short time (~1 min), at about 400°C.[4] The contact between the electrodes and the Q2DEG seems to be obtained by the diffusion of Ge down past the heterointerface. The Schottky barrier gate is often defined with a small recess, the depth of which depends on the specific application of the device.

In the depletion mode the thickness of the doped layer under the gate must be of the order of the depletion width of the Schottky barrier or larger, while in the enhancement mode the layer is much thinner so that the barrier depletes the electron channel without external voltages applied and the device is normally *off.* The modes of operation are therefore analogous to those of the MOSFET, because the AlGaAs resembles an *insulator.* In HEMT structures, however, interface states are virtually absent. Also, the metal gate and the channel are separated by only a few tens of nanometers and, since the dielectric constant of $Al_xGa_{1-x}As$ is larger than that of SiO_2, this fact contributes to larger transconductances.

To avoid conduction through the $Al_xGa_{1-x}As$, which has inferior transport properties, the device geometry has to be chosen so that the depletion region due to the Schottky gate and the one generated at the interface, due to the band edge discontinuity, will overlap. In normally-on devices, the depletion by the Schottky gate should just merge with the interface depletion. Normally-on structures are more suitable for high-speed analog applications, such as low-noise microwave amplifiers.

In normally-off devices, no current flows through the device unless a positive voltage is applied to the gate. This type of structure is suitable for use in high-speed digital integrated circuits because of low power dissipation.

2.2 Quantization

The motion of electrons in the quantum channel can be characterized by an envelope wavefunction[5]

$$\psi(x,y,z) = \zeta_j(z)\exp(i\mathbf{k}\cdot\mathbf{r}), \tag{2–1}$$

where z is the direction perpendicular to the heterointerface, and $\mathbf{r}$ and $\mathbf{k}$ are the two-dimensional position vector and the wavevector for motion parallel to the interface. The function $\zeta_j(z)$ (for the jth subband) satisfies the Schrödinger equation[5,6]

$$-\frac{\hbar^2}{2m^*}\frac{d^2\zeta_j(z)}{dz^2} + V(z)\zeta_j(z) = E_j\zeta_j(z), \tag{2–2}$$

where the effective potential $V(z)$ is given approximately by

$$V(z) = -e\phi(z) + V_b(z) + V_{ex}(z), \tag{2–3}$$

where $\phi(z)$ is the electrostatic potential, $V_b(z)$ is the step function describing the interface barrier, and $V_{ex}(z)$ is the local exchange correlation potential.

The electrostatic potential is given by Poisson's equation,

$$\frac{d^2\phi(z)}{dz^2} = -\frac{e}{\epsilon}\left[\sum_j N_j|\zeta_j(z)|^2 + N_A(z) - N_D(z)\right], \tag{2–4}$$

where N_A and N_D are the concentrations of acceptors and donors, and N_j represents the electron concentration, in subband j, which in equilibrium is

$$N_j = \frac{D}{e}k_B T \ln\left[1 + \exp\left(\frac{E_F - E_j}{k_B T}\right)\right] \tag{2–5}$$

for spin degeneracy 2 and valley degeneracy 1. The two-dimensional density of states is $D = em^*/(\pi\hbar^2)$. A self-consistent iteration of the Schrödinger and Poisson equations leads to the determination of the subband energy levels E_j and of the corresponding wavefunctions ζ_j.[5]

An analytic solution for the energy levels is possible if $V(z)$ is approximated by an infinite barrier at the interface ($z = 0$) and by $V(z) = eF_s z$ for $z > 0$, where F_s is an effective electric field assumed to be constant throughout the layer. This model leads to the Airy equation with the solution[7]

$$\zeta_j(z) = Ai\{(2m^*eF_s/\hbar^2)^{1/3}[z - (E_j/eF_s)]\}, \tag{2–6}$$

$$E_j(z) = (\hbar^2/2m^*)^{1/3}\left[\frac{3}{2}\pi eF_s\left(j + \frac{3}{4}\right)\right]^{2/3}. \tag{2–7}$$

Variational solutions can be obtained by approximating the z-dependent part of the envelope wavefunction with a trial function. For instance, the following expressions can be used for the first and the second subband.[7]

$$\zeta_1(z) = (b_0^3/2)^{1/2} z \exp(-b_0 z/2), \tag{2–8}$$

$$\zeta_2(z) = \left[\frac{3b_1^2}{2(b_0^2 - b_0 b_1 + b_1^2)}\right]^{1/2}[z - (b_0 + b_1)z^2/6]\exp(-b_1 z/2), \tag{2–9}$$

where the parameters b_0 and b_1 are determined by minimizing the total energy of the system for each value of the charge. Both the Airy function and the variational solutions imply that the electron wavefunction is zero at the heterointerface. The self-consistent method which iterates the Schrödinger and Poisson equations until convergence is obtained is more accurate and takes into account the penetration of the wavefunction into the $Al_xGa_{1-x}As$. With this last method it is much easier to include higher-order subbands and to treat cases in which the approximation of a triangular well is not justified.

2.3 Hydrodynamic Transport Equation Model

In this section we describe the solution of the HEMT transport problem by use of the first four moments of the Boltzmann equation. These equations link four interdependent variables: electron concentration, current, average energy, and energy flux. The coefficients of these equations are approximated as functions of the average energy, where the functional dependencies can be determined from steady-state homogeneous Monte Carlo calculations or experimental results. In such a manner the important hot-electron properties are included. This procedure was first proposed for MESFETs.[8]

Poisson's equation completes the set of differential equations, introducing the potential as another dependent variable for self-consistency. All equations are solved numerically using finite-difference schemes for both time and space in the approach presented here, with separate treatment for the bulk and the quantum well.

The four moment equations have been given in a very convenient form by Stratton.[9] Neglecting generation and recombination, they are[10]

$$\frac{\partial n}{\partial t} = \frac{1}{e} \nabla \cdot \mathbf{j}, \tag{2–10}$$

$$\mathbf{j} + \frac{\partial}{\partial t}(\tau_{\mathrm{HF}}\mathbf{j}) = -e\mu n \nabla \psi + e\nabla(Dn), \tag{2–11}$$

$$\frac{\partial nE}{\partial t} = -\mathbf{j} \cdot \nabla \psi - nB - \nabla \cdot \mathbf{s}, \tag{2–12}$$

$$\mathbf{s} + \frac{\partial}{\partial t}(\tau_{\mathrm{HF}}\mathbf{s}) = \mu_E nE \nabla \psi - \nabla(D_E nE), \tag{2–13}$$

where $\mathbf{s}$ is the energy flux, and the coefficients μ, D, B, μ_E, D_E, and τ_{HF} are the mobility, diffusivity, energy dissipation factor, flux mobility, flux diffusivity, and high-frequency factor, respectively, all taken as functions of the average energy, E. Separate longitudinal and transverse diffusion parameters should be used. In order to simplify the numerical formulation, the longitudinal parameter is commonly applied strictly in the direction parallel to the heterointerface and the transverse one in the perpendicular direction.

The energy flux equation (2–13) may be further simplified. By assuming a given energy distribution, μ and μ_E can be expressed in terms of averages of the momentum relaxation time τ, as[10]

$$\mu = \frac{e}{m} \frac{\langle \tau\epsilon \rangle}{\langle \epsilon \rangle}, \tag{2–14}$$

and

$$\mu_E = \frac{e}{m}\frac{\langle\tau\epsilon^2\rangle}{\langle\epsilon\rangle^2} = \frac{\langle\tau\epsilon^2\rangle}{\langle\tau\epsilon\rangle\langle\epsilon\rangle}\mu. \tag{2–15}$$

In a similar way, the flux diffusivity can be expressed as

$$D_E = \frac{\langle\tau\epsilon^2\rangle}{\langle\tau\epsilon\rangle\langle\epsilon\rangle}D, \tag{2–16}$$

for both longitudinal and transverse components. One can then rewrite Eq. 2–13:

$$\mathbf{s} + \frac{\partial}{\partial t}(\tau_{HF}\mathbf{s}) = -\alpha[-\mu nE\nabla\psi + \nabla(DnE)], \tag{2–17}$$

where the coefficient α is defined by

$$\alpha = \frac{\langle\tau\epsilon^2\rangle}{\langle\tau\epsilon\rangle\langle\epsilon\rangle}. \tag{2–18}$$

The high-frequency parameter τ_{HF} is on the order of 0.1 ps, and since transient delays on the order of 5 to 10 ps are expected, τ_{HF} may be neglected without introducing much error. The transport equations then become

$$\frac{\partial n}{\partial t} = \nabla\cdot[-\mu n\nabla\psi + \nabla(Dn)] \tag{2–19}$$

$$\frac{\partial nE}{\partial t} = -\mathbf{j}\cdot\nabla\psi - nB + \nabla\cdot\alpha[-\mu nE\nabla\psi + \nabla(DnE)] \tag{2–20}$$

Widiger *et al.*[10] used the dependence on the average energy of mobility, longitudinal and transverse diffusivity, and power dissipation, for both the bulk and the quantum well, as obtained from Monte Carlo simulations. The low field mobility was taken from the experimental data in Ref. 11, for the two-dimensional case, and for the bulk system was taken as 40,000 $cm^2/V\cdot s$, corresponding to an impurity concentration of 2×10^{16} cm^{-3} at 77 K. The transport parameters, as functions of energy, were calculated by a steady-state Monte Carlo simulation with polar-optical and deformation-potential scattering, and an ensemble technique to include electron-electron scattering. Carrier screening effects were calculated for a uniform electron concentration of 10^{17} cm^{-3}. Since screening is important only for low electron energies, and electron concentrations are typically on the order of 10^{17} cm^{-3} at low energies, such an assumption is not expected to introduce appreciable error.

Transport in the HEMT can take place in both the bulk GaAs and in the quantum well. One expects the conduction primarily in the lowest quantum subband near the source, since the fields there are small, whereas conduction in the bulk will dominate near the pinchoff point. The model introduced by Widiger *et al.*[10] considers only the lowest subband and the bulk system. The $Al_xGa_{1-x}As$ is considered to be completely depleted, and therefore no conduction takes place there.

The transport equations in the quantum well are similar to Eqs. 2–19 and 2–20, used for three-dimensional transport, with the addition of coupling terms accounting for transfer of electrons to the bulk system. These coupling terms are analogous to generation-recombination terms. The equations for one subband are[10]

$$\frac{\partial n_I}{\partial t} = \nabla_I \cdot [-\mu_I n_I \nabla_I \psi + \nabla_I(D_I n_I)] + \frac{1}{e} J_c, \tag{2–21}$$

$$\frac{\partial n_I E_I}{\partial t} = -\mathbf{j}_I \cdot \nabla_I \psi - n_I B_I + \nabla_I \cdot \alpha[-\mu_I n_I E_I \nabla_I \psi + \nabla_I(D_I n_I E_I)] - S_c, \tag{2–22}$$

where the subscript I denotes the quantity defined at the interface, and the J_c and S_c terms couple the subband to the other subbands or the bulk system. This coupling is introduced in the bulk system through the interface boundary conditions. The differential operator ∇_I operates only on the interface plane. The coupling coefficients in Eqs. 2–21 and 2–22 are simply the perpendicular current and energy flux of the bulk system ending at the interface, determined from Eqs. 2–19 and 2–20 as $J_c = (\mathbf{j}_\perp)_{\text{int}}$ and $S_c = (\mathbf{s}_\perp)_{\text{int}}$. It is easy to verify that total charge and energy are conserved, by integrating Eqs. 2–19 through 2–22. A relation for the rate of transfer between the quantum and bulk systems is also needed to specify the coupling between them, since the inclusion of the coupling coefficients only guarantees conservation.

Assuming that the relative concentrations of the quantum system and of the bulk at the adjoining edge obey local quasi-equilibrium, the bulk concentration at the interface n_{int}, in the nondegenerate case, and n_I can be related as[10]

$$n_I = N_{Ic}\left[1 - \exp\left(-\frac{E_1 - E_0}{kT_e}\right)\right] \ln\left[1 + (n_{\text{int}}/N_c)\exp\left(-\frac{E_1 - E_0}{kT_e}\right)\right], \tag{2–23}$$

where N_c and N_{Ic} are the effective density of states of GaAs in three and two dimensions, respectively, and T_e is the electron temperature, which here is represented with the average energy. E_0 and E_1 are the energy levels of the first two subbands. It has been assumed in this formulation that the potential of the bulk system, at the edge of the quantum system, coincides with the second subband energy level, E_1.

To accurately calculate E_0 and E_1 one should use the self-consistent procedure introduced in the previous section. Widiger *et al.*[10] used the approximate expressions

$$E_0 = (\hbar^2/2m)^{1/3}(0.88)(eF_0)^{2/3}, \tag{2–24}$$

$$E_1 = (\hbar^2/2m)^{1/3}[(0.28)(eF_0)^{2/3} + (0.94)(eF_1)^{2/3}] + E_0, \tag{2–25}$$

where F_0 is the confining field at the interface and F_1 is the field screened by the quantum-well charge $F_1 = F_0 - en_I/\epsilon$.

The potential and the electric field are obtained from Poisson's equation,

$$\nabla^2\psi = e(n - N_D)/\epsilon. \tag{2–26}$$

This equation applies to both the GaAs and the $Al_xGa_{1-x}As$ layers, which in general have different dielectric constants and donor concentrations. The difference between the perpendicular displacement in the $Al_xGa_{1-x}As$ at the interface and at the GaAs quantum-well bulk interface equals the interface charge as

$$\epsilon_1 \frac{\partial\psi}{\partial r_\perp}\bigg|_{\text{int}^-} - \epsilon_2 \frac{\partial\psi}{\partial r_\perp}\bigg|_{\text{bulk}^+} = en_I, \tag{2–27}$$

where ϵ_1 and ϵ_2 are the dielectric constants of $Al_xGa_{1-x}As$ and GaAs, respectively. At the heterointerface the continuity rules require that the parallel fields are equal on the two sides and the potential is continuous, i.e. $\psi_{\text{int}^-} = \psi_{\text{int}^+}$. To relate the potential at the edge of the second quantum level (the edge of the bulk system) to that in the $Al_xGa_{1-x}As$, the quantum energy is included, i.e. $e\psi_{\text{bulk}^+} = e\psi_{\text{int}^-} - E_1$. Widiger *et al.*[10] also include surface states through appropriate boundary conditions for the electric field.

The general procedure for solving the system of equations is shown in Fig. 2–3. An explicit finite difference scheme has been used over a two-dimensional variable-width positional mesh. Poisson's equation is solved at each time step with a simple "red/black" successive-over-relaxation (SOR) method. The basic device topology adopted in the simulations is the same one shown in Fig. 2–2. The size of the structure is 3 μm from source to drain with a 1-μm gate length. The insulation layer is $Al_xGa_{1-x}As$ with a 50-nm width, a doping of 5×10^{17} cm^{-3}, and $x = 0.32$. The device is simulated at 77 K.

Figures 2–4 to 2–11 show the simulation results of four different biasing potentials at steady state. The first of each pair of figures (Figs. 2–4, 2–6, 2–8, and 2–10) shows the electron concentration in both the quantum well and the bulk. The horizontal axis denotes the position between the source and the drain, with the gate boundaries between the 1- and 2-μm marks. The axis entending out of the figure is the distance into the bulk from the interface. The second figures of each pair (Figs. 2–5, 2–7, 2–9, and 2–11) are composed of a top portion, showing the average elec-

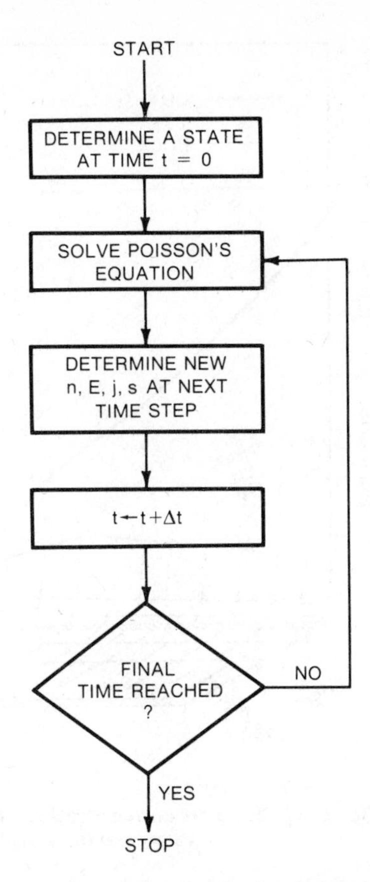

Fig. 2–3. Iteration scheme for the hydrodynamic model.

tron velocity parallel to the surface for both the bulk and the quantum well, and a bottom portion which shows the parallel electric field and the average electron energy at the interface. Energy-field and velocity-field relationships obtained from Monte Carlo calculations for a homogeneous semiconductor have been used to predict the average electron velocity and the average energy, given in the top and the bottom portion of the figure, respectively.

These figures show that the current at the source is carried in the quantum system, while, because of electron heating and a reduced confining field, the current is carried in the bulk system at the pinch-off point. High parallel fields cause, at pinch-off, a dramatic increase in the average velocity to well above the maximum steady-state velocity, an effect known as *velocity overshoot.* The moment equation approach described above is known to overestimate the overshoot effect in cases of very steep fields, as in Figs. 2–5 and 2–7. Such an overestimate arises from a neglect of the exact band structure which limits electron velocities in GaAs to about

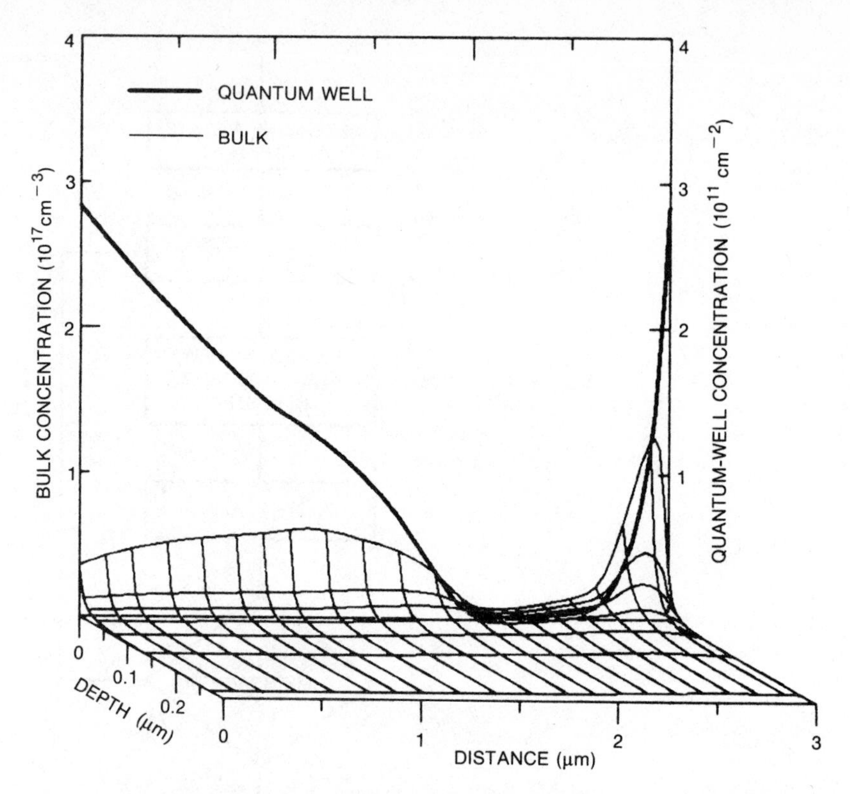

Fig. 2–4. Electron concentration in the quantum well and the bulk for $V_g = -0.06$ V and $V_d = 0.9$ V, at 77 K.

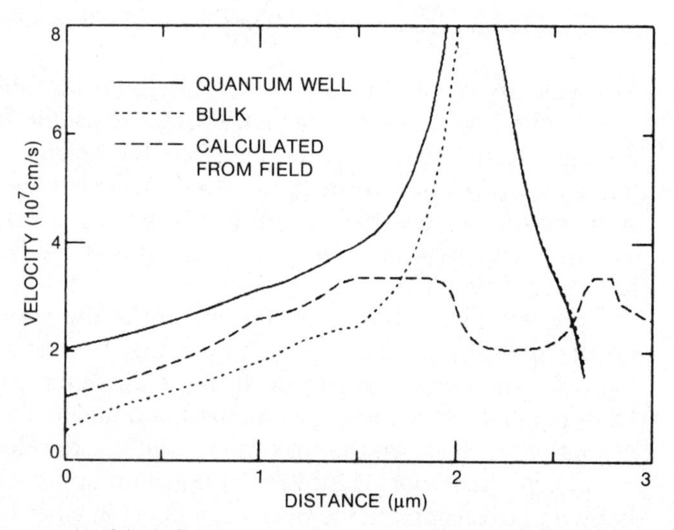

(A) Comparison of the bulk velocity at the surface, the quantum-well velocity, and the velocity predicted from the electric field by Monte Carlo results.

Fig. 2–5. Velocities, fields, and energies for $V_g = -0.06$ V and $V_d = 0.9$ V, at 77 K.

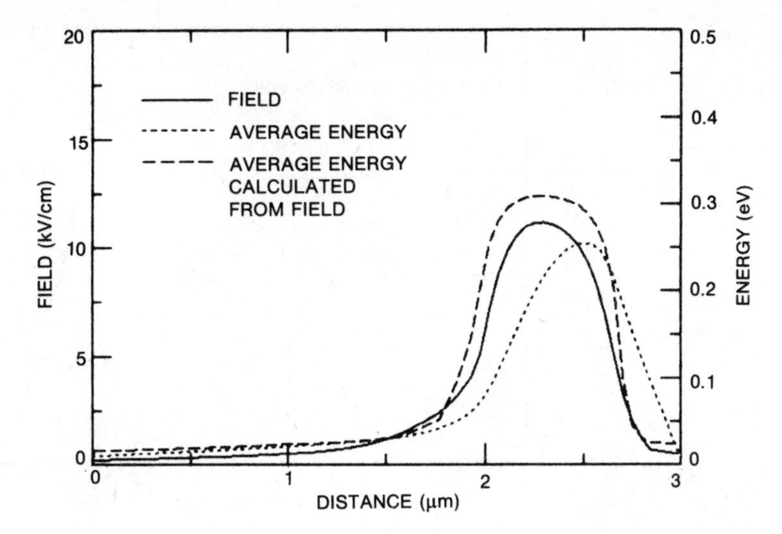

(B) Comparison of the electric field, the average energy and the average velocity predicted from the field.

Fig. 2–5 *continued*

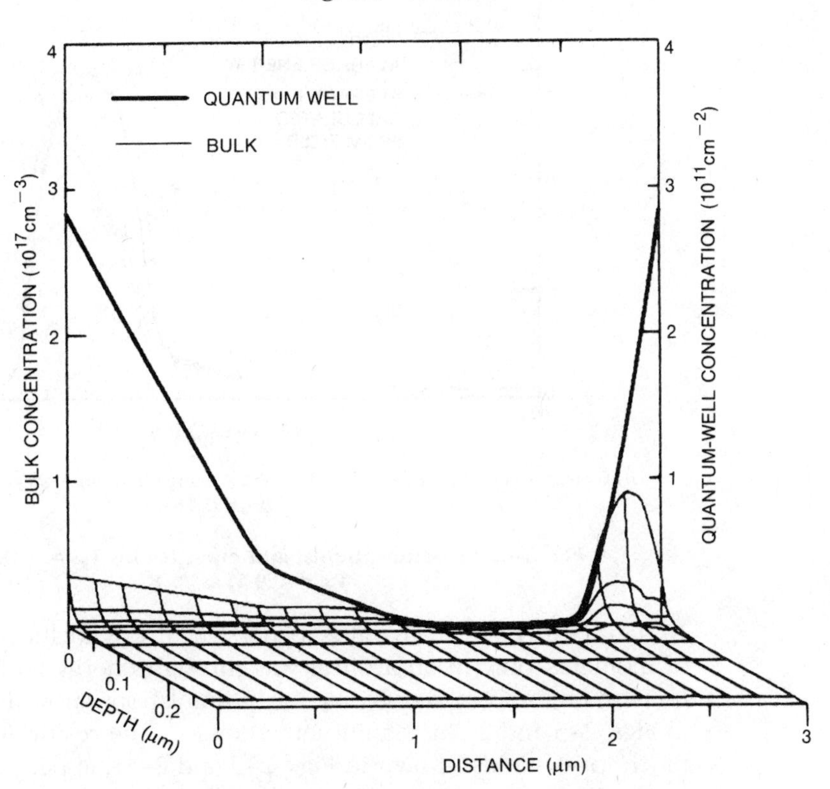

Fig. 2–6. Electron concentration for V_g = −0.21 V and V_d = 0.9 V, at 77 K.

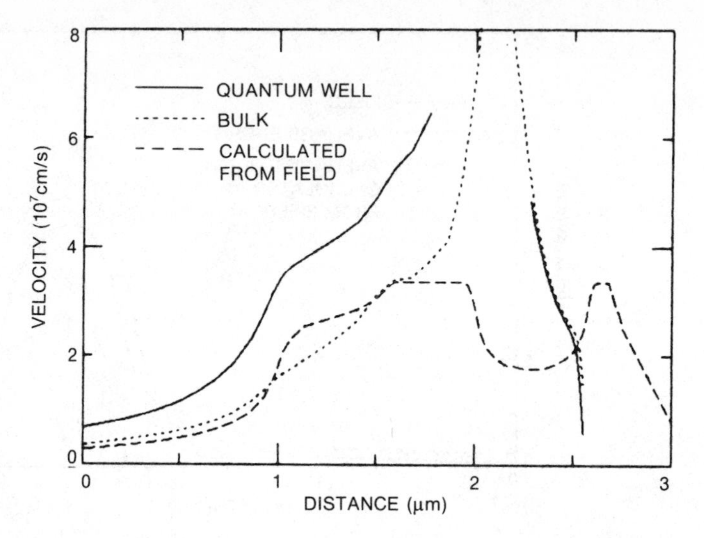

(A) Comparison of the bulk velocity at the surface, the quantum-well velocity, and the velocity predicted from the electric field by Monte Carlo results.

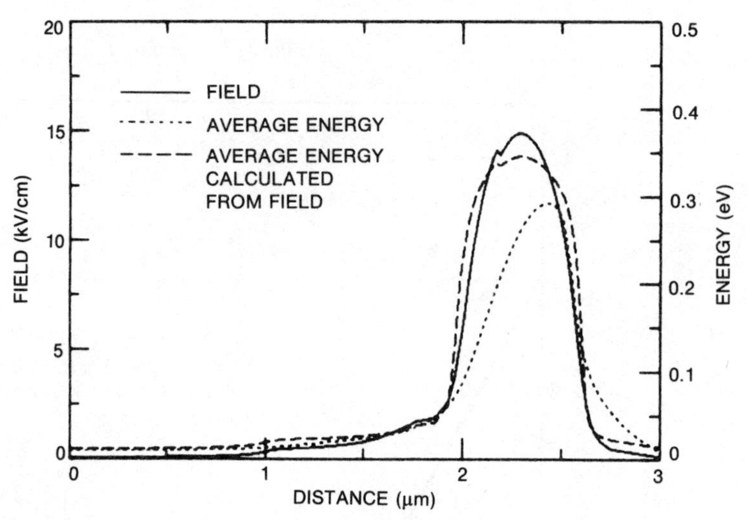

(B) Comparison of the electric field, the average energy, and the average velocity predicted from the field.

Fig. 2–7. Velocities, fields, and energies for V_g = −0.21 V and V_d = 0.9 V, at 77 K.

10^8 cm/s. Also, the many-valley band structure is included in the method of moments only in an artifical way through energy and momentum relaxation times. Complete Monte Carlo simulations show that the overshoot in Figs. 2–5 and 2–7 is significantly reduced. The results for the velocity at much lower fields, shown in Figs. 2–9 and 2–11, appear instead to be accurate. By comparing Figs. 2–8 and 2–9 with Figs. 2–10 and 2–11, one can

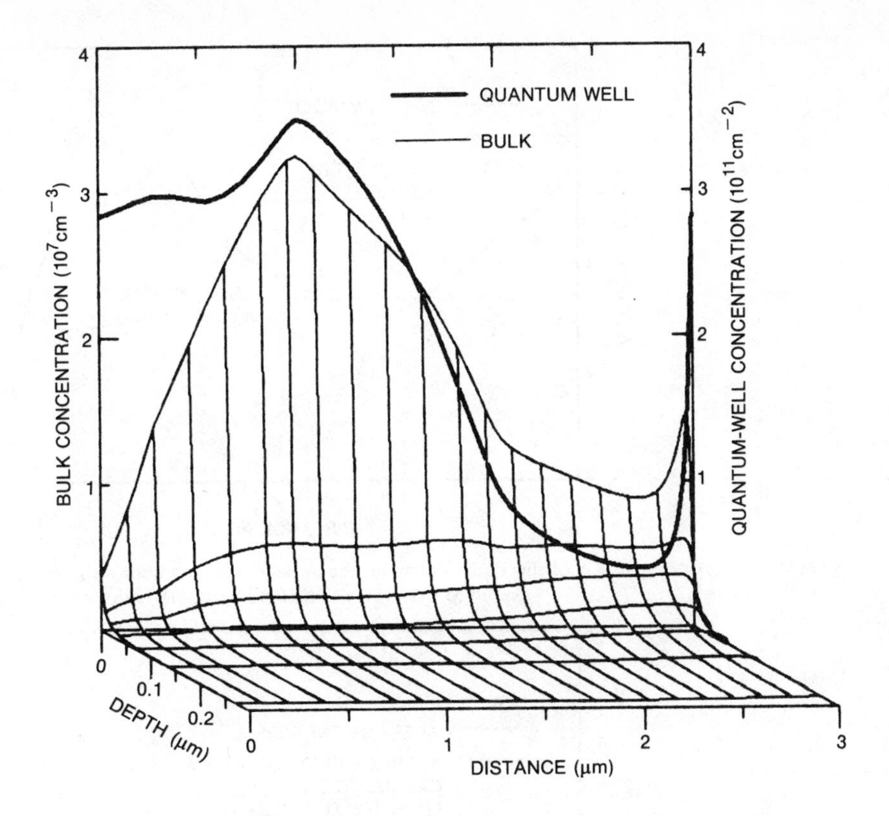

Fig. 2–8. Electron concentration for V_g = 0.39 V and V_d = 0.9 V, at 77 K.

see that, under current saturation conditions with a fixed gate bias, the various quantities in the source and gate regions are independent of the drain bias, while those at the pinch-off point, particularly at overshoot velocity, are highly dependent on the drain bias. A velocity saturation mechanism at the pinch-off point clearly does not exist for all applied voltages.

Simulations of both switch-on and switch-off conditions have been performed by Widiger *et al.*[10] and by Kizilyalli *et al.*[12], using as initial and final states the previously discussed steady-state solutions. In the following example[12] for a 1-μm gate HEMT at 300 K, switch-on starts at V_d = 0.9 V and V_g = −0.21 V (Fig. 2–12) and switches to V_g = 0.39 V with V_d fixed (Fig. 2–13). The results for source and drain currents during the transients are shown in Fig. 2–14 (switch-on) and in Fig. 2–15 (switch-off). The device is considered to be in steady-state conditions when the currents have stabilized and are within ±5 percent of each other. The characteristic switch-off time is significantly faster than the characteristic switch-on time. In the turn-on transient it is observed that the steady-state electric field and the electron concentration at the source of the device establish rap-

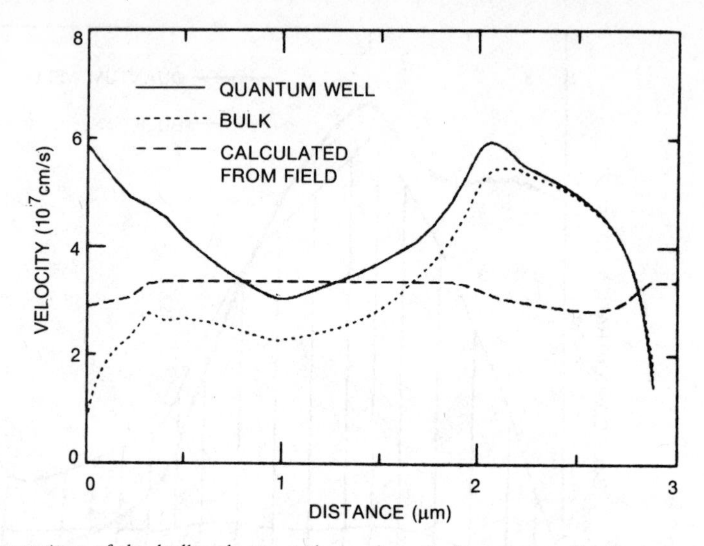

(A) Comparison of the bulk velocity at the surface, the quantum-well velocity, and the velocity predicted from the electric field by Monte Carlo results.

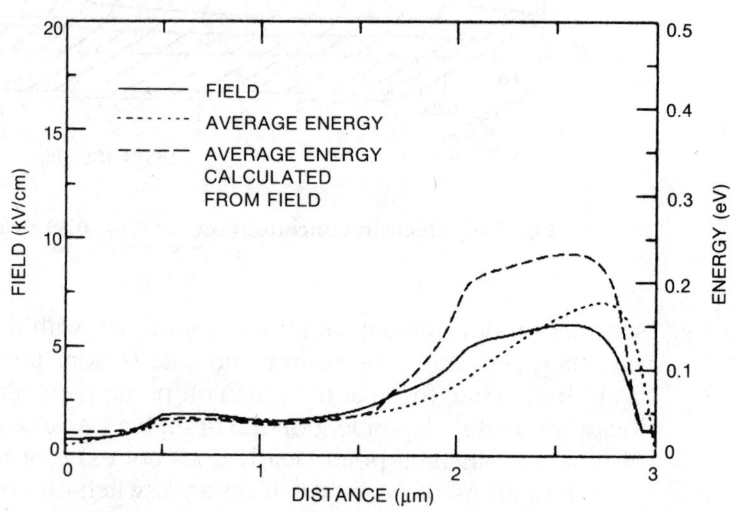

(B) Comparison of the electric field, the average energy, and the average velocity predicted from the field.

Fig. 2–9. Velocities, fields, and energies for V_g = 0.39 V and V_d = 0.9 V, at 77 K.

idly. Therefore, the source current converges to its steady-state values within picoseconds. The drain current is established only after the electrons that have entered the device from the source contact at time 0 reach the drain edge of the gate. The switch-on time is therefore determined approximately by the ratio of the device length to the average transit velocity.

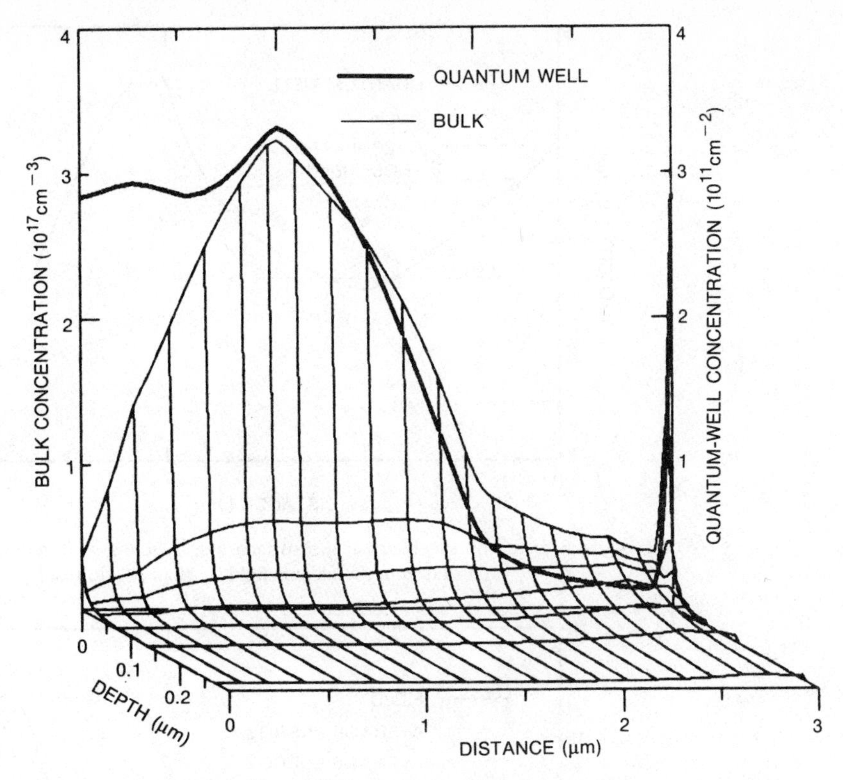

Fig. 2–10. Electron concentration for V_g = 0.39 V and V_d = 1.5 V, at 77 K.

To switch the device off, however, it is sufficient to separate the electrons spatially under the gate by lowering the gate potential and driving the electrons to their respective contacts. Electrons need not transit the full device length to switch off.

The switching time in practice includes also the time required to charge a capacitive load. This type of delay will indeed dominate the delays of LSI-type circuits. For switching purposes the capacitance can be estimated to be approximately 2 pF/mm. With a 0.6-V potential change required to switch a device, and a drain driving current of 300 mA/cm, an additional 4-ps switch-on delay for each fan-out gate is found. Such delays depend critically on the transconductance.

The model presented in this section only treats the active region of the device. The contacts undoubtedly play an important role in establishing the device characteristics. By choosing a boundary condition which constrains the electron concentration at the source and drain contacts to a fixed value, in Refs. 10 and 12 the contacts are given a zero resistance, which surely contributes to the excellent performance calculated. A high

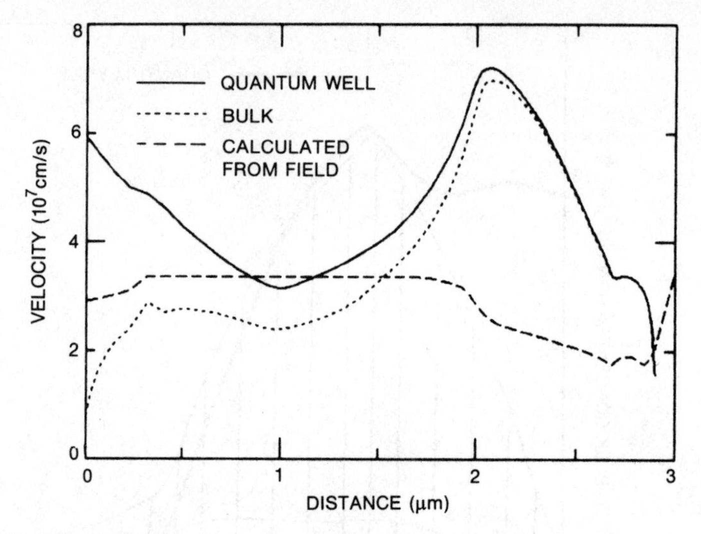

(A) Comparison of the bulk velocity at the surface, the quantum-well velocity, and the velocity predicted from the electric field by Monte Carlo results.

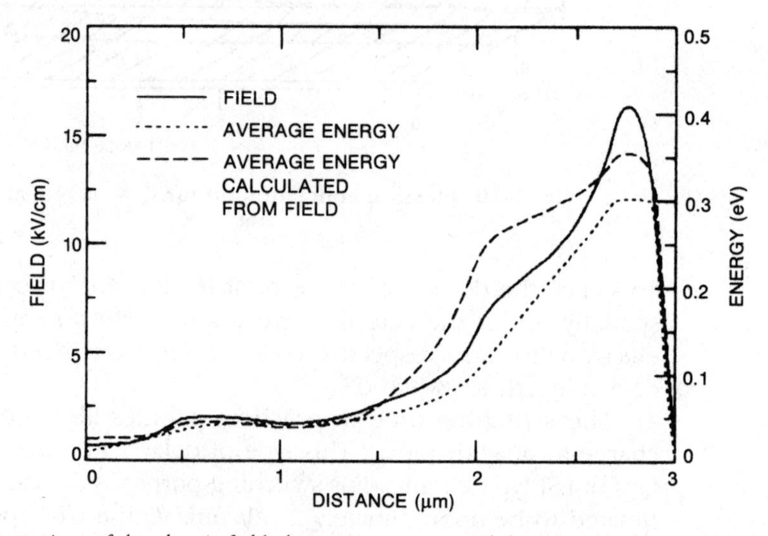

(B) Comparison of the electric field, the average energy, and the average velocity predicted from the field.

Fig. 2–11. Velocities, fields, and energies for V_g = 0.39 V and V_d = 1.5 V, at 77 K.

source resistance will reduce transconductance, whereas a high drain resistance will cause saturation to occur at higher drain voltages. Although the $Al_xGa_{1-x}As$ is completely depleted at equilibrium for the simulated device, hot electrons can transfer from the GaAs to the $Al_xGa_{1-x}As$ if the average energies are comparable to the heterojunction barrier (on the order

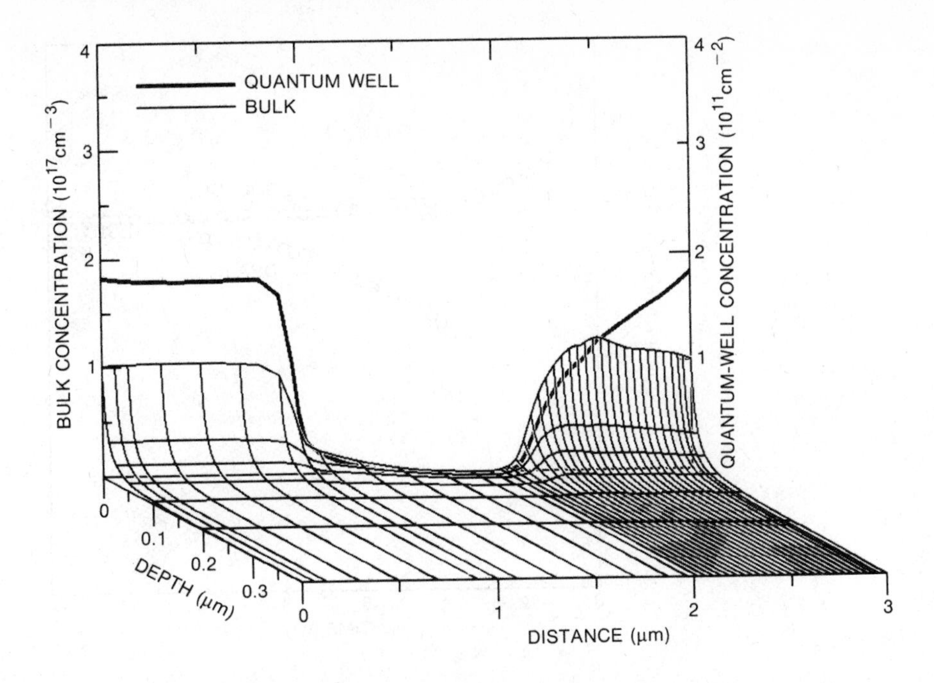

Fig. 2–12. Electron concentration for a 1 μm gate HEMT, for V_g = −0.21 V and V_d = 0.9 V, at 300 K.

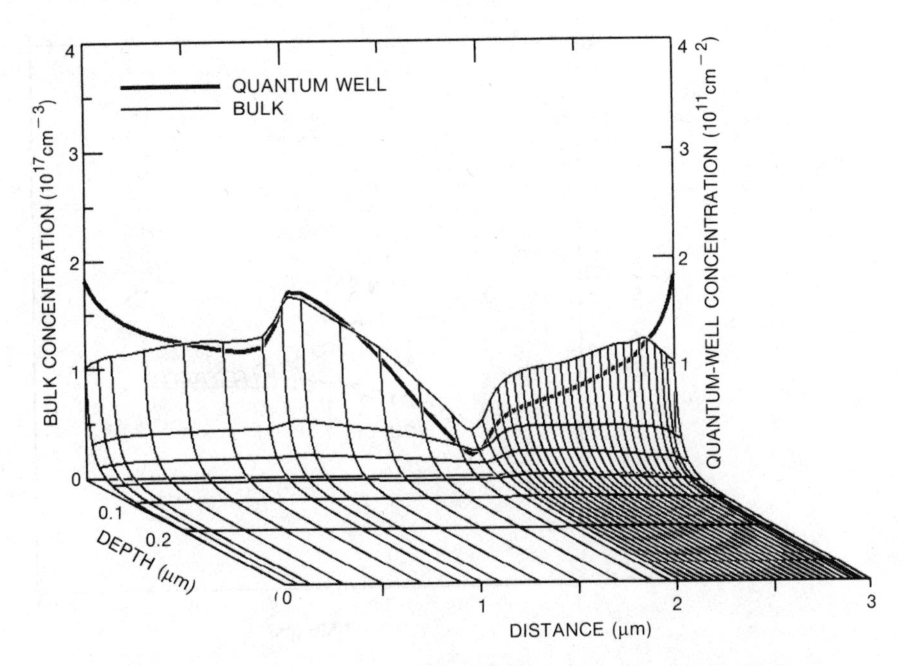

Fig. 2–13. Electron concentration for a 1 μm gate HEMT, for V_g = 0.39 V and V_d = 0.9 V, at 300 K.

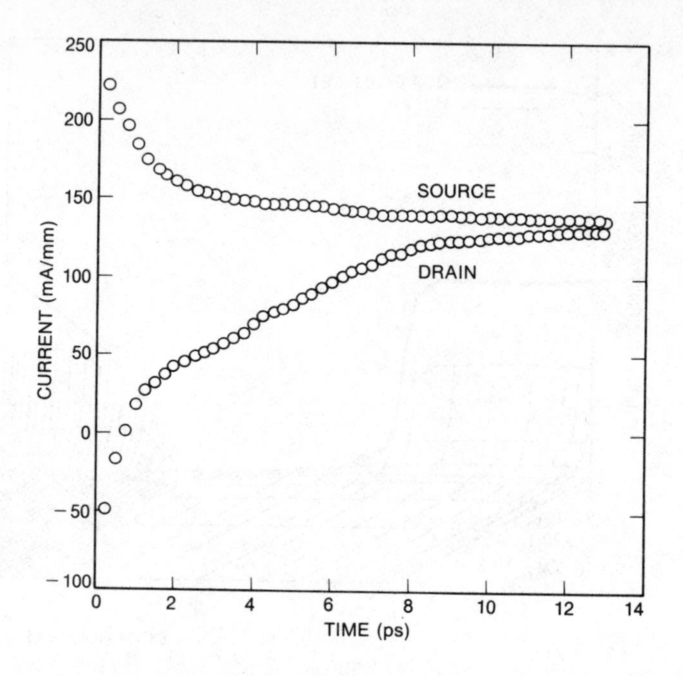

Fig. 2–14. Switch-on currents at source and drain. The gate potential is switched from −0.21 to 0.39 V with V_d = 0.9 V, at 300 K.

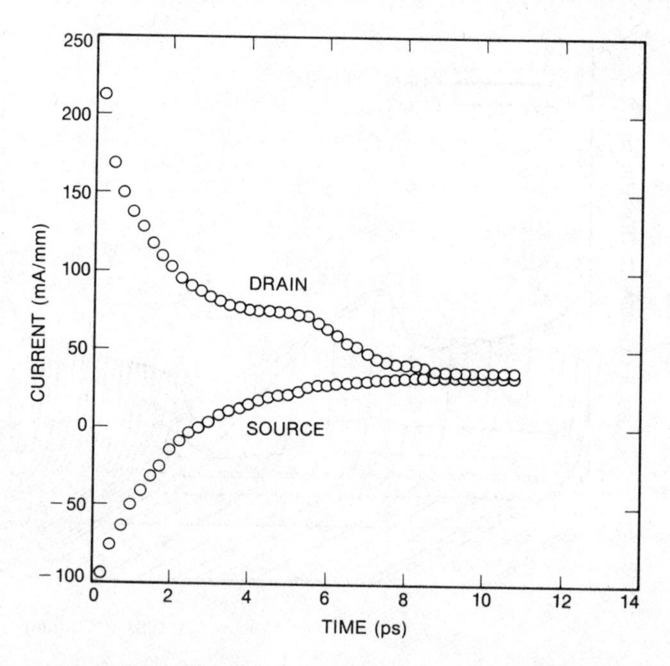

Fig. 2–15. Switch-off currents at source and drain. The gate potential is switched from 0.39 to −0.21 V with V_d = 0.9 V, at 300 K.

of 0.3 eV). The steady-state results indicate that such energies are present only on the drain side of the gate. If electron transfer near the drain occurs, the speed of propagation will decrease because of the poorer conduction properties of the $Al_xGa_{1-x}As$.

2.4 Scaling Properties of the HEMT

The hydrodynamic model presented in the previous section has been used to study the scaling properties of a HEMT structure at 300 K.[12] The device topology in this study has been the same as shown in Fig. 2–2. To derive a scaling scheme for the HEMT we have to consider the following parameters: the total device length (L_{SD}), the gate length (L_G), the $Al_xGa_{1-x}As$ thickness (w), the doping concentration (N_D), the active layer thickness, and the applied biases. The horizontal dimensions (L_{SD} and L_G) are the most important parameters for the study of hot electron transport and its effect on scaling properties. The active layer thickness can be ignored in this analysis. The ratio of the transconductance to the sum of gate and interconnect capacitances gives the basic switching speed, and since the transconductance increases with decreasing the $Al_xGa_{1-x}As$ thickness, w, the smallest possible values for w are desirable, even at the expense of high gate capacitances. The lower bound for w is determined by the breakdown properties of the material. A constant value $w = 50$ nm and a constant doping $N_D = 5 \times 10^{17}$ cm^{-3} has been chosen by Kizilyalli *et al.*[12] for the $Al_xGa_{1-x}As$. For the scaling of operating voltages there are two possible approaches: constant voltage and constant field scaling. The switching speeds are almost identical for the two approaches, so constant voltage scaling has been chosen since significantly higher transconductances are obtained. In these calculations, the gate is always one third of L_{SD} and is symmetrically placed between source and drain. Only horizontal dimensions are varied in the investigation. The drain voltage has been kept constant at 0.9 V, while, for the switching analysis, the gate voltage is instantaneously changed from low level, $V_{GL} = -0.21$ V, to high, $V_{GH} = 0.39$ V. Devices with gate lengths between 2.0 and 0.5 μm have been considered.

The results for the current switching time as a function of length are given in Fig. 2–16. As shown in the previous section, the switch-off time is always smaller than the switch-on one. The characteristic switching times are clearly reduced when the HEMT is scaled down, since the length an electron must traverse is reduced, while average transit velocities and electric fields are increased.

An increase in the drain-source current is expected as the horizontal dimensions of the device are reduced. A linear scaling law would predict a

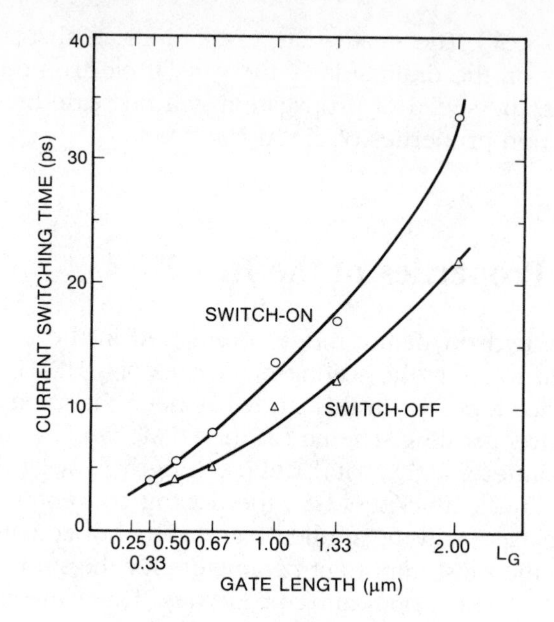

Fig. 2–16. Current switching time versus gate length.

relation $I \propto L^{-1}$. The current levels for high and low gate voltages are shown in Fig. 2–17, where the drain voltage is kept at 0.9 V. The dependence of currents on gate length is not linear, but roughly estimated to be $I_H = I_L^{(0)} L^{-0.7}$ and $I_L = I_L^{(0)} L^{-1.2}$, where the superscript 0 labels the 1.0 μm gate length device. The current-gate length product obtained for various gate lengths is given in Fig. 2–18. The results show that a linear scaling is applicable for devices with a gate length larger than 2 μm. When

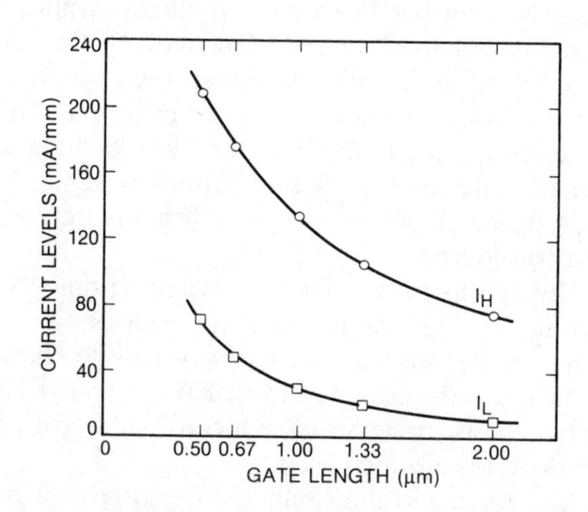

Fig. 2–17. High and low current levels versus gate length.

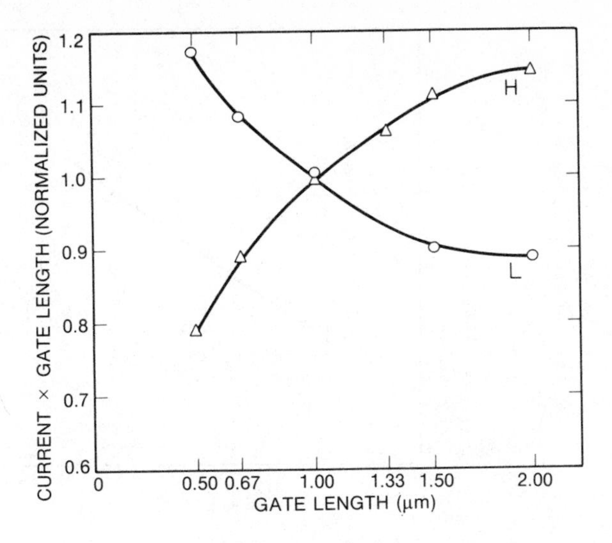

Fig. 2–18. Current-gate length product versus gate length.

the gate is smaller than 2 μm, the transport is dominated by two competing hot-electron effects: the velocity saturation (mobility reduction due to increased average electron energies) and overshoot effects.

The transconductance can be calculated from the steady-state drain current by changing the gate voltage in discrete steps. The plot of Fig. 2–19 shows the transconductance as a function of gate voltage, for various device dimensions. The curves are uniform over most gate voltages, increas-

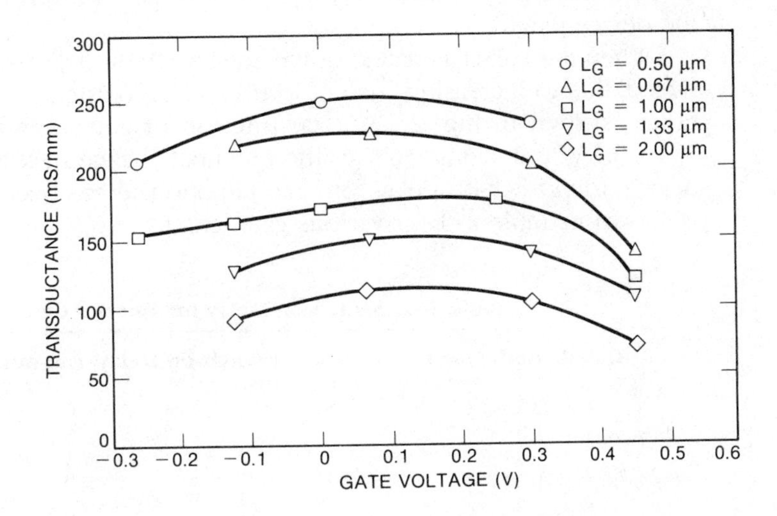

Fig. 2–19. Transconductance versus gate length.

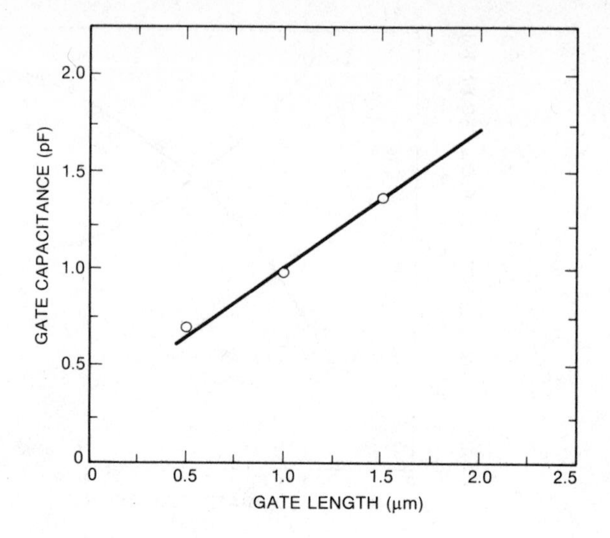

Fig. 2–20. Gate-to-source capacitance versus gate length.

ing slightly at first and decreasing for gate voltages above approximately 0.4 V. The sharp reduction of transconductance, which is experimentally observed at higher gate voltages and attributed to conduction in the $Al_xGa_{1-x}As$, is not seen here, because the model does not take into account the real-space–transfer effect. The model cannot be extended to devices with gate lengths smaller than 0.5 μm, where the transconductance is expected to saturate and not increase with further reduction of the device length.

When the HEMT is scaled down, the switch-on delay is significantly reduced, due to increasing current levels and decreasing gate-source capacitance, shown in Fig. 2–20 as a function of gate length. Using again $V_D = 0.9$ V, and taking 0.6 V as the potential change necessary to switch a device, the switch-on delay for each fan-out gate has been calculated and is shown in Table 2–1 for various gate lengths.

Table 2–1. Switch-on Delay for Fan-out Gates

Gate Length (μm)	Switch-on Delay/Fan-out Gate (ps)
0.5	2.0
1.0	4.5
1.33	7.1
1.50	8.2
2.0	14.0

2.5 Ensemble Monte Carlo Models

A more precise study of the electron transport properties in the quantum channel of the HEMT requires a detailed inclusion of the subband levels and wavefunctions. At this level of accuracy, one also must solve the Boltzmann transport equation and take into account the scattering in each subband. A stochastic solution of the resulting complicated system of equations may be obtained using a complete self-consistent Monte Carlo method. In an ensemble Monte Carlo (EMC) simulation, the dynamics of the individual particles inside the device are accurately modeled, reproducing stochastically the particle movements with sequences of free flights, interrupted by instantaneous scattering events. The choice of the scattering mechanism is done by random selection, according to the appropriate rates corresponding to the particle energy at the end of each flight. Self-consistency is achieved by solving Poisson's equation periodically during the simulation time, to update the electric fields. Since, in an EMC simulation, the effect of the scattering mechanisms on the particle flow can be studied in great detail, hot-electron and nonlinear effects are naturally included. It can be shown that a Monte Carlo method gives a stochastic solution of the Boltzmann equation.[13] An extensive account of the Monte Carlo techniques for semiconductor transport can be found in Refs. 13, 14, and 15.

For the analysis of the transport in the quantum channel, one has to calculate the two-dimensional scattering rates. It can be shown that for the $Al_xGa_{1-x}As/GaAs$ system, the square of the scattering matrix element is[16,17]

$$|M_{mn}|^2 = \int |M(Q,q)|^2 \, |I_{mn}(q)|^2 \, dq, \tag{2–28}$$

where $M(Q,q)$ is the corresponding three-dimensional matrix element, Q and q are the phonon wavevector components parallel and perpendicular to the interface. The indices m and n indicate the subbands. If $n = m$, the scattering is intrasubband, and if $n \neq m$, it is intersubband. The overlap integral is given by

$$|I_{nm}(q)|^2 = \int \zeta_n(z) \, \zeta_m(z) \exp(iqz) \, dz, \tag{2–29}$$

where $\zeta_n(z)$ is the envelope wavefunction discussed in Sec. 2–2.

The important scattering mechanisms to be considered in the channel are polar-optical phonon scattering, deformation-potential acoustic phonon scattering, and ionized impurity scattering. Using the Fermi golden rule and the matrix elements given by Eq. 2–28, the intrasubband

and intersubband scattering rates can be calculated for each of the subbands which are included in the model. The ionized impurity scattering rates in a modulation-doped structure are normally much smaller than phonon scattering rates, and may often be neglected, particularly when an undoped spacer layer is present at the interface.

Using the scattering rates obtained, steady-state and transient behavior of single–quantum-well structures in homogeneous conditions can be studied by Monte Carlo methods. In the case of steady-state analysis, it is sufficient to follow the trajectory of one particle for a long period of time, and the steady-state transport properties, such as average drift velocity and population of each subband and each valley, are calculated from the history of the electron motion. The steady-state calculation is typically continued until the scattering events exceed 20,000 in order to obtain a good convergence of the estimated quantities. For the analysis of transient transport phenomena, an ensemble of electrons must be simulated at the same time and the initial distribution of electrons plays an important role.

Calculated steady-state velocity-field curves for the Q2DEG confined in the potential well, including five subbands, are shown in Figs. 2–21A and 2–21B, for 77 K and 300 K, respectively.[17] These figures also show the distribution of electrons in the subbands and in the L-valley, which is treated as three-dimensional. The initial two-dimensional carrier density is taken to be $N_s = 5 \times 10^{11}$ cm^{-2}.

A large number of electrons populates the first subband at low electric fields. With the increase of the applied field, more and more electrons transfer to higher subbands, until at still higher fields intervalley transfer begins, to the L-valley at first and then to the X-valley. The velocity characteristics are compared with those for bulk GaAs with $N_D = 10^{17}$ cm^{-3}. The critical field where the peak velocity is observed for the Q2DEG is shifted to lower fields compared with that for bulk GaAs. The shift is caused by the reduced impurity scattering for the Q2DEG, and we can expect high velocity values at low fields, which is significant for device applications. The mobility obtained at 77 K for 500 V/cm is 2.5×10^4 cm^2/V·s.[17]

The transient drift velocity response at 77 and 300 K is shown in Figs. 2–22A and 2–22B, respectively.[17] The velocity for the Q2DEG exhibits the peak value after a time period which is nearly equal to that of bulk GaAs with a doping $N_D = 10^{18}$ cm^{-3}, for an electric field of 10 kV/cm at 77 K. The peak value for the Q2DEG is higher than for bulk, and the velocity response is much steeper, due to reduction in ionized impurity scattering. With an increasing field, the transient behavior for the Q2DEG is very close to that for the three-dimensional gas. Therefore, it is concluded that the velocity advantages are significant only at low and intermediate fields.

The inclusion of a complete Q2DEG dynamics in a full simulation of an HEMT, as outlined so far, is rather complicated because of the large inho-

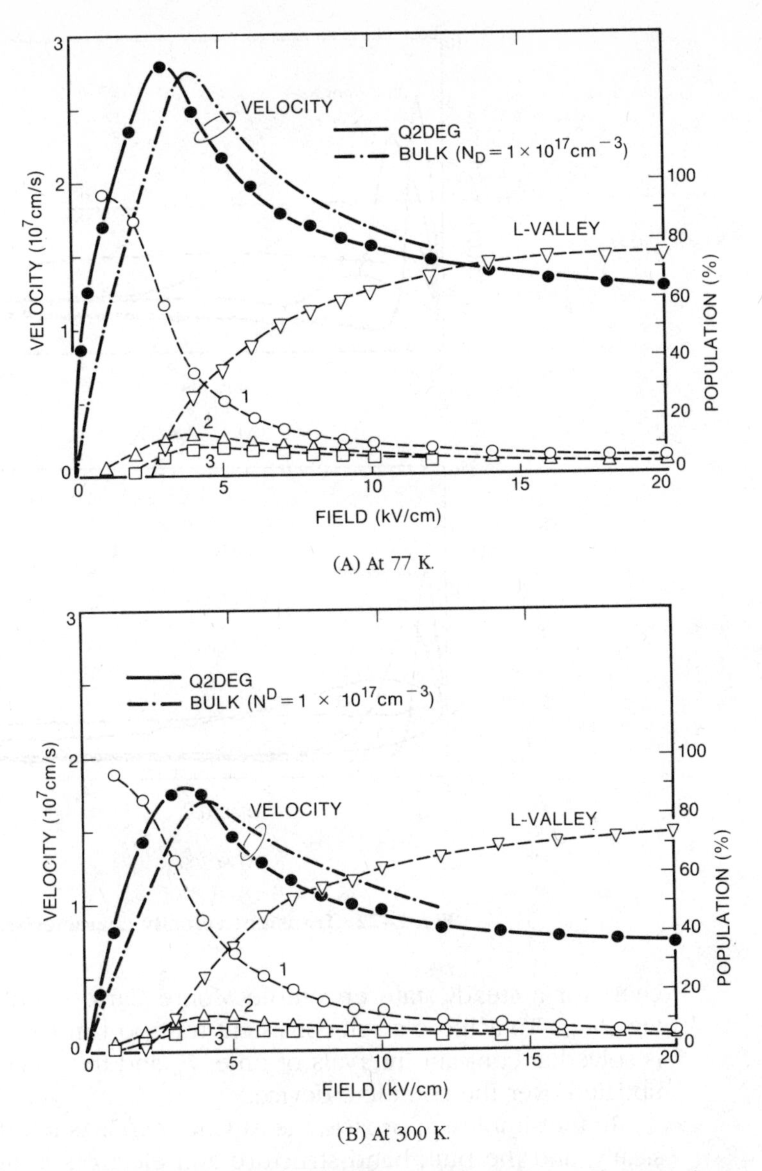

(A) At 77 K.

(B) At 300 K.

Fig. 2–21. Steady-state velocity and population in the first three subbands and in the L-valley, versus field. The dash-dotted line represents calculated velocities for three-dimensional transport.

mogeneity along the channel. This would require the self-consistent solution of the Schrödinger equation for a large number of potential distributions. Since an ensemble Monte Carlo model for device simulation already requires a very considerable computational effort, some approximations are in order for the transport in the conduction channel. A schematic flow-

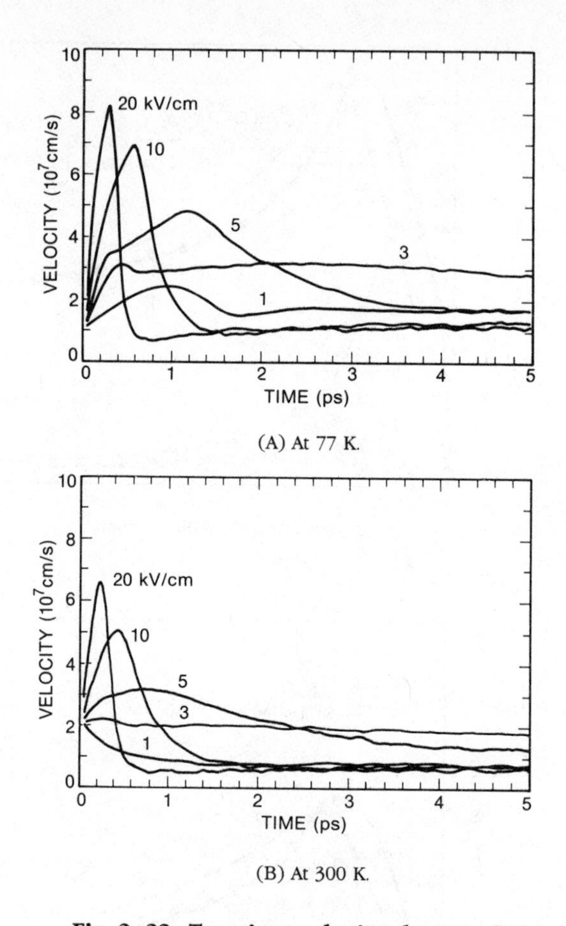

Fig. 2–22. Transient velocity characteristics.

chart for a steady-state ensemble Monte Carlo simulation is shown in Fig. 2–23. To achieve a self-consistent transport model, Poisson's equation is solved at constant intervals of time, T_s, and the electric field values are updated over the complete device.

In the simplest approach, the $Al_xGa_{1-x}As$/GaAs interface is treated classically, and the bulk band structure and electron scattering mechanisms are assumed in the GaAs channel.[18] Although such a model cannot be completely accurate, it is valuable under high-field conditions, when the size quantization effects are less important, and it allows an analysis of the trends under the gate and in the pinch-off region. The real space transfer effect, which accounts for hot electrons transferring from the GaAs into the $Al_xGa_{1-x}As$ region, can also be included in the Monte Carlo simulation in a natural way. This is not possible in a hydrodynamic model.

A relatively simple way to include quantum effects is to assume an approximate solution for the subband wavefunctions, as obtained from the

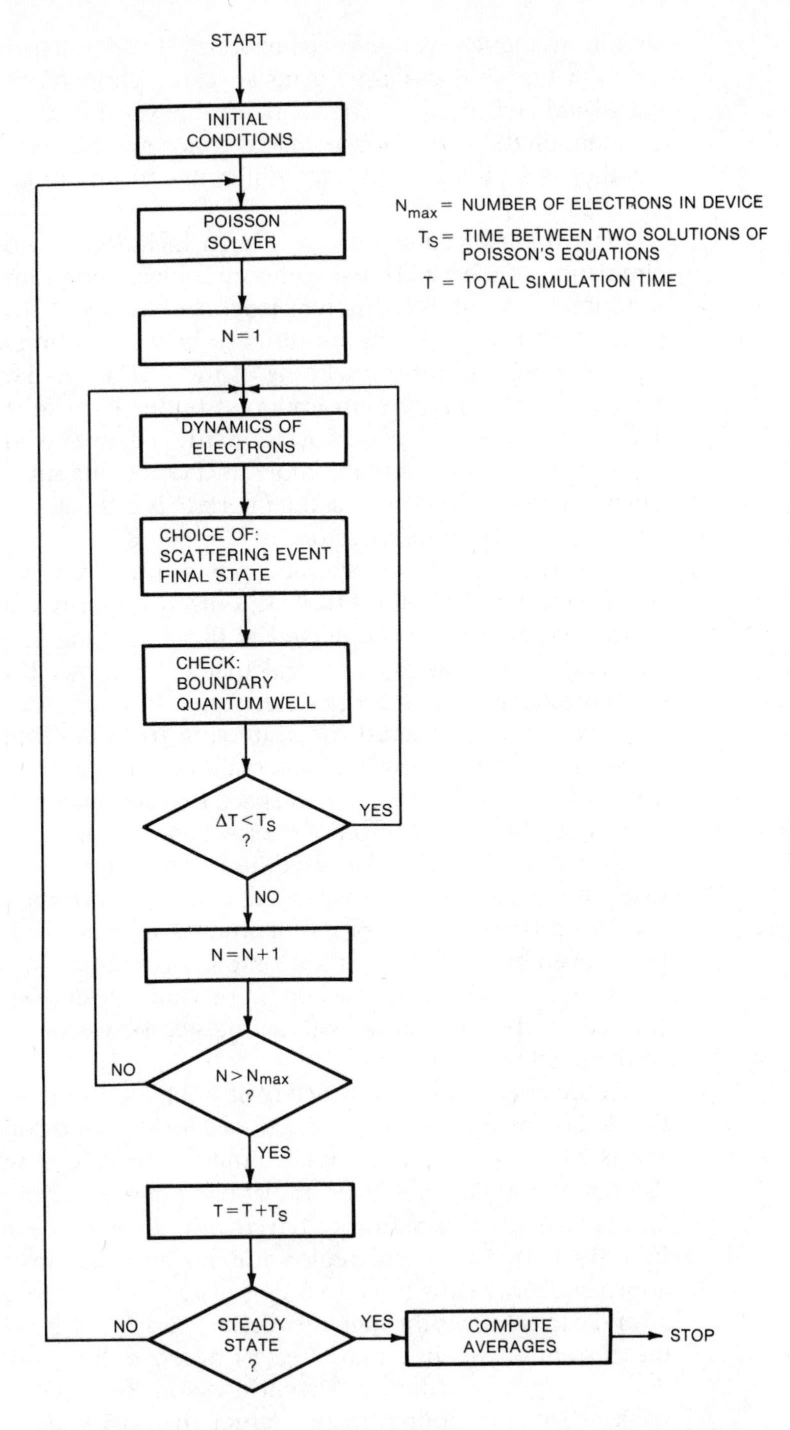

Fig. 2–23. Flowchart for the ensemble Monte Carlo simulation of the HEMT.

infinite triangular-well approximation.[19] If variational results are used, as given by Eqs. 2–8 and 2–9 for instance, the phonon scattering rates can be calculated analytically.[20] The shape of the wavefunctions is, in general, dependent on the local surface charge concentration and it is easy to create a number of scattering tables for different concentrations, to be used in the calculations.

The first two subbands have been included in a device Monte Carlo simulation of the HEMT, using the variational approximation for the wavefunctions.[21] A submicrometer structure has been investigated, with the gate length $L_G = 0.3\ \mu m$, the distance between source and drain contacts $L_{SD} = 0.7\ \mu m$, and the thickness of the $Al_xGa_{1-x}As$ layer $w = 60$ nm, of which 10 nm constitute an undoped buffer layer at the heterointerface. The concentration of donors considered in the $Al_xGa_{1-x}As$ layer is $N_D = 2 \times 10^{17}\ cm^{-3}$, and the mole fraction of aluminum $x = 0.3$. The conduction band discontinuity at the interface is estimated to be $\Delta E_c = 0.25$ eV, and the simulation temperature is $T = 300$ K.

In this model,[19] acoustic and polar-optical phonon scatterings have been considered for the Q2DEG. Ionized impurity and remote impurity scatterings have been neglected at this operating temperature. In each subband, scattering rates are calculated for intrasubband and intersubband mechanisms, and for hot electrons whose energy is greater than ΔE_c, rates are also calculated for scattering from confined states to three-dimensional states, as well as intervalley scattering to the L satellite valley. When electrons experience real space transfer and cross the interface, two basic laws, the conservation of energy and the conservation of momentum parallel to the interface, are used. In addition, particles are assumed to remain in the same k-space valley (Γ, L, or X), since the intervalley transfer involves a large amount of momentum change and is less likely to occur. For a given bias condition, the Monte Carlo simulation is continued until a steady-state condition is reached. Approximately 20,000 electrons are considered for the simulation, with a time step between two solutions of Poisson's equation of 10 fs.

The current-voltage characteristics obtained in this way are shown in Fig. 2–24. The upper curve is calculated for flat-band conditions at the gate, and is given as a reference upper limit for the case in which current flows through the $Al_xGa_{1-x}As$ layer. In flat-band condition, the applied gate voltage V_g cancels the Schottky barrier Φ_B due to the metal contact, i.e. $V_g + \Phi_B = 0$. The useful region for the normal transistor operation lies approximately below the curve $V_g + \Phi_B = -0.4$ V, since in this case virtually no current flows through the $Al_xGa_{1-x}As$ layer. When $V_g + \Phi_B = -0.3$ V, the current curve shows an effect of negative differential mobility. At low drain voltages virtually no current flows in the $Al_xGa_{1-x}As$, but when the voltage becomes approximately larger than 0.5 V, the total drain current

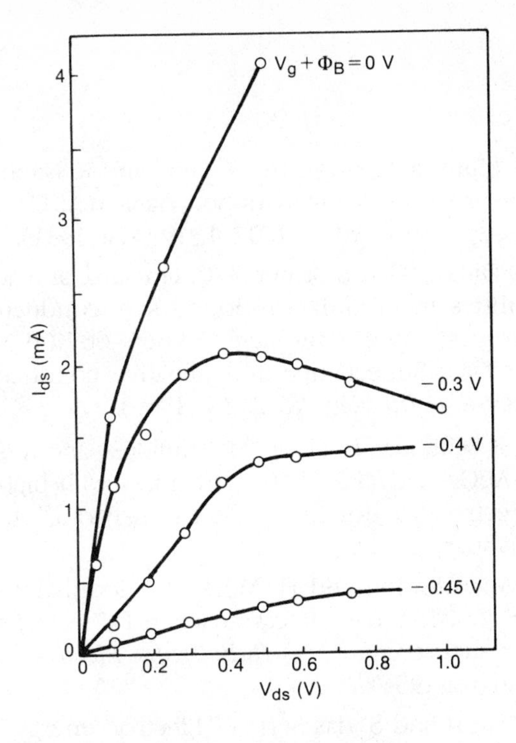

Fig. 2–24. I_{ds} versus V_{ds} for the 0.3 μm gate HEMT calculated with ensemble Monte Carlo simulation. The curves are scaled for a 0.1 mm wide device.

starts decreasing since more and more carriers flow through the $Al_xGa_{1-x}As$ layer. This is due partly to real space transfer and partly to the reduced barrier at the Schottky gate.

A systematic investigation of the device with Monte Carlo methods is limited by the need for very large computational resources. However, only in a Monte Carlo model it is possible to fully include hot electron effects, in particular real space transfer. Since the conduction in the $Al_xGa_{1-x}As$ layer is naturally included without any restrictive assumptions, it is also possible to evaluate directly the potentially degrading effects of parallel conduction.

2.5.1 Note Added in Proof

Since the time this chapter was written, Monte Carlo methods have been developed by D. H. Park and K. F. Brennan.[22]

References

1. T. Mimura, S. Hiyamizu, T. Fujii, and K. Nanbu, "A new field effect transistor with selectively doped GaAs/n-$Al_xGa_{1-x}As$ heterojunctions," *Jpn. J. Appl. Phys.,* 20(5): L317–L319 (May 1981).
2. R. Dingle, H.L. Stormer, A.C. Gossard, and W. Wiegman, "Electron mobilities in modulation-doped semiconductor heterojunction superlattices," *Appl. Phys. Lett.,* 33:665–667 (1978). See also L. Esaki and R. Tsu, "Superlattice and negative conductivity in semiconductors," *IBM Res., Int. Rep. RC 2418* (1969).
3. E.F. Schubert, K. Ploog, H. Dambkes, and K. Heime, "Selectively doped n-$Al_xGa_{1-x}As$/GaAs heterostructures with high-mobility two-dimensional electron gas for field effect transistors," *Appl. Phys. A,* 33(1): 63–76 (January 1984).
4. P.M. Solomon and H. Morkoç, "Modulation-doped GaAs/AlGaAs heterojunction field-effect transistors (MODFET's), ultrahigh-speed devices for supercomputers," *IEEE Trans. Electron Dev.,* ED31 (8): 1015–1027 (August 1984).
5. F. Stern and S. das Sarma, "Electron energy levels in GaAs-$Ga_{1-x}Al_xAs$ heterojunctions," *Phys. Rev. B,* 30(2): 840–848 (July 15, 1974).
6. K. Yokoyama and K. Hess, "Intersubband phonon overlap integrals for AlGaAs/GaAs single-well heterostructures," *Phys. Rev. B,* 31(10): 6872–6874 (May 15, 1985).
7. T. Ando, A.B. Fowler, and F. Stern, "Electronic properties of two-dimensional systems," *Rev. Mod. Phys.,* 54(2): 437–669 (April 1982).
8. R.K. Cook and J. Frey, "Two-dimensional numerical simulation of energy transport effects in Si and GaAs MESFET's," *IEEE Trans. Electron Dev.,* ED-29(6): 970–977 (June 1982).
9. R. Stratton, "Diffusion of hot and cold electrons in semiconductor barriers," *Phys. Rev.,* 126(6): 2002–2014 (June 15, 1962).
10. D.J. Widiger, I.C. Kizilyalli, K. Hess, and J.J. Coleman, "Two-dimensional transient simulation of an idealized high electron mobility transistor," *IEEE Trans. Electron Dev.,* ED-32(6): 1092–1102 (June 1985).
11. M. Keever, W. Kopp, T.J. Drummond, H. Morkoç, and K. Hess, "Current transport in modulation-doped $Al_xGa_{1-x}As$/GaAs heterojunction structures at moderate field strengths," *Jpn. J. Appl. Phys.,* 21(10): 1489–1495 (October 1982).
12. I. C. Kizilyalli, K. Hess, J. L. Larson, and D. J. Widiger, "Scaling properties of high electron mobility transistors," *IEEE Trans. Electron Dev.,* ED-33(10): 1427–1433 (October 1986).

13. W. Fawcett, A. D. Boardman, and S. Swain, "Monte Carlo determination of electron transport properties in gallium arsenide," *J. Phys. and Chem. of Solids,* 31(9): 1963–1990 (September 1970).
14. C. Jacoboni and L. Reggiani, "The Monte Carlo method for the solution of charge transport in semiconductors with applications to covalent materials," *Rev. Mod. Phys.,* 55(3): 645–705 (July 1983).
15. R. W. Hockney and J. W. Eastwood, *Computer Simulation Using Particles,* New York: McGraw-Hill (1981).
16. P. J. Price, "Electron transport in polar heterolayers," *Surf. Sci.,* 113(1–3): 199–210 (January 1982).
17. K. Yokayama and K. Hess, "Monte Carlo study of electronic transport in $Al_xGa_{1-x}As/GaAs$ single-well heterostructures," *Phys. Rev. B,* 33(8): 5595–5606 (April 15, 1986).
18. T. Wang and K. Hess, "Calculation of the electron velocity distribution in high electron mobility transistors using an ensemble Monte Carlo method," *J. Appl. Phys.,* 57(12): 5336–5339 (June 15, 1985).
19. U. Ravaioli and D. K. Ferry, "Monte Carlo study of the quasi two-dimensional electron gas in the high electron mobility transistor" *Superlattices and Microstructures,* 2(1): 75–78 (January 1986).
20. J. P. Polonovsky and K. Tomizawa, "Phonon scattering of quasi two-dimensional electron gas in a single heterostructure," *Jpn. J. Appl. Phys.,* 24(12): 1611–1618 (December 1985).
21. U. Ravaioli and D. K. Ferry, "Monte Carlo investigation of the high electron mobility transistor," in *High Speed Electronics,* eds. B. Kallback and H. Beneking, Berlin: Springer-Verlag, 136–139 (1986).
22. D. H. Park and K. F. Brennan, "Theoretical analysis of a pseudomorphic HEMT using an ensemble Monte Carlo simulation," *IEEE Trans. Electron Dev.,* ED-36:1254–1263 (1989).

Chapter 3

JFET GaAs Logic

R. Zuleeg, J. K. Notthoff and G. L. Troeger

3.1 Introduction

The single-gate GaAs JFET on semi-insulating substrate was first fabricated by a zinc-diffused gate into an epitaxially grown n-type channel region and dates back to 1968.[1] In a parallel development the MESFET and the JFET were utilized in the design of digital integrated circuits. While depletion-mode MESFETs are now primarily used for high-speed buffered FET logic (BFL),[2] Schottky-diode FET logic (SDFL),[3] and direct-coupled FET logic (DCFL),[4] their potential as candidates for VLSI and eventually GSI (giant-scale integration) is prohibitive because of the high power dissipation per logic gate. The technology which will reign supreme in this high-complexity IC design is expected to be DCFL. Either enhancement-mode MESFETs (E-MESFETs) or enhancement-mode JFETs (E-JFETs) have been utilized in the design of DCFL circuits. These devices are fabricated by reducing the channel height so that a normally-off enhancement-mode FET results with a positive pinch-off, or threshold, voltage. The important difference in these DCFL circuits is that the logic swing of a circuit using MESFETs is limited by the forward turn-on voltage of the Schottky-barrier transistor gate to about 0.5 V, whereas for the E-JFET this voltage is increased to about 1 V. This gives a larger noise immunity for the E-JFET circuits and places stringent fabrication requirements on threshold voltage control and uniformity for the E-MESFET to obtain a good processing yield.

Although the E-JFET has a higher noise margin than the E-MESFET, because of the possible higher logic swing, the demonstrated complementary design of E-JFET circuits with enhancement-mode n- and p-channel devices offers even further increase in noise margin and very low power dissipation per gate.[5] This complementary design lends itself to the fabrication of a low-power static random-access memory (SRAM) of high yield capability owing to the high noise margin and excellent threshold voltage control by a double ion-implanted pn-junction gate. In addition, the complementary memory cell can be made immune to single-event upset (SEU) from energetic protons and heavy ions, a design innovation already exercised for Si CMOS SRAMs.[6,7]

The speed-power performance of the homojunction GaAs E-JFET technology will be presented and discussed. It should be noted that complementary circuit design is only practical in GaAs with the E-JFETs because the E-MESFET has a low Schottky barrier height for the p-channel device which prohibits practical circuit designs and operation. The performance of complementary E-JFETs in static RAM designs of 1K (1 kilobit) and 4K will be presented and the prospects of 16K and 64K will be addressed.

Low-power operation of E-JFET gates in DCFL is advantageous for the design of gate arrays and microprocessors. A 4-bit microprocessor (MDC2901) was designed, fabricated, and tested which emulates the Si-based AMD2901.[8] With the exception of the 16- by 4-bit register file, which was implemented using complementary cells, the entire chip was designed with resistive-load E-JFETs.

The design of a 1500-gate array was completed using E-JFET drivers with resistive load only and one personalization for a 5-bit ALU in parallel with a 5-by-5 multiplier will also be described.[9] Further expansion of this approach to 6000- and 10,000-gate arrays is planned and simulated performance characteristics will be presented and discussed.

A 32-bit microprocessor, based on the RISC architecture,[10] was implemented in E-JFET technology.[11] With improvements in the GaAs JFET currently in development, the goal is an execution cycle of 5 ns, which corresponds to a clock rate of 200 MHz. A status report on this development will be presented.

As a logical continuation of GaAs homojunction complementary circuit design, the heterojunction complementary circuit design is being explored as a possible second generation of GaAs integrated circuits. Exploratory heterojunction complementary circuit designs, utilizing molecular beam epitaxy (MBE) of AlGaAs on GaAs, are under development. These are based on two-dimensional electron gas (2DEG) and two-dimensional hole gas (2DHG) properties. As a consequence of the increased electron and hole mobilities of the heterojunction concept, these heterojunction complementary circuits promise the highest switching speed achievable with the least power dissipation for future digital integrated circuits. Experimental design structures are available from publications.[12,13,14,15]

In Sec. 3.2 the voltage-current and temperature characteristics of the JFET will be presented. The equivalent circuit for modeling, as described, is utilized in circuit performance predictions. Logic design of integrated circuits, employing the E-JFET, completed or presently under development, is addressed in Sec. 3.3. The all-ion implantation planar process for integrated-circuit fabrication is described in Sec. 3.4. This section also documents the achieved processing yields of a pilot line and discusses yield issues of this technology for VLSI circuit complexity. Prospects for VLSI, and eventually GSI, are analyzed in Sec. 3.5 together with applications of the future.

3.2 JFET Theory and Model

3.2.1 Voltage-Current Characteristics

The steady-state $I_D V_D$-V_G characteristics for a uniformly doped channel have been derived by Shockley[16] assuming that electron drift velocity is proportional to the electric field. According to this theory the current saturates for

$$V_D \geq V_{sat} = V_G - V_T, \tag{3–1}$$

when the channel is pinched off at the drain. For ion-implanted channels the MOSFET-like relation

$$I_D = 2K[(V_G - V_T)V_D - V_D^2/2] \tag{3–2}$$

with

$$K = w\mu\epsilon\epsilon_0/2a_{eff}L \tag{3–3}$$

where μ is the mobility, and[17]

$$a_{eff} = R_p + 2\sigma_p\sqrt{2/\pi} \tag{3–4}$$

appears more appropriate,[18] since the width of the depletion layer varies little along most of the channel and with applied gate voltage, its boundary being pinned more or less to the peak of the dopant concentration. As in Schockley's theory, the current saturates for $V_D \geq V_{sat} = V_G - V_T$, at

$$I_{sat} = K(V_G - V_T)^2. \tag{3–5}$$

The parameters w, L, and $a_{eff} = a$ are defined in Fig. 3–1, which presents a cross section of a JFET.

Deviations of voltage-current characteristics from the square-law relations are encountered, however, when series resistance values of the source contact and the region between the source and gate edge are appreciable.[19] Equation 3–5 must then be modified, because the effective gate voltage is now

$$V_G^1 = V_G - R_S I_{sat}, \tag{3–6}$$

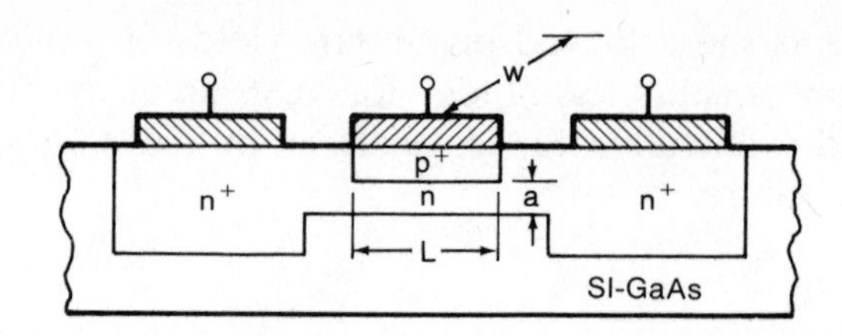

Fig. 3–1. Cross section of ion-implanted enhancement-mode JFET with channel width *W*, channel length *L*, and channel height *a*.

where R_S is the total ohmic series resistance between source and gate. The reduced drain saturation current I_{DS}^* is approximately

$$I_{DS}^* \cong \frac{I_{sat}}{1 + 2R_S K(V_G - V_T)} . \tag{3–7}$$

A uniform reduction of saturation current arising from equal resistances for all transistors would be tolerable. However, nonuniform alloying over a large wafer area causing resistance variations would be objectionable since it would result in a spread of saturation current.

Optimization of device performance characteristics is accomplished through tailoring of ion-implantation profiles. The smallest channel height gives the highest drain saturation current I_{DS}, for a given gate voltage, V_G, according to Eqs. 3–3 and 3–5. This requires rather large channel doping, however, to maintain the desired threshold voltage. The design goal is $a = 100$ nm so that, for a channel doping of 1.5×10^{17} cm^{-3}, $V_T = V_B - V_P = 0.2$ V. With $L = 2.5\ \mu$m and $\mu = 4000$ cm^2/V · s, one obtains $I_{DS}/w = 80(V_G - V_T)^2$ mA/mm and $g_m/w = (1/w)(dI_{DS}/dV_G) = 160(V_G - V_T)$(mA/V)/mm. Thus for $V_G = +1$ V we have $I_{DS} = 51$ mA/mm and $g_m = 128$ mS/mm. Since the gate capacitance per unit length of this device is $C_G/w = \epsilon\epsilon_0 L/a \cong 2.5$ pF/mm, the estimated intrinsic *RC* time-constant response is

$$\tau = 2.2C_G/g_m \cong 50 \text{ ps} .$$

For a JFET with $w = 10\ \mu$m and a 2.5-kΩ load resistance, the extrinsic *RC* time-constant response is then

$$\tau = 2.2C_G R_L = 150 \text{ ps} .$$

Experimental values of 150 to 200 ps have been obtained with complex integrated circuits.

The important parameters in controlling the current-voltage characteristics are primarily the channel doping, N, and channel height a, with w and L being well defined by photolithographic delineations. The channel doping affects *I-V* characteristics through the threshold voltage

$$V_T = V_B - V_P , \tag{3–8}$$

where

$$V_P = (q/\epsilon\epsilon_0)\int_0^a dx' \int_0^{x'} N(x)\,dx \tag{3–9}$$

is the pinch-off voltage, and V_B is the gate versus channel built-in potential. For a uniformly doped epitaxial layer,

$$V_P = Nqa^2/2\epsilon\epsilon_0 . \tag{3–10}$$

Equation 3–3, with Eqs. 3–5, 3–8, and 3–10, then shows that

$$\frac{\delta I_{DS}}{I_{DS}} = \frac{2V_P}{V_G - V_T}\left(\frac{\delta N}{N}\right) + \left(\frac{4V_P}{V_G - V_T} - 1\right)\frac{\delta a}{a} \tag{3–11}$$

For the typical values $V_P = 1.0$ V, $V_G = +1.0$ V, and $V_T \cong 0$ V, this gives numerically

$$\frac{\delta I_{DS}}{I_{DS}} = 2\frac{\delta N}{N} + 3\frac{\delta a}{a}. \tag{3–12}$$

Variations of built-in voltage and of mobility are neglected, since they are only second-order effects.

With a 10-percent change of doping N and a 10-percent change of a, the fractional change of drain saturation current I_{DS} is 0.5, or 50 percent. Thus a tight control of both a and N is required over large wafer areas to achieve a good uniformity of electrical-device characteristics and reasonable yields in integrated-circuit fabrication.

Changes of N and a that are less than 10 percent over small wafer areas can be obtained with standard techniques of diffusion and epitaxy. It is, however, not the yield of devices and circuit per wafer but rather the yield of good wafers which renders the standard technology marginal in getting good overall yields in the integrated-circuit fabrication. Implantation of silicon ions is now standard practice for channel regions, and typical measurements on GaAs devices indicate that a drain saturation current variation of only 10 percent can be obtained over large wafer areas. This amounts to a control of N and a within a 2-percent variation. These tolerances can be maintained for an optimized enhancement-mode structure, and ion implantation offers the necessary tolerances to ensure device uniformity and thus provide better yields in integrated-circuit fabrication than does the standard technology of diffusion and epitaxy.

Uniformity of performance characteristics for ion-implanted devices depends primarily on the results of the capping and annealing procedure and secondarily on the ohmic contact and series resistance control obtainable with the presently used fabrication technologies.

Radiation effects in GaAs E-JFETs arising from neutron fluences and total doses of ionizing radiation have been assessed.[20] It was established that the degradation of the electrical characteristics of enhancement-mode gallium arsenide junction field-effect transistors exposed to fast neutrons ($E > 10$ keV) or ionizing radiation (CO^{60}) is caused substantially by changes in mobility and free-carrier concentration. Circuit operation of devices with channel impurity concentrations of about 10^{17} cm^{-3} will not be impaired by fast neutron fluences of 10^{15} neutrons/cm^2 and ionizing radiation doses of 10^8 rad (GaAs). In addition, single-event upset (SEU) to heavy ions and protons, and logic upset to pulsed ionizing radiation, was measured with E-JFET static random-access memories (SRAMs) and are available from the literature.[21]

3.2.2 Temperature Characteristics

A temperature-dependent semiempirical GaAs JFET device model was developed to interpret the experimental data and implemented in a circuit simulator. The model incorporates the temperature dependence of the threshold voltage, transconductance, subthreshold current, substrate leakage current, gate current, and implanted load resistor conductivity.[22]

The dominant factor affecting operation at both low and high temperatures is the threshold voltage shift of the JFETs, which is approximately -1.5 mV/°C as shown in Fig. 3–2. The performance factor, K, and threshold voltage, V_T, are both temperature dependent and change R_{on}, the on-resistance in the linear region, as well as I_{DS} of the FET. These are expressed by the relations

$$I_{DS}(T) = \frac{\epsilon\epsilon_0\mu(T)}{2a(T)}\left(\frac{W}{L}\right)[V_G - V_T(T)]^2, \tag{3–13}$$

and

$$R_{on} = \frac{1}{2K(T)\,[V_G - V_T(T)]}. \tag{3–14}$$

An accurate physical model for temperature dependence of V_T requires determining the channel-to-substrate built-in voltage, involving an iterative method and such parameters as trap concentration and occupation as a function of temperature.[23] These are not easily measurable. For the temperature under consideration here, however, the temperature dependence of the threshold voltage can be accurately approximated by a linear function with the addition of a second-order term as follows:

$$V_T(T) = V_{TO} + \Delta T + bT^2, \tag{3–15}$$

where V_{TO} is the nominal threshold voltage at $T = 300$ K. Typically $\Delta \cong -1.33$ mV/°C and $b \cong 1.2$ μV/°C^2 for n-channel JFETs and $\Delta \cong$

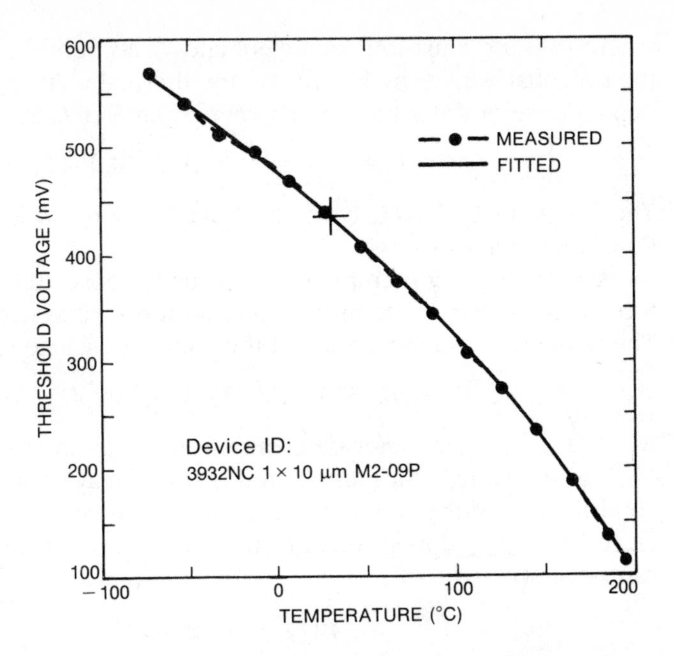

Fig. 3–2. Threshold voltage V_T as a function of temperature for a GaAs E-JFET. (*After Hyun* et al.[22])

−2.33 mV/°C for p-channel JFETs. Fig. 3–2 compares the measured and calculated threshold voltages for an n-channel JFET.

The performance parameter, K, of Eq. 3–3 is temperature dependent, where $\epsilon\epsilon_0$ is the permittivity of GaAs, μ is the mobility, and $a_{\text{eff}} = a$ is the channel height. In Eq. 3–3, mobility is the most important parameter determining the temperature dependence of K. The effective channel height is also a function of temperature but its effect is less significant. For moderate doping conditions, the electron mobility in the channel can be described by[24]

$$\frac{1}{\mu} = \frac{1}{\mu_0}\left(\frac{T}{300}\right)^a + 4 \times 10^{-21} N_d \left(\frac{300}{T}\right)^{1.5}, \qquad (3\text{–}16)$$

where μ_0 is the low field mobility at $T = 300$ K and N_d is the donor concentration.

Because of its exponential increase with temperature, the subthreshold current becomes an important parameter in determining the operation of GaAs FET circuits at high temperatures. The subthreshold current is given by

$$I_{ds} = I_0 \exp\left(\frac{V_{gs} - V_T}{mV_{\text{th}}}\right)(1 - V_{ds}/V_{se}), \qquad (3\text{–}17)$$

where m is the emission coefficient and V_{se} accounts for the modulation of the potential barrier by V_{ds} and V_{th}, the thermal voltage.[25] The temperature dependence of the saturation current parameter I_0 can be expressed as

$$I_0 = I_0(T = 300\ \text{K})\,(T/300)^{1.5-a}\,. \tag{3–18}$$

The exponent $1.5 - a$ is typically about 0.3, hence I_0 is an increasing function of temperature.

At relatively high temperatures (usually above 100°C), the shunting effect of the substrate dominates the subthreshold current characteristics. The temperature dependence of the shunt resistance can be expressed as

$$R_{\text{sh}} \cong R_{\text{sho}}(300/T)^{3/2}(T/300)^{2.3}\exp(E_g/2kT)\,, \tag{3–19}$$

where R_{sho} is a parameter determined by the geometry of the device.

The resistance of ion-implanted resistors may be modeled as a parallel combination of the implanted-region resistance and the substrate resistance. For lightly doped resistors, the carrier mobility expression, Eq. 3–16, can be simplified to

$$\mu(T) = \mu_0(300/T)^a\,, \tag{3–20}$$

where a is dependent on the doping concentration ($a = 2.3$ for intrinsic GaAs) and μ_0 is the mobility at $T = 300$ K. Then, the temperature dependence of the implanted-region resistance, R_i, may be described approximately by

$$R_i = R_0(T/300)^a\,, \tag{3–21}$$

where R_0 is the nominal resistor value at 300 K. The expression for the substrate resistance, R_{sh}, may be roughly expressed as

$$R_{\text{sh}} \cong R_{\text{so}}\exp(E_g/2kT)\,. \tag{3–22}$$

Neglecting the temperature dependence of R_{so} ($R_{\text{so}} \propto 1/T$), the total resistance may be rewritten as

$$R_T = \frac{R_0(T/300)^a R_{\text{so}}\exp(E_g/2kT)}{R_0(T/300)^a + R_{\text{so}}\exp(E_g/2kT)}\,. \tag{3–23}$$

The calculated curve R_T versus T, using Eq. 3–23, is shown in Fig. 3–3. As can be seen, R_T has a maximum value at $T = T_m$. Thus, the value of R_{so} can be obtained from Eq. 3–23 by letting $\partial(R_T)/\partial T = 0$ at $T = T_m$, which yields

$$R_{\text{so}} = \frac{E_g R_0}{kT_m(a+1)}(T_m/300)^a\exp(-E_g/2kT_m)\,. \tag{3–24}$$

Hence the input parameters are a, R_0 and T_m, which can be easily extracted from the measured data.

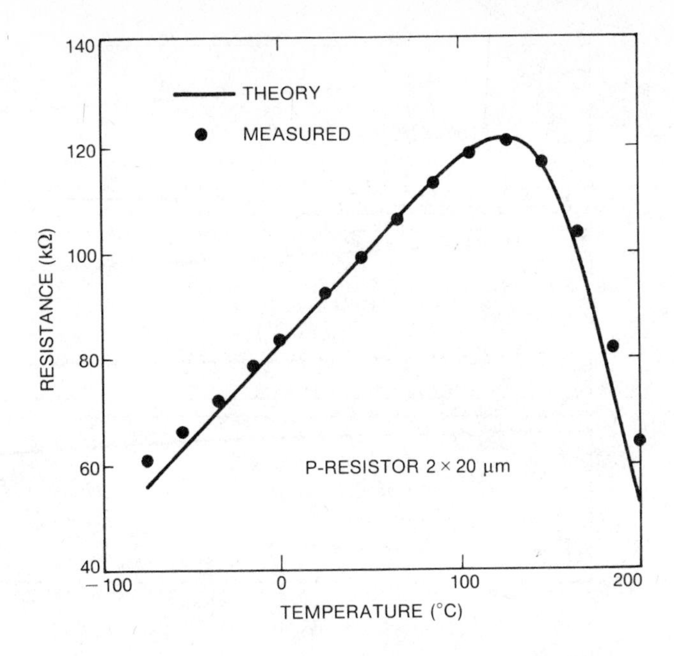

Fig. 3–3. GaAs implanted resistor measured and calculated temperature characteristic. (*After Hyun* et al.[22])

The temperature variation of logic levels, i.e. logic output "low," V_{OL}, and "high," V_{OH}, for the three-JFET DCFL gate configuration was simulated over the temperature range of −55°C to +125°C. The results are presented in Fig. 3–4 for complementary, resistive-load, and enhancement/depletion-mode (E/D) JFET inverters. While the first two versions offer a tolerable reduction of logic swing up to +125°C, the E/D inverter suffers a severe degradation above +100°C, and therefore has a limited range for high-temperature operation.

3.2.3 Modeling and Equivalent Circuits

Advancing the JFET technology from LSI to VLSI circuit complexity cannot be achieved without taking full advantage of process, parameter, and circuit modeling. An overview of semiconductor device modeling is shown in Fig. 3–5. This flowchart encompasses the entire modeling activity, as applied to the JFET technology at the McDonnell Douglas Microelectronics Center. Extensive modeling is used in integrated-circuit practice to predict and optimize circuit performance and yield. One can start from first principles of physics or one can resort to empirical models which are derived from curve fitting to experimental data and data extraction from device

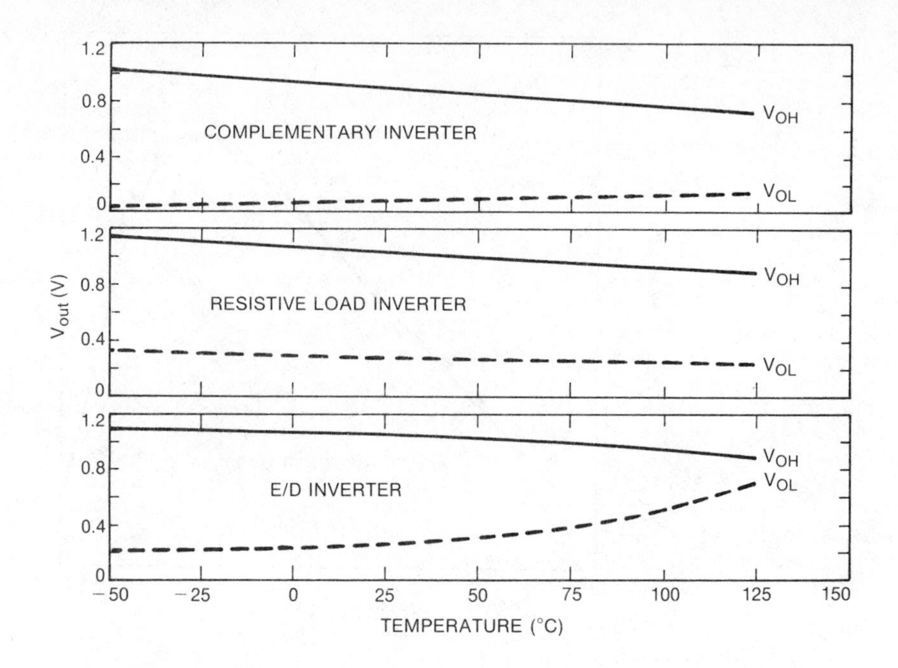

Fig. 3–4. Logic levels of three E-JFET inverters as a function of temperature.

scaling experiments. As described in the previous section, the GaAs JFET voltage-current characteristics and their temperature dependence have been analyzed using a semiempirical model. With the aid of these voltage-current characteristics and the equivalent circuit of Fig. 3–6, a modified SPICE2 simulation was developed. The two-diode model was selected, and the additional circuit elements, with respect to FET terminology, are self-explanatory. Special analytical treatments were found mandatory for R_{sh}, the shunt resistance in parallel with the channel conductance, and the current generators I_B and I_{sub}, which relate to currents due to breakdown and leakage through the semi-insulating substrate material.

As an example of simulation utility the 4-bit ALU performance, as measured and as simulated, for two different load resistances is presented in Fig. 3–7. With a load resistance value of 2.5 kΩ, the circuit of lot 1 measured an add time of 2.2 ns for a power dissipation of 55 mW. For this load resistance this should be compared with the simulated values of 2.1 ns for a 50-mW power performance. The trade-off between speed and power is evident from the circuit from lot 2. With increased load resistance, i.e. 4.0 kΩ, the power dissipation is reduced to 32 mW, with a corresponding add-time increase to 2.9 ns. The good agreement of simulation with experimental results demonstrates the accuracy of our modified JFET SPICE2

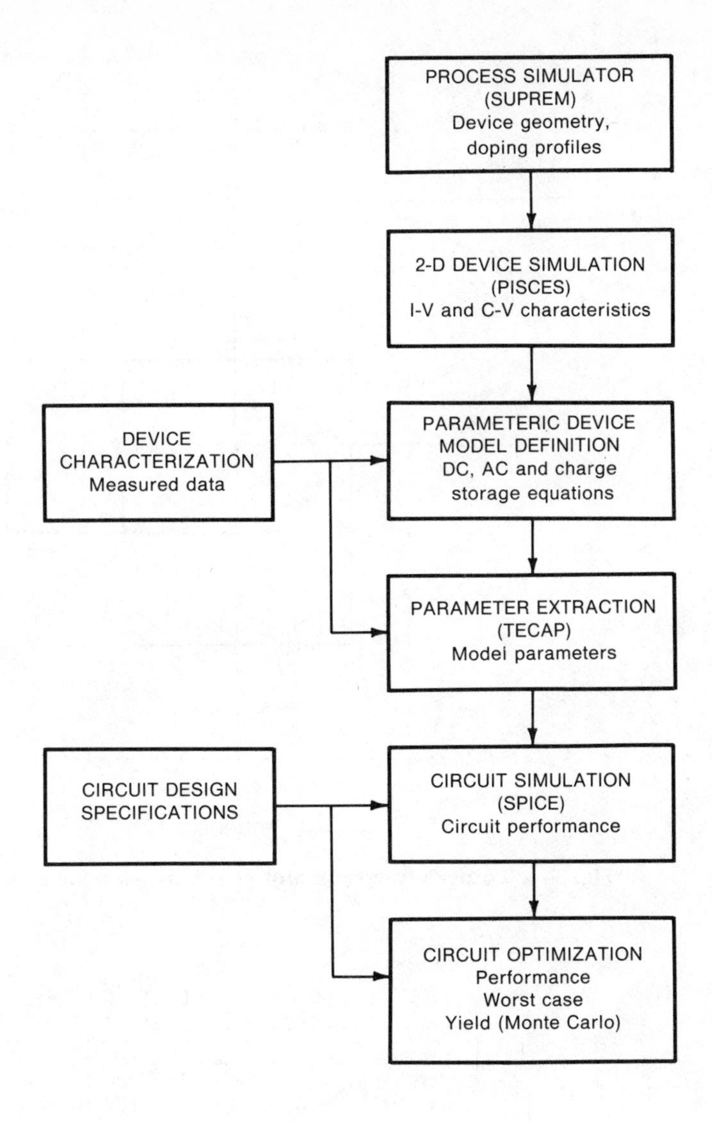

Fig. 3–5. Overview of semiconductor device modeling.

model and the voltage- and current-dependent circuit elements of Fig. 3–6, which were used for this simulation. The result establishes a confidence level for application of this model to predict the performance of circuits with higher complexity.

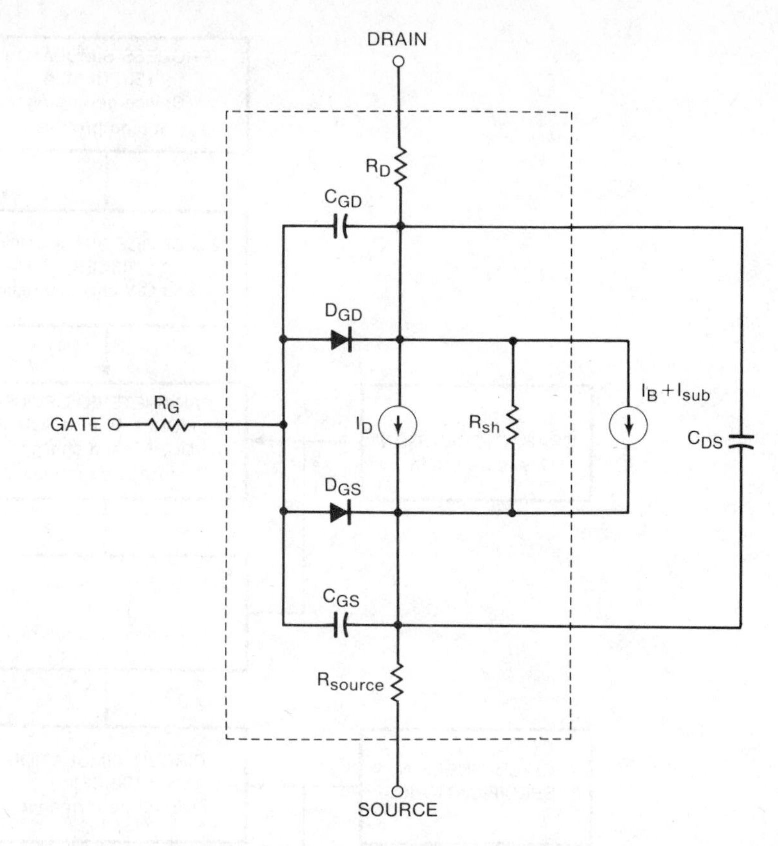

Fig. 3–6. Equivalent circuit and elements used for modeling of E-JFETs.

	LOT 1	LOT 2	SIMULATION
ADD TIME (ns)	2.2	2.9	2.1
POWER (mW)	55.0	32.0	50.0
LOAD VALUE (kΩ)	2.5	4.0	2.5

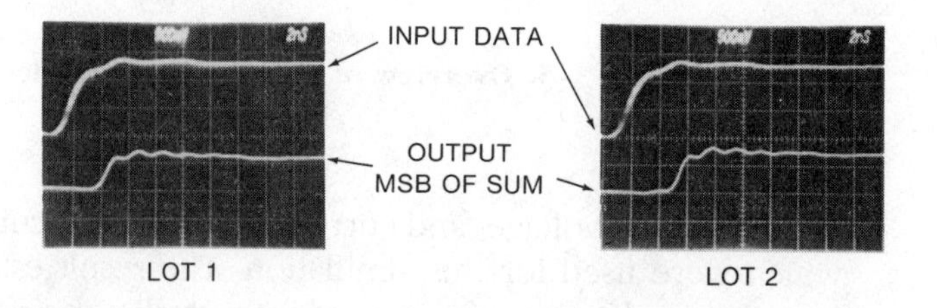

Fig. 3–7. Comparison of simulated and measured speed-power performance of 4-bit ALU.

3.3 Logic Design and Integrated Circuits

Logic and IC design with GaAs junction field-effect transistors is similar to that with other GaAs FET technologies. However, many important differences exist in the circuit design. After a look at the basic logic building blocks, similarities with other logic families, circuit approaches and design issues are explored here. Important design considerations for large circuits are discussed, and circuit yield analysis techniques and temperature effects are touched on.

3.3.1 Logic Design

NOR gates, NAND gates, and cross-coupled inverter-pair flip-flops are the basic building blocks of E-JFET logic. The preferred gate circuit is the NOR gate for two reasons: better switching speed and higher fan-in capability. A NAND gate of the same power dissipation and logic-voltage levels requires larger inverter devices to maintain the logic low level (see Fig. 3–8); this results in lower switching speed. For this reason, NAND gates with more than three inputs are rarely used in E-JFET logic.

E-JFET NOR gates can be constructed with different load elements and circuit configurations. Figure 3–9 shows a collection of direct-coupled field-effect transistor logic (DCFL) two-input NOR gate circuits. Figure 3–9A represents the most common circuit with a resistive load, Fig. 3–9B has a

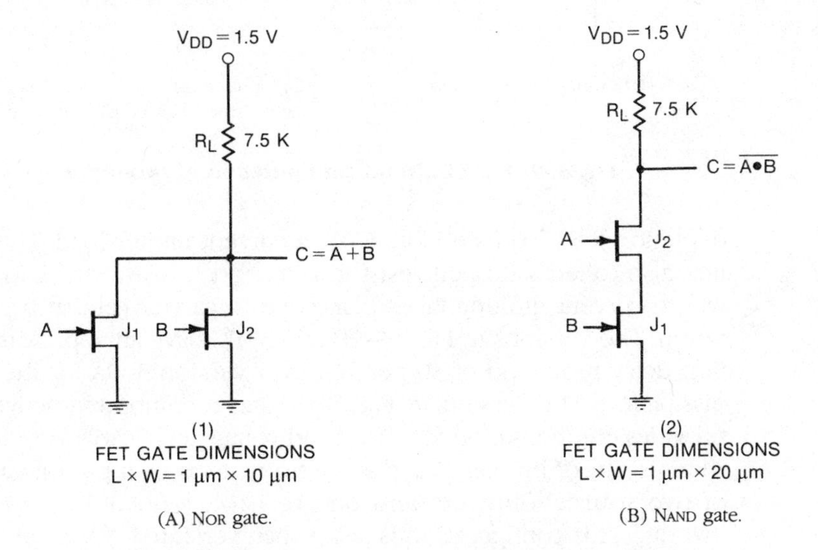

(A) NOR gate.

(B) NAND gate.

Fig. 3–8. E-JFET gates implemented with a resistive load.

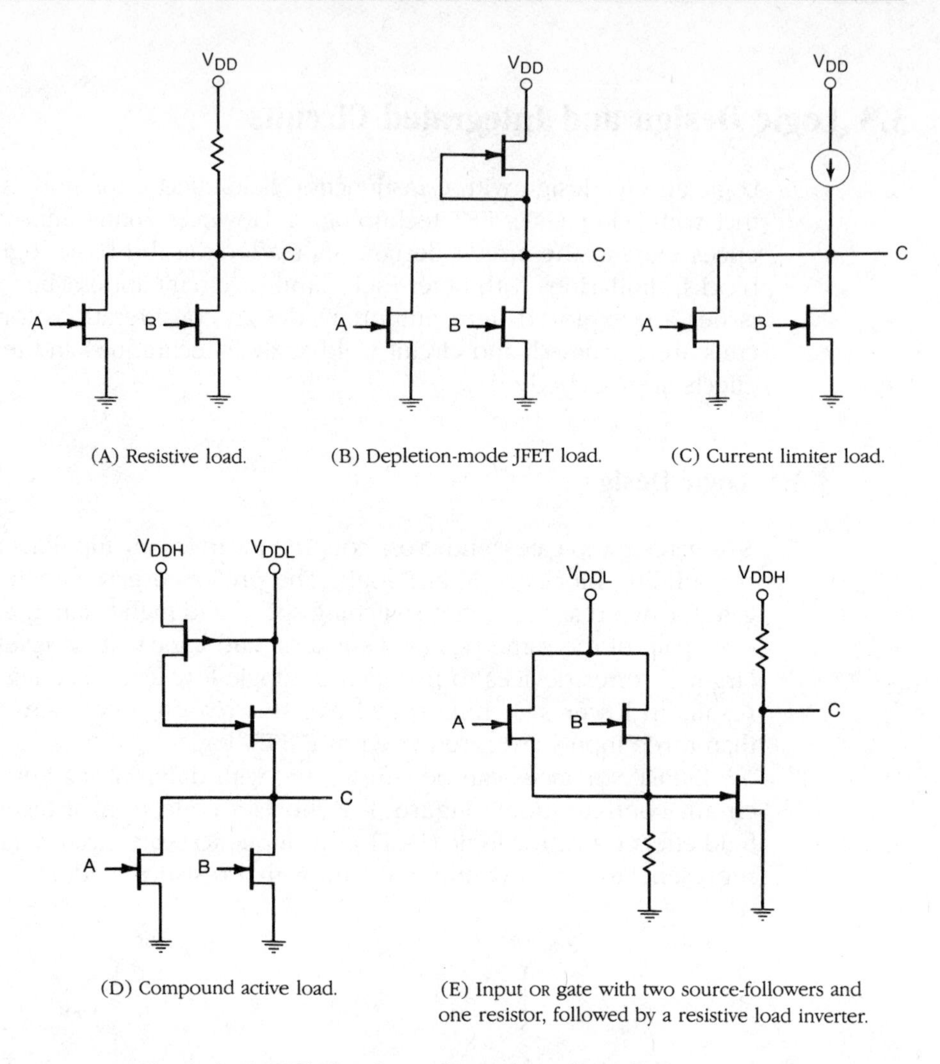

(A) Resistive load.

(B) Depletion-mode JFET load.

(C) Current limiter load.

(D) Compound active load.

(E) Input OR gate with two source-followers and one resistor, followed by a resistive load inverter.

Fig. 3–9. E-JFET circuit configuration of two-input NOR gates.

depletion JFET load, and Fig. 3–9C a current limiter load. The current limiter, also called saturated resistor, is a short n-resistor of 1 to 2 μm length where current limiting takes place due to carrier (electron) velocity saturation. The versions in Figs. 3–9B and 3–9C have an approximate propagation delay reduction of 30 percent over version 3–9A for the same power dissipation. The version in Fig. 3–9D has a compound active load which provides compensation for threshold-voltage (V_T) variations. Version 3–9E, reminiscent of bipolar ECL gate circuitry, uses an input OR gate composed of two source-followers and one resistor, followed by a resistive-load inverter. This configuration is faster than version 3–9A for fan-out above 3, but requires more components and layout area. Version 3–9F is the same

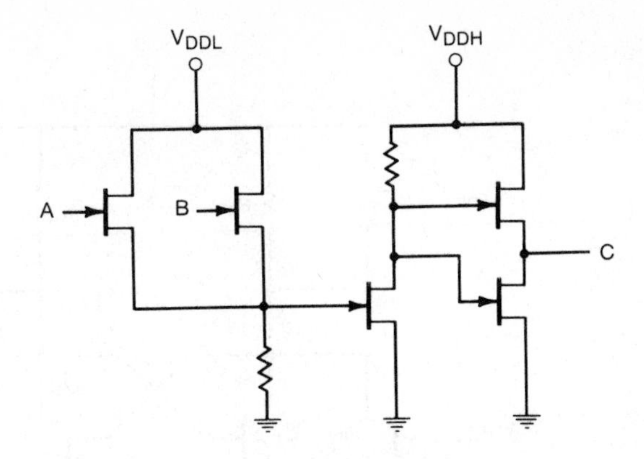

(F) Same as (E) with added totem pole stage.

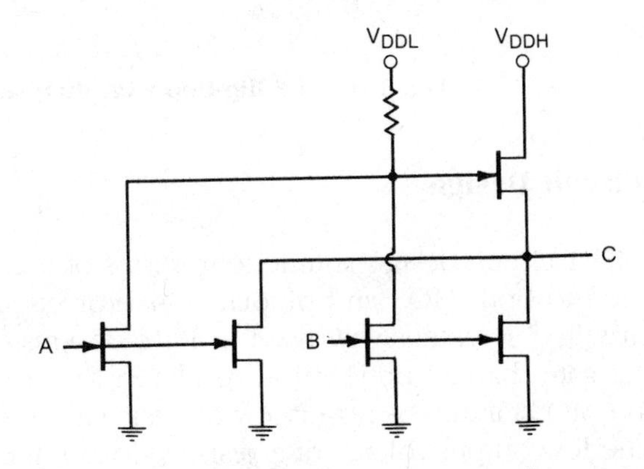

(G) Version (A) with quasi-complementary totem pole output stage.

Fig. 3–9. *continued*

version with a totem-pole stage added, useful where large fan-outs or parasitic capacitances must be driven. Another buffered-gate version[26] is shown in Fig. 3–9G. A quasi-complementary or totem-pole output stage has been integrated with the gate of version 3–9A. The output source follower's drain terminal is connected to a lower supply voltage, V_{DDL} (1.2 to 1.5 V), than V_{DDH}, which is kept in a range of 2.0 to 3.0 V, depending on speed requirements. This gate provides near-equal rise and fall times and propagation delays, an important consideration when minimum skew of waveform edges is critical. Type D, R-S, or JK flip-flops are constructed by interconnecting NOR gates. The setting or resetting of an R-S flip-flop or latch can be accomplished by the conventional pulling down of the high node and letting the low node rise, or pulling up the low node by a source-follower JFET (see Fig. 3–10). Both methods can be used simultaneously to minimize propagation delay.

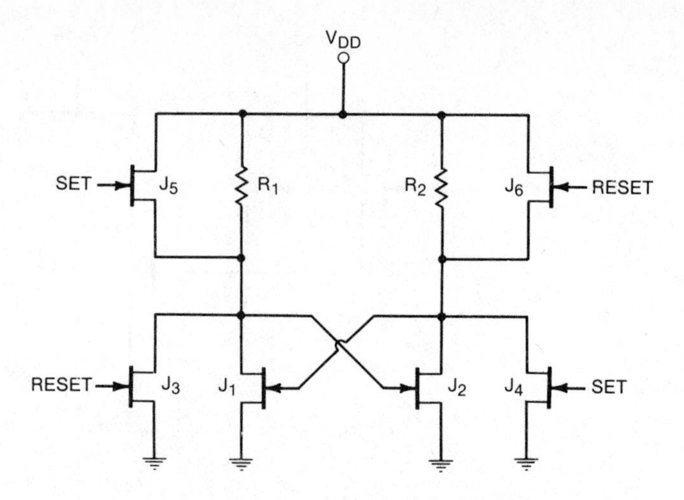

Fig. 3–10. R-S flip-flop with dual set-reset inputs.

3.3.2 Circuit Design

E-JFET circuit design is unique in terms of logic levels, but circuitwise related to both MOS and bipolar transistor circuit design. Complementary circuitry[5] is extensively used in RAM designs. However, the existence of the gate-channel pn-junction results in forward-gate currents not present in CMOS circuits. An overview of basic circuit design issues is given here. The low output voltage of a resistive-load E-JFET inverter is given by

$$V_{OL} = R_{on} I_{DS} \tag{3–25}$$

where $R_{on} = 1/2K(V_G - V_T)$. Commonly, the load resistor (for an active load element) is sized to produce a V_{OL} equal or close to the threshold voltage (V_T) of the FET, thus

$$V_{OL} = V_T, \tag{3–26}$$

which, for worst-case conditions, should be

$$V_{OL1} \leqq V_{T2},$$

where V_{OL1} is the output voltage of inverter 1, V_{T2} the threshold voltage of the E-JFET driven by it. The value of V_{OH}, the high output voltage when the inverter FET is turned off, is given by

$$V_{OH} = V_{DD} - I_R V_R = V_{GS(on)}, \tag{3–27}$$

where $V_{GS(on)}$ is the gate-source voltage of the FET driven by the inverter. V_{OH} can be determined graphically by plotting I_G versus V_{GS}, and a load line given by the load resistor of the inverter. Figure 3–11 shows the inter-

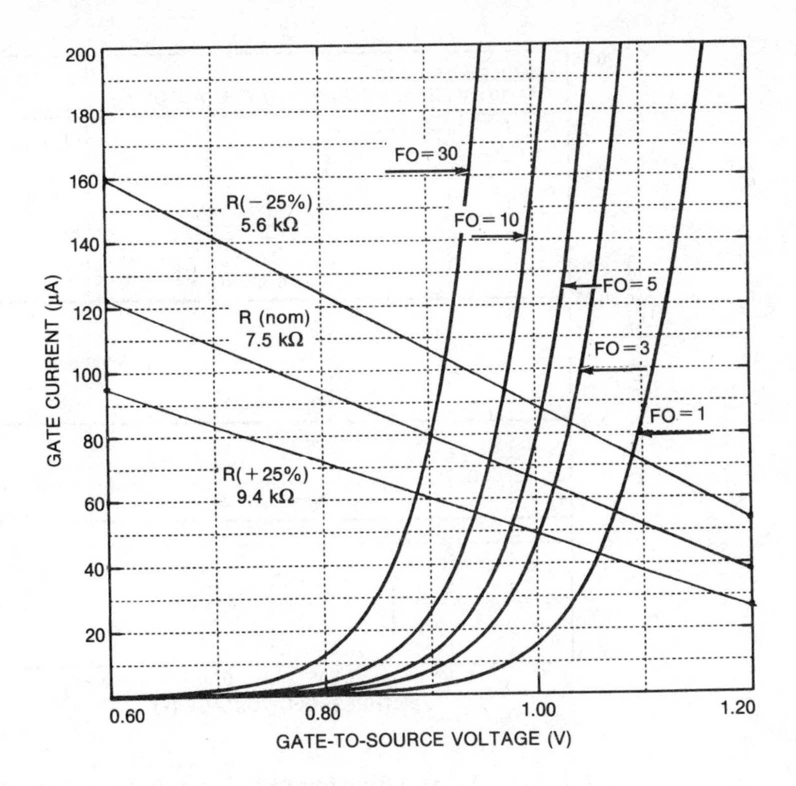

Fig. 3–11. V_{OL} of resistive load inverter (Fig 3-8A) as a function of fan-out and load resistor value.

cept of the load line with the I_G versus V_{GS} curve for fan-outs of 1, 3, 5, 10, and 30. The load lines represent a nominal, a 25 percent lower, and 25 percent higher than nominal resistor value. The change of V_{OH} as function of fan-out is an important parameter to be considered in circuit design; the V_{OH} "window" is widened when JFET parameter ranges, instead of fixed nominal values, are applied. The standard deviations based on empirical data, which are commonly used in JFET circuit simulation, are $\sigma_{VT} = 50$ mV, $\sigma_{K'} = 8\ \mu A/V^2$, and $\sigma_R = 5$ percent. The mean values of these three parameters may vary 100 mV for V_T, from 75 to 110 $\mu A/V^2$, and ± 10 percent for the resistor. While these mean parameter values may vary much less within a lot of wafers being processed, present lot-to-lot variations, depending upon material and processing conditions, are closer to the values listed above.

An illustration of how parametric variations affect V_{OL} is given in Fig. 3–12, where V_{OL} is plotted as a function of FET threshold voltage. In this plot the gate voltage-drain current gain parameter, K', is automatically adjusted as a function of V_T. This relationship, based on measured data, is

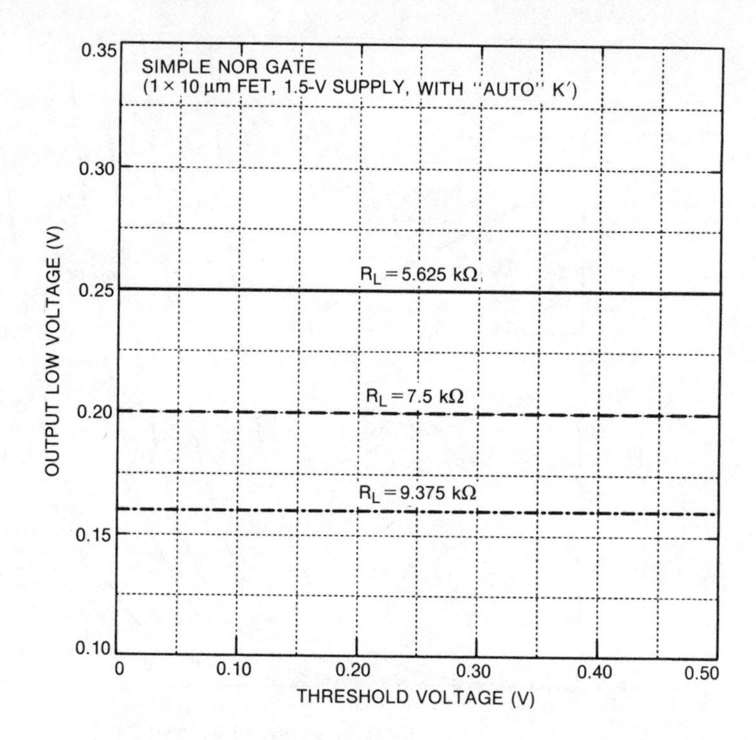

Fig. 3–12. V_{OL} as a function of V_T for various load resistors.

shown in Fig. 3–13, where K' increases as V_T increases and compensates for the increase in the value of the drain-source on resistance, R_{on}. If a flat K' is used, V_{OL} changes are on the order of 10 mV for a 200-mV change in V_T. The range of load resistor value stipulated in the example shown in Fig. 3–14 is nominally 7.5 kΩ, the lower limit value is 7.5 kΩ × 0.75 = 5.625 kΩ, and the high limit value is 7.5 kΩ × 1.25 = 9.375 kΩ.

The distribution of device parameters leads to a distribution of V_{OH} and V_{OL} values, and distribution of the inverter or gate performance in switching mode. The four parameters that represent gate switching performance are rise time t_r, fall time t_f, delay from rising input edge to falling output edge at 50 percent of logic swing, t_{pdRF}, and delay from falling input edge to rising output edge, t_{pdFR}. A good circuit design keeps t_r and t_f equal, with ±10 percent an acceptable tolerance goal, and t_{pdRF} and t_{pdFR} equal with 20 percent a reasonable goal for signal waveform skew. These goals can be readily accomplished for a small circuit by iterative circuit design which establishes first a nominal case, then examines several worst cases where the device parameters are at the edge of their respective design "windows." The windows are based on experimental results commonly

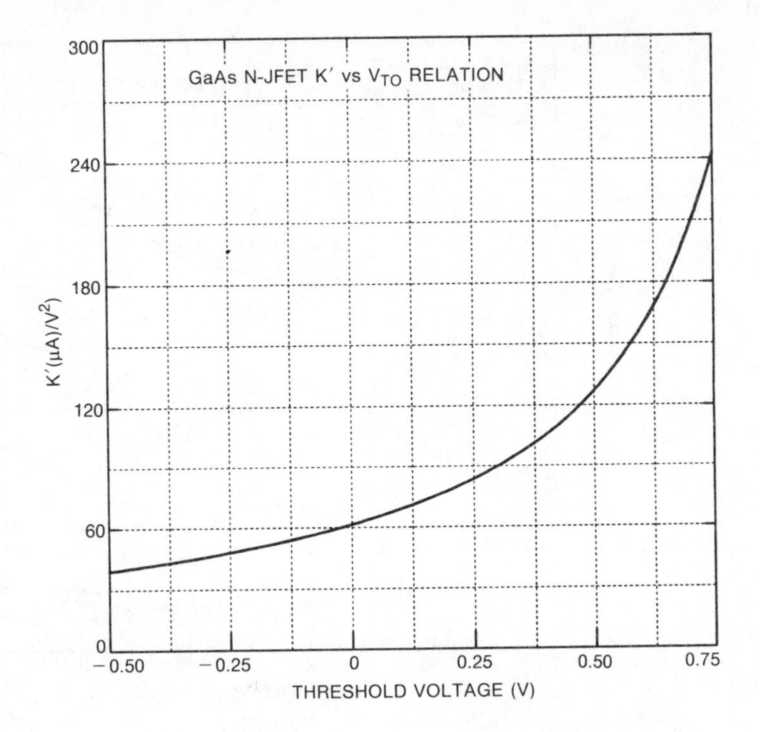

Fig. 3–13. *K′* values as a function of threshold voltage.

based on the maximum variation of the mean value plus or minus three times the standard deviation over the chip area. Based on the values discussed earlier, windows used typically in E-JFET circuit design are 0.05 V to 0.55 V for V_T, 50 $\mu A/V^2$ to 125 $\mu A/V^2$ for K' and 0.75 to 1.25 times the nominal resistor value. The compatibility of different circuits to interface with each other and guarantee functionality can be analyzed by establishing an input-output protocol that each circuit must obey. In the absence of such a protocol, statistical analysis using multiple computer runs based on Monte Carlo type device parameter variation is the best approach. For a design aimed at 100 percent circuit yield, a rectangular distribution of parameter variations is useful; for circuit yields less than 100 percent, a Gaussian distribution is recommended.

For complex circuits, there is a multiplicity of worst cases which is difficult to derive by hand. The best design tool here is the Monte Carlo analysis where certain failure criteria are specified. The parameter values of the failing circuits are saved, subsequently examined, and corrections or redesigns of the circuit are made. This is an iterative procedure which can predict the circuit performance distribution and circuit yield.

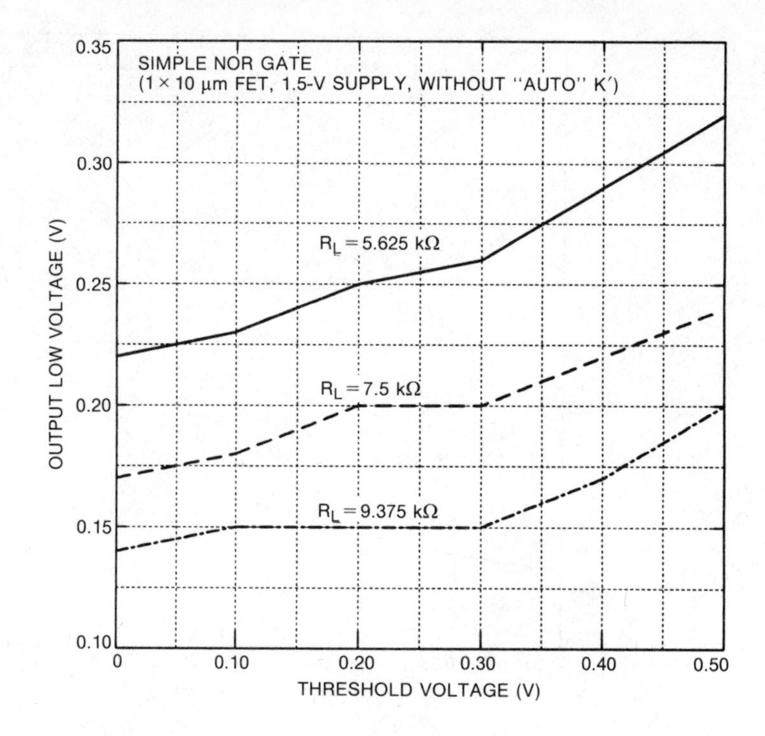

Fig. 3–14. V_{OL} as a function of V_T for various load resistors.

In contrast to MOS circuit design, where parameter windows are relatively small, GaAs designs require a more detailed and statistical analysis to obtain high circuit yield.

When designing a circuit for a temperature range (commonly a −55°C to 125°C range is specified for military applications), the design procedure discussed above is repeated for the two end-points of the range. If required, the design is changed or adjusted and then reconciled with the 25°C design. As discussed in Sec. 3.2.2, device parameters change with temperature, the dominant one being V_T, which decreases approximately 1.5 mV/°C. Correct subthreshold drain-source current modeling is very important at elevated temperatures in gate circuits with large fan-in or fan-out and low-current curcuits, such as low-power memory cells, registers, and linear circuits, that have high-impedance circuit nodes.

Design considerations for E-JFET circuits are equally applicable to E-MESFET circuits, which are now preferred over D-MESFET circuits because of lower power dissipation and higher circuit density and complexity per chip. While the V_{OH} values for E-JFET circuits are in the range of 0.9 to 1.2 V, they are 0.5 to 0.6 V for the E-MESFET due to the lower diode forward-voltage characteristics of the Schottky gate. Consequently, parameter windows for the E-MESFET must be reduced, particularly V_T mean and standard deviations, to approximately one half of the E-JFET windows. This requires tight material and process control, and for this reason the E-JFET technology may be preferable for VLSI and beyond.

E-JFET circuit designs can also be implemented with p-channel E-FETs which allows design analogues to silicon CMOS with the difference that E-JETs consume gate current when forward biased. However, the gate current, typically 1 μA/μm of FET width at V_{GS} = 1.0 V, is small and makes it possible to construct useful logic gates with power dissipation of 1.0 to 40 μW, depending on switching speed required.

Figure 3–15 shows simulated two-input complementary gate switching characteristics as a function of power dissipation; Figure 3–16, the power dissipation as function of power supply voltage. The switching performance of DCFL with GaAs E-JFETs for the complementary and resistive load inverter is shown in Fig. 3–17. These speed-power relations were obtained with 99-stage inverter chains and fan-in = fan-out = 1. With 0.8-V supply voltage, the complementary circuitry displays a switching energy of

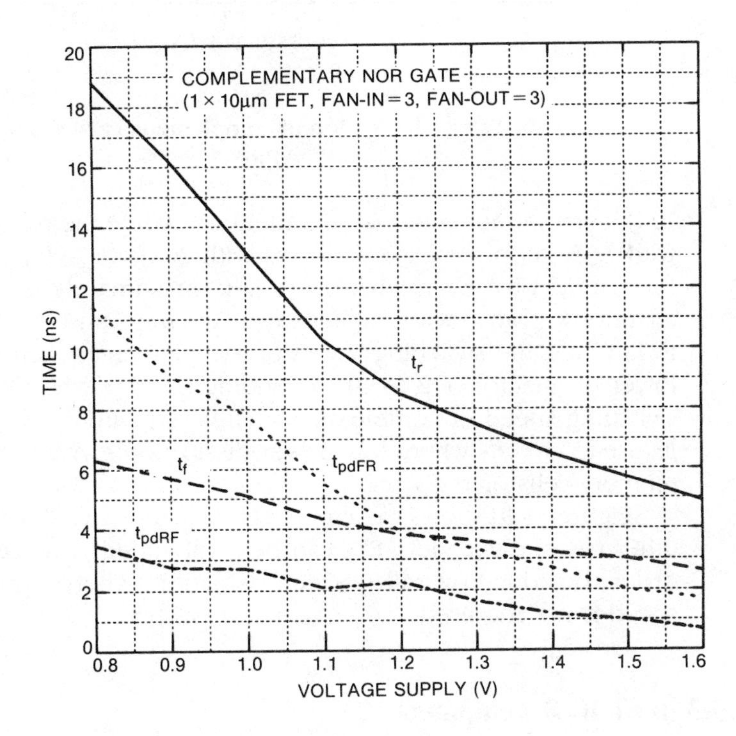

Fig. 3–15. Response times of complementary NOR gate as a function of supply voltage.

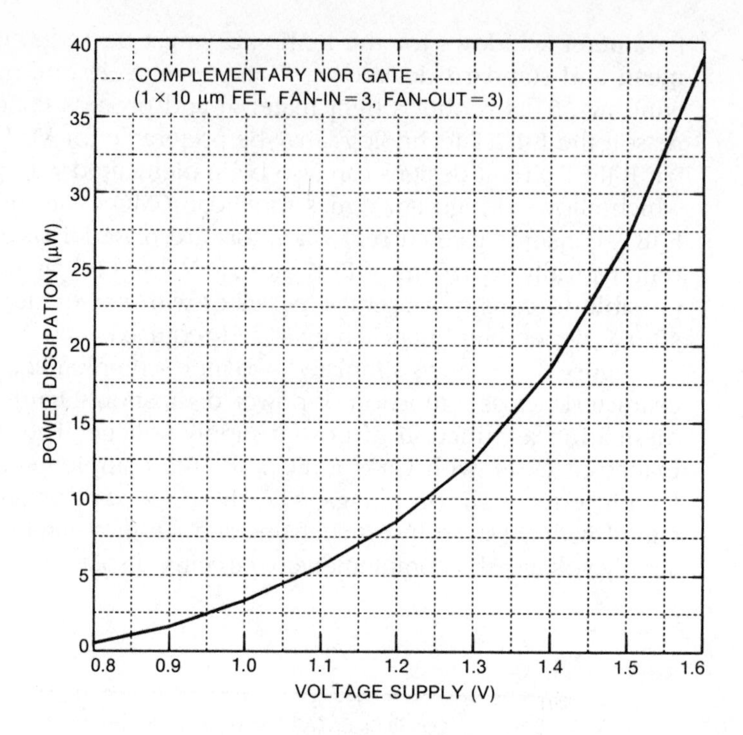

Fig. 3–16. Power dissipation of complementary NOR gate as a function of supply voltage.

10 fJ, corresponding to 330 ps and 30 μW, while the resistive load inverter with 1.5-V supply voltage performs with 20 fJ, i.e. 200 ps and 100 μW.

The K' of p-channel E-JFETs is approximately one tenth of the n-channel K', but device capacitances are the same for p- and n-channel FETs. Thus, the switching speed of complementary inverters and gates is lower by a factor of 2 to 10, depending upon power dissipation. The low switching speed of complementary E-JFET (C-EJFET) logic has restricted its use to circuits where low power dissipation is mandatory, such as static memory cells or registers.

Figure 3–18 shows the circuit schematic of a C-EJFET memory cell. In other cases, p-channel FETs can be "embedded" in n-channel circuits in order to reduce overall power dissipation, where optimized switching speed is not required.

3.3.3 JFET IC Development

Since the inception of E-JFET logic in 1978, steady progress has been made. E-JFET IC development is carried out by the McDonnell Douglas

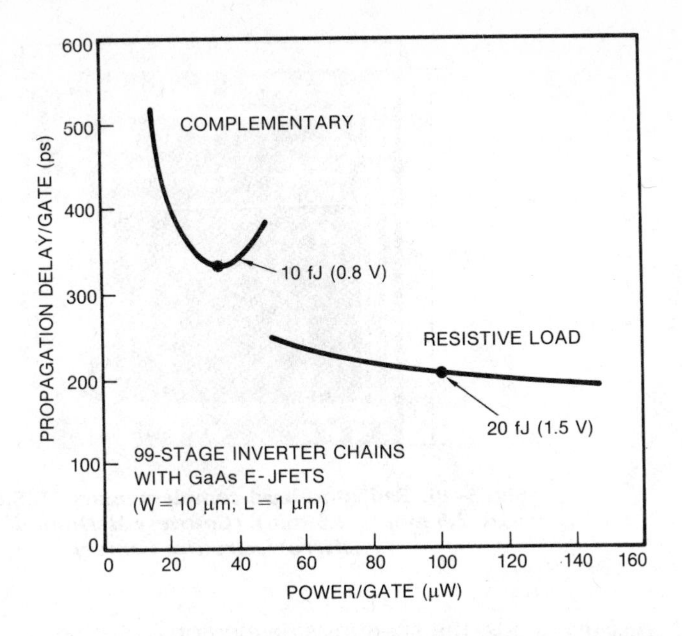

Fig. 3–17. Switching performance of DCFL with GaAs E-JFETs for complementary and resistive-load inverters.

Microelectronics Center (MDMC) in the United States and by Sony Corporation in Japan. Initial demonstration circuits included frequency dividers, inverter chains, 16-bit shift registers and small SRAMs, 32- × 1- and 64- × 1-bit. In 1983, the first 256-bit SRAM was fabricated and tested. The next step was the development of a 1K SRAM which debuted[27] in 1984. While the peripheral circuitry consisting of column and row decoding, I/O circuits and read/write control were designed with n-channel E-JFETs and

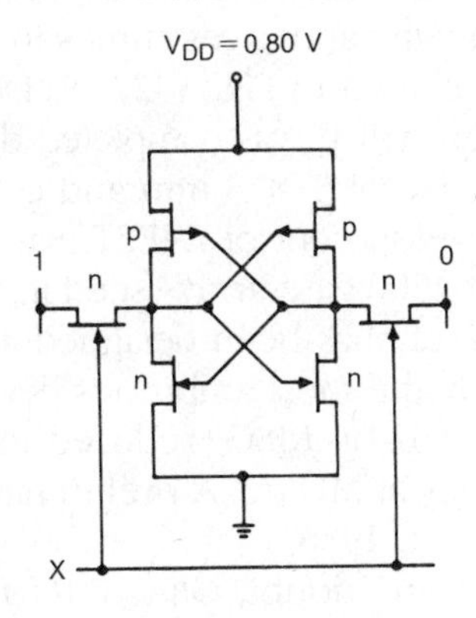

Fig. 3–18. Complementary E-JFET static RAM cell.

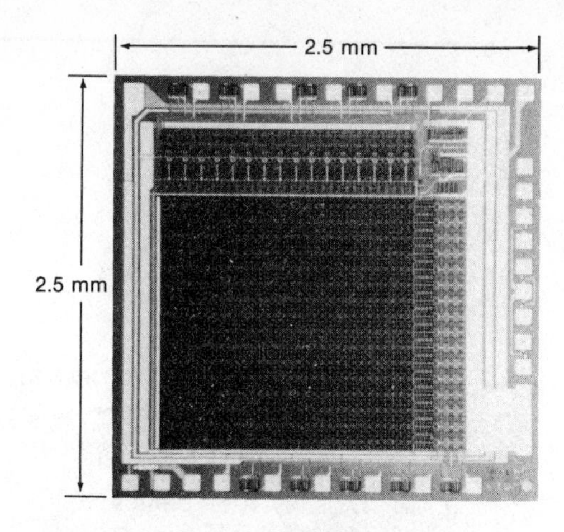

Fig. 3–19. Radiation-hard complementary 1K SRAM (chip size 2.5 mm × 2.5 mm). ***(Courtesy McDonnell Douglas Microelectronics Center)***

resistive loads, the six-transistor memory cell used a complementary flip-flop design. The complementary cell offers higher noise margin and circuit yield than the corresponding resistive-load version. The access and write time for the 1K SRAM is in the range of 5 to 10 ns, typically. Power dissipation at 100-MHz operation is 50 mW. Figure 3–19 is a photograph of the 1K SRAM chip, which is 2.5 mm × 2.5 mm.

During 1985, a 4-bit, bit-slice microprocessor was built by MDMC and tested. It showed the feasibility of producing GaAs E-JFET circuits with complexities of several thousand transistors. A photograph of this 4-bit microprocessor is shown in Fig. 3–20. Sony demonstrated an 8- × 8-bit multiplier in 1986, which is shown in Fig. 3–21.

A 4K SRAM was developed at MDMC during 1986 with parts showing full functionality and access times in the 6- to 12-ns range. A photograph of the chip is shown in Fig. 3–22. McDonnell Douglas proceeded with a 16K SRAM design which was completed during the summer of 1987. The design has a chip size of 5 × 8 mm and contains 105,096 JFETs.

The development of E-JFET gate arrays was begun during 1985 with a three-input 1500-gate array (see Fig. 3–23). A larger version of three-input 6000-gate array has been designed and was evaluated during 1987.

Probably the most ambitious development effort is the design and fabrication of a 32-bit RISC (reduced instruction set computer) microprocessor underway at MDMC. A preliminary chip demonstrated the feasibility of a 32-bit ALU in 1986.

Without any doubt, GaAs E-JFET IC development is progressing at a fast pace and is accelerating, as users recognize the advantages of this technology, which can be summarized as low power dissipation, speeds

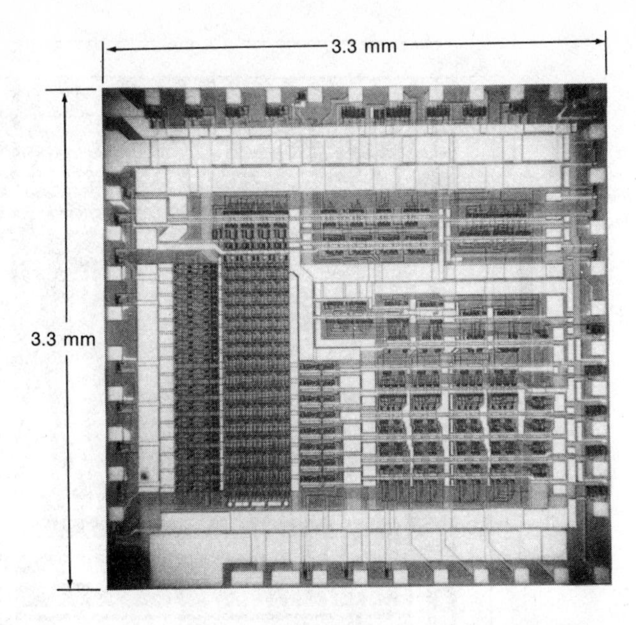

Fig. 3–20. Photomicrograph of MD 2901 4-bit GaAs microprocessor (chip size 3.3 mm × 3.3 mm). ***(Courtesy McDonnell Douglas Microelectronics Center)***

equal to or better than silicon ECL high layout density, high circuit yield, and excellent radiation hardness.

3.4. Processing and Yield

Effective utilization of any device technology for individual circuits or as part of a large system requires a fabrication process that can produce functional circuits in sufficient quantities to meet user demands. This section describes the JFET process that has been developed at MDMC that is producing high circuit yields and is compatible with production processing methods.[5,28] Using this process a pilot line facility has been established capable of 120 wafer starts per single-shift five-day week. Fully functional circuits that have been produced on the line include 1K, 4K, and 16K SRAMs, 4-bit slice microprocessors, 16-bit ALUs, and 1500- and 6000-gate arrays, and a 32-bit microprocessor.

3.4.1 Processing

The GaAs wafers used for the JFET pilot line process are 3 in in diameter and are grown by undoped LEC techniques. Qualification procedures in-

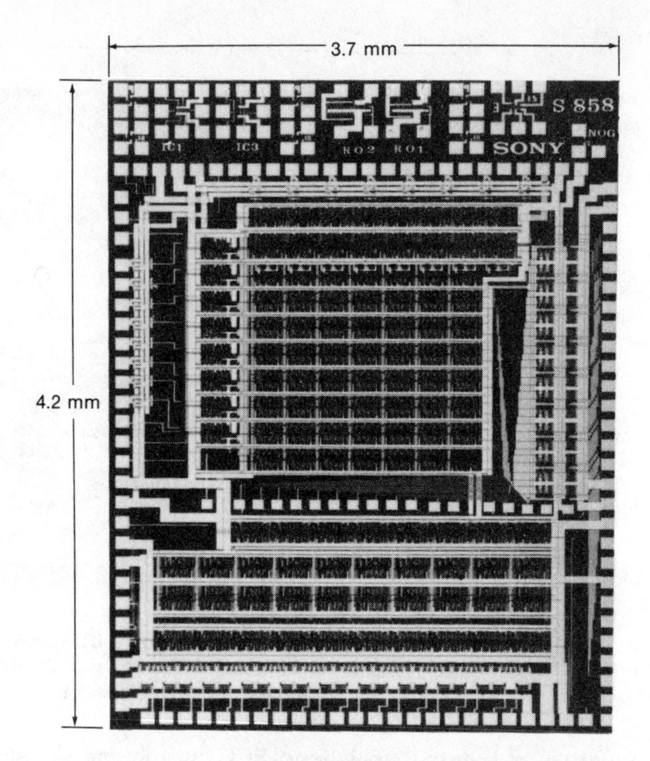

Fig. 3–21. 8- × 8-bit multiplier/accumulator (chip size 4.2 mm × 3.7 mm). ***(Courtesy Sony Corp.)***

clude surface finish, surface flatness, wafer diameter and thickness, flat length, implant resistivities, and JFET device characteristics. After cleaning the wafer surface to remove surface contaminants and polishing damage, a Si_3N_4 layer is deposited to protect the surface during processing and to form the annealing cap. Since dopants are implanted through the nitride layer to form the n-type and p-type regions of JFETs, resistors, and diodes, nitride thickness uniformity and repeatibility are essential to control of device properties. The nitride thickness uniformity achieved in pilot line operation is ± 0.5 percent across single wafers and ± 1 percent from wafer to wafer. Pin holes and particles are less than $0.5/cm^2$.

Figure 3–24 shows the ion-implant sequence for the complementary JFET process. Photoresist is used as the mask for each of the selective implants. The implant species are Si^+ for n-type implants and Mg^+ for p-type implants. Annealing of the wafers to remove the implant damage and activate the dopants is done in a furnace.

For the n-channel E-JFETs, Si^+ ions are implanted at 150 keV with doses up to $1 \times 10^{13}/cm^2$, to form the channel, and Mg^+ ions at energies near 50 keV and doses of $5 \times 10^{13}/cm^2$ for the p^+ gate. Figure 3–25 shows

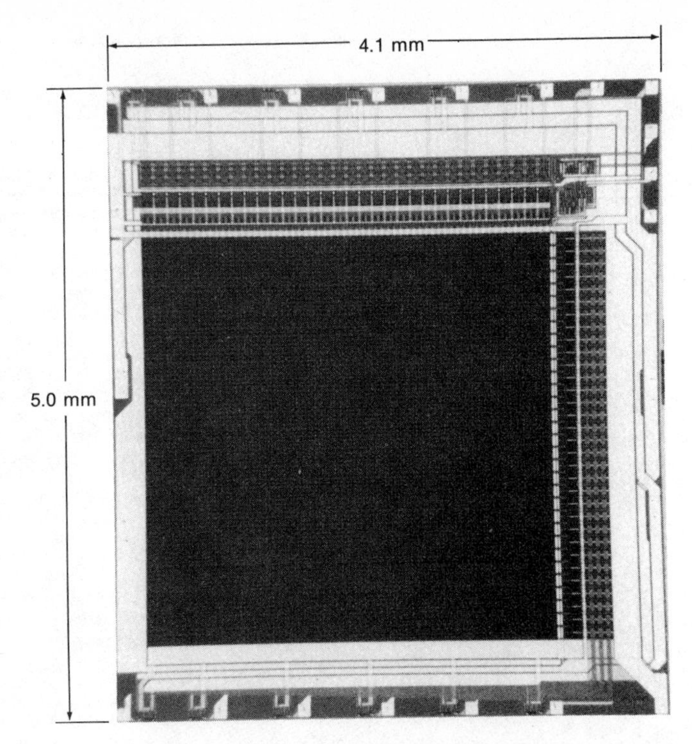

Fig. 3–22. 4K × 1 SRAM with E-JFETs (chip size 5 mm × 4.1 mm).
(Courtesy McDonnell Douglas Microelectronics Center)

typical impurity profiles for the gate and channel implants. The built-in voltage generated by the p^+ and n regions depletes the carriers in the channel and causes the device to be normally off. The threshold voltage of the device can be adjusted to the desired value by changing the gate implant energy which varies the depth of the p^+n junction. Similarly, p-channel E-JFETs are produced by using a shallow n^+ implant for the gate and a deep p implant for the channel. The threshold voltage of the p-channel device can also be adjusted to the desired value by varying the gate implant.

Typical *I-V* characteristics of 1-μm × 10-μm n-channel and p-channel E-JFETs are shown in Fig. 3–26. Transconductances of n-channel JFETs measured between $V_{GS} = 0.9$ V and $V_{GS} = 1.0$ V were greater than 100 mS/mm with $V_T = 0.2$ V. The highest value of g_m measured under these conditions was 192 mS/mm. Transconductances of p-channel JFETs measured between $V_{GS} = -0.9$ V and $V_{GS} = -1.0$ V were greater than 5 mS/mm with $V_T = -0.2$ V. The highest value of p-channel g_m measured under these conditions was 14 mS/mm. Gate diode forward *I-V* curves for p^+n diodes and n^+p diodes, presented in Fig. 3–27, show the 1-V turn-on voltage of pn junctions that allow DCFL circuits to operate with 1-V logic levels.

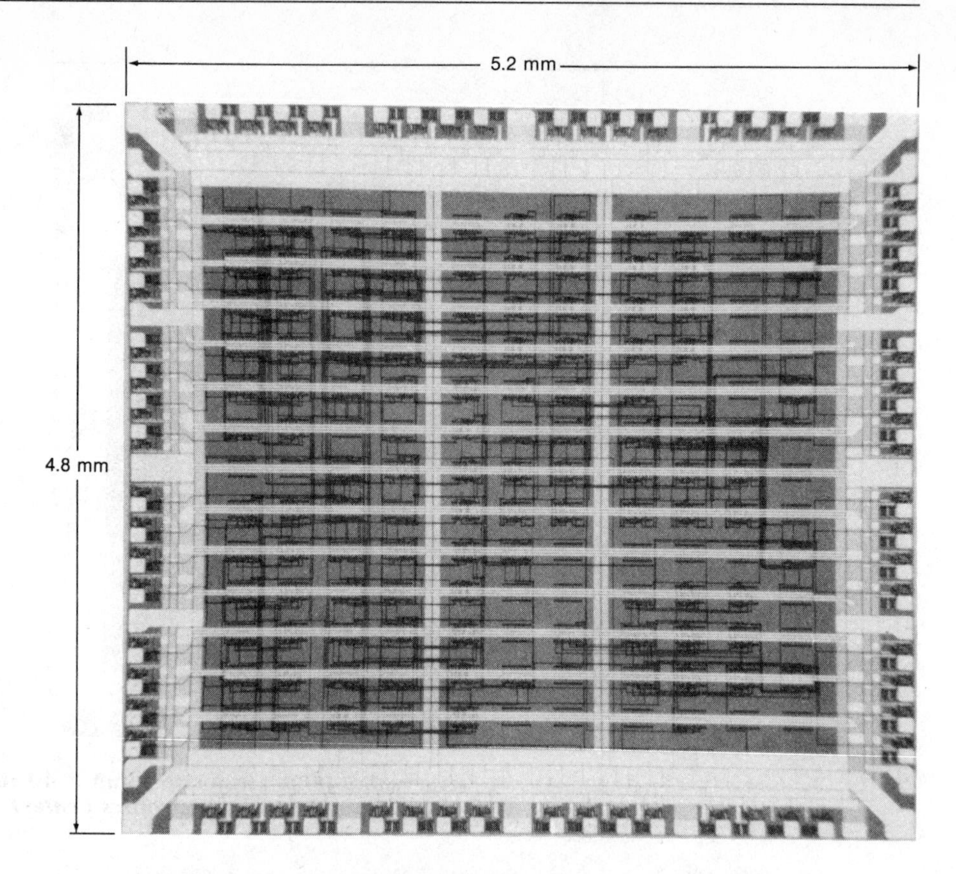

Fig. 3–23. 1.5K array personalized as 5-bit ALU in parallel with a 5- × 5-bit multiplier (chip size 5.2 mm × 4.8 mm). ***(Courtesy McDonnell Douglas Microelectronics Center)***

The remainder of the process defines the ohmic contacts and the two-level interconnects. The GeAuNi ohmic contacts to the n^+ regions are patterned by nitride-assisted PR lift and then alloyed. The first interconnect layer, which also forms the ohmic contact to the p^+ region, is sputtered PtAu patterned with ion milling. Next a SiO_2 or Si_3N_4 dielectric layer is deposited for the crossover insulator, vias etched, and the second-level interconnect sputtered and ion milled. Typical values of contact resistance are $1 \times 10^{-6}\ \Omega \cdot cm^2$ and $2 \times 10^{-6}\ \Omega \cdot cm^2$ for the n^+ and p^+ contacts, respectively. The interconnect resistivity is 150 mΩ per square for the first level and 60 mΩ per square for the second level.

No isolation implants are used during the process. Isolation resistance is measured on processed wafers with the n^+ to n^+ and p^+ to p^+ test structures shown in Fig. 3–28. The gap between the doped layers is 3 μm and the width is 100 μm. Typical resistance values are greater than $10^9\ \Omega$ for the n^+ to n^+ pattern and $10^7\ \Omega$ for the p^+ to p^+ patterns.

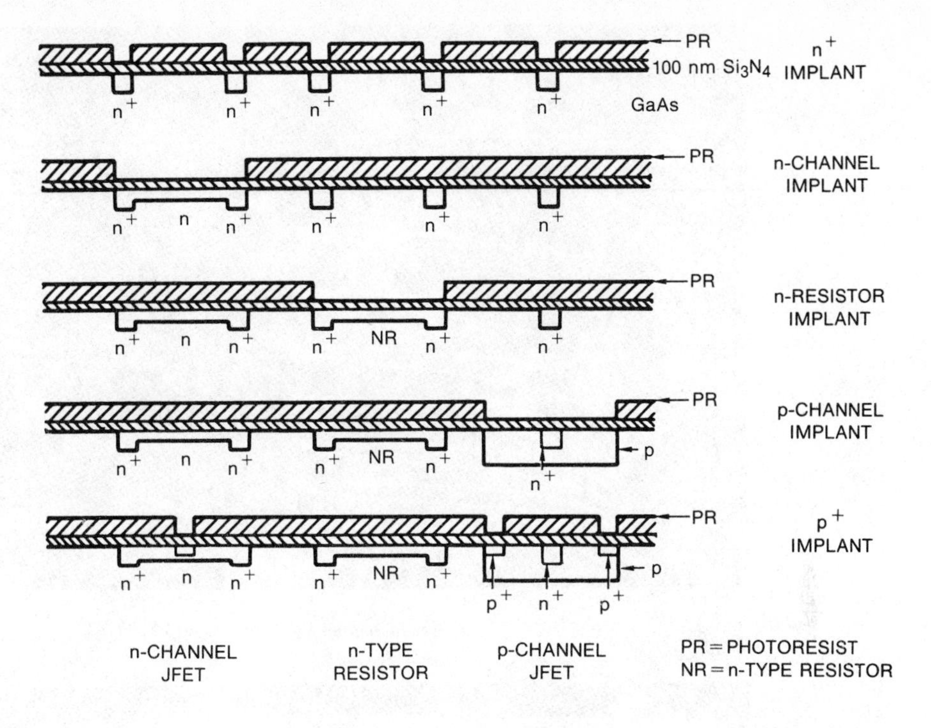

Fig. 3–24. Ion-implant sequence for complementary E-JFET process.

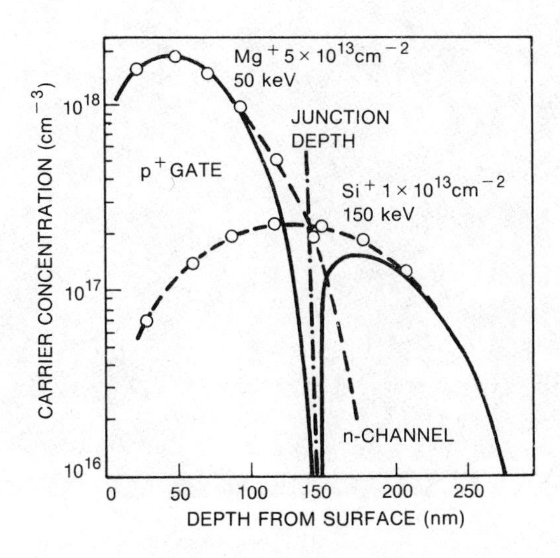

Fig. 3–25. Carrier concentration profiles of gate and channel implants for n-channel E-JFET.

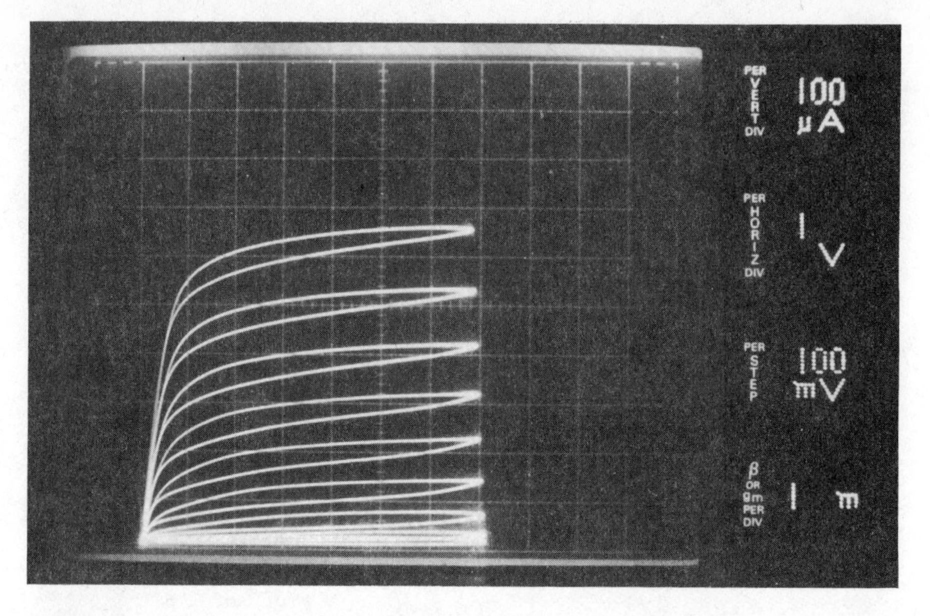

(A) For n-channel V_{GS} = 0 to +1.0 V.

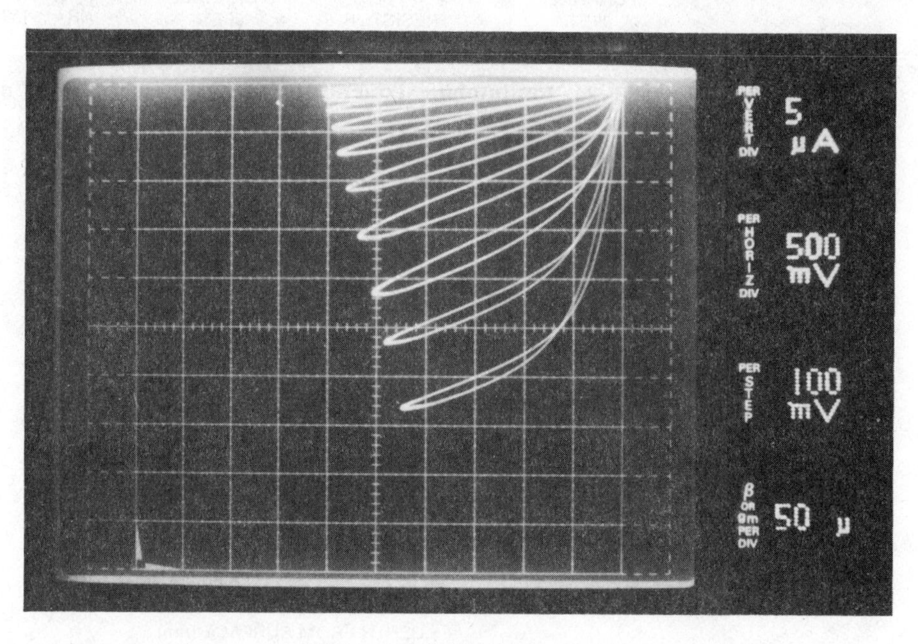

(B) For p-channel V_{GS} = 0 to −1.0 V.

Fig. 3–26. Curve tracer plots of room temperature drain current-voltage relations for L = 1-μm and w = 10-μm E-JFETs with p and n channels.

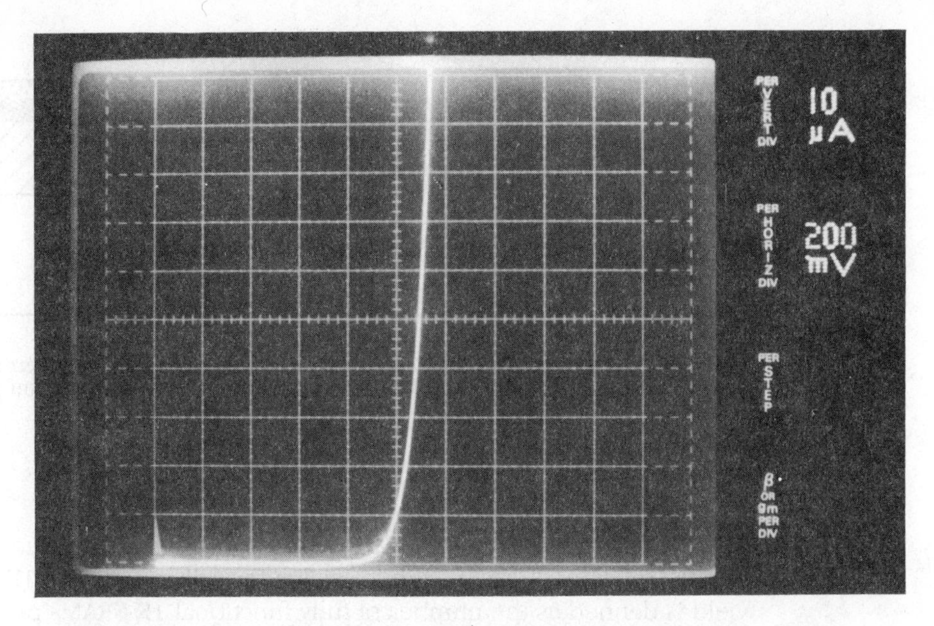

(A) For p$^+$n diodes.

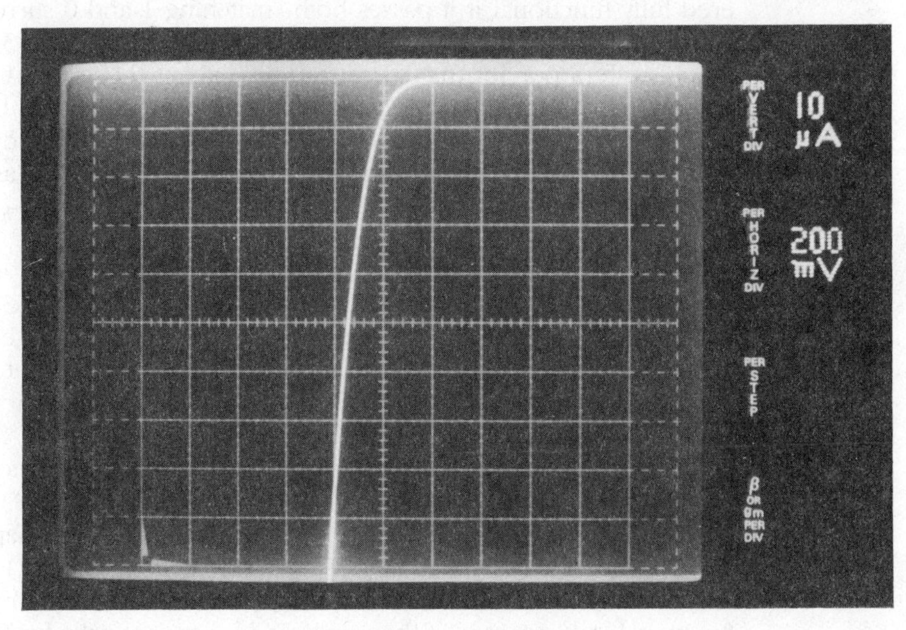

(B) For n$^+$p diodes.

Fig. 3–27. Room temperature current-voltage plots of p- and n-channel E-JFETs for $L = 1\ \mu m$ and $w = 10\ \mu m$.

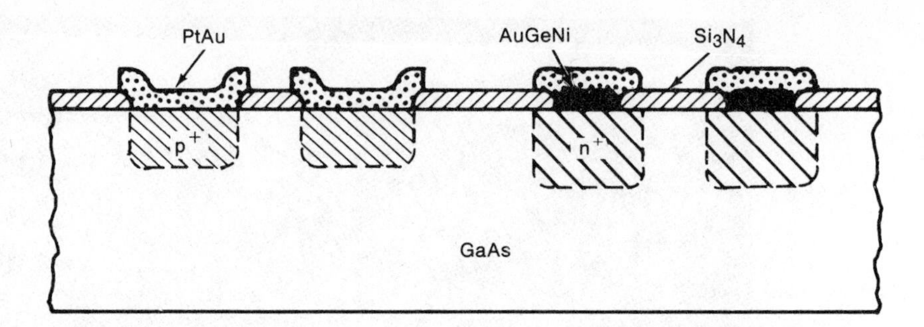

Fig. 3–28. Substrate isolation test structures. The gap between n^+ and p^+ contacts is 3 μm and the width of both structures is 100 μm.

3.4.2 Yield Results

Circuits fabricated on the MDMC GaAs pilot line have produced excellent yields. Figure 3–29 shows the lot yields for fully functional 1K SRAMs. Lot yield is defined as the number of fully functional 1K SRAMs per lot divided by the total number of 1K SRAMs completed per lot. A 1K SRAM is considered fully functional if it passes both "matching 1 and 0 surrounded by 8 complement" and GALPAT tests. For the lots shown in Fig. 3–29, each of the 10 wafers per lot that were started (except those that broke), were completed and included in the yield calculation. The single best lot yield is 27.5 percent. The highest 1K SRAM yield on an individual wafer is 47 percent. This is a significant result for GaAs digital ICs, as the chip is 2.5 mm × 2.5 mm and contains over 7000 n-channel and p-channel transistors.

Figure 3–29 also shows a promising trend: the yield is improving as more lots are processed and experience is gained. This is the result of the primary benefits of the pilot line program: process a large enough number of wafers to determine the yield limiting factors, institute correction, and verify that the corrections solve the problem without introducing harmful side effects.

Figure 3–30 presents a wafer map of yield for a complementary E-JFET 4K SRAM which is 3.8 mm × 4.4 mm and consists of 18,833 n-channel E-JFETs, 8192 p-channel E-JFETs, and 414 resistors. The map shows the yield of individual 4K SRAMs passing GALPAT. Two of the 114 SRAMs are fully functional, for a wafer yield of 2 percent, with many chips near fully functional. A histogram of the numbered functional cells for SRAM on this wafer, Fig. 3–31, shows that most of the SRAMs are clustered at the high yield end of the histogram with a small number at the low yield end and very few in between. This type of distribution is typically observed when random effects, such as particulates, on threshold voltage variations cause

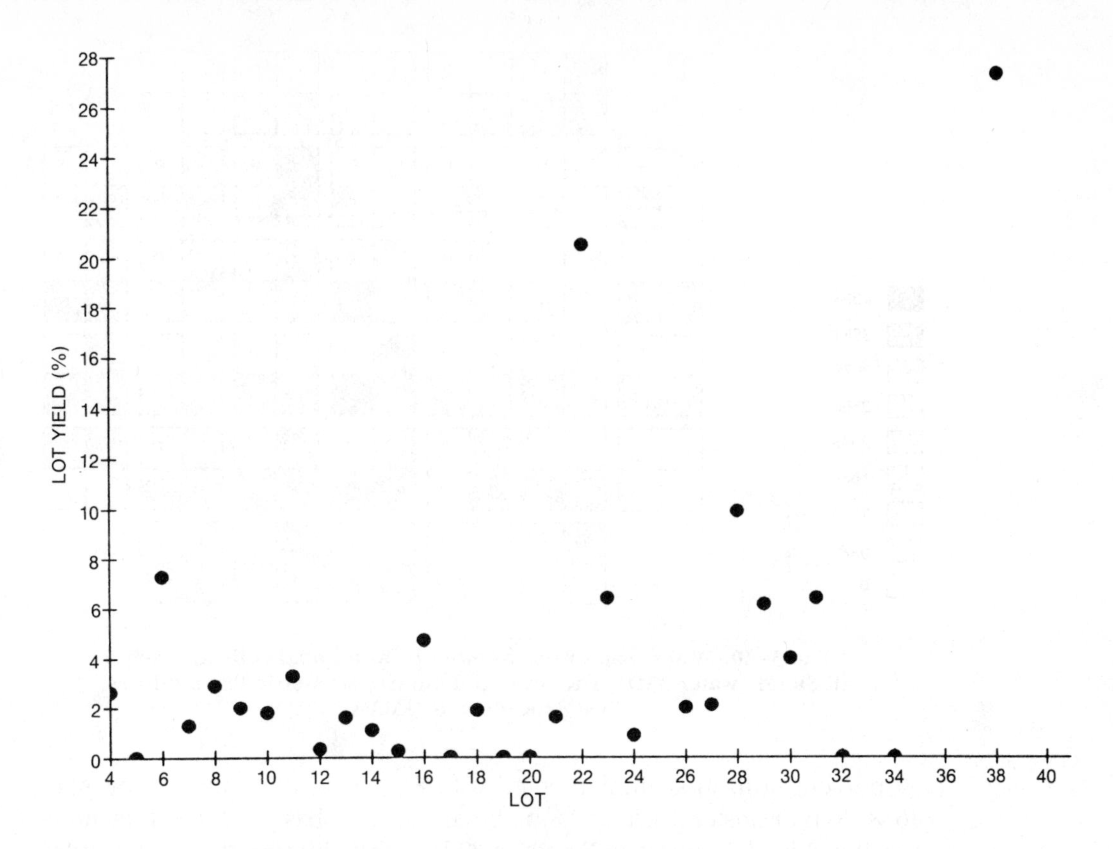

Fig. 3–29. Completed wafer lot yield of complementary 1K SRAMs versus lot number.

individual cells to fail at the test conditions. The causes of the cell losses are under investigation. Recent evaluations indicate that less than half of the losses of the 4K SRAM cells are due to particulates while the remainder is due to material or process parameter nonuniformity. Improvements in these yield limiting factors are expected to raise yield significantly.

Other fully functional circuits fabricated on the pilot line include a GaAs emulation of the 2901 4-bit slice processor (3.3 mm × 3.3 mm, 1860 transistors, and best wafer yield of 8 percent), a 1.5K gate array personalization of a 5-bit ALU in parallel with a 5 × 5 multiplier (4.8 mm × 5.2 mm, 4576 transistors, and best wafer yield of 5 percent); and most recently a 32-bit ALU (4.5 mm × 5.5 mm, 5005 transistors, and best wafer yield of 1.3 percent). Recent design upgrades and process improvements have significantly increased yields. Best wafer yields of fully functional SRAMs are now 79 percent, 41 percent, and 9 percent for the 1K SRAMs, 4K SRAMs, and 16K SRAMs, respectively. In addition a 6080 gate array with

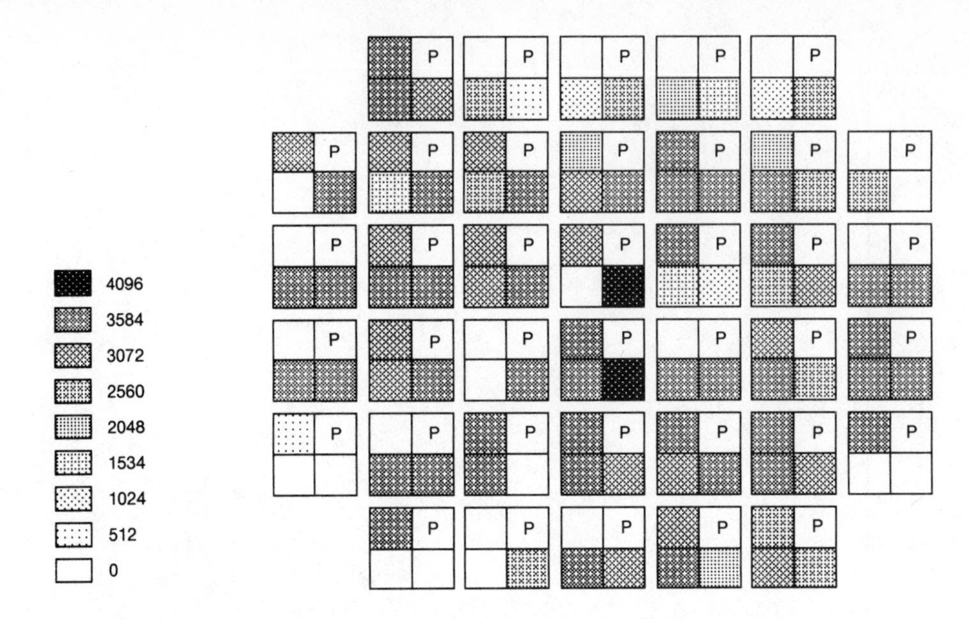

Fig. 3–30. Wafer map of the number of functional cells for each 4K SRAM (wafer 33D). The letter p denotes parametric test modules. Test procedure is GALPAT.

77 percent utilization implemented with 8- × 8-bit multiplier, 16-bit ALU, 16 × 16-bit register stack, and 8-bit binary counter has produced fully functional yields of 14 percent. Finally, a 32-bit CPU with the RISC architecture has achieved wafer yields of 6 percent for fully functional chips using more than 4000 test vectors with greater than 85-percent fault coverage. These results indicate the high yield potential of E-JFET circuits for fabrication of LSI and VLSI chips.

3.4.3 Yield Issues for VLSI

Yield in IC fabrication can be assigned to limitations due to design, i.e. the sensitivity of a particular design to the normal process variations of a parameter such as threshold voltage, and uniform or random processing defects. For small-size circuits (SRAMs) of low complexity, the design yield is high. Figures 3–32 and 3–33 give design yields close to 100 percent for a 1000-gate chain with complementary gates and 90 percent for resistive-load gates, for our present process parameter uniformity, σ_{th}, and design window.[29] Therefore, the yield of 256-bit and 1K static RAMs is determined by Poisson statistics and can be described by

$$Y = Y_0 \exp(-DA), \tag{3-28}$$

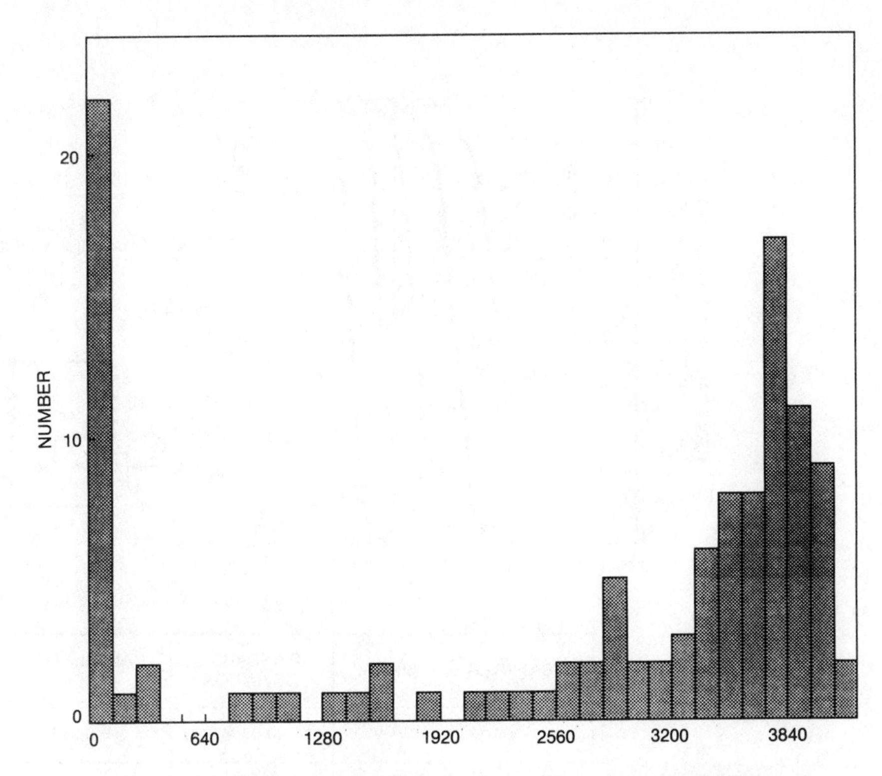

Fig. 3–31. Histogram of the number of functional cells for each 4K SRAM on wafer 33D. Test procedure is GALPAT.

where D is the defect density and A the area. Experimental results are shown in Fig. 3–34 for the 256-bit, the 1K, and the 4K SRAMs processed on our pilot line. The experimental results for the 256-bit and 1K SRAMs reveal a yield percentage related to about 10 defects/cm^2. However, the 4K SRAM with about 1 to 2 percent yield falls way below this theoretical curve, which predicts 25 percent. Failure analysis of the SRAM nonfunctional bits, rows, and columns identified both process and design weaknesses that limited yield. After correcting design deficiencies and upgrading the process by reducing random particulates and improving wafer cleaning procedures during the contact and interconnect formation steps, the increased circuit yields reported at the end of Sec. 3.4.2 were achieved. The defect density for the high JFET density SRAMs of 10 defects/cm^2 has now been achieved for the LSI 4K SRAM and the VLSI 16K SRAM. For combinatorial logic circuits, such as the 6080 gate array and the 32-bit RISC CPU, with lower JFET densities the defect density from Eq. 3–28 is 2.8 defects/cm^2.

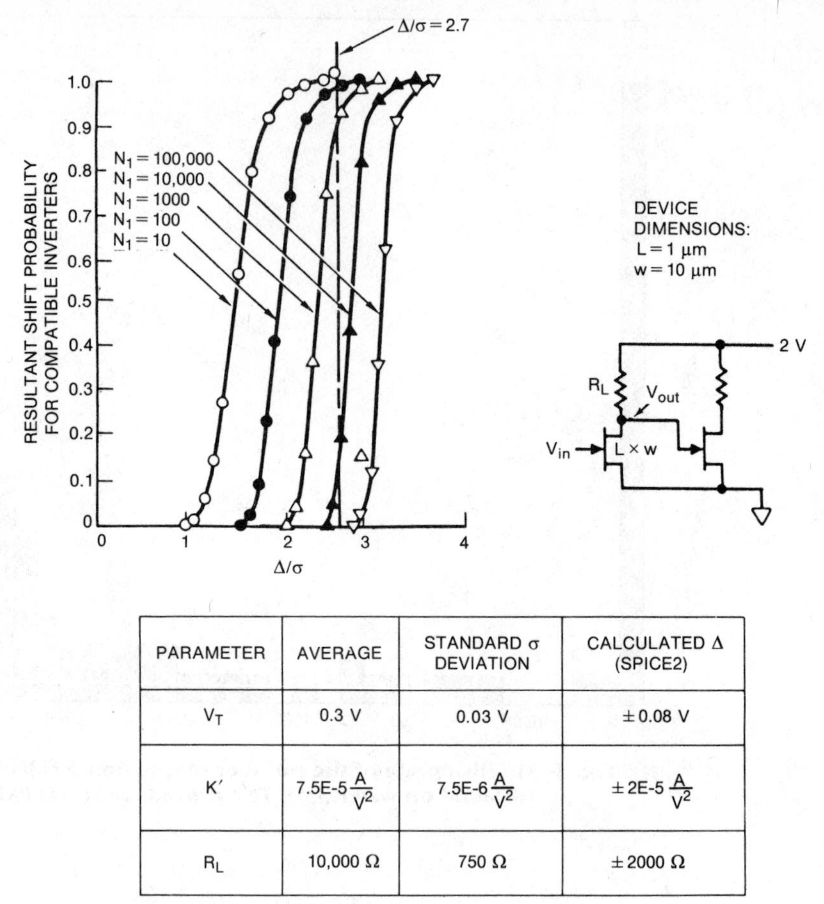

PARAMETER	AVERAGE	STANDARD σ DEVIATION	CALCULATED Δ (SPICE2)
V_T	0.3 V	0.03 V	± 0.08 V
K′	7.5E-5 $\frac{A}{V^2}$	7.5E-6 $\frac{A}{V^2}$	± 2E-5 $\frac{A}{V^2}$
R_L	10,000 Ω	750 Ω	± 2000 Ω

$\Delta/\sigma = 2.7$ for these parameters calculated from SPICE2 simulations.

Fig. 3–32. Resultant shift probability for *n* compatible resistive-load inverters as a function of Δ/σ.

It is our observation and interpretation, for the high circuit complexities above 1K SRAMs, that the yield is no longer dominated by the number of defects per square centimeter but becomes limited by material uniformity and quality and process-related material parameters. Primarily, the control of threshold voltage over large chip areas is the key parameter for yield. For instance, if we change the standard deviation in Fig. 3–32 and Fig. 3–33 to 40 mV, instead of the listed 30 mV, then the 1000-gate circuits of the resistive-load design have a yield close to zero while the complementary gate design still has 90 percent yield. The significance of the threshold voltage control becoming a yield controlling parameter becomes

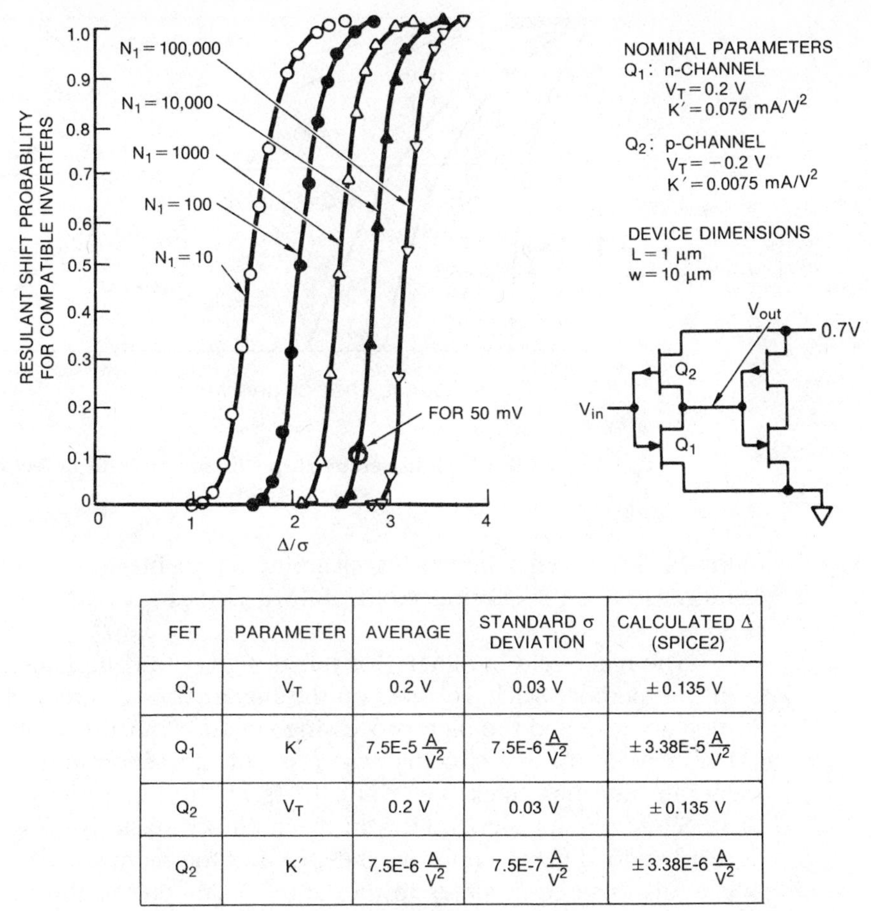

FET	PARAMETER	AVERAGE	STANDARD σ DEVIATION	CALCULATED Δ (SPICE2)
Q_1	V_T	0.2 V	0.03 V	±0.135 V
Q_1	K′	7.5E-5 $\frac{A}{V^2}$	7.5E-6 $\frac{A}{V^2}$	±3.38E-5 $\frac{A}{V^2}$
Q_2	V_T	0.2 V	0.03 V	±0.135 V
Q_2	K′	7.5E-6 $\frac{A}{V^2}$	7.5E-7 $\frac{A}{V^2}$	±3.38E-6 $\frac{A}{V^2}$

$\Delta/\sigma = 4.5$ for these parameters calculated from SPICE2 calculations.

Fig. 3–33. Resultant shift probability for *n* compatible complementary inverters as a function of Δ/σ.

more evident when we take the 10,000-gate complexity for the complementary circuitry with an increased threshold voltage standard deviation of 50 mV, which reduces the yield to 5 percent. This design sensitivity to parameter control is believed to compete at and above the 4K SRAM complexity level with the yield limitations due to the process defect density. Both areas need improvement to guarantee reasonable yields in the range of 5 to 20 percent for VLSI GaAs ICs of our technology.

As processes and designs continue to be improved, the large SRAMs eventually will not be design limited. Yields of 4K SRAMs could then be around 20 to 25 percent, and still an acceptable 3 percent for 16K SRAMs

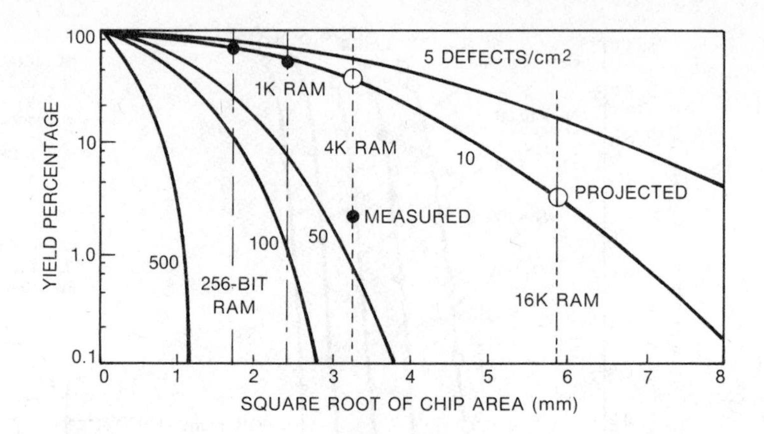

Fig. 3–34. Circuit yield versus chip dimension with defect density as a parameter.

for 10 defects/cm^2. Improving cleanliness to achieve 5 defects/cm^2 could increase the processing yield of 16K RAMs to a comfortable value of 15 percent.

The high yield in SRAMs, for our design with complementary circuitry in the memory stack, is based on the higher noise margin of the complementary gate and the high processing yield for our threshold voltage control. The peripheral circuitry, which is only a fraction of the gates utilized in the memory stack, i.e. 2449 JFETs in the peripheral circuitry of a 4K SRAM versus 24,576 JFETs in the memory stack, is designed with the resistive-load gate to improve the speed of the memory. The resistive-load gate also has good yields in the large SRAMs due to the low number of gates required.

The key issue, other than particle reduction, for high yield VLSI GaAs JFET circuits is threshold voltage control. Improvement of the uniformity of process parameters such as nitride thickness, gate length, implant dose and energy dielectric stress, and annealing temperature will benefit V_T control. However, eventually GaAs material properties, which include EL2 average value and uniformity, donor and acceptor compensation levels, EPD, sawing and polishing subsurface damage, gettering sites, and stoichiometry will be the limiting factors in threshold voltage control and ultimate yield.

3.5 Prospects for VLSI

The captive technology of future GaAs ICs was at first assigned to D-MESFETs for producing commercial circuits.[30] While this technology is mature to produce ICs with BFL, SDFL, or modified logic with level shifting, at complexity levels of MSI and VLSI, the power dissipation constraints per chip may exclude its application for VLSI. The E-MESFET technology on the other hand has VLSI capability, owing to the low power dissipation per gate, but as pointed out[30] has difficulties in meeting the threshold voltage control to a standard deviation of about 25 mV, because of the limited logic voltage swing of typically 0.5 V. The voltage difference between logical 0 and 1 states demands a 20 times standard deviation of threshold voltage for a good enough margin to produce LSI circuits with practical yields. The control of this parameter is extremely difficult over large wafer areas with the process of self-aligned gate (SAG) and self-aligned implanted n-layer technology (SAINT). On the other hand, the E-JFET circumvents these difficulties. Because of the junction gate structure, the logic swing is doubled to about 0.8 to 1.0 V. This automatically relaxes the control of threshold voltage standard deviation to about 50 mV for LSI and 35 mV for VLSI. Experimental results with E-JFET devices over 2-in wafers have demonstrated[31] a standard deviation of 30 mV and over 3-in wafers a range of 40 to 50 mV, which, therefore, qualifies this technology for producing VLSI circuits with acceptable yields. The yield figures of fully functional 1K SRAMs of 39 percent and 4K SRAMs of 2 percent with a 3-in pilot line process are verifications of this prediction. The E-JFET technology with resistive load for combinational logic and especially the complementary E-JFET technology for low-power operation, such as static RAMs and shift registers, are therefore a sound technology basis for VLSI circuit production, such as 4K, 16K, and eventually 64K static RAMs and 6000- and 10,000-gate arrays. The existing JFET technology, as described in this chapter, is the first GaAs technology to claim operation over the full military temperature range; it possesses a built-in radiation hardness (to a total dose ionizing radiation of 10^8 rad [GaAs]) and offers other advantages in harsh radiation environments. With its low power consumption per gate, it offers high integration complexities with acceptable yields, but in its standard planar design does not produce the highest switching speed possible with GaAs material. Self-aligned gate structures—now under exploration and development—are viable solutions to bring in line with the high electron mobility in GaAs material the promise of high speed.

Progress in GaAs JFET integration technology is compared in Fig. 3–35 with that of Si, following Moore's law of doubling device complexity per year for almost two decades.[32] Although the GaAs integration technology started about 15 years later than the Si one, it is gratifying to observe a doubling-squared progression in integration complexity. With Si technology entering the VLSI circuit field about 1975 and now even boasting GSI (giant-scale integration) parts, such as 1M (1 megabit) and 4M SRAMs, GaAs integration has just reached the VLSI boundary with the 16K static RAM. It is a challenge for the GaAs technology to advance fabrication and material technology to culminate in GSI in the early nineties.

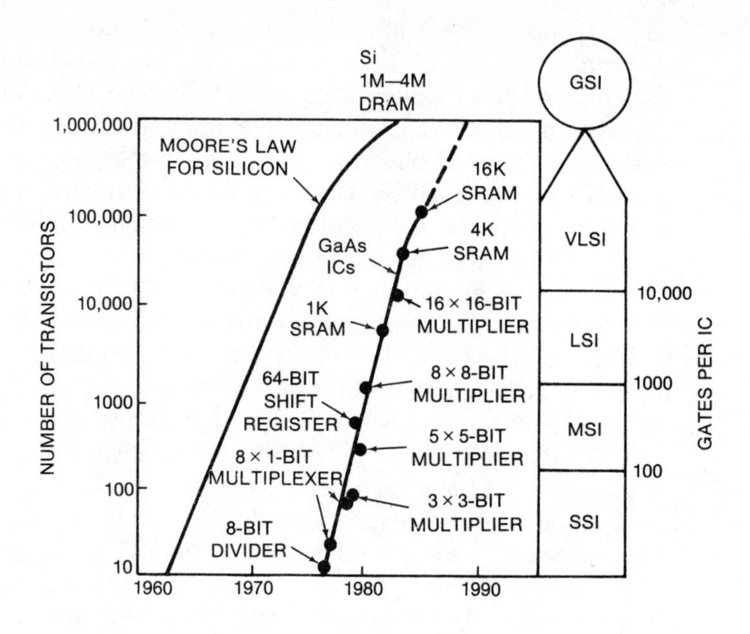

Fig. 3–35. Maturing of GaAs IC technology in comparison to Si IC progress.

To fully exploit GaAs in digital logic circuits requires special design approaches. Because the GaAs gates switch extremely fast, the ratio of on-chip propagation delays to off-chip communication delays is greater than in Si. Even in today's supercomputers, which use very high speed ECL logic in small-scale integration (SSI), chip-to-chip communication is a more important factor in the computer's cycle time than on-chip gate delays. This suggests that higher levels of integration and novel computer architectures are necessary before GaAs technology will pay real dividends in high-performance computers.

References

1. R. Zuleeg, *Proc. IEEE,* 56:879 (1968).
2. R. L. van Tuyl *et al., IEEE J. Solid State Cir.,* SC-12:485 (1977).
3. R. C. Eden *et al., IEEE J. Solid State Cir.,* SC-14:221 (1979).
4. A. W. Livingston and P. J. T. Meller, *1980 IC Symp.,* paper no. 10.
5. R. Zuleeg *et al., IEEE Electron Dev. Lett.,* EDL-5:21 (1984).
6. R. Zuleeg *et. al., IEEE Trans. Nucl. Sci.,* NS-31:1121 (1984).
7. S. E. Diehl *et al., IEEE Trans. Nucl. Sci.,* NS-30:4501 (1983).

8. W. A. Geideman *et al., GOMAC Tech. Dig.,* 169 (1986).
9. T. P. Nicalek *et al., IEEE Cust. ICs Conf. Proc.,* 434 (1985).
10. J. L. Hennessy, *IEEE Trans. Computers,* C-33:1221 (1984).
11. T. L. Rasset *et al., Computer,* 19:60 (October 1986).
12. R. A. Kiehl and A. C. Gossard, *IEEE Electron Dev. Lett.,* EDL-5:521 (1984).
13. T. Mizutani *et al., Electron. Lett.,* 21:1116 (1985).
14. N. C. Cirillo *et al., IEEE Electron Dev. Lett.,* EDL-6:645 (1985).
15. T. Mizutani *et al., GaAs IC Symp. Tech. Dig.,* 107 (1986).
16. W. Shockley, *Proc. Inst. Rad. Eng.,* 40:1374–1382 (1952).
17. G. W. Taylor *et al., IEEE Trans. Electron Dev.,* ED-26:172–182 (1979).
18. R. Zuleeg, J. K. Notthoff, and K. Lehovec, *IEEE Trans. Electron Dev.,* ED-25:628–639 (June 1978).
19. R. C. Eden, B. M. Welch, and R. Zucca, *IEEE J. Solid State Cir.,* SC-131:419–425 (1978).
20. R. Zuleeg, J. K. Notthoff, and K. Lehovec, *IEEE Trans. Nucl. Sci.,* NS-24:2305–2308 (1977).
21. R. Zuleeg, Chapter 11 in *VLSI Electronics, Microstructure Science, Vol. 11,* New York: Academic Press (1985).
22. C. H. Hyun *et al., GaAs IC Symp. Tech. Dig.,* 119–122 (1986).
23. C. H. Chen, M. Shur, and A. Peczalski, *IEEE Trans. Electron Dev.,* ED-33(6): 792–798 (June 1986).
24. J. S. Blakemore, *J. Appl. Phys.,* 53(10): R123–R181 (October 1982).
25. R. J. Brewer, *Solid-State Electron,* 18:1013–1017 (1975).
26. R. Zuleeg and J. K. Notthoff, "Symmetrical input NOR/NAND gate circuit," U.S. Patent No. 4,038,563 (July 26, 1977).
27. T. P. Nicalek and J. K. Notthoff, *Proc. Cust. Int. Cir. Conf.,* 434–435 (1985).
28. G. L. Troeger *et al., IEDM Tech. Dig.,* 497–500 (1979).
29. C. H. Vogelsang *et al., GaAs IC Symp. Tech. Dig.,* 149–152 (1983).
30. R. C. Eden *et al., IEEE Spectrum,* 20:30–37 (1983).
31. G. L. Troeger and J. K. Notthoff, *GaAs IC Symp. Tech. Dig.,* 78–81 (1983).
32. G. E. Moore, *IEDM Tech. Dig.,* 11–12 (1975).

Chapter 4

Schottky Barriers on GaAs

W. Mönch

4.1 Introduction

The rectifying behavior of metal-semiconductor contacts was first reported by Braun[1] in 1873. In his studies he used natural as well as synthetic metal sulfides such as chalcopyrite, iron pyrite, galena, and fahlore. Only a few years later, selenium was added to this list by Adams and Day[2] and Siemens.[3] Although the physics of unipolar conduction in such junctions was not understood until 1938, "cat's whisker" rectifiers, which consisted of a metal point pressed against a crystal, usually of lead sulfide, and, later on, plate rectifiers based on first cuprous oxide and then selenium were invented and successfully applied in the rapidly growing field of broadcasting. Even after the quantum theory of semiconductors had been formulated by Wilson,[4] the positive sign of the Hall coefficient had to be correctly determined for cuprous oxide.[5] Furthermore, chemical and physical depletion at interfaces in rectifiers made of cuprous oxide and selenium, respectively, needed to be distinguished so that Schottky could eventually publish his "Halbleitertheorie der Sperrschicht" (Semiconductor Theory of Barrier Layers).[6] He realized that, in metal-selenium contacts, the potential barrier originates from a space charge on the semiconductor side of the interface. The space charge is made up of charged dopants, or, in other words, from a layer which is depleted of the mobile majority carriers. The rectifying behavior of such Schottky barriers, as they are called, is determined by their barrier heights, i.e. the energy distance at the interface between the Fermi level and the conduction-band bottom, when the semiconductor is doped n-type, and the valence-band top, respectively, for p-type doping.

After Schottky's basic studies on metal-semiconductor contacts, the main task remaining has been to find a physical model which explains all the chemical trends observed with the barrier heights of such junctions.

This had turned out to be a real problem, since most of the models hitherto proposed were "linear" in character, i.e. the barrier height was more or less linearly related to a property of the metal or of the semiconductor, while the experimental data obviously exhibit more complicated trends, as will be shown for GaAs in Sec. 4.2. In these circumstances, many "rules" have been suggested which only "explained" specific subsets of experimental data (see, for example, Refs. 7 through 11). Quite recently, however, Mönch[12] has proposed a new model which combines two of the earlier "linear models." These are the virtual gap-state (ViGS) model of metal-induced gap states, the principles of which were proposed by Heine[13] already in 1965, and the defect model by Wieder,[14] Spicer *et al.*,[15] and Williams *et al.*[16] While none of these can describe all the experimental data, the new ViGS-plus-defects model was shown to explain all the chemical trends of the barrier heights observed with Schottky barriers on silicon[12] and gallium arsenide.[17]

A detailed discussion of the ViGS-model of metal-induced gap states and the defect model will be given in Sec. 4.6 and 4.7, respectively. First of all, however, the GaAs data base used will be presented in Sec. 4.2. Section 4.3 then discusses the well-known Schottky-Mott rule, which should be found only when no interface states are present in the gap, as will be explained in Sec. 4.4. Since the identification of the metal-induced gap states by the virtual gap states (ViGS) of the complex semiconductor bandstructure is based on results of adsorption studies, these data will be presented and analyzed in Sec. 4.5. Finally, the ViGS-plus-defects model will be developed and applied to Schottky barriers in Secs. 4.8 through 4.10. Section 4.11 then deals with metal composition and reactions at the interface.

4.2 The Data Base

This chapter will not attempt to review all of the many studies which have been published on GaAs Schottky barriers. The reasons are twofold. First, discussions on the comparability of barrier heights obtained by different experimental techniques will be avoided. Therefore, only data determined from forward-bias current-voltage (I-V) characteristics measured with Schottky diodes will be considered. This means that all investigations which, for example, employed photoemission, and thus thin metal overlayers, are not included in the present discussion. Some of these will be mentioned in Sec. 4.11. Second, the preparation of the GaAs substrates prior to the metal deposition and the fabrication process of the Schottky diode itself were reported to affect the barrier heights of the final devices. This then means that all the experimental conditions should be kept identical when the barrier heights of different deposited metals are compared.

The influence of, for example, post-fabrication annealing treatments will be briefly discussed in Sec. 4.11.

Those two criteria are met by the systematic investigations of GaAs Schottky contacts by Waldrop[18, 19] and by Spicer and coworkers.[20] Both groups started from clean GaAs surfaces, which were prepared *in situ* and were held at room temperature during the evaporation of the metals. Waldrop prepared his [001]-oriented samples by chemical etching, followed by a momentary heating to approximately 550°C in UHV (ultrahigh vacuum), which removes the inevitable native oxide, approximately 1 nm thick. Newman *et al.*[20] used {110} surfaces prepared by cleavage *in situ*.

Assuming thermionic transport across the barrier, the current-voltage characteristics read (see, for example, Ref. 21)

$$j = j_s \exp(e_0V/nk_BT)[1 - \exp(-e_0V/k_BT)], \tag{4-1}$$

where the saturation current density is given by

$$j_s = A^*T^2 \exp[-e_0(\Phi_B - \Delta\Phi)/k_BT]. \tag{4-2}$$

Here, A^* is the effective Richardson constant (8.16 $A \cdot cm^{-2}K^{-2}$ for n-type and 74.4 $A \cdot cm^{-2}K^{-2}$ for p-type GaAs),[18] $\Delta\Phi$ is the image force correction (approximately 0.04 eV for the dopings used[18]), and n is the "ideality" factor which amounts to 1.02 for an ideal junction.[21]

The experimental barrier heights and ideality factors reported by Waldrop[18, 19] and by Newman *et al.*[20] are listed in Table 4–1. In addition, data reported for Ru-GaAs {100} contacts by Ludwig and Heymann[22] are also included. They electrochemically deposited their ruthenium films. The table clearly shows that the ideality factors are generally much smaller and close to the "ideal" value $n_{id} = 1.02$ for n-type than for p-type doping. Therefore, the barrier heights given for the samples doped n-type are considered to be the more reliable ones. Nevertheless, the barrier heights determined with substrates of both types of doping add up to almost the band gap energy. Only Pb, Cr, Fe, and above all Mg are showing somewhat larger deviations. This observation will be discussed in Sec. 4.10.

The experimental data on metal contacts to GaAs and other semiconductors have been compiled by Brillson.[23] There the reader may also find the results of the many photoemission studies published up to 1982.

4.3 The Schottky-Mott Rule of Barrier Heights

The first model, trying to explain the chemical trends of barrier heights observed in metal-semiconductor contacts, was already proposed by Mott[24] and Schottky[25] in 1938 and 1940, and is sketched in Fig. 4–1. In a *Gedankenexperiment*, a metal-semiconductor contact may be formed by

Table 4–1. Barrier Heights and Ideality Factors Determined from *I-V* Characteristics of GaAs Schottky Barriers

	{100}*					{100}‡	
Metal	Φ_{Bn} (eV)	n	Φ_{Bp} (eV)	n	$\Phi_{Bn} + \Phi_{Bp}$ (eV)	Φ_{Bn} (eV)	n
Mg	0.62	1.03	0.55	1.44	1.17		
Al	0.85	1.07	0.61	1.06	1.46	0.85	1.05
Ti	0.83	1.03	0.56	1.37	1.39		
Cr	0.77	1.04	0.56	1.09	1.33	0.67	1.06
Mn	0.81	1.07	0.56	1.18	1.37	0.72	1.05
Fe	0.72	1.02	0.60	1.23	1.32		
Co	0.76	1.08	0.61	1.19	1.37		
Ni	0.77	1.15	0.62	1.13	1.39	0.77	1.06
Cu	0.96	1.05	0.45	1.24	1.41	0.87	1.05
Ru	1.09†	1.03†					
Pd	0.91	1.03	0.50	1.21	1.41	0.85	1.05
Ag	0.90	1.03	0.50	1.21	1.40	0.89	1.05
Sn						0.77	1.06
Tb	0.84	1.04					
Dy	0.85	1.09					
Er	0.85	1.05					
Au	0.89	1.03	0.50	1.15	1.39	0.92	1.05
Pb	0.80	1.03	0.52	1.13	1.32		
Bi	0.77	1.03	0.61	1.08	1.38		

*Refs. 18 and 19, except for Ru.
†Ref. 22.
‡Ref. 20.

approaching a metal and a semiconductor until they finally form an intimate, abrupt contact. The semiconductor is assumed to be doped n-type and, what is more important, to exhibit no surface states in the bandgap between valence-band top W_{vs}, and conduction-band bottom, W_{cs}. The bands are thus flat up to the surface. Since the work functions of the metal and the semiconductor generally differ, but the Fermi levels are equal in thermal equilibrium, an electric field exists in the vacuum gap between both surfaces. Consequently, the metal and the semiconductor have to carry a surface charge of equal density, but of opposite sign, i.e. charge neutrality requires

$$Q_{\text{met}} + Q_{sc} = 0 . \tag{4–3}$$

In the example chosen in Fig. 4–1, the sign is negative for the metal charge, Q_{met}, but positive for the semiconductor charge, Q_{sc}.

In principle, the electric field will enter into both the metal and the semiconductor. This penetration of the electric field is neglected in the metal since its screening length typically amounts to less than 0.1 nm due

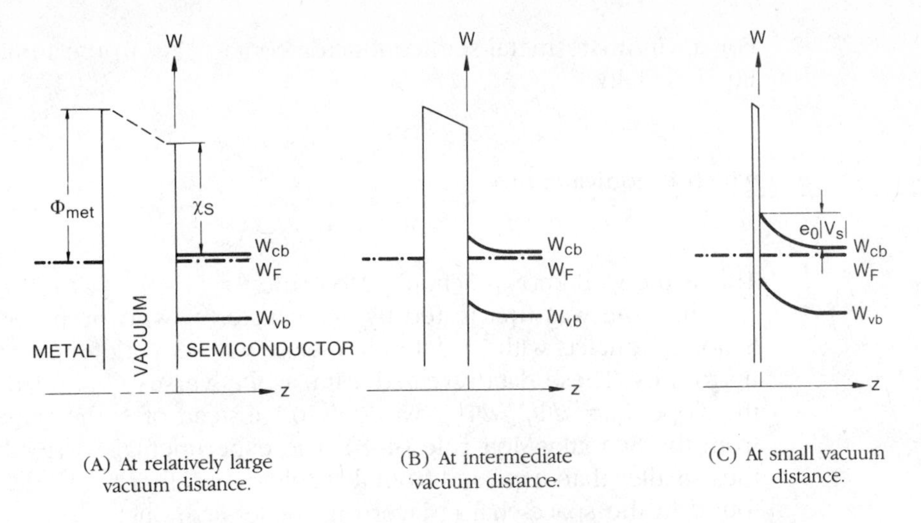

Fig. 4–1. Energy-band diagram of a metal and a semiconductor forming a metal-semiconductor contact with no interface states being present (Schottky-Mott model).

to the large electron density. This is quite different in nondegenerately doped semiconductors, where the density of mobile carriers is many orders of magnitude lower and the screening length, i.e. the Debye length, is correspondingly larger. For the example chosen in Fig. 4–1, the positive charge at the semiconductor surface results from a depletion of mobile electrons which leaves behind a layer of positively charged, static donors. The potential drop across such space-charge layers may be calculated by solving Poisson's equation (see, for example, Refs. 26 and 27), and in a depletion layer the space-charge density is found to be related to the surface band-bending $e_0|V_s|$ by

$$Q_{sc} = +(2\epsilon_s\epsilon_0 N_D e_0|V_s|)^{1/2} \tag{4–4}$$

to better than 10 percent for $e_0|V_s| > 5k_BT$.

Let us now assume that the metal and the semiconductor form a parallel-plate capacitor. The voltage drop across the vacuum gap of width d then is $(\Delta\Phi/e_0 - |V_s|)$, where $\Delta\Phi$ is the difference in the metal and the semiconductor work functions, and the surface charge density is

$$q/A = \frac{\epsilon_0}{e_0}\frac{\Delta\Phi - e_0|V_s|}{d}. \tag{4–5}$$

Combining Eqs. 4–4, 4–5, and the charge-neutrality condition (4–3), one obtains

$$\frac{(\Delta\Phi - e_0|V_s|)^2}{e_0|V_s|} = \frac{2e_0^2\epsilon_s N_D}{\epsilon_0}d^2. \tag{4–6}$$

For an intimate metal-semiconductor contact, i.e. in the limit of $d \cong 0$, Eq. 4–6 yields

$$\Delta\Phi - e_0|V_s| = 0\,, \tag{4–7}$$

which is equivalent to

$$\Phi_{Bn} = \Phi_{\text{met}} - \chi_s\,. \tag{4–8}$$

This is the well-known Schottky-Mott rule.[24,25]

This rule was first tested by Schweickert,[28] who prepared selenium Schottky contacts with 21 different metals, whose work functions varied by up to 3 eV. These data revealed a linear Φ_{Bp} versus Φ_{met} relationship but the slope $S_\Phi = d\Phi_{Bp}/d\Phi_{\text{met}}$ was only 0.1 instead of 1, the value expected from the Schottky-Mott rule (4–8). The experimental barrier heights are thus smaller than expected from that rule or, in other words, less charge is found in the space-charge layer on the semiconductor side than on the metal side of the junction. This means that on the semiconductor side of such contacts another reservoir of charge must exist in addition to the space-charge layer. Bardeen[29] was the first to introduce electronic interface states in this context.

Before Bardeen's model will be discussed in Sec. 4, the barrier heights compiled in Table 4–1 are also plotted against the work functions of the metals used. This is done in Fig. 4–2. The straight line in this figure gives

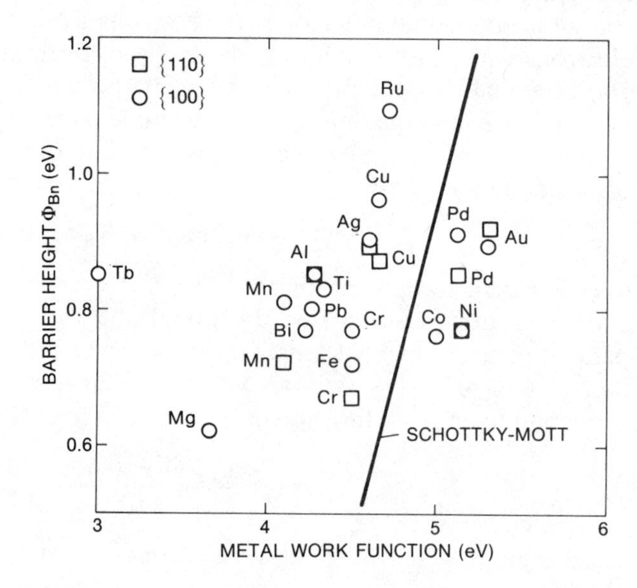

Fig. 4–2. The "GaAs data base" and the Schottky-Mott rule. *(Experimental barrier heights after Waldrop*[18,19] *(○) and Newman* et al.[20] *(□); metal work functions after Michaelson.*[76]*)*

the barrier heights expected from the Schottky-Mott rule (4–8). Here, an electron affinity of 4.05 eV was used. This value was determined with cleaved GaAs {110} surfaces[30] and also is the average of the more widely differing values evaluated for {100}-oriented surfaces of different stoichiometries and reconstructions (see Ref. 31). Obviously, the Schottky-Mott rule does not explain the chemical trends observed with the barrier heights of GaAs Schottky contacts.

4.4 Interface States in Schottky Barriers

Bardeen[29] suggested that electronic interface states lying in the region where the conduction band of the metal overlaps the bandgap of the semiconductor might absorb charge and, by this, shield the depletion layer. The wavefunctions of such interface states will rapidly decay into the semiconductor. Thus, an electric double layer of atomic dimensions δ_{eff} will form. This is schematically shown in Fig. 4–3. The charge-neutrality

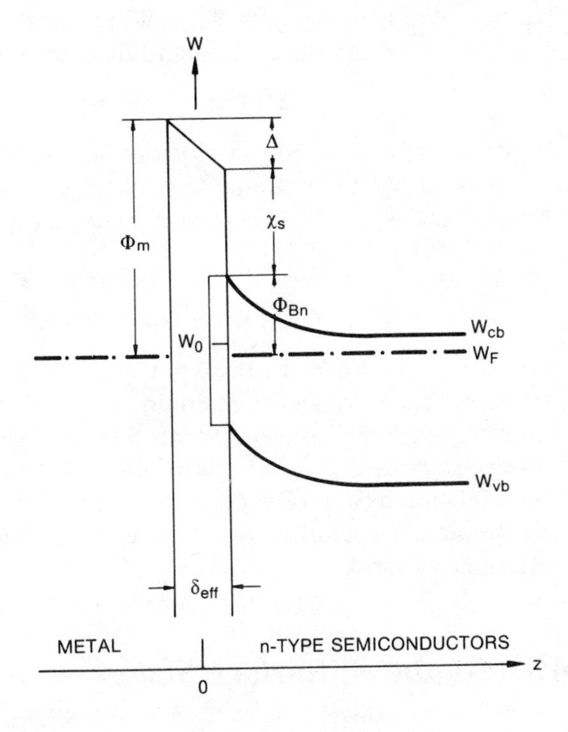

Fig. 4–3. Energy-band diagram of a metal-semiconductor contact with interface states.

condition (4–3) now has to be replaced by

$$Q_{met} + Q_{ss} + Q_{sc} = 0\,, \tag{4–9}$$

where Q_{ss} is the excess charge in the interface states. Cowley and Sze[32] and, later, Tejedor *et al.*[33] have evaluated the Φ_{Bn} versus Φ_{met} relation for Schottky barriers containing interface states. They postulated a continuum of interface states of constant density D_{vs} across the semiconductor bandgap. The character of the interface states was assumed to be acceptorlike above, and donorlike below, an energy, W_0, i.e. $Q_{ss} = 0$ for $W_F = W_0$. Therefore, W_0 is called the charge-neutrality level (CNL) of this continuum of interface states. With these assumptions, one obtains

$$\Phi_{Bn} = S_\Phi(\Phi_{met} - \chi_s - D_j) + (1 - S_\Phi)(W_{cs} - W_0) \tag{4–10}$$

with

$$S_\Phi = \left(1 + \frac{e_0^2}{\epsilon_0} D_{vs}\delta_{eff}\right)^{-1}. \tag{4–11}$$

The second term in the right-hand side of Eq. 4–11 is the ratio of the potential drop Δ across the interface double-layer to the deviation of the barrier height from $\Phi_{B0} = W_c - W_0$ when the Fermi level passes through the CNL of the interface states and their excess charge is zero:

$$\Delta/(\Phi_{Bn} - \Phi_{B0}) = e_0^2 D_{vs}\delta_{eff}/\epsilon_0\,. \tag{4–12}$$

D_j accounts for a surface-specific dipole contribution to the work function.

Two simple, but extreme, cases will be considered: $S_\Phi = 1$, or $D_{vs} = 0$, and $S_\Phi = 0$, or $D_{vs} = \infty$. Equation 4–10 then yields

$$\Phi_{Bn} = \Phi_{met} - \chi_s - D_j \quad \text{for } S_\Phi = 1 \tag{4–13}$$

$$= W_c - W_0 = \Phi_{B0} \quad \text{for } S_\Phi = 0\,. \tag{4–14}$$

As expected, the Schottky-Mott rule (4–8) is obtained for a metal-semiconductor contact containing no interface states. Large densities of interface states will pin the Fermi level at their charge-neutrality level independently of the metal work function. According to Eq. 4–11, a value $S_\Phi = 0.1$, for example, needs 2×10^{14} interface states per $eV \cdot cm^2$ when the dipole-layer width is assumed as $\delta_{eff} = 0.24$ nm, which is a typical interatomic distance.

4.5 Adsorbate-induced Surface States

Clean, and well-cleaved, GaAs {110} surfaces, for example, do not have any surface states within the bulk bandgap and, according to the charge-neutrality condition (4–3), the bands are thus flat up to the surface.[30,34–40]

For a review, the reader is referred to Ref. 31. Schottky contacts on such oriented surfaces, on the other hand, show depletion layers, i.e. band bending, and thus exhibit interface states within the bandgap. Now the more complex charge-neutrality condition (4–9) applies. Until now, no detailed theoretical calculations were carried out to explain the chemical trends observed. Thus, simple concepts are needed. These may be achieved from investigations of the charge rearrangement at adsorbate-covered semiconductor surfaces, i.e. from submonolayer physics. For this purpose, the adsorption of cesium, sulfur and chlorine on GaAs {110} surfaces will be considered and Cl_2: GaAs {110} will be discussed as an example.

4.5.1 Adsorbate-induced Surface Dipoles and Surface States

Figure 4–4 displays the variations of the work function of cleaved GaAs {110} surfaces, measured by using a Kelvin probe, as a function of chlorine

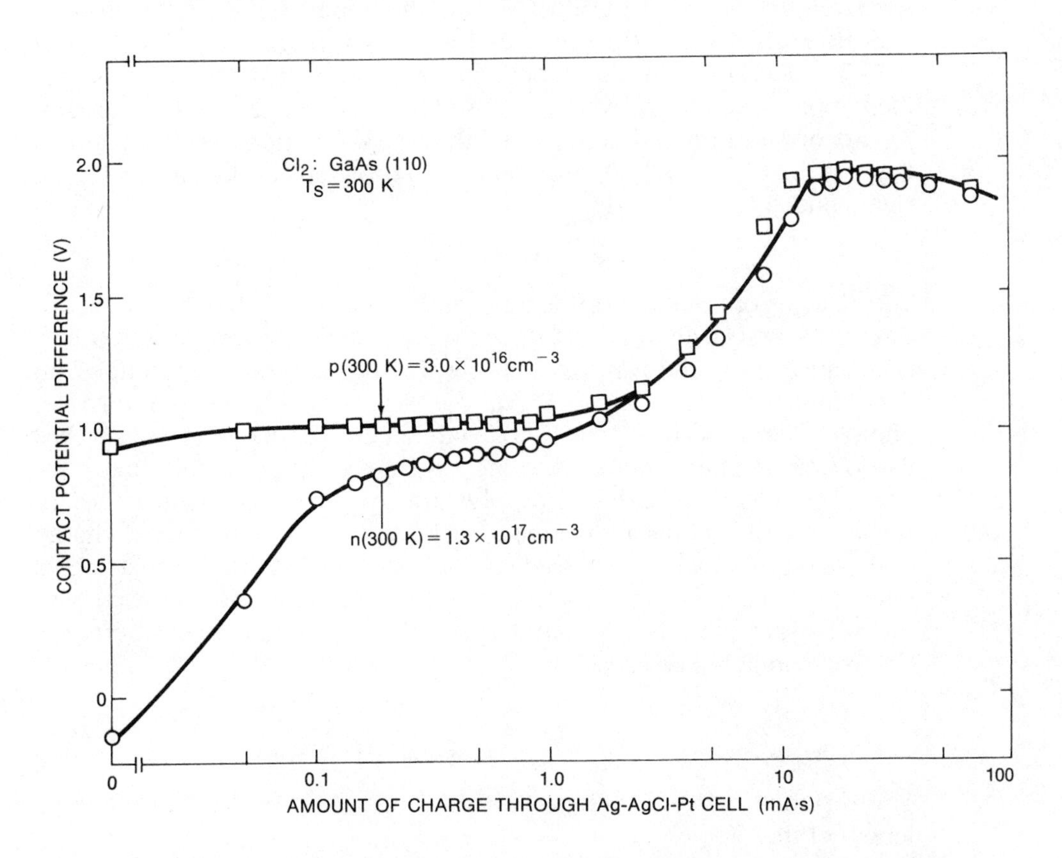

Fig. 4–4. Variations of the work function of cleaved GaAs {110} surfaces exposed to Cl_2 emitted from an electrolytic Ag-AgCl-Pt solid-state cell. (*After Troost* et al.[41])

exposure.[41] Chlorine molecules were provided by an electrolytic solid-state cell which consisted of an AgCl pellet between an Ag cathode and a perforated Pt anode. According to Faraday's law, the amount of charge passed through the cell is proportional to the number of chlorine molecules released at the anode.

The work function of a semiconductor may be written as

$$\Phi = W_{vac} - W_F = I + (W_{vs} - W_F), \tag{4–15}$$

where $I = W_{vac} - W_{vs}$ is the ionization energy. Provided the bands are flat up to the surface, i.e. $W_{vs} = W_{vb}$, the difference $\Phi_p - \Phi_n$ of the work functions measured with samples doped p- and n-type, respectively, amounts to

$$\Phi_p - \Phi_n = W_g - (W_p + W_n), \tag{4–16}$$

where $W_p = W_F - W_{vb}$ and $W_n = W_{cb} - W_F$ are the positions of the Fermi level relative to the respective band edges for p- and n-type doping, respectively, which may be evaluated from the doping densities. In Fig. 4–4, the difference $\Phi_p - \Phi_n$ deviates by 0.17 eV from the value 1.43 − (0.145 + 0.024) eV. This deviation is attributed to a surface band-bending at the sample doped n-type which is caused by cleavage-induced surface states.

Adsorbate-induced changes of the work function, as displayed in Fig. 4–4 for Cl_2: GaAs {110}, may be due to variations of both terms on the right-hand side of Eq. 4–15, i.e.

$$\Delta\Phi = \Delta I - \Delta(W_{vs} - W_F). \tag{4–17}$$

The changes of the ionization energy and of the surface band-bending are caused by adsorbate-induced surface dipoles and adsorbate-induced surface states within the gap, respectively. To separate the contributions on the right-hand side of Eq. 4–17, ΔI was determined by using ultraviolet photoemission spectroscopy (UPS). As the upper panel of Fig. 4–5 reveals, the ionization energy is found to increase as a function of chlorine exposure. The adsorbate-induced dipoles are thus oriented so that the adsorbed chlorine atoms are charged negatively. The positive countercharge of these dipoles then has to reside in the wavefunction tails of the adatoms into the semiconductor. This charge transfer between the adsorbate and the substrate may be estimated if the dipole layer is described as a charged parallel-plate capacitor:

$$Q_{ad} = -\frac{\epsilon_0 \Delta I}{e_0 t_{eff}}. \tag{4–18}$$

The distance between the two plates is approximated by an effective length of the dipoles:[13]

$$t_{eff} = r_{cov} + t_{vs}/\epsilon_s, \tag{4–19}$$

where $r_{cov} = 0.099$ nm is the covalent radius of chlorine, $t_{vs} \cong 0.2$ nm is

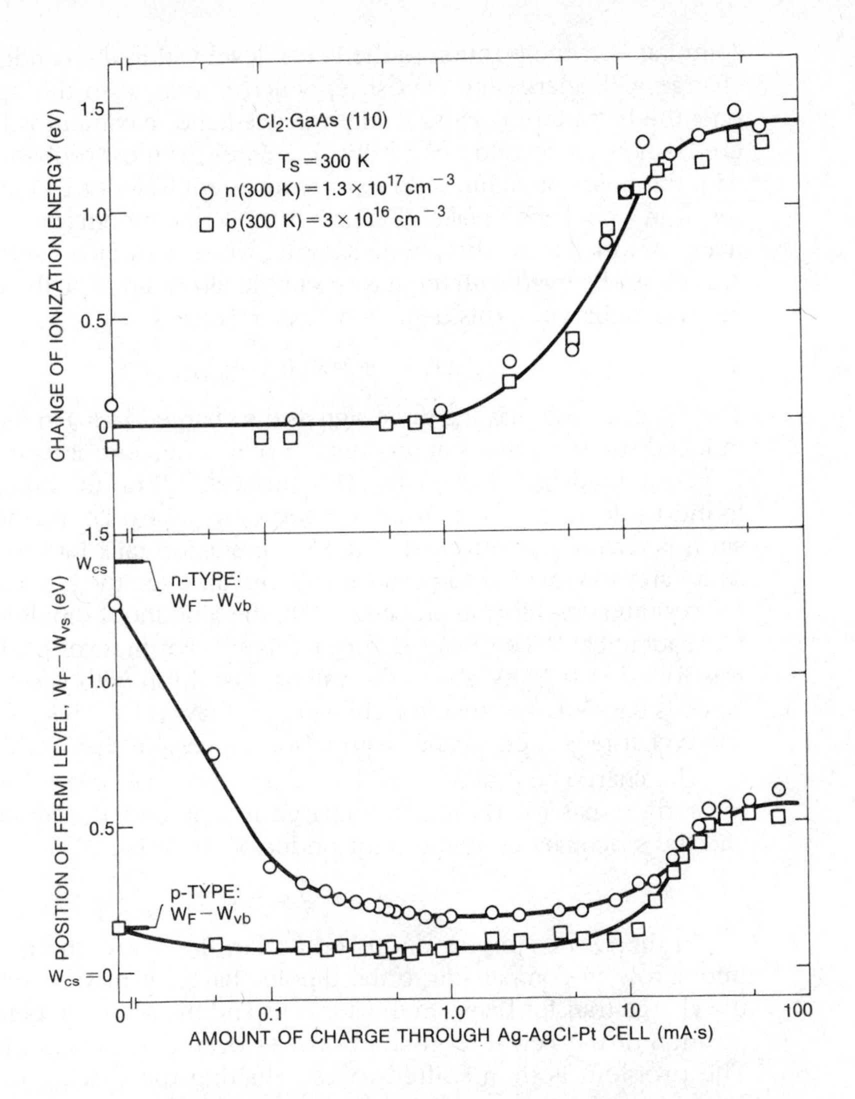

Fig. 4–5. Variation of the ionization energy and of the surface band-bending of cleaved GaAs {110} surfaces exposed to Cl_2 emitted from an electrolytic Ag-AgCl-Pt solid-state cell. *(After Troost* et al.[41])

the decay length of surface states in GaAs (see Sec. 6), and ϵ_s is the dielectric constant of GaAs. For $\Delta I \cong 1$ eV, one then obtains

$$Q_{ad} \cong -5 \times 10^{13}\ e_0/\text{cm}^2 . \qquad (4\text{–}20)$$

The position of the Fermi level, relative to the top of the valence band, as determined from the measured changes of the work function and of the ionization energy, is plotted as a function of the chlorine exposure in the lower panel of Fig. 4–5. The Fermi level shows a behavior which is called "pinning," i.e. the position of the Fermi level within the bandgap does not change with increasing exposure, which is related to the uptake. Here, only the first pinning close to the valence-band maximum will be considered, since the second one (at $W_{vs} + 0.55$ eV) is most probably correlated with the onset of etching. Up to exposures equivalent to 10 mA · s, which result in a chlorine uptake of approximately one monolayer, the bands remain almost flat for the p-type sample, while a surface band-bending of 1.24 eV is observed with the n-type sample. From Eq. 4–4, the corresponding space charge in this depletion layer results as

$$Q_{sc} = +2 \times 10^{12}\, e_0/\text{cm}^2 . \qquad (4\text{–}21)$$

This space charge has the same sign as the charge (4–20) in the adsorbate-induced surface states, but amounts to only 4 percent of its magnitude.

Cesium adsorbed on GaAs {110} surfaces,[42–44] on the other hand, was found to decrease the ionization energy by 3.5 eV, i.e. the adsorbed cesium is charged positively and the wavefunction tails into the GaAs substrate are now carrying a negative charge. Although the signs are different for cesium and chlorine on GaAs {110}, the amount of dipole charge is almost identical.[45] The main difference is the Fermi-level position, which was found at 0.73 eV above the valence-band top with cesium. This distance is 0.5 eV larger than for chlorine on GaAs {110}. This means that the space charge is even smaller with adsorbed cesium than with chlorine.

The charge-neutrality condition, at an adsorbate-covered semiconductor surface, has to consider the charge in adatom-induced surface states and the space charge in the semiconductor, i.e. it is

$$Q_{ss}^{ai} + Q_{sc} = 0 . \qquad (4\text{–}22)$$

In the two examples discussed so far, the space charge Q_{sc} plays a minor role in comparison to the dipole charge. It may be supposed that the charge transfer between the adatoms and the substrate determines the position of the Fermi level at adsorbate-covered semiconductor surfaces. The problem is then shifted to establishing the driving force for the charge transfer between the adatoms and the substrate and to determining the nature of the adsorbate-induced surface states.

4.5.2 Charge Transfer at Adsorbate-covered Surfaces

The covalent bonds between the adsorbates, cesium or chlorine, and the gallium-arsenide substrate exhibit a partly ionic character. Following Pauling's concept,[46] the ionicity of a covalent bond in a single-bonded diatomic molecule *A-B* is determined by the difference $X_A - X_B$ of the atomic electronegativities. Gallium arsenide is characterized by an electronegativity of $X_{GaAs} = 2$, which is the geometric mean[47] of the electronegativities of gallium and arsenic. For the two adsorbates considered here, the electronegativity differences $X_{GaAs} - X_{Cs} = 1.2$ and $X_{GaAs} - X_{Cl} = -1.2$ have opposite signs, but the same absolute value. Therefore, Cs and Cl are predicted to become positively and negatively charged, respectively, to the same extent when bonded to GaAs. And indeed, this is what has been observed, as discussed in the preceding section. For a rough estimate of the total charge transfer between adsorbate and substrate, each adatom will be assumed to form a single bond with the substrate. The charge transfer within an isolated adatom-substrate molecule is then given by

$$\Delta q_1 = 0.16|X_{ad} - X_{sub}| + 0.035|X_{ad} - X_{sub}|^2 .$$

This scale is a revised version[48] of Pauling's ionicity of single bonds in covalent diatomic molecules. With increasing coverage, however, these adatom-substrate dipoles mutually interact via their electrostatic fields,[49] which gives a depolarization. At saturation coverage, which is 0.6 of a monolayer for Cs on GaAs, the dipole moment is reduced to approximately 0.4 of its initial value.[44,45] For Cs and Cl adsorbed on GaAs {110}, this then gives

$$Q_{ad} \cong \pm 0.4\Delta q_1 e_0 \times 0.6\sigma_{110} = \pm 5.3 \times 10^{13} e_0/\text{cm}^2 , \qquad (4\text{–}23)$$

where $\sigma_{110} = 8.85 \times 10^{14}\ \text{cm}^{-2}$ is the total density of sites in a GaAs {110} plane. This value (4–23) is in good agreement with the charge density (4–20) in adatoms, estimated from the change in ionization energy experimentally observed.

4.5.3 Virtual Gap States of the Complex Band Structure of Semiconductors

The complex band structure of a semiconductor will be explained for a one-dimensional solid, i.e. a linear chain. For details, the reader is referred to the early papers by Maue[50] and Goodwin[51] and the book by Garcia-Moliner and Flores.[52]

Let us assume a linear chain with lattice constant a and the simple periodic potential

$$V = V_0 + 2V_1 \cos\left(\frac{2\pi}{a} z\right) . \qquad (4\text{–}24)$$

The resulting band structure is schematically given in Fig. 4–6. In the valence band and the conduction band, which are separated by an energy gap of width $2|V_1|$, the wavefunctions are Bloch waves with real wavevector $\mathbf{k}$. Schrödinger's equation also allows solutions with complex wavevectors. Such solutions are physically meaningless in the bulk since they would diverge for $z \to +\infty$ or $-\infty$ depending on the sign of the imaginary part of the wavevector. For a finite sample, however, surface states are conceivable. The wavefunctions of these are exponentially decaying into the semiconductor. The energy level of such a surface state would then be found within the "bulk" bandgap.

Solving Schrödinger's equation for complex wavevectors,

$$k_{vs} = \pi/a + i\gamma\,, \tag{4–25}$$

gives the dispersion relation for energies within the bandgap

$$W - V_0 = W_1 \pm \sqrt{V_1^2 - 4W_1\left(\frac{\hbar^2}{2m_0}\gamma^2\right)}\,, \tag{4–26}$$

with

$$W_1 = \frac{\hbar^2}{2m_0}\left(\frac{\pi}{a}\right)^2.$$

The wavefunctions

$$\psi_{vs} \propto e^{\gamma z}\cos\left(\frac{\pi}{a}z + \phi\right), \quad z < 0\,, \tag{4–27}$$

are exponentially decaying along the chain, and their phase varies across the gap as

$$\sin 2\phi = +\frac{1}{V_1}\sqrt{V_1^2 - (W - V_0 - W_1)^2}\,. \tag{4–28}$$

The density of states is obtained as

$$D_{vs}(W)\,dW = \frac{1}{\pi a^2}\frac{d\phi}{dW}\,dW = \frac{1}{\pi a^2}[V_1^2 - (W - V_0 - W_1)^2]^{-1/2}\,dW\,. \tag{4–29}$$

These states, which are solutions of Schrödinger's equation for complex wavevectors, are called the virtual gap states (ViGS) of the complex band structure. Such solutions may also be found for three-dimensional semiconductors.[53] For a linear chain, which is considered here, the variations of the imaginary part of the wavevector and of the density of states of the ViGS are schematically shown in Fig. 4–6.

With a finite chain, a real surface state is found when its wavefunction (Eq. 4–27), which is exponentially decaying into the chain, can be fitted to a tail exponentially decreasing into vacuum. With the matching plane at $z = +a/2$ and for the bulk potential (Eq. 4–24), this boundary condition can be satisfied only when the Fourier coefficient opening the gap at the boundary of the Brillouin zone is positive, i.e. $V_1 > 0$. As Fig. 4–7 shows, the potential has to be attractive at the surface for a real surface state to

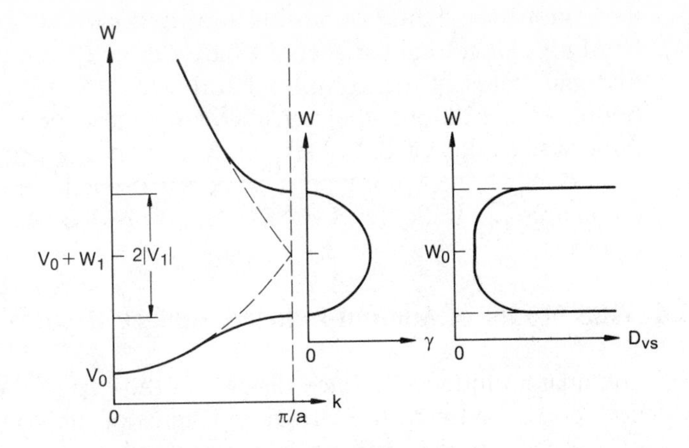

Fig. 4–6. Complex band structure and density of virtual gap states for a linear chain.

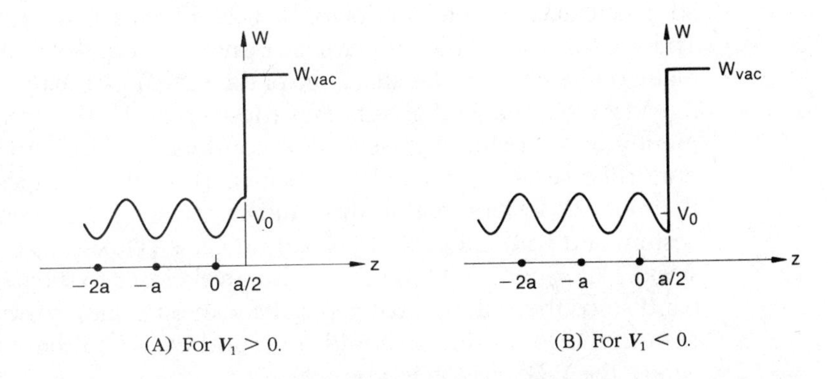

Fig. 4–7. Energy diagram of a finite linear chain for cosine potentials of different sign in the bulk.

exist (see also Ref. 54). The energy of this real surface state is given by

$$W_{ss} - V_0 = W_1\left(\frac{W_{vac} - V_0}{W_{vac} - V_0 - V_1}\right)^2 - V_1\left(\frac{W_{vac} - V_0}{W_{vac} - V_0 - V_1}\right), \tag{4–30}$$

and the respective imaginary part of the wavevector γ_{ss} is then obtained by inserting W_{ss} from Eq. 4–30 into Eq. 4–26.

For a Fourier coefficient $V_1 > 0$, the energy of the surface state shifts across the bandgap, as a function of the depth of the potential well $W_{vac} - V_0$. The wavefunctions of the surface states are derived from the bulk bands and, therefore, they are predominantly acceptorlike closer to the conduction-band bottom but mostly donorlike nearer to the valence-band top. Since real surface states are derived from the continuum of virtual gap states of the complex band structure according to the specific boundary conditions, their character may also be thought to vary in the same way as the ViGS. Consequently, the energy level where they change from donor to acceptor type is called the branch point or, more intuitively, the charge-neutrality level (CNL). Here, it is denoted W_0.

4.5.4 ViGS Model of Adatom-induced Surface Dipoles

The adatom-induced surface dipoles in the gap of the semiconductor are now correlated with the virtual gap states of the complex band structure of the semiconductor. In this, we are following an early suggestion of Heine.[13] He proposed that the wavefunctions of the adatoms are decaying into the semiconductor and identified these tails with the virtual gap states. They then take up the charge chemically transferred between the specific adsorbate and the semiconductor. This is shown schematically in Fig. 4–8A. The tails in the semiconductor are given by Eq. 4–27. As has been explained in Sec. 4.5.3, the character of the virtual gap states, and thus also of the adsorbate-induced gap states, varies across the bandgap from predominantly acceptorlike (close to the conduction band) to predominantly donorlike (close to the valence band). Thus, the chemical trends of the Fermi-level position within the bandgap, which have been observed with cesium and with chlorine adsorbed on GaAs {110} surfaces, can be understood: Chlorine should pin the Fermi level closer to the top of the valence band since there the virtual gap states exhibit donor character, while with cesium the Fermi level should be found closer to the conduction band where the ViGS are of acceptor type.

For a more detailed analysis, the positions of the Fermi level, relative to the top of the valence band, as determined with Cs, Cu, Ag, Au, S and Cl adsorbed on GaAs {110} surfaces[41,43,55,56] are plotted versus the difference $X_{ad} - X_{GaAs}$ of the Pauling electronegativities in Fig. 4–9. Here, the difference in electronegativities represents the chemical charge transfer from the adsorbate into the adsorption-induced gap states which is not otherwise accessible. The experimental results do indeed follow the chemical trend predicted from the present model. The data points for Cs, S, and Cl are exactly fitted by a straight line ($r = 0.999$) and they are thus predicting the charge-neutrality level of the adsorbate-induced gap states to lie at $X_{ad} - X_{GaAs} = 0$ by 0.46 eV above W_{vs}. This value agrees remarkably well with the CNL of the ViGS as calculated by Tersoff[57] for GaAs. Earlier estimates by Tejedor and Flores,[58] who used a one-dimensional model similar to the one discussed in Sec. 4.5.3, gave $W_0 = W_{vs} + 0.55$ eV. This close agreement between the values calculated and determined from experimental

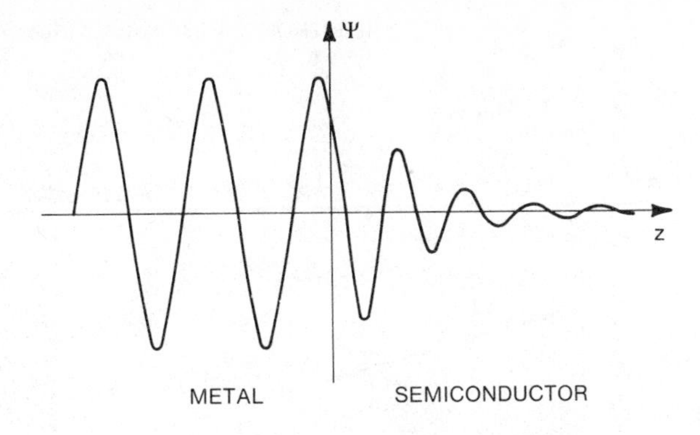

(A) Metal wavefunction.

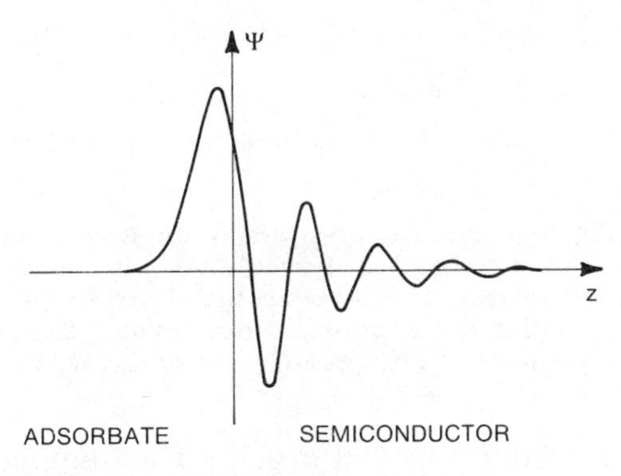

(B) Adsorbate wavefunction.

Fig. 4–8. Tailing of adsorbate and metal wavefunctions into a semiconductor.

data is strongly supporting the model presented here. It is also worth mentioning that the results reported for Cu and Ag are also close to that line defined by the data points for Cs, S, and Cl together with the theoretical CNL. For Au the experimental data point is off from that straight line by 0.15 eV. This is most probably caused by disorder which is introduced in the surface region due to intermixing of gold and substrate atoms. Such problems will be discussed in Secs. 4.8 and 4.9.

4.6 ViGS Model of Metal-induced Gap States in Schottky Barriers

At clean metal surfaces, the wavefunctions of the electrons are tailing into vacuum. When a metal is overlayed with an insulator or semiconductor

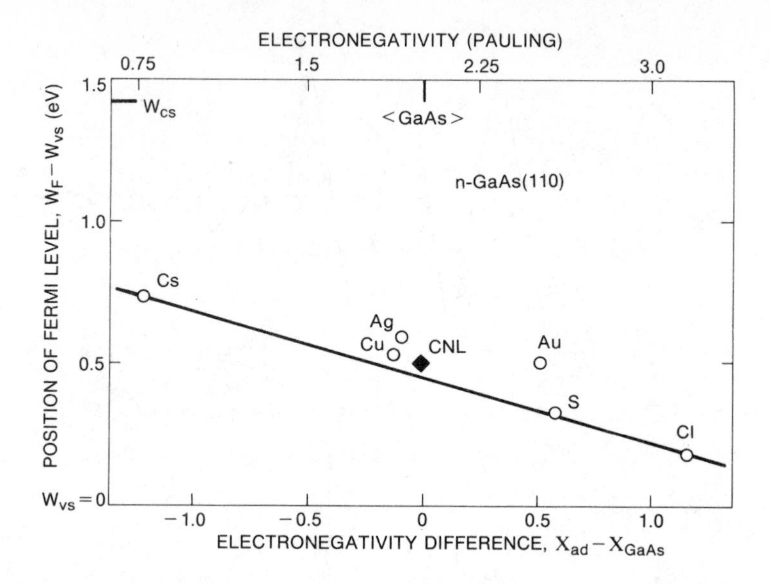

Fig. 4–9. Pinning positions of the Fermi level relative to the top of the valence band at cleaved GaAs {110} surfaces covered with various adsorbates plotted versus the electronegativity of the adatoms. ***(Data after Troost* et al.,[41] *Spicer* et al.,[43] *Blömacher, Koenders, and Mönch,*[55] *Chin* et al.[56]; *CNL of the ViGS of GaAs after Tersoff*[57]*)***

the wavefunctions will match across the interface at energies where the conduction band of the metal overlaps the valence bands of the insulator or semiconductor. However, the metal wavefunctions will tail into the semiconductor or insulator in that energy range where the metal conduction band is facing a gap in the semiconductor or insulator band-structure. Since the valence bands of semiconductors generally exhibit a larger width than metal conduction bands, semiconductor valence-electrons are expected to tail into the metal at energies below the metal conduction band as well.

With respect to the barrier heights in Schottky contacts, the tailing of the metal wavefunctions into the semiconductor is of interest at energies where the top of the metal conduction band overlaps the bandgap between valence and conduction band of the semiconductor. Heine[13] has pointed out that these tails of the metal wavefunctions are described by the virtual gap states of the complex band structure of the semiconductor. This is schematically explained in Fig. 4–8B. These metal-induced gap states, as they also have been called,[59] are an intrinsic property of the semiconductor, and they are identical with the tailing states of the adatom wavefunctions discussed in the preceding section (Fig. 4–6). Therefore, the model, where the chemically driven charge transfer across the interface populates the virtual gap states of the semiconductor and by this determines the interface position of the Fermi level within the semiconductor bandgap, can also be applied to metal-semiconductor contacts.

Figure 4–10 shows what is expected from this model. Let us assume that the density of the virtual gap states, and thus of metal-induced gap states, is constant across the bandgap. At least for one-dimensional models, this is a fair assumption as Eq. 4–29 and Fig. 4–6 show. With the Fermi level at the interface being close to the conduction band, the charge Q_{vs}, which may be transferred to the virtual gap states, is negative and large. When the Fermi level is moved towards and then below the charge-neutrality level of the ViGS the amount of charge Q_{vs} first decreases, then vanishes for $W_F = W_0$ and eventually becomes increasingly positive. The charge transferred from the metal into the virtual gap states has not been calculated and, therefore, it is again described by the difference $X_{met} - X_{GaAs}$ of the electronegativities of the metals and the semiconductor, which is GaAs here. Inversion of the W_F versus $(X_{met} - X_{GaAs})$ diagram then gives the Φ_{Bn} versus $(X_{met} - X_{GaAs})$ plot expected from the ViGS model of adsorbate-induced gap states.

In Fig. 4–11, the barrier heights listed in Table 4–1 are plotted versus the difference $X_{met} - X_{GaAs}$. Here, the electronegativities given by Miedema[60,61] are used since they were derived from an analysis of chemical trends of various properties of metal alloys and intermetallic compounds. The Pauling and the Miedema scales are related by

$$X_{Mied} = 1.93X_{Paul} + 0.87, \qquad (4\text{–}31)$$

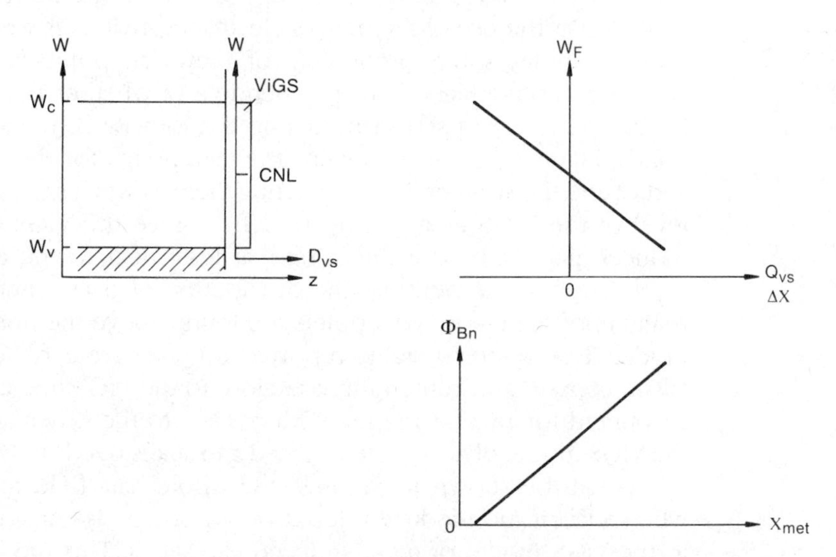

Fig. 4–10. Variation of the charge in virtual gap states as a function of Fermi-level position within the bandgap, and expected barrier height as a function of metal electronegativity.

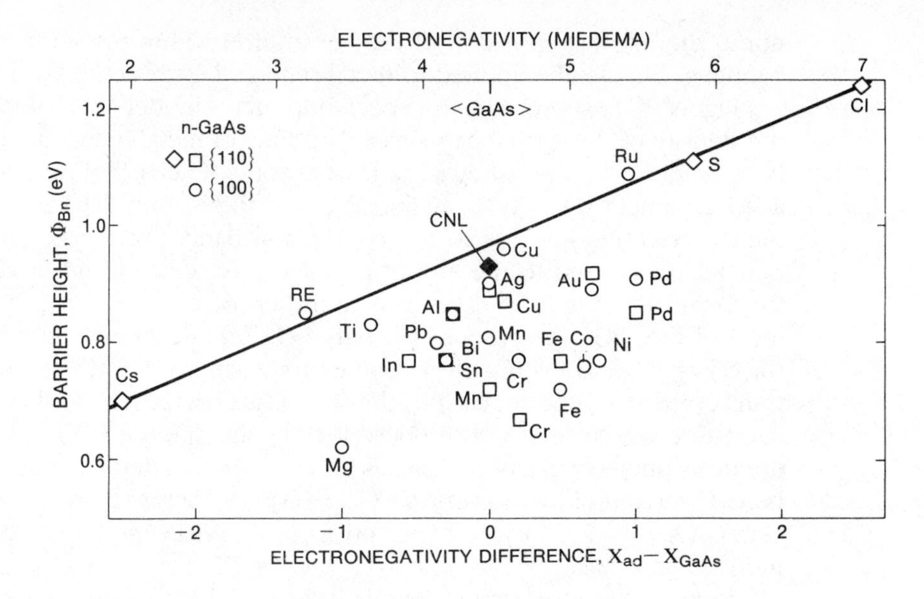

Fig. 4–11. Application of the ViGS model of metal-induced gap states to the "GaAs data-base": barrier heights plotted versus the difference of metal and GaAs electronegativities and comparison with the "ViGS-line" from Fig. 4-9.

with $r = 0.92$. Since metal-semiconductor bonds are to be described here, it is felt that the use of Miedema's electronegativities is well justified. However, except for some minor shifts of a few data points along the abscissa, the conclusions reached are independent of whether the Miedema or the Pauling scales are used in modeling the ionicity of metal-semiconductor bonds. Figure 4–11 also contains the data points for the adsorbates Cs, S, and Cl and the straight line connecting them as well as the charge-neutrality level of the ViGS as given by Tersoff.[57] Since the ViGS-model of metal-induced gap states is a linear one, at least near to the charge-neutrality level, it cannot account for the distribution of data points. But—what is really important—no data points are found above the line defined by that model. The barrier heights reported for the rare-earth metals, titanium, silver, copper, and ruthenium, are close to the ViGS line and are thus well accounted for by that model. With respect to the other Schottky contacts, the ViGS model of adsorbate-induced gap states needs to be supplemented.

As will be shown in Sec. 4.9, the whole data field may be explained when a fabrication-induced defect of donor-type is considered in addition to the ViGS model of metal-induced gap states. This model, however, differs from the Unified Defect Model of Schottky barriers,[62] which will be presented in the next section.

Before the discussion turns to the influence of defects in Schottky barriers, the few theoretical studies which were devoted to calculate electronic properties of metal-semiconductor interfaces will be considered. Unfortunately, the chemical trends of the charge distribution at metal-semiconductor interfaces were not systematically examined. Louie *et al.*[62,63] have investigated the electronic structure of the Al-contacts to four semiconductor surfaces, namely {111} surfaces of Si and {110} surfaces of GaAs, ZnSe, and ZnS. They used a self-consistent pseudopotential approach and modeled the Al by jellium of appropriate electron density. For all four interfaces, they found a transfer of electronic charge across the interface from the metal to the semiconductor or, in other words, an interface dipole. On the semiconductor side, they obtained metal-induced gap states, which are decaying exponentially from the interface into the semiconductor and contain the charge transferred. The sign of the charge transfer is easily understood, since Al is less electronegative than the semiconductors. For example, $X_{Al} - X_{GaAs} = -0.4$ on the Pauling scale and $X_{Al} - X_{GaAs} = -0.25$ on the Miedema scale.

The decay lengths and the density of states of the MIGS and the barrier heights $\Phi_{Bp} = W_F - W_{vs}$, as given by Louie *et al.*,[63] are listed in Table 4–2. Their values are compared with the data evaluated earlier by Louis *et al.*[64] and Tejedor *et al.*,[33] who have used a one-dimensional model, and the energies of the charge-neutrality level of the ViGS, as computed by Tersoff.[57,66] First of all, both the decay lengths and the densities of states are found to correlate with the widths of the bandgaps, as to be expected from Eqs. 4–26 and 4–29 for a simple one-dimensional model. Second, the theoretical model of Louie *et al.*[63] gives the position of the Fermi level,

Table 4–2. Decay Length γ^{-1}, Density of States D_{vs}, and Barrier Height Φ_{Bp} and Position of Charge-Neutrality Level $W_0 - W_{vs}$, Respectively, of Metal-Induced Gap States in Al-Semiconductor Contacts and of Virtual Gap States of the Semiconductors

		MIGS by Al†			ViGS‡,§		
	W_g* (eV)	γ^{-1} (nm)	D_{vs} (10^{14} eV^{-1} cm^{-2})	Φ_{Bp} (eV)	γ^{-1} (nm)	$D_{vs}(W_0)$ (10^{14} eV^{-1} cm^{-2})	$W_0 - W_{vs}$§ (eV)
Si	1.12	0.3	4.5	0.52	0.41	—	0.35
GaAs	1.43	0.28	5.0	0.63	0.24	3.7	0.5
ZnSe	2.7	0.15	2.0	2.48	0.10	2.28	1.7
ZnS	3.56	0.10	1.4	3.18	0.09	1.96	1.4

*Ref. 65.
†Ref. 63.
‡Refs. 33 and 64.
§Refs. 57 and 66.

in the Al-semiconductor contacts considered, above the charge-neutrality levels of the ViGS as evaluated by Tersoff.[57,66] This finding is in agreement with the concept presented in Sec. 4.5.4, that the chemically driven charge transfer across an adsorbate-semiconductor interface into adsorbate-induced gap states (which are identified as the ViGS of the complex band structure of the semiconductor) determines the interface position of the Fermi level in the semiconductor bandgap.

Figures 4–9 and 4–11 give a linear relationship between the barrier height, when it is determined by the ViGS of the semiconductor, and the metal electronegativity. For the slope

$$S_X = d\Phi_{Bn}/dX_{met}, \quad (4\text{–}32)$$

one obtains $S_X = 0.22$ from Figs. 4–9 and 4–11. Louis *et al.*,[64] as well as Louie *et al.*,[63] have estimated this slope parameter as 0.1, which value is a factor of two smaller than the experimental one. In this respect, more detailed theoretical investigations are highly desirable. We will return to this problem in Sec. 4.10.

4.7 Unified Defect Model

In 1979, Spicer *et al.*[15] noted that the deposition of Cs, Al, Ga, and In, as well as the adsorption of oxygen on GaAs {110} surfaces, induce almost the same positions of the Fermi level within the bandgap, $W_F - W_{vs} = 0.79$ eV and 0.55 eV on samples doped n- and p-type, respectively. Similar results were reported for thin overlayers on other compound semiconductors, such as InP and GaSb. These findings were interpreted by the Unifed Defect Model.[67] It assumes that native defects are formed in the semiconductor during the deposition of the metal and that the energy levels of these defects are then "pinning" the Fermi level. Vacancies,[68] as well as antisite defects,[69] which are considered to be more typical defects in III-V compound semiconductors, have been proposed in this context. The Unified Defect Model has become very popular. Its basic implications will be briefly outlined. For a more extended review the reader is referred to Ref. 70.

As the numbers given above indicate, the deposition of metals results in the formation of depletion layers. When the semiconductor is doped n-type, for example, the space charge, which is given by Eq. 4–4, is positive. The Unified Defect Model assumes that this space charge is balanced by an excess charge, Q_{ss}, located only in those defects. This means that the charge-neutrality condition

$$Q_{sc} + Q_{ss} = 0, \quad (4\text{–}33)$$

requires no charge transfer across the interface. At the metal interface to a semiconductor doped n-type, the charge Q_{ss} may in principle be provided by acceptor-type defects. The occupancy of discrete interface states of acceptor character at energy W_{ss}^{A} is then given by

$$Q_{ss}^{A} = -e_0 N_{ss}^{A}\left[g_A \exp\left(\frac{W_{ss} - W_F}{k_B T}\right) + 1\right]^{-1}, \qquad (4\text{–}34)$$

where their total density is N_{ss}^{A} and g_A is their degeneracy. By inserting Eqs. 4–4 and 4–34 in the charge-neutrality condition (Eq. 4–33), the variation of the Fermi level within the bandgap may be calculated as a function of the density of interface defects. This is shown in Fig. 4–12 for typical parameters. This curve shows the behavior that was experimentally observed for the surface band-bending, for example, at GaAs {110} surfaces covered with Al, Ga, In, and also Ge. It, therefore, was concluded that the density of surface states or defects and of adatoms are interrelated by[71,72]

$$N_{\text{def}} = N_{ss}^{A,D} = q^{A,D} N_{\text{ad}}. \qquad (4\text{–}35)$$

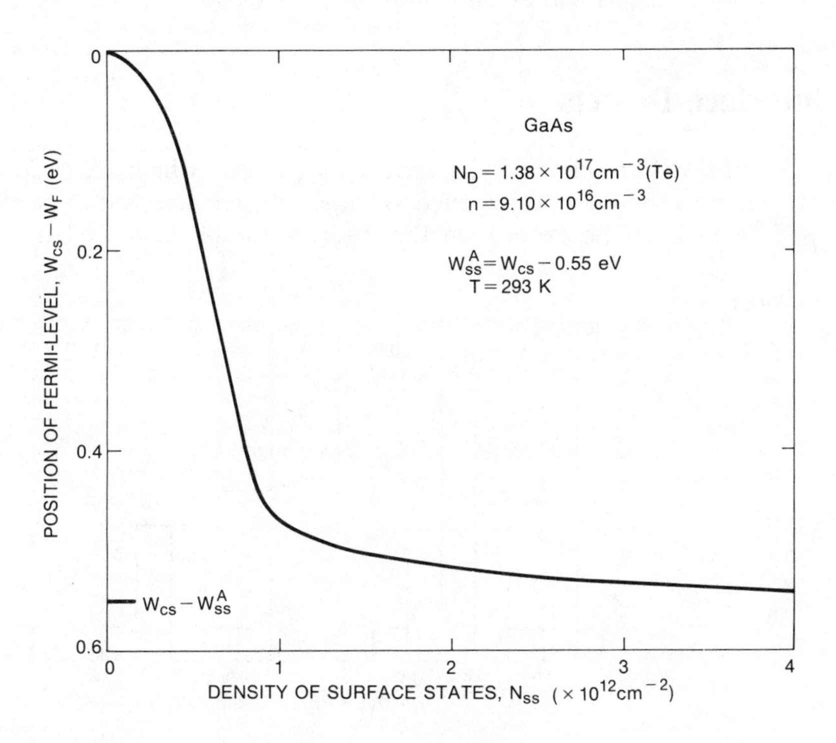

Fig. 4–12. Surface band-bending as a function of density of discrete surface states.

The superscript D denotes defects of donor character. For GaAs {110}, the experimental data could be well described by[71,72]

$$q^A = q^D = 0.06 \pm 0.04 . \tag{4–36}$$

In the Unified Defect Model, charge transfer across the interface, which is caused by chemical bonding, was not considered. While this might be tolerable for Ge, In, Ga, and even for Al on GaAs, it certainly is incorrect for Cs, S, and Cl. Furthermore, the rather new data for chlorine and sulfur adsorbed on GaAs clearly show that there is no unique pinning position of the Fermi level, which is independent of the adsorbates.

In addition, the data set listed in Table 4–1 gives no evidence for a preferred barrier height. In Fig. 4–13 the barrier heights are classified in intervals of a tenth of an electronvolt. No "clustering" at specific values is discernible, but the same broad distributions between 0.65 and 1.09 eV remain for both {100}- and {110}-oriented substrates.

For these reasons, it is concluded that defect levels do not *primarily* determine the barrier heights at semiconductor interfaces and adsorbate-covered surfaces. This, however, does not mean that defects do not play a role at all, as will be shown in the next section.

4.8 Interface Defects

The ViGS model of metal-induced gap states, which are occupied according to the chemically driven charge transfer across the interface, cannot explain all the experimentally observed variations of the barrier heights in

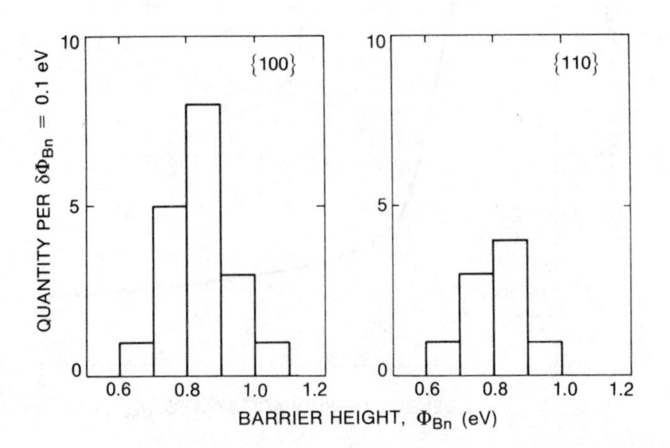

Fig. 4–13. Distribution of the "GaAs data base" after classification in intervals of barrier heights of 0.1 eV

Schottky contacts, as was demonstrated by Fig. 4–11. This model will now be supplemented by a defect.

Let us assume that we have additional defects of donor type close to the interface or surface. The charge-neutrality conditions (4–9) have to be modified by adding the charge density Q_{sd} in the defects:[45]

$$Q_{\text{met}} + Q_{ss} + Q_{sd} + Q_{sc} = 0. \tag{4–37}$$

Figure 4-14 now explains the influence of such defects on the position of the Fermi level or, what is equivalent, on the barrier height in Schottky contacts when the metals are varied. The donor-type defect is assumed to have an energy level above the charge neutrality level of the ViGS. As long as that much negative charge is transferred to the semiconductor to keep the Fermi level above the defect level, the defects are all neutral and the Fermi-level position is only determined by the virtual gap states. With increasing metal electronegativity, however, the Fermi level will approach the defect level and the defects gradually become charged positively. With the electronegativity of the metal increasing further, the Fermi level gets pinned at the position of the defect level. When all the defects are eventually charged, the virtual gap states will again take up charge and will again determine the position of the Fermi level as a function of electronegativity of the metal. The line drawn in full now runs in parallel to the dashed one, which is followed when no defects are present (see Fig. 4–10). The shift is directly proportional to the density of defects. The lower panel of Fig. 4–14 displays the variation of the barrier height as a function of the metal electronegativity when the Schottky barriers contain defects of donor type. The dashed line would again be followed with no donor-type defects present at all. This means that the introduction of donor-type defects at the interface of Schottky contacts reduces the barrier heights, which are larger than $\Phi'_{Bn} = W_{cs} - W_{sd}$ at an interface free of defects.

This problem has been theoretically approached by Zhang *et al.*[73] At a jellium-silicon interface, they replaced the first Si layer, which is closest to the boundary, by a layer of "donor" atoms and then self-consistently calculated the charge distribution at the interface by using an *ab initio* pseudopotential approach. They found the Fermi level shifted towards the conduction-band bottom by 0.28 eV, which corresponnds to an equivalent lowering of the barrier height for samples doped n-type. They concluded that "the change in the Schottky-barrier height arises from the charge transfer from the 'donor atoms' to the metal," a conclusion that is equivalent to the charge-neutrality condition (4–37). Substantial shifts of the Fermi level need large densities of interfacial defects, i.e. $\gtrsim 10^{14}$ defects per square centimeter.[73–75]

The same simple analysis may be carried out for interface defects of acceptor type. Their incorporation results in an increase of the barrier height for $\Phi''_{Bn} < W_{cs} - W_{sa}$ in defect-free junctions.

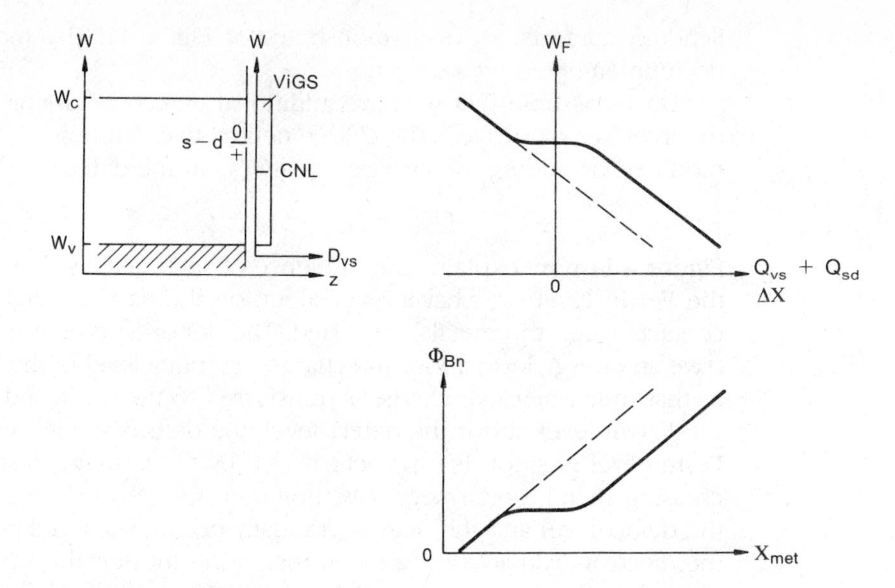

Fig. 4–14. Variation of surface charge in virtual gap states and discrete surface defects of donor type as a function of Fermi-level position within the bandgap, and expected barrier heights as a function of metal electronegativity.

4.9 Metal-induced Gap States plus Fabrication-induced Defects

The ViGS model of metal-induced gap states, which are then occupied according to the chemically driven charge transfer across the interface, only gives an upper bound for the barrier heights observed with GaAs Schottky contacts. Barrier heights lower than predicted from that model, however, are to be expected when interface defects of donor type are present, as was shown in the preceding section. All the chemical trends reported for GaAs Schottky barriers may thus be understood by that ViGS model of metal-induced gap states plus an interface defect of donor type. From the data plotted in Fig. 4–11 or Fig. 4–15, it follows that the contacts with rare-earth metals titanium, copper, silver, and ruthenium contain very low densities of such interface defects while the barriers based on nickel, cobalt, iron, and chromium are most heavily defected. The dashed line in Fig. 4–15 gives the Φ_{Bn} versus X_{met} curve for the largest density of defects needed to explain the present data set. The donors are assumed to have an energy level at 0.65 eV below the bottom of the conduction band. This

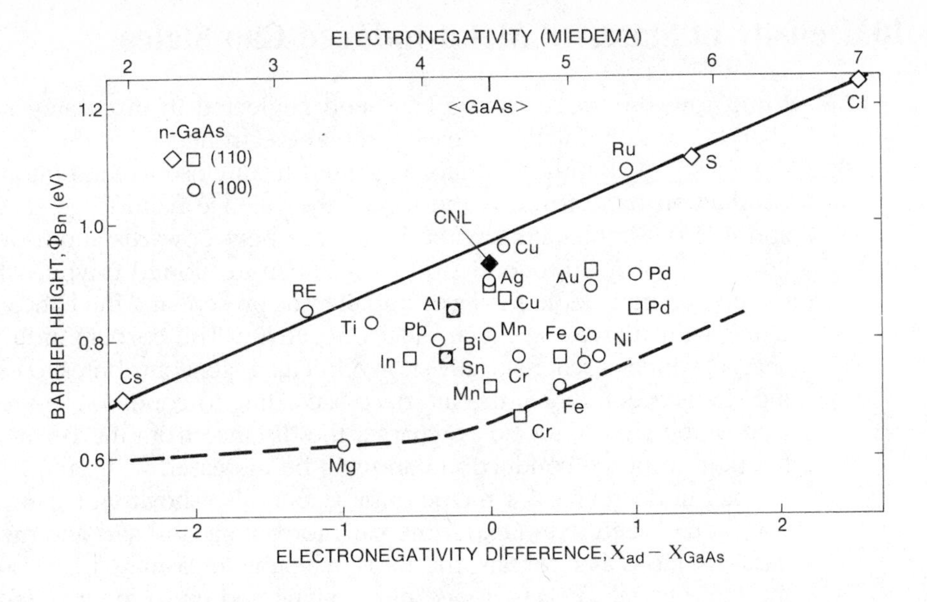

Fig. 4–15. Application of the ViGS-plus-defects model to the "GaAs data base": barrier heights plotted versus the metal electronegativity and the "ViGS line" as well as a "ViGS-plus-defects" curve.

value is very uncertain since it is based on the Mg data-point only. It may well be that the donor level is still closer to the conduction-band edge.

These interface donors are fabrication-induced. This is indicated by the variations of the barrier heights observed with the differently oriented GaAs substrates when the same metals were used. For Al, Ag, and Au, the barrier heights were found to be identical for both substrate orientations. Discrepancies, which are larger than the limits of experimental error, are observed with manganese, chromium, iron, and palladium. The case of iron is of special significance, since here the order of the barrier heights with respect to the orientation is clearly inverted.

The conclusion, that the ViGS model of metal-induced gap states, plus fabrication-induced interface defects, can explain the chemical trends observed with the barrier heights of Schottky barriers, is also supported by an analysis of data reported for metal- and silicide-silicon contacts.[12] Here again, the fabrication-induced defects were found to be of donor type. At present, the nature of these defects cannot be speculated on, since even their energy level cannot be exactly determined from the experimental data available so far.

4.10 Density of States of Metal-induced Gap States

Until now, the space charge has been neglected in discussing interface charge-neutrality. With increasing metal electronegativity the position of the Fermi level at the interface is shifted from close to the bottom of the conduction band towards the top of the valence band (Figs. 4–9, 4–11, and 4–15). Simultaneously, the bands are bent upwards and a depletion layer builds up in a semiconductor substrate doped n-type, while for p-type doping a depletion layer initially was present and the bands are flattened with increasing metal electronegativity. The corresponding space charge, which is schematically shown in Fig. 4–16, contributes to establishing charge neutrality at the interface according to condition 4–9 or 4–37. Due to the sign of the space charge, the distance from the Fermi level to the bottom of the conduction band will be increased with samples doped p-type but decreased for n-type doping, as is also shown in Fig. 4–16. This is what has been found experimentally for both adsorbates and metal contacts. As Table 4–1 reveals, the barrier heights measured with contacts of the same metal to GaAs doped n- or p-type add up to the gap energy minus (50 ±40) meV on the average. No chemical trends are discernible. As for the adsorbates, the difference of the barrier heights measured with substrates of both types of dopings amounts to 0.1 eV for chlorine, as shown in Fig. 4–5,[41] while with cesium[43] and with sulfur[55] the position of

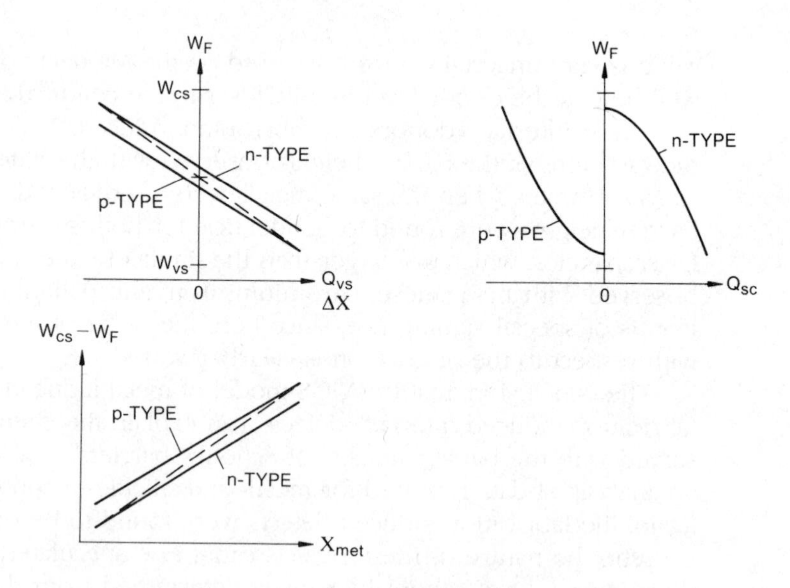

Fig. 4–16. Charge balance at a defect-free metal-semiconductor interface.

the Fermi level at the surface was found to be independent of the type of doping to within ±40 meV, the limits of experimental error.

From these data, the density of states may be estimated for the metal-induced gap states or, what is equivalent, for the virtual gap states of the complex GaAs band-structure. Except for extremely low barriers in highly defected contacts, the fabrication-induced donors will always be well above the Fermi level. Therefore, for all the metal-contacts to GaAs, with the exception of Mg-GaAs, the charge Q_{sd} in those defects remains the same when the doping is changed from p- to n-type. In the charge neutrality condition 4–37, only variations of the space-charge density and the charge in metal-induced gap states have to be considered. The density of states in question thus results from

$$D_{vs}(W_F) = (Q^n_{sc} + |Q^p_{sc}|)/[W_g - (\Phi_{Bn} + \Phi_{Bp})]. \tag{4–38}$$

As an estimate, doping levels of 5×10^{17} cm^{-3} donors and acceptors, respectively, will be assumed and the barrier heights are chosen as $\Phi_{Bn} = 0.83$ eV and $\Phi_{Bp} = 0.55$ eV, which are typical. Using the approximation (4–4) for the space charge in depletion layers, the density of states is

$$D_{vs} \cong 8.6 \times 10^{13}\ \text{eV}^{-1}\text{cm}^{-2}. \tag{4–39}$$

This value is approximately a factor of 5 lower than those densities of states evaluated by Louie *et al.*[63] as well as by Louis *et al.*[33,64] from their theoretical models (see Table 4–2).

The density of states of the metal-induced gap states also enters into the slope parameter S_Φ, as has been discussed in Sec. 4.4. According to Eq. 4–11, a lower density of interface states results in a larger slope parameter S_Φ. The work function of a solid and its electronegativity are directly related as

$$\Phi_{\text{met}} = 0.88X_{\text{Mied}} + 0.61\,, \tag{4–40}$$

when the work functions compiled by Michaelson[76] and the Miedema electronegativities are used. When the widths of the dipole layer are assumed to be the same, then the deviation of the theoretical value $S_X = 0.1$ from the expermiental $S_X = 0.22$ is easily understood from the lower density of states derived for the metal-induced gap states from the experimental data.

4.11 Metal Composition at the Interface

The development of GaAs-based semiconductor devices requires stable metal-GaAs contacts which can be reproducibly fabricated. This need has motivated numerous studies on the thermal aging of metal films deposited on GaAs. Both the metallurgy and changes of the barrier height, as a function of annealing treatment, have been studied. Such investigations were partly guided by the success which has been achieved with silicide-silicon contacts, where epitaxial silicides exhibiting low densities of interface states could be grown (see, for example, Ref. 77). With metal-GaAs contacts, however, the conditions are quite different. Even the prediction of reaction products is extremely difficult since the ternary phase diagrams are not known.

The ViGS model of metal-induced gap states needs the chemically driven charge transfer across the metal-semiconductor interface for a prediction of the barrier height. This then requires that the chemical composition at the metal side of the junction is known. The present analysis of the data base characterizes the contact by the electronegativity of the pure metal, although it is well known that even at room temperature chemical reactions may occur at metal-GaAs interfaces. The results of a systematic study by Waldrop *et al.*[78,79] will be considered. Their samples were (100) oriented and were chemically etched, which still leaves several monolayers of native oxide on the surface. Right before metal deposition, the oxide films were removed *in situ* by heating the samples to 550°C. The surfaces were then found to be atomically clean. X-ray photoemission spectroscopy (XPS) was used to find interfacial reactions via chemical shifts in the Ga(3d) and As(3d) core-level signals. The following metals were deposited: Sn, Ag, Au, Al, Fe, Cr, and Ti[78] and W, Ta, Ir, Mo, and Re.[79] Sn and Ag were found to form abrupt interfaces. As with the other metals studied, chemical reactions were observed at the interface. For the refractory metals, the interfacial layers have a thickness of approximately 1 nm and contain an arsenide of the metal deposited and metallic gallium. With gold overlayers, an intermetallic Au-Ga compound seems to form (see also Ref. 80). The spectra taken with Ti contacts indicate the presence of both As- as well as Ga-intermetallic compounds.

Photoemission spectroscopy is quite sensitive in detecting chemical shifts of core levels which are indicative of chemical reactions. Very often, however, it is difficult to determine the composition of such interface compounds exactly. For this purpose, transmission electron and X-ray diffraction are much more efficient, but these techniques lack the sensitivity of XPS. Using soft X-ray photoelectron spectroscopy (SXPS), nickel was found to react with cleaved GaAs {110} surfaces.[81] As with the refractory metals on {100} surfaces, the dissociation of GaAs and the formation of a nickel arsenide and metallic gallium were detected. The composition of that arsenide could not be identified. On the other hand, no interfacial reaction products could be detected with Ni deposited at room temperature on (001) and (111) oriented substrates when electron diffraction was employed.[82] After an annealing treatment at as low a temperature as 100°C, however, the intermetallic compound Ni_2GaAs was clearly identified. An increase of the annealing temperature to 600°C resulted in a decomposition of the Ni_2GaAs into two phases, namely NiGa and NiAs. The electronic properties of the contacts are apparently correlated with the composition at the surface. With Ni_2GaAs present a barrier height of 0.84 eV was measured, while after the formation of NiGa and NiAs the contacts were ohmic. This finding seems to imply that, in addition to the formation of those two phases, more complex reactions have occurred during the high-temperature treatments.

The Ni contacts to GaAs were annealed in a forming gas atmosphere. Detailed studies with Pd on GaAs surfaces, for example, revealed that anneals in ultrahigh vacuum (UHV) or in forming gas may produce quite different overlayer compositions.[83,84] After deposition at room temperature, the ternary phase $PdGa_{\cong 0.3}As_{\cong 0.2}$ was detected by using electron diffraction.[84] When the samples were subsequently annealed in UHV at temperatures above 350°C, the diffraction patterns revealed the presence of PdGa only, while $PdAs_2$ was additionally observed after an identical treatment under forming gas. The $PdAs_2$ is segregated at the surface[85] and it can be easily removed by, for example, ultrasonic cleaning.[83]

These few examples have clearly demonstrated that the interfacial compositions very much depend on the conditions under which the metal-GaAs contacts are fabricated. This again justifies the reduction of the data base in Sec. 4.2 to results which have been obtained by two groups, who, in each case, have characterized, prepared, and measured all their GaAs Schottky contacts in the same way.

4.12 Concluding Remarks

The chemical trends in the barrier heights of metal-GaAs contacts can be explained by the ViGS model of metal-induced gap states, which are occupied according to the chemically driven charge transfer across the interface, and by additional fabrication-induced interface-defects of donor type. The experimental studies at submonolayer coverage of monovalent adatoms such as cesium and chlorine have contributed considerably to the development of this model. The basic idea of the ViGS model of adsorbate-induced gap states has been formulated some 20 years ago, but chemical trends have not been theoretically worked out since then. Quite recently, however, this model has been shown to explain the chemical trend of the Schottky-barrier slope-parameter $S_p = d\Phi_{Bn}/d\Phi_{\text{met}}$ as observed with 19 different semiconductors and insulators.[87]

The fabrication-induced defects need to be identified and the control of their density is more desirable. With GaAs, as with Si, these defects are of donor type. Here, experimental methods other than the ones hitherto in use should be employed for the characterization of those defects. Magnetic resonance techniques might be applicable in the near future, since their sensitivity has steadily increased during recent years.

In the present version of the ViGS model of metal-induced gap states, the charge transfer across the interface is approximated by the difference of the electronegativities of the overlayer and the substrate. As long as no detailed theoretical approaches to the chemical trends of the barrier heights in Schottky contacts are available, the electronegativities have to be used in this regard. Besides, the electronegativities are purely empirical in nature, and their trends across the periodic table of the elements have been justified by more elaborate theoretical approaches.[86] However, the chemical compositions at the interfaces should be known in any case since detailed theoretical models will also need such data as an input. Here, a wide field is open for experimental research.

With a few exceptions, the interaction of metals and semiconductors results in an intermixing at the interface already at room temperature. The formation of compounds between overlayer and substrate atoms is enhanced at elevated temperatures. This concept has led to silicide-silicon instead of metal-silicon contacts. Some of the silicides show epitaxial growth on silicon and thus give almost perfect interfaces, which exhibit barrier heights as predicted from the ViGS model of metal-induced gap states. For GaAs and other III-V compound semiconductors, it may be more difficult to achieve epitaxial growth of metallic compounds between a metal and one of the substrate components. However, such interfaces might also give almost perfect Schottky barriers on GaAs.

4.12.1 Note Added in Press

After the completion of this chapter at the end of April 1987, new experimental results were published on the growth mode of metals on semiconductors at low and room temperatures and on the formation of metal-semiconductor barriers at low temperatures. The analysis of these data contributed considerably to the understanding of adatom-induced surface states and surface dipoles, and of the primary mechanism determining barrier heights in Schottky contacts, which is the continuum of metal-induced gap states. For experimental and theoretical details, the reader is referred to Refs. 88 to 90 and the papers cited therein.

References

1. F. Braun, "Über die Stromleitung durch Schwefelmetalle," *Pogg. Ann.*, 153:556 (1874).

2. W. G. Adams and R. E. Day, "The action of light on selenium," *Proc. Roy. Soc.,* 25:113 (1876).
3. F. Braun, "Versuche über Abweichungen vom Ohm'schen Gesetz in metallisch leitenden Körpern," *Sitzungsber. Naturforsch. Gesell.* (Leipzig), p. 49 (1876).
4. A. H. Wilson, "The theory of electronic semi-conductors I, II," *Proc. Roy. Soc. A,* 133:458 (1936) and 134:277 (1937).
5. O. Fritsch, "Elektrisches und optisches Verhalten von Halbleitern X," *Ann. Phys.,* 22:375 (1935).
6. W. Schottky, "Halbleitertheorie der Sperrschicht," *Naturwissenschaften,* 26:843 (1938).
7. C. A. Mead, "Metal-semiconductor surface barriers," *Solid-State Electron.,* 9:1023 (1966).
8. J. M. Andrews and J. C. Phillips, "Chemical bonding and structure of metal-semiconductor interfaces," *CRC Crit. Rev. Solid State Sci.,* 5:405 (1975).
9. I. O. McCaldin, T. C. McGill, and C. A. Mead, "Correlation for III-V and II-VI semiconductors of the Au Schottky barrier energy with anion electronegativity," *J. Vac. Sci. Technol.,* 13:802 (1976).
10. L. N. Brillson, "Transition in Schottky barrier formation with chemical reactivity," *Phys. Rev. Lett.,* 40:260 (1978).
11. G. Ottaviani, K. N. Tu, and J. W. Mayer, "Interfacial reaction and Schottky barrier in metal-silicon systems," *Phys. Rev. Lett.,* 44:884 (1980).
12. W. Mönch, "On the role of virtual gap states and defects in metal-semiconductor contacts," *Phys. Rev. Lett.,* 58:1260 (1987).
13. V. Heine, "Theory of surface states," *Phys. Rev.* 138:A1689 (1965).
14. H. Wieder, "Perspectives on III-V compound MIS structures," *J. Vac. Sci. Technol.,* 15:1498 (1978).
15. W. E. Spicer, P. W. Chye, P. R. Skeath, and I. Lindau, "New and unified model for Schottky barrier and III-V insulator interface states formation," *J. Vac. Sci. Technol.,* 16:1422 (1979).
16. R. H. Williams, R. R. Varma, and V. Montgomery, "Metal contacts to silicon and indium-phosphide," *J. Vac. Sci. Technol.,* 16:1418 (1979).
17. W. Mönch, "Chemical trends in Schottky barriers," *Phys. Rev. B,* 37:7129 (1988).
18. J. R. Waldrop, "Electrical properties of ideal metal contacts to GaAs: Schottky-barrier height," *J. Vac. Sci. Technol. B,* 2:445 (1982).
19. J. R. Waldrop, "Rare-earth metal Schottky-barrier contacts to GaAs," *Appl. Phys. Lett.,* 46:864 (1985).

20. N. Newman, M. van Schilfgaarde, T. Kendelewicz, M. D. Williams, and W. E. Spicer, "Electrical study of Schottky barriers on atomically clean GaAs (110) surfaces," *Phys. Rev. B,* 33:1146 (1986).

21. E. H. Rhoderick, *Metal Semiconductor Contacts,* Oxford: Claredon (1980).

22. M. Ludwig and G. Heymann, "Ruthenium-induced surface states on n-GaAs surfaces," *J. Vac. Sci. Technol. B,* 4:485 (1986).

23. L. N. Brillson, "The structure and properties of metal-semiconductor interfaces," *Surf. Sci. Rep.,* 2:123 (1982).

24. N. F. Mott, "Note on the contact between a metal and an insulator or semiconductor," *Proc. Cambridge Phil. Soc.,* 34:568 (1938).

25. W. Schottky, "Abweichungen von Ohm'schen Gesetz in Halbleitern," *Phys. Z.,* 41:570 (1940).

26. A. Many, Y. Goldstein, and N. B. Grover, *Semiconductor Surfaces,* Amsterdam: North-Holland (1965).

27. D. R. Frankl, *Electrical Properties of Semiconductor Surfaces,* Oxford: Pergamon (1967).

28. H. Schweikert, "Über Selengleichrichter," *Verh. Phys. Ges.,* 20(3): 99 (1939). The data reported at that meeting were presented in Ref. 25.

29. J. Bardeen, "Surface states and rectification at a metal semiconductor contact," *Phys. Rev.,* 71:717 (1947).

30. J. van Laar, A. Huijser, and T. L. van Rooy, "Electronic surface properties of Ga and In containing III-V compounds," *J. Vac. Sci. Technol.,* 14:894 (1977).

31. W. Mönch, "On the surface physics of III-V compound semiconductors" in *Festkörperprobleme* (Advances in Solid State Physics), ed. P. Grosse, 24:229, Braunschweig: Vieweg (1984).

32. A. M. Cowley and S. M. Sze, "Surface states and barrier height of metal-semiconductor systems," *J. Appl. Phys.,* 36:3212 (1965).

33. C. Tejedor, F. Flores, and E. Louis, "The metal-semiconductor interface: Si(111) and zincblende (110) junctions," *J. Phys. C: Solid State Phys.,* 10:2163 (1977).

34. J. van Laar and J. J. Scheer, "Influence of volume doping on Fermi-level position at GaAs surfaces," *Surf. Sci.,* 8:342 (1967).

35. A. Huijser and J. van Laar, "Work function variations of GaAs cleaved single crystals," *Surf. Sci.,* 52:202 (1975).

36. W. Gudat, D. E. Eastman, and J. J. Freeouf, "Empty surface states on semiconductors," *J. Vac. Sci. Technol.,* 13:250 (1976).

37. W. E. Spicer, I. Lindau, P. E. Gregory, C. M. Garner, P. Pianetta, and P. W. Chye, "Synchrotron radiation studies of electronic structure and sur-

face chemistry of GaAs, GaSb and InP," *J. Vac. Sci. Technol.,* 13:780 (1976).

38. W. Mönch and H.J. Clemens, "Surface states at clean, cleaved GaAs (110) surfaces," *J. Vac. Sci. Technol.,* 16:1238 (1979).
39. A. Huijser, J. van Laar, and T.L. van Rooy, "Electronic surface properties of UHV-cleaved III-V compounds," *Surf. Sci.,* 62:472 (1977).
40. J.R. Chelikowsky, S.G. Louie, and M.L. Cohen, "Relaxation effects on the (110) surface of GaAs," *Phys. Rev. B,* 14:4724 (1976).
41. D. Troost, L. Koenders, L.-Y. Fan, and W. Mönch, "Pinning of the Fermi level close to the valence-band top by chlorine adsorbed on cleaved GaAs (110) surfaces," *J. Vac. Sci. Technol. B,* 5:1119 (1987).
42. T. Madey and J. Yates, "Electron-stimulated desorption and work function studies of clean and cesiated (110) GaAs," *J. Vac. Sci. Technol.,* 8:39 (1971).
43. W.E. Spicer, P.E. Gregory, P.W. Chye, I.A. Babalola, and T. Sukegawa, "Photoemission study of the formation of Schottky barriers," *Appl. Phys. Lett.,* 27:617 (1975); *J. Vac. Sci. Technol.,* 13:233 (1976).
44. H.J. Clemens, J. von Wienskowski, and W. Mönch, "On the interaction of cesium with cleaved GaAs (110) and Ge (111) surfaces," *Surf. Sci.,* 78:648 (1978).
45. W. Mönch, "Virtual gap states and Fermi level pinning by adsorbates at semiconductor surfaces," *J. Vac. Sci. Technol. B,* 4:1085 (1986).
46. L. Pauling, *The Nature of the Chemical Bond,* Ithaca: Cornell University Press (1939).
47. R.T. Sanderson, *Chemical Periodicity,* New York: Van Nostrand Reinhold (1975).
48. N.B. Hannay and C.P. Smyth, "The dipole moment of hydrogen fluoride and the ionic character of bonds," *J. Am. Chem. Soc.,* 68:171 (1946).
49. J. Topping, "On the mutual potential energy of a plane network of doublets," *Proc. Roy. Soc. London* (Ser. A), 114:67 (1927).
50. A.W. Maue, "Die Oberflächenwellen in der Elektronentheorie der Metalle," *Z. Physik,* 94:717 (1935).
51. E.T. Goodwin, "Electronic states at the surface of cyrstals," *Proc. Cambridge Phil. Soc.,* 35:205 (1939).
52. F. Garcia-Moliner and F. Flores, *Introduction to the Theory of Solid Surfaces,* Cambridge: Cambridge University (1979).
53. Y.-Ch. Chang, "Complex band structure of zinc-blende materials," *Phys. Rev. B,* 25:605 (1982).

54. J. Zak, "Symmetry criterion for surface states in solids," *Phys. Rev. B*, 32:2218 (1985).
55. M. Blömacher, L. Koenders, and W. Mönch, "Electronic properties of sulfur-covered GaAs (110) surfaces," *J. Vac. Sci. Technol. B*, 6:1416 (1988).
56. K.K. Chin, S.P. Pan, D. Mo, P. Mahowald, N. Newman, I. Lindau, and W.E. Spicer, "Electronic structure and Schottky-barrier formation of Ag on n-type GaAs (110)," *Phys. Rev. B*, 32:918 (1985).
57. J. Tersoff, "Calculation of Schottky barrier heights from semiconductor band structures," *Surf. Sci.*, 168:275 (1986).
58. C. Tejedor and F. Flores, "A simple approach to heterojunctions," *J. Phys. C: Solid State Phys.*, 11:L19 (1978).
59. S.G. Louie and M.L. Cohen, "Electronic structure of a metal-semiconductor interface," *Phys. Rev. B*, 13:2461 (1976).
60. A.R. Miedema, F.R. de Boer, and P.F. de Chatel, "Empirical description of the role of electronegativity in alloy formation," *J. Phys. F: Metal Phys.*, 3:1558 (1973).
61. A.R. Miedema, P.F. de Chatel, and F.R. de Boer, "Cohesion in alloys—fundamentals of a semi-empirical model," *Physica*, 100 B:1 (1980).
62. S.G. Louie and M.L. Cohen, "Electronic structure of a metal-semiconductor interface," *Phys. Rev. B*, 13:2461 (1976).
63. S.G. Louie, J. Chelikowsky, and M.L. Cohen, "Ionicity and the theory of Schottky barriers," *Phys. Rev. B*, 15:2154 (1977).
64. E. Louis, F. Yndurain, and F. Flores, "Metal-semiconductor junction for (110) surfaces of zinc-blende compounds," *Phys. Rev. B*, 13:4408 (1976).
65. Landolt-Börnstein, *Numerical Data and Functional Relationships in Science and Technology*, Vols. 17a and b, ed. K.H. Hellwege, Heidelberg: Springer (1982).
66. J. Tersoff, "Theory of semiconductor heterojunctions: The role of quantum dipoles," *Phys. Rev. Lett.*, 30:4874 (1984).
67. W.E. Spicer, I. Lindau, P. Skeath, and C.Y. Su, "Unified defect model and beyond," *J. Vac. Sci. Technol.*, 17:1019 (1980).
68. M.S. Daw and D.L. Smith, "Vacancies near semiconductor surfaces," *Phys. Rev. B*, 20:5150 (1979).
69. R.E. Allen and J.D. Dow, "Unified theory of point-defect electronic states, core excitons and intrinsic electronic states at semiconductor surfaces," *J. Vac. Sci. Technol.*, 19:383 (1981).
70. W. Mönch, "Chemisorption-induced defects at interfaces of compound semiconductors," *Surf. Sci.*, 132:92 (1983).

71. H. Gant and W. Mönch, "On the chemisorption of Ge on GaAs (110) interfaces," *Applications Surf. Sci.,* 11/12:332 (1982).
72. W. Mönch and H. Gant, "Chemisorption-induced defects on GaAs (110) surfaces," *Phys. Rev. Lett.,* 48:512 (1982).
73. S. B. Zhang, M. L. Cohen, and S. G. Louie, "Interface potential changes and Schottky barriers," *Phys. Rev. B,* 32:3955 (1985).
74. A. Zur, T. C. McGill, and D. L. Smith, "Fermi-level position at a semiconductor-metal interface," *Phys. Rev. B,* 28:2060 (1983).
75. C. B. Duke and C. Mailhiot, "A microscopic model of metal-semiconductor contacts," *J. Vac. Sci. Technol. B,* 3:1170 (1985).
76. B. Michaelson, "The work function of the elements and its periodicity," *J. Appl. Phys.,* 48:4729 (1979).
77. P. E. Schmid, M. Liehr, F. K. Le Goues, and P. S. Ho, "Schottky-barrier and electronic states at silicide-silicon interfaces," *Mat. Res. Soc. Symp. Proc.,* 54:469 (1986).
78. J. R. Waldrop and R. W. Grant, "Interface chemistry of metal-GaAs Schottky-barrier contacts," *Appl. Phys. Lett.,* 34:630 (1979).
79. J. R. Waldrop, S. P. Kowalczyk, and R. W. Grant, "Refractory metal contacts to GaAs," *J. Vac. Sci. Technol.,* 21:607 (1982).
80. W. G. Petro, T. Kendelewicz, I. Lindau, and W. E. Spicer, "Au-GaAs (110) interface," *Phys. Rev. B,* 34:7089 (1986).
81. T. Kendelewicz, M. D. Williams, W. G. Petro, I. Lindau, and W. E. Spicer, "Interfacial chemistry and Schottky-barrier formation of Ni/InP(110) and Ni/GaAs(110) interfaces," *Phys. Rev. B,* 32:3758 (1985).
82. A. Lahar, M. Eizenberg, and Y. Komem, "Interfacial reactions between Ni films and GaAs," *J. Appl. Phys.,* 60:991 (1986).
83. X.-F. Zeng and D. D. L. Chung, "Structural characterization of the interfacial reactions between palladium and gallium arsenide," *J. Vac. Sci. Technol.,* 21:611 (1982).
84. T. S. Kuan, T. L. Freeouf, P. E. Batson, and E. L. Wilkie, "Reactions of Pd on (100) and (110) GaAs surfaces," *J. Appl. Phys.,* 58:1519 (1985).
85. J. O. Olondafe, P. S. Ho, H. J. Hovel, J. E. Lewis, and J. M. Woodall, "Contact Reactions in Pd/GaAs junctions," *J. Appl. Phys.,* 50:955 (1979).
86. W. Mortier, S. K. Gosh, and S. Shankar, "Electronegativity equalization method for the calculation of atomic charges in molecules," *J. Am. Chem. Soc.,* 108:4315 (1986).
87. W. Mönch, "On the present understanding of Schottky contacts," in *Festkörperprobleme* (Advances in Solid State Physics), ed. P. Grosse, 26:67, Braunschweig: Vieweg (1986).

88. W. Mönch, "Mechanisms of Schottky barrier formation in metal-semiconductor contacts," *J. Vac. Sci. Technol. B,* 6:1270 (1988).
89. W. Mönch, "Mechanisms of barrier formation in Schottky contacts, in metallization and metal-semiconductor interfaces," *Proc. of a NATO Advanced Research Workshop,* ed. J. P. Betra, New York: Plenum (1989), in press.
90. W. Mönch, "Physics of Schottky contacts," *Rep. Prog. Phys.,* in press.

Chapter 5

Picosecond Optical Characterization of GaAs Integrated Circuits

K. J. Weingarten, M. J. W. Rodwell, and D. M. Bloom

5.1 Introduction: Emerging Applications for High-Speed, Internal-Note Testing

The development of advanced GaAs devices and integrated circuits has been spurred by a number of applications, including microwave and millimeter-wave radar and communication systems, fiber-optic digital data transmission at gigahertz rates, high-speed data acquisition, and the constant push for faster digital logic in high-speed computers and signal processors; the ICs developed for these applications are creating new demands on high-speed electronic instrumentation.

One demand is for increased instrument bandwidth. GaAs MESFETs have been demonstrated with maximum frequency of oscillation, f_{max}, in excess of 110 GHz,[1] while pseudomorphic InGaAs/AlGaAs modulation-doped field-effect transistors[2] have shown power-gain bandwidth products which extrapolate to give $f_{max} \cong 200$ GHz, resonant tunnelling diodes have exhibited oscillation at 56 GHz,[3] and heterojunction bipolar transistors are expected to show similar performance. Because the maximum frequency of oscillation of these devices is often greater than the 100-GHz bandwidth of commercial millimeter-wave network analyzers, f_{max} is estimated by extrapolation from measurements at lower frequencies. Used as switching elements, propagation delays and transition times of 1 to 10 ps should be expected for these devices, times well below the resolution of commercial sampling oscilloscopes. In either case the device bandwidth exceeds that of the measurement instrument.

A second demand is for noninvasive access to the internal signals within high-speed integrated circuits. GaAs digital integrated circuits of MSI (medium-scale integration) complexity and 1- to 5-GHz clock rates are now available commercially, as are GaAs monolithic microwave integrated circuits (MMICs) of SSI (small-scale integration) complexity and 1- to

26-GHz bandwidths. More complex LSI (large-scale integration) digital circuits are under development, and experimental SSI digital circuits operating with 18-GHz clock rates[4] have been demonstrated. In contrast to silicon LSI integrated circuits operating at clock rates in the tens and hundreds of megahertz, the development of GaAs high-speed circuits is hampered both by poorly refined device models and by layout-dependent circuit parasitics associated with the high frequencies of operation. A test instrument providing noninvasive measurements within the integrated circuit would permit full characterization of complex high-speed ICs.

These issues have motivated a number of researchers to investigate alternative test techniques, both to increase measurement bandwidth and to allow internal testing of ICs. This chapter reviews a variety of new general test techniques for ICs, then describes in more detail the principles, capabilities, and IC measurement results of direct electro-optic sampling, a measurement technique that allows for internal-node voltage measurements in GaAs ICs with picosecond time resolution, corresponding to bandwidths in excess of 100 GHz.

5.2 Electrical Test Methods

The capabilities and characteristics of conventional test methods are determined by two features—the IC probe that connects the test instrument to the circuit and the test instrument itself. The IC probe has its own intrinsic bandwidth, or ability to transmit high-frequency signals, that may limit the test method. In addition, the probe also determines an instrument's ability to probe *internally* to the IC due to its size (limiting its spatial resolution) and influence on circuit performance (loading of the circuit from its characteristic and parasitic impedances). The test instrument sets the available bandwidth given perfect IC probes or packaged circuits, and defines the type of electrical test, such as measuring the time or frequency response.

5.2.1 Electrical IC Probes

Connection of a test instrument to an IC begins with the external connectors, typically 50-Ω coaxial cable with a microwave connector, such as SMA, K-connector, APC-2.4, APC-3.5, or APC-7. Connections to an IC are then made with a transition from the coaxial cable to some type of probe with a contact size comparable to an IC bond pad. Low-frequency signals are often connected with needle probes. At frequencies greater than several hundred megahertz these probes have serious *parasitic* impedances, due to shunt capacitance from fringing fields and series inductance from the long, thin needle. These parasitic impedances and the relatively large probe size compared with IC interconnects limit their effective use to low-frequency external input or output circuit responses at the bond pads.

For signals greater than several hundred megahertz, the recently developed Cascade Microtech probe[5,6] has demonstrated well-defined, repeatable IC connections to microwave frequencies of 50 GHz. These probes consist of transitions from a coaxial connector to a 50-Ω transmission line which tapers to bond pad size contacts. They offer excellent frequency response and have a potential frequency range to millimeter-waves (50 to 100 GHz). However, their size limits test points to IC bond pads and their 50-Ω chartacteristic impedance limits their use to input or output sections of the IC.

Thus, electrical probes suffer from a measurement dilemma. Good high-frequency probes use transmission lines to control the line impedance from the coaxial transition to the IC bond pad and to reduce parasitic impedances. The low characteristic impedance of such lines (typically 50 Ω) limits their use to input/output connections. High-impedance probes suitable for probing intermediate circuit nodes have significant parasitic impedances at microwave frequencies, severely perturbing the circuit operation and affecting the measurement accuracy. In both cases, the probe size is large compared with IC interconnect size, limiting their use to test points the size of bond pads.

5.2.2 Electrical Test Instruments

Several types of conventional test instruments exist for measuring high-speed electrical signals: sampling oscilloscopes, spectrum analyzers, and network analyzers. Sampling oscilloscopes measure the time response of repetitive signals with a resolution as short as 20 ps or a corresponding bandwidth of 20 GHz. Combined with transmission line probes this instrument gives either time-domain reflectometry measurements or signal waveforms of an IC's external response, but has neither the time resolution required for state-of-the-art GaAs ICs nor the ability to measure the internal node response of MSI or LSI ICs. The recent introduction of a new type of sampling oscilloscope based on Josephson junction technology[7] offers a time resolution of about 5 ps or a bandwidth as high as 70 GHz. As with slower sampling oscilloscopes, 50-Ω connectors limit its use to external characterization of ICs.

Spectrum and network analyzers measure the response of circuits in the frequency domain, with a frequency range of typically ~26 GHz and limited extension of this range through the millimeter-wave frequencies (300 GHz). A spectrum analyzer measures the power spectrum of a signal; that is, it measures the magnitude of the power content as a function of frequency. Typical measurement applications of spectrum analyzers include measuring harmonic and intermodulation distortion of amplifiers or nonlinear circuits, general RF/microwave signal, modulation, and noise analysis, and less frequently the magnitude of the frequency response of circuits. Network analyzers measure the vector response (magnitude and phase) as a function of frequency of the small-signal, *linear* characteristics of microwave circuits, usually expressed in terms of the scattering parameters. Typical measurement applications of network analyzers include measuring the frequency response of circuits, such as filters and other passive elements, active elements such as amplifiers, and device characteristics of high-speed transistors. The scattering parameters of a transistor, for example, are used for direct determination or extrapolation of f_{max}, the device's maximum frequency of oscillation, a figure of merit of the device's operation speed. The frequency range of these instruments can be extended to millimeter-wave frequencies (300 GHz for the spectrum analyzer, ~100 GHz for the network analyzer) using external source multipliers, mixers and waveguide connectors, but the frequency coverage is limited to 1.5:1 waveguide bandwidths and the waveguide connectors require a hybrid mount of the IC in a waveguide package, preventing wafer testing of the IC.

By Fourier transforming the small-signal frequency-domain two-port device parameters, network analyzers can provide equivalent time-domain measurements of a network's small-signal step or impulse response. However, for large-signal responses the network is no longer linear and the principle of superposition cannot be applied; thus the network analyzer cannot measure the large-signal, time waveform response (due to amplifier saturation, for example). Although spectrum analyzers can measure the harmonic spectrum magnitude of saturated or large-signal circuit responses, the phase of the harmonics is not measured, and thus the large-signal time waveforms again cannot be inferred. Both instruments also rely on 50-Ω connectors and IC probes, limiting their ability to probe ICs to the external response. For network analysis, de-embedding the device parameters from the connector and circuit fixture response becomes progressively more difficult at increasing frequencies, an important issue for millimeter-wave circuit testing.

5.3 Nonelectrical, Novel Test Techniques

With the objective of either increased bandwidth or internal IC testing with high spatial resolution, or both, a number of new test techniques have been introduced and demonstrated. As background for this chapter, a review of a number of these techniques with representative references is given.

5.3.1 SEM Voltage Contrast (Electron-Beam) Testing

Motivated by the need to test logic level signals in VLSI silicon ICs, the voltage contrast scanning electron microscope (SEM) was developed in the late 1960s for detecting voltages on IC metal lines.[8–11] This technique uses an electron beam from a SEM to stimulate secondary electron emission from surface metallization. For a metal conductor at ground or negative potential, the electrons have sufficient energy to be collected by a detector shielded by an energy analyzer grid. Metal lines at a positive potential retard the emitted electrons, lowering their energy and reducing the number of electrons detected. The detected signal is small for IC voltage levels; to improve time resolution the signal is sampled with electron beam pulses and averaged to improve the signal-to-noise ratio.

Commercial SEMs have sensitivities of 1 to 10 mV, bandwidths up to 2 GHz or time resolution of ~1 ns, and a spatial resolution as small as 20 nm. Compared with typical operating speeds of commercial silicon VLSI (clock rates of ~10 to 100 MHz) this technique has good time resolution, good sensitivity, and excellent spatial resolution. The systems, based on standard SEMs with thermionic cathodes, chop the electron beam to increase their time resolution. For decreasing electron-beam pulse duration, the average beam current also decreases, degrading measurement sensitivity and limiting practical systems to a time resolution of several hundred picoseconds. To overcome this limitation, a group at IBM Research has implemented a photocathode triggered by an intense picosecond optical pulse to generate short, high-current electron pulses.[12] This approach shows promise for achieving time resolution approaching 10 ps with the SEM probe.

The major drawbacks of SEM testing are its complexity, its vacuum system, its relatively high cost, and the necessary sample preparation. Electron-beam testing of GaAs ICs also may suffer from problems due to charging of the deep levels (impurity-induced charge traps) in the GaAs substrate, perturbing circuit operation.

5.3.2 Photoemissive Sampling

Photoemissive sampling, similar to E-beam testing, is based on analyzing secondary electron emission from IC conductors. The technique uses intense, energetic light from a pulsed laser focused on an IC conductor to generate photoelectrons. An extraction/retarding grid combination placed in close proximity to the conductor energy analyzes the electrons, with a resulting secondary electron emission varying with the conductor potential. The feasibility of this new approach is made possible by picosecond pulse width, and high peak power lasers, and offers a potential improvement in time resolution and sensitivity over the SEM probe. The technique is not available commercially, but a number of researchers have demonstrated systems with time resolution as short as 7 ps with good spatial resolution and millivolt sensitivity.[13–15] However, scaling the response time to picoseconds has required increasing the extraction field to levels approaching the field strengths associated with surface inversion of the semiconductor substrate; this may severely increase the invasiveness of this approach. As with the scanning electron microscope, extraction fields may also cause backgating of the MESFETs on GaAs circuits through charging of deep levels.

5.3.3 Charge Sensing

Silicon, due to its centrosymmetric crystal structure, exhibits no bulk second-order optical nonlinearities, such as the electro-optic effect, to its use as a basis for an optical measurement system. Third-order effects, such as voltage-dependent second-harmonic generation or the optical Kerr effect, are in general very weak and result in impractical systems in terms of measurement sensitivity and implementation. An alternative technique, charge sensing in devices, was recently developed as a measurement technique for IC devices in silicon.[16,17] This system is based on interferometrically detecting the presence of charge at IC device junctions. Charge or free carriers in the semiconductor contribute to the index of refraction of the material as described by the plasma-optical relation

$$n = n_0\left(1 - \frac{\omega_p^2}{\omega^2}\right)^{1/2},$$

$$\omega_p^2 = \frac{q^2 N}{\epsilon_s m_e^*} + \frac{q^2 P}{\epsilon_s m_h^*},$$

where n_0 is the bulk index of refraction, ω is the optical probe frequency, ω_p is the plasma resonance frequency, N is the electron concentration in the conduction band, P is the hole concentration in the valence band, ϵ_s is the permittivity of the substrate material, and m_e^* amd m_h^* are the electron and hole conductivity effective masses. The change in charge density is detected with a compact optical interferometer using a 1.3 μm wavelength semiconductor laser as the probe, shown schematically in Fig. 5–1. The Wollaston prism separates the input beam into a probe beam and a reference beam. The presence of charge in the probe beam path results in a

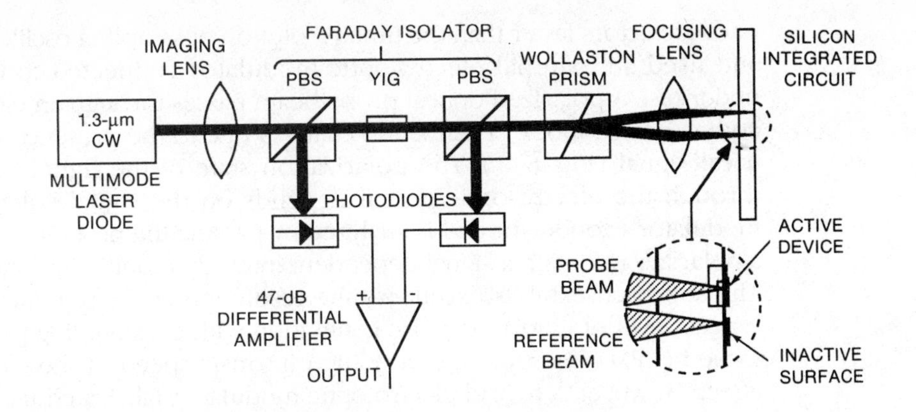

Fig. 5–1. Charge-sensing system schematic. ***(After Heinrich, Bloom, and Hemenway*** [16] ***and Heinrich* et al.**[17]**)**

change in output intensity at the photodiodes proportional to the charge concentration. PBS denotes a polarizing beamsplitter cube.

This technique has demonstrated single-shot detection of a silicon bipolar-junction transistor's switching signal in a 200-MHz bandwidth (Fig. 5–2). Since the plasma-optical effect occurs in all semiconductor materials, this technique is applicable to GaAs ICs and shows promise for studying device characteristics and charge dynamics.

5.3.4 Electro-Optic Sampling

Electro-optic sampling uses an electro-optic light modulator to intensity modulate a probe beam in proportion to a circuit voltage. The technique was initially developed to measure the response of photoconductors and

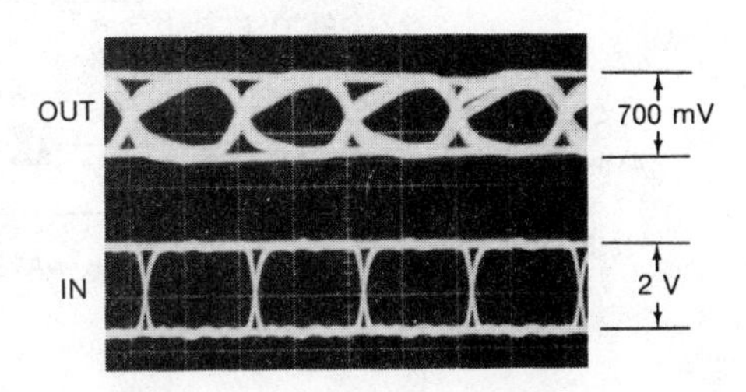

Fig. 5–2. Eye diagram of the stored base charge in a silicon bipolar-junction transistor, as monitored by the optical charge-sensing system (upper trace). The corresponding base voltage is shown in the lower trace. The input signal is 25-megabaud pseudo-random Manchester-coded data. Time scale is 20 ns per division.

photodetectors faster than the time resolution of sampling oscilloscopes[18–21] and used an external electro-optic modulator connected to the device under test. A polarized optical probe beam passes through an electro-optic crystal, whose index of refraction changes due to the presence of an electrical signal (Fig. 5–3). The polarization state of the light after passing through the electro-optic crystal depends on the signal voltage on the modulator's conductor and is analyzed by passing the probe beam through a polarizer to obtain a signal-dependent intensity modulation of the probe. The approach exhibits excellent time resolution (<0.5 ps) due to the advanced level of ultrashort pulse generation with the colliding-pulse mode-locked (CPM) laser and the very high intrinsic speed of the electro-optic effect. However, a hybrid electro-optic modulator, fabricated using lithium tantalate, requires connecting the test point on the IC to the transmission line of the modulator, restricting its use to external test points. Unless carefully designed, the system bandwidth will be degraded by the hybrid connection between the modulator and the device under test, due to the loading of the test point from the relatively low characteristic impedance of the transmission line, the parasitic impedances associated with bond wires, and the parasitic elements associated with evanescent mode excitation at the interface between dissimilar transmission-line substrates.

Using an electro-optic needle probe (Figs. 5–4 and 5–5),[22] the technique has been extended to internal node probing of ICs. The needle, a

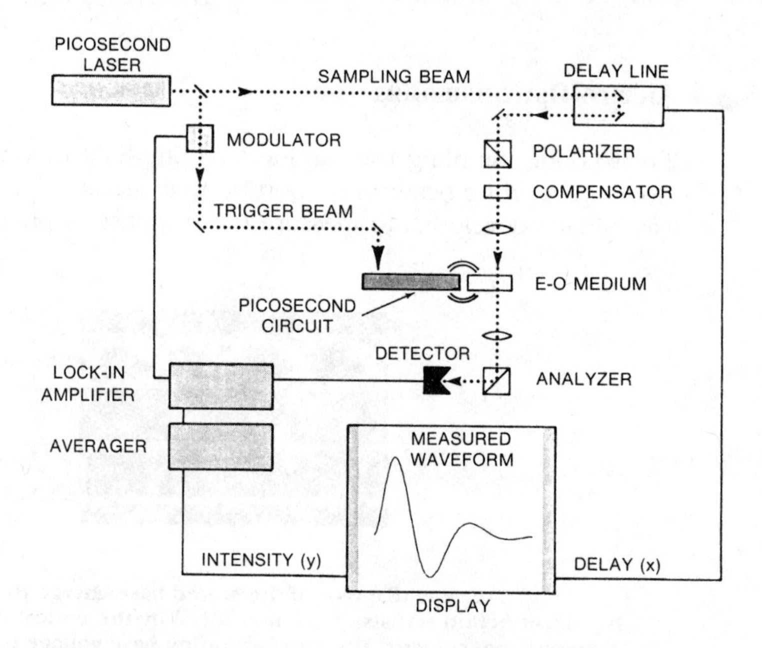

Fig. 5–3. Electro-optic sampling system. ***(After Valdmanis, Mourou, and Gabel.[18] Courtesy AT&T Bell Laboratories)***

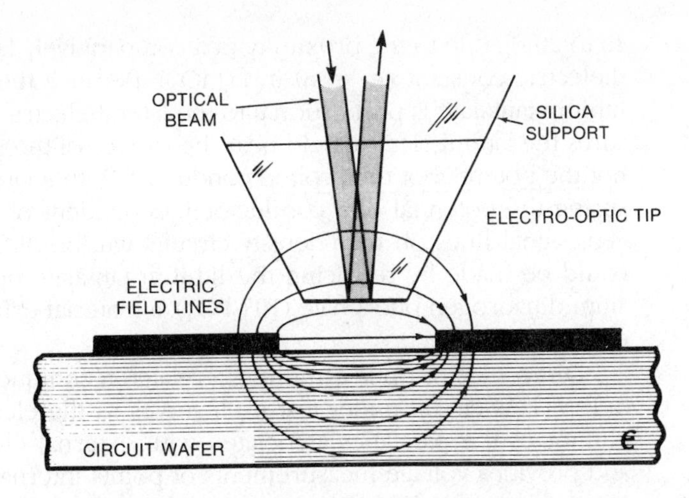

Fig. 5–4. Electro-optic needle probe for on-wafer measurements. ***(After Valdmanis and Pei.[22] Courtesy AT&T Bell Laboratories)***

fused silica rod with a 40- × 40-μm tip of an electro-optic material (lithium tantalate, $LiTaO_3$), is brought into the vicinity of a conductor, introducing fields within the probe tip. As with the hybrid electro-optic scheme, the fields perturb the polarization of the probing beam, and the polarization change is analyzed by a polarizer. The $LiTaO_3$ electro-optic material is transparent to visible wavelength light, allowing for use of ultrashort sampling pulses from a CPM laser. Because the probe relies on no optical or electrical properties of the circuit under test, circuits of any substrate material can be tested without sample preparation. The probe exhibits some circuit invasiveness through capacitive loading on the order of

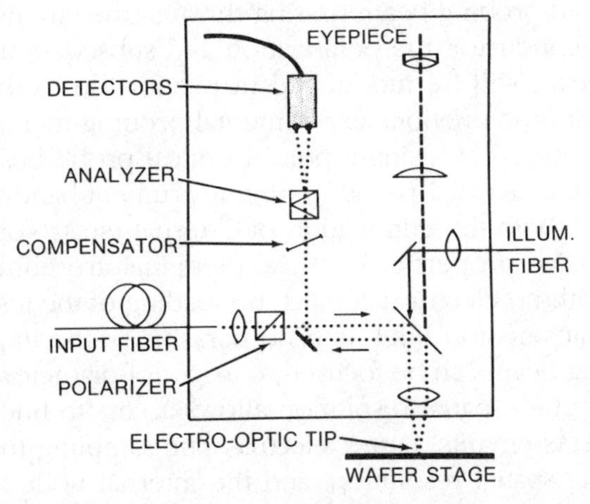

Fig. 5–5. Sampling system for on-wafer measurements with the electro-optic needle probe. ***(Courtesy AT&T Bell Laboratories)***

10 to 20 fF, due to the proximity of a comparatively large probe tip of high dielectric constant ($\epsilon_r = 40$ in $LiTaO_3$). Because the polarization shift in lithium tantalate is proportional to the lateral electric field, the probe measures the lateral electric field near the surface of the integrated circuit, and not the potential of the probed conductor. With a lateral field probe, measuring the potential of a conductor independent of the potential of adjacent signal lines on high-density circuits will be difficult; a voltage probe could be made by replacing the lithium tantalate tip with a tip of potassium dihydrogen phosphate (KH_2PO_4), a material exhibiting a longitudinal electro-optic effect.

Direct electro-optic sampling, a related technique, where the substrate of the GaAs circuit under test itself serves as the electro-optic modulator, eliminates the parasitics associated with external electro-optic elements, and provides voltage measurements of points internal to the IC with picosecond time resolution and micrometer spatial resolution. This method and its principles, capabilities, and results are described in the following sections.

5.4 Principles of Direct Electro-Optic Sampling in GaAs ICs

Gallium arsenide is itself an electro-optic material. Through the electro-optic effect, the electric fields associated with conductor voltages within a GaAs circuit induce small anisotropic changes in the optical index of refraction of the substrate.[23,24] With the use of a suitably oriented and polarized probing beam passing through the circuit's substrate in the vicinity of a conductor, the polarization and subsequently the intensity of the probe beam will be modulated in proportion to the conductor's potential. As with the previous experimental probing methods, good time resolution is achieved by using a pulsed optical probe beam, with pulse durations on the order of 1 ps, permitting instrument bandwidths greater than 100 GHz.

With the elimination of external electro-optic elements, associated invasive properties of those elements are eliminated, permitting probing with no electrical contact, no loading of the test point with low-impedance transmission lines, and no parasitic probe impedance. Because the probing beam can be focused to a spot of diameter of several micrometers, the probe's spatial resolution allows access to finely spaced conductors in LSI GaAs circuits. Direct electro-optic sampling thus provides the bandwidth, the spatial resolution, and the internal node access necessary for characterization of high-speed GaAs integrated circuits.

5.4.1 The Electro-Optic Effect

The electro-optic effect is an anisotropic variation in a material's dielectric constant (and hence index of refraction) occurring in proportion to an applied electric field. The effect, whose origin lies in small quadratic terms in the relationship between an applied field and the resulting material polarization, is present in a variety of noncentrosymmetric crystals. Among these are GaAs, InP, and AlAs, used for high-speed semiconductor devices, and lithium niobate ($LiNbO_3$), lithium tantalate ($LiTaO_3$) and potassium dihydrogen phosphate (KH_2PO_4), used for nonlinear optical devices. Centrosymmetric crystals do not exhibit the electro-optic effect; notable among these materials are silicon and germanium.

The change in refractive index of these materials with electric field can be used for optical phase-modulation, and, from this, polarization-modulation or intensity-modulation.[25,26] Lithium niobate and lithium tantalate electro-optic modulators are used in both lasers, and fiber-optic systems; direct electro-optic sampling uses the electro-optic effect in GaAs to obtain voltage-dependent intensity modulation of a probe beam.

5.4.2 Electro-Optic Voltage Probing in a GaAs Crystal

The principal axes of a GaAs IC fabricated on standard (100)-cut material are shown in Fig. 5–6. The **X**, **Y**, and **Z** axes are aligned with the ⟨100⟩ directions of the GaAs cubic Bravais lattice, while the **Y**′ and **Z**′ axes are aligned with the $[01\bar{1}]$ and [011] directions, parallel to the cleave planes along which a GaAs wafer is scribed into individual ICs. Because the $[01\bar{1}]$ and [011] directions, parallel to the IC edges, are also the eigenvectors of

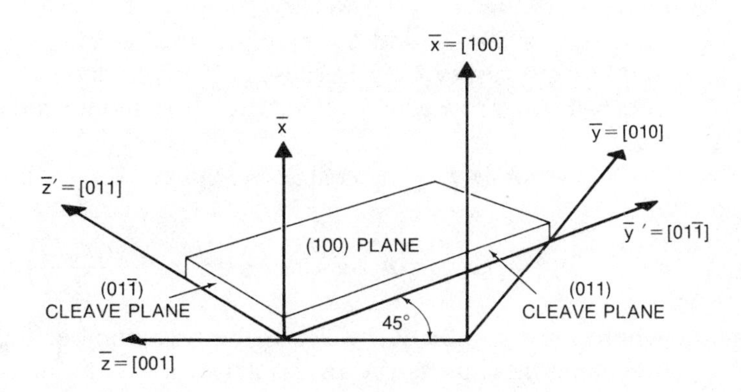

Fig. 5–6. Principal axes and cleave planes in (100)-cut gallium arsenide.

the electro-optic effect, the **X**, **Y′** and **Z′** axes are the natural coordinate system for describing electro-optic sampling in GaAs.

In GaAs, the refractive indices $n_{y'}$ and $n_{z'}$ in the **Y′** and **Z′** directions are[27]

$$n_{y'} = n_0 - \frac{n_0^3 r_{41} E_x}{2}, \tag{5–1A}$$

$$n_{z'} = n_0 + \frac{n_0^3 r_{41} E_x}{2}, \tag{5–1B}$$

where E_x is the component of the circuit electric field in the **X** direction, n_0 is the zero-field refractive index, and r_{41} is the electro-optic coefficient for GaAs. Given a beam propagating in the **X** direction, these field-dependent refractive indices will result in differential phase modulation of the beam components having electric fields $e_{y'}$ and $e_{z'}$ polarized in the **Y′** and **Z′** directions.

The wave equations for propagation in the **X** direction are

$$\frac{\partial^2 e_{y'}}{\partial x^2} = \left(\frac{n_{y'}^2}{c^2}\right)\frac{\partial^2 e_{y'}}{\partial t^2}, \tag{5–2A}$$

and

$$\frac{\partial^2 e_{z'}}{\partial x^2} = \left(\frac{n_{z'}^2}{c^2}\right)\frac{\partial^2 e_{z'}}{\partial t^2}, \tag{5–2B}$$

where c is the speed of light in vacuum. Sinusoidal optical fields thus propagate with phase velocities $c/n_{y'}$ and $c/n_{z'}$, and are given by

$$e_{y'}(x,t) = A_{y'} \cos(\omega t - k_{y'}x + \phi_{y'}), \tag{5–3A}$$

$$e_{z'}(x,t) = A_{z'} \cos(\omega t - k_{z'}x + \phi_{z'}), \tag{5–3B}$$

where $k_{y'} = \omega n_{y'}/c$ and $k_{z'} = \omega n_{z'}/c$ are the wavenumbers along the **Y′** and **Z′** axes, respectively. Because of the field-dependent refractive indices (Eq. 5–1) the propagating *e*-fields are phase modulated by the electric field,

$$e_{y'}(x,t) = A_{y'} \cos\left[\omega t - kx\left(n_0 - \frac{n_0^3 r_{41} E_x}{2}\right) + \phi_{y'}\right], \tag{5–4A}$$

$$e_{z'}(x,t) = A_{z'} \cos\left[\omega t - kx\left(n_0 + \frac{n_0^3 r_{41} E_x}{2}\right) + \phi_{z'}\right], \tag{5–4B}$$

where now $k = 2\pi/\lambda_0 = \omega/c$ is the wavenumber for λ_0, the free-space wavelength. Thus as the two polarization components propagate they undergo a differential phase shift proportional to the x component of the electric field resulting in a change in the beam's polarization.

Consider the electro-optic amplitude modulator shown in Fig. 5–7. At the point of entry to the GaAs wafer, at $x = 0$, a circularly polarized probe beam has electric fields given by

$$e_{y'}(0,t) = A\cos(\omega t), \quad (5\text{–}5A)$$

$$e_{z'}(0,t) = A\cos(\omega t - \pi/2). \quad (5\text{–}5B)$$

After propagating through a distance w, the thickness of the substrate, the relative phase of the two field components is shifted in proportion to the electric field

$$e_{y'}(w,t) = A\cos(\omega t - \phi_0 + \Delta\phi/2), \quad (5\text{–}6A)$$

$$e_{z'}(w,t) = A\cos(\omega t - \phi_0 - \pi/2 - \Delta\phi/2), \quad (5\text{–}6B)$$

where

$$\phi_0 = \frac{2\pi n_0}{\lambda_0} w$$

is the total phase shift through the substrate in the absence of an electric field and

$$\Delta\phi = \frac{2\pi}{\lambda_0} n_0^3 r_{41} E_x w = \frac{2\pi}{\lambda_0} n_0^3 r_{41} V \quad (5\text{–}7)$$

is the change in phase between the **Y**′ and **Z**′ polarizations due to the electro-optic effect. The electric field expressions of Eq. 5–5 then no

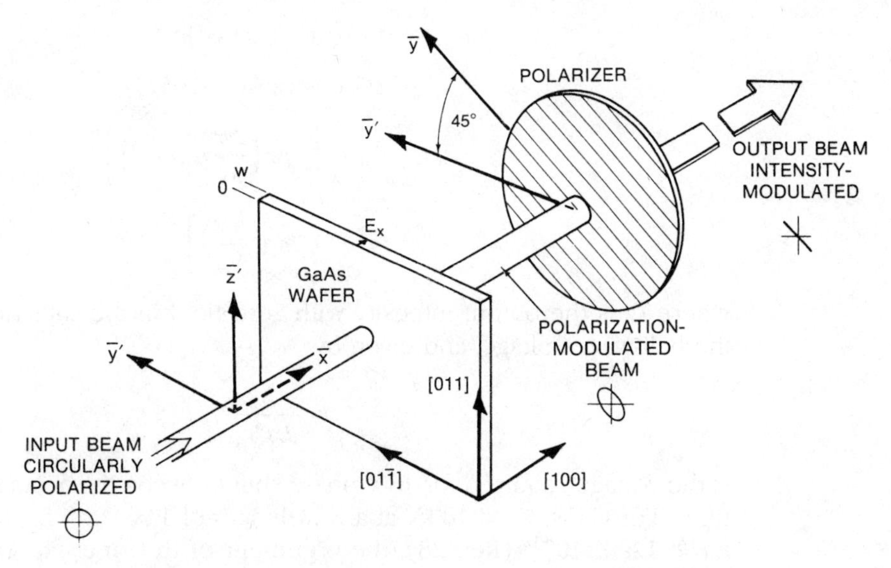

Fig. 5–7. Gallium arsenide electro-optic intensity modulator.

longer represent circularly polarized light due to this additional shift, $\Delta\phi$; the polarization emerging from the substrate has changed from circular to slightly elliptical. The product wE_x of the substrate thickness and the x component of the electric field is the potential difference V between the front and back surfaces of the GaAs wafer where the probing beam enters and exits the wafer. The change in beam polarization is thus a function only of the potential difference, V, across the wafer at the probed point, and is independent of the particular field direction and distribution giving rise to V.

To measure this voltage-induced polarization change, the beam emerging from the GaAs is passed through a polarizer oriented parallel to the [010] (**Y**) direction. With the polarizer oriented at 45° to the **Y**′ and **Z**′ axes, its output field $e_y(t,z)$ is the difference of $e_{z'}$ and $e_{y'}$:

$$e_y(t, z = \text{polarizer}) = \frac{e_{z'} - e_{y'}}{\sqrt{2}}. \tag{5–8}$$

Substituting in Eq. 5–6, we find that e_y is amplitude modulated by $\Delta\phi$:

$$\begin{aligned} e_y(t, z = \text{polarizer}) &\propto \cos(\omega t - \phi_0 + \Delta\phi/2) \\ &\quad - \cos(\omega t - \phi_0 - \pi/2 - \Delta\phi/2) \\ &\propto \cos(\omega t - \phi_0 + \pi/4)\cos(\Delta\phi/2 - \pi/4) \end{aligned} \tag{5–9}$$

The intensity of the output beam, detected by a photodiode, is proportional to the square of e_y:

$$\begin{aligned} P_{\text{out}} &\propto \{\text{time average of}\} e_y^{\,2} \\ &= 2P_0 \cos^2(\pi/4 - \Delta\phi/2) \\ &= P_0\left[1 + \sin\left(\frac{2\pi}{\lambda_0} n_0^3 r_{41} V\right)\right] \\ &= P_0\left[1 + \sin\left(\frac{\pi V}{V_\pi}\right)\right], \end{aligned} \tag{5–10}$$

where P_0 is the output intensity with zero field in the substrate. V_π, called the half-wave voltage, and given by

$$V_\pi = \frac{\lambda_0}{2n_0^3 r_{41}},$$

is the voltage required for 180° phase shift between the **Y**′ and **Z**′ polarizations. For GaAs, $V_\pi \cong 10$ kV at a wavelength of 1.064 μm for $n_0 = 3.6$, and $r_{41} \cong 1.4 \times 10^{-12}$ (Ref. 28); the argument of the sine expression is thus small for typical voltages V encountered on integrated circuits, and Eq. 5–11 can be approximated by

$$P_{\text{out}} \cong P_0\left(1 + \frac{\pi V}{V_\pi}\right). \tag{5–11}$$

Thus, for substrate voltages to several hundred volts, the output beam intensity is nearly linear and in direct proportion to the voltage across the substrate. The intensity of the output beam, detected by a photodiode, is thus a measure of the voltage across the substrate of the IC. To make useful measurements of the voltages in microwave and high-density digital GaAs ICs, the simplified probing geometry of Fig. 5–7 must be adapted to the conductor geometries found on these circuits.

5.4.3 Probing Geometries in GaAs ICs

While the simplified probing geometry of Fig. 5–7 provides modulation of the probe beam intensity in proportion to the voltage across the wafer, the arrangement is not readily applied to circuit measurements. This transmission-type arrangement would require separate lenses for focusing and collecting the probe beam, precisely aligned on opposite sides of the wafer. Also, high-density interconnections on the circuit side of digital ICs and backside metallization on many microwave ICs would obstruct passage of the beam through the wafer. Reflection-type probing geometries, as shown in Fig. 5–8, provide better access to the wafer, using only a single lens and using the IC metallization for reflection. The frontside geometry is suitable for probing microstrip transmission lines of MMICs. The backside geometry permits very tight focusing of the probing beam to a diameter limited by the numerical aperture of the focusing lens, and provides a probe beam modulation sensitive to the probed conductor's voltage but independent of nearby signal conductors,[29] a necessity for testing high-density ICs.

For microstrip transmission lines typically used in MMICs, the fields extend from the conductor to the ground plane, roughly the distance of the substrate thickness; the probe beam is focused from the top of the substrate through the fringing fields of the conductor to a beam spot size diameter approximately one tenth of the substrate thickness.

Other MMICs may use planar transmission lines such as coplanar waveguides (CPW) to provide controlled impedance connections between circuit elements at microwave frequencies. MSI/LSI circuits typically use thin metal interconnects (3 to 10 μm); these conductors act as lumped elements with series inductance and shunt capacitance per unit length, and the electric field distribution is a strong function of the proximity of a conductor to ground. For probing these types of interconnects and for planar transmission lines, the characteristic extent of the fields into the substrate

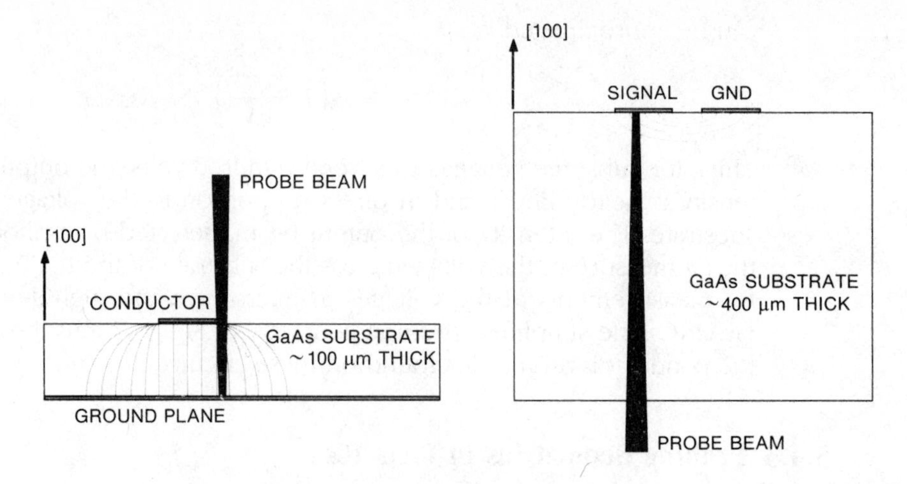

(A) Frontside geometry is used for probing microstrip transmission lines on MMICs.

(B) Backside geometry is used for probing planar transmission lines on MMICs and wire interconnects on digital ICs.

Fig. 5–8. Reflection-mode probing geometries for electro-optic sampling of GaAs integrated circuits.

is the distance between the signal conductor and ground. For typical ICs the substrate thickness (400 to 500 μm unthinned) is much greater than signal-ground spacings, concentrating the electric fields close to the substrate surface so that the back of the substrate is at an equipotential with respect to individual conductors on the IC surface. The optical probe, focused through the back of the substrate to a spot diameter less than or equal to the conductor width and reflected from the signal conductor, senses the signal voltage on that conductor with little sensitivity to the presence of other nearby conductors.

In both cases the optical probe is reflected from metallization on the IC. This reflection geometry has the advantage of doubling the sensitivity of the measurement by doubling the interaction length in the GaAs. However, the signal is also proportional to the amount of light reflected; if the IC metallization varies, the signal strength also varies. In general, the signal can be normalized to the amount of reflected light; however, metallization such as ohmic contacts, which may be very rough, will suffer from poor sensitivity if they scatter a large portion of the probe beam. For backside probing the back of the wafer should be sufficiently polished to allow for passage of the probe beam with negligible scattering, and for frontside probing the ground plane should be sufficiently reflective.

In this reflection-mode probing, the incident and reflected beams, centered on the microscope lens for optimum focusing, are separated by their polarizations (Fig. 5–9). The linearly polarized incident probe beam

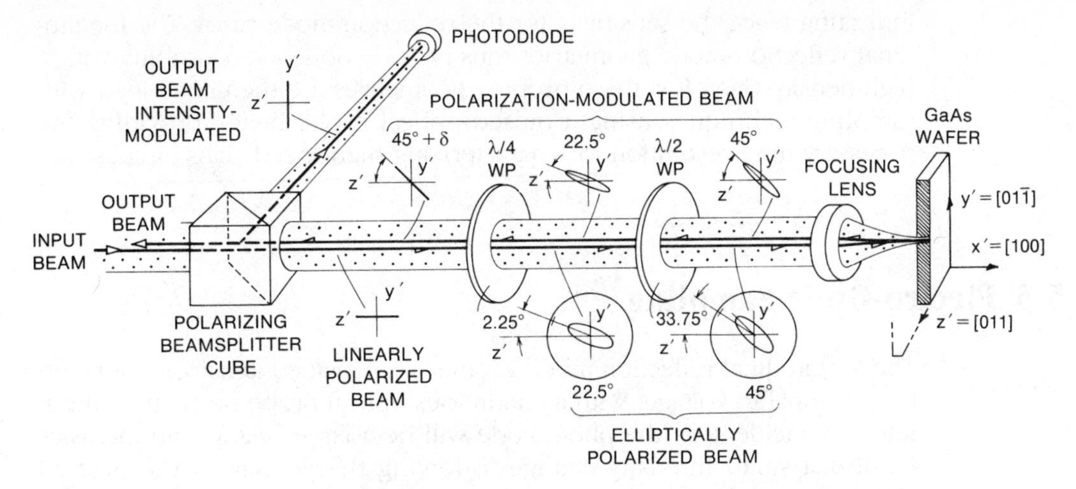

Fig. 5–9. Coaxial arrangement for separation of incident and reflected beams in reflection-mode probing.

from the polarizing beamsplitter passes through a quarter-waveplate, an optical birefringent element used to adjust beam polarizations. The birefringent axes of the quarter-wave plate are oriented at 22.5° to the axis of the beamsplitter, producing an elliptical polarization. A rotatable half-wave plate oriented at 33.5° then aligns the major axis of this elliptical polarization at 45° to the [011] direction of the GaAs substrate (i.e. oriented at 45° to the substrate cleave planes), and the probe beam is focused by a microscope objective next to a conductor (from the front) or on the conductor (from the back). The reflected beam passes back through the lens and the wave plates, producing a linear polarization at 45° to the axes of the polarizing beamsplitter; the polarization component at 90° orientation is then directed by the beamsplitter onto a photodiode. Through the electro-optic effect, the probed conductor voltage perturbs the polarization of the returning beam, changing the angle of the linearly polarized light prior to the polarizer, and thus the intensity incident upon the photodiode. The resulting intensity modulation is identical with the analysis in Sec. 4.2, and the current in the photodiode varies in proportion to the probed intensity as in Eq. 5–11:

$$I \cong I_0\left(1 + \frac{\pi V}{V_\pi}\right),$$

where I_0 is the photocurrent with $V = 0$, and where V_π is now given by

$$V_\pi = \frac{\lambda_0}{4n_0^3 r_{41}} \cong 5 \text{ kV},$$

indicating twice the sensitivity for the reflection-mode cases. The longitudinal reflection-mode geometries thus provide noncontact probing within high-density GaAs ICs; these probing geometries are then combined with sampling techniques using a pulsed optical probe beam, providing the necessary time resolution for characterizing high-speed GaAs circuits.

5.5 Electro-Optic Sampling

The longitudinal reflection-mode geometries provide intensity modulation proportional to voltage. With a continuous optical probe beam, the output intensity incident on the photodiode will be a large steady-state intensity, I_0, plus a small intensity change following the voltage of the probed conductor; microwave-frequency or picosecond rise-time signals on the probed conductor will result in microwave-frequency or picosecond rise-time modulation of the probe beam. Detection of this modulation would require a photodiode/receiver system with bandwidth comparable to that of the detected signal; the bandwidth of the electro-optic probing system would be limited to that of the photodiode, the receiver system, and the oscilloscope displaying the signal. With the bandwidth of commercial sampling oscilloscopes limited to ~20 GHz (see Sec. 2.2), and the bandwidth of efficient infrared photodiodes currently limited to ~20 GHz, the probing system would be limited to a bandwidth of ~14 GHz, insufficient for probing many high-speed and microwave GaAs circuits. In addition, because of the very small modulation provided by the electro-optic effect, direct detection of a probe beam intensity-modulated at microwave bandwidths would result in an extremely low signal-to-noise ratio, and thus very poor instrument sensitivity.

Mode-locked laser systems in conjunction with optical pulse compressors can generate extremely short optical pulses; pulses as short as 8 fs[30] have been generated at visible wavelengths, while subpicosecond pulse widths have been generated at the infrared wavelengths[31,32] where GaAs is transparent. Sampling techniques, using a pulsed optical probe to achieve time resolution set by the optical pulse duration and the circuit-probe interaction time, permit instrument bandwidths exceeding 100 GHz. Two related methods, synchronous sampling and harmonic mixing, are used in electro-optic sampling. In synchronous sampling, equivalent-time measurements of the voltage waveforms on probed conductors are generated in a manner similar to the operation of a sampling oscilloscope. In harmonic mixing, the electro-optic sampler measures the amplitude and phase of sinusoidal voltages on probed conductors, thus emulating a microwave network analyzer.

5.5.1 Equivalent-Time (Synchronous) Sampling

In equivalent-time sampling, an optical probe pulse with repetition rate f_0 samples a repetitive voltage waveform. If the waveform repeats at *exactly* Nf_0, an integer multiple of the probe repetition rate, an optical pulse interacts with the waveforms every Nth period at a fixed point within its repetition period. Thus, over many optical pulse and voltage waveform repetitions, these pulses sample the voltage waveform at a single point in time within the cycle. Each optical pulse thus undergoes an equal modulation in its intensity; the resulting change in the average intensity of the probe beam is proportional to the signal and detected by a photodiode receiver whose frequency response can be much less than the optical pulse repetition frequency.

To detect the entire time waveform, the waveform frequency is increased by a small amount, Δf (Fig. 5–10). The probe pulses are then slowly delayed with respect to the waveform, sampling successively delayed points, so that the average intensity at the photodiode changes in proportion to the waveform, but *repeating* at a rate Δf, as shown below.

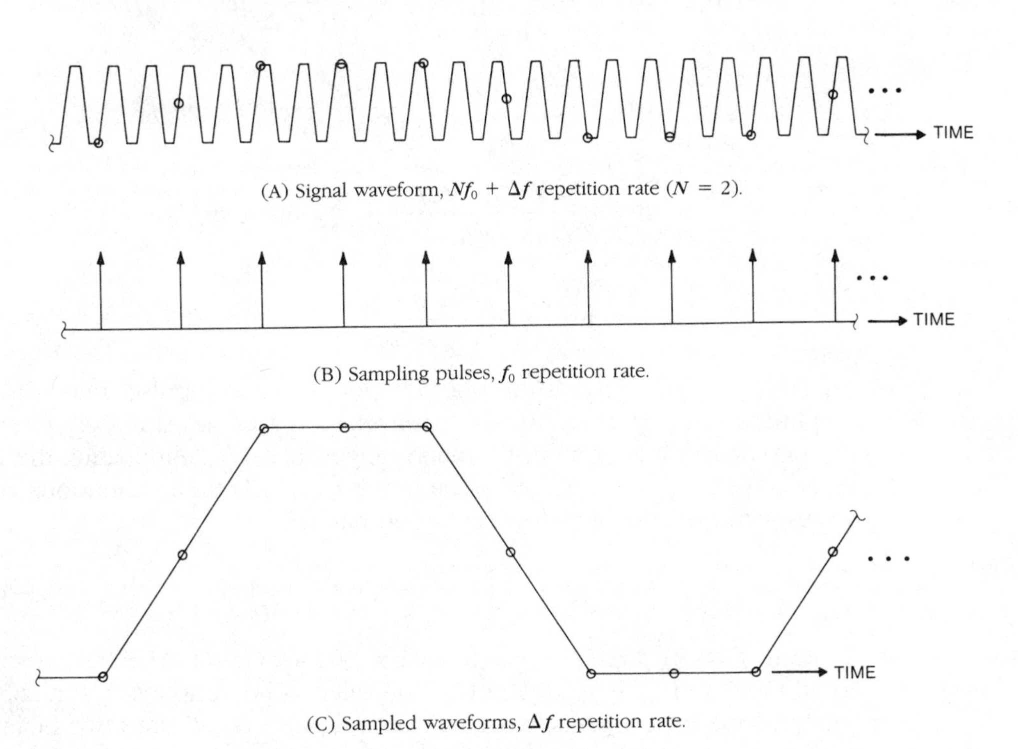

TIME
(A) Signal waveform, $Nf_0 + \Delta f$ repetition rate ($N = 2$).
TIME
(B) Sampling pulses, f_0 repetition rate.
TIME
(C) Sampled waveforms, Δf repetition rate.

Fig. 5–10. Equivalent-time sampling. Typically, $f_0 \cong 82$ MHz, $0 \leqq N < 500$, and $\Delta f \cong 10$ Hz.

After Eq. 5–11, the photodiode current $i(t)$ is

$$i(t) \cong I(t)\left(1 + \frac{\pi V(t)}{V_\pi}\right),$$

where $I(t) = rP(t)$ is the photocurrent with $V(t) = 0$, r is the photodiode responsivity, $P(t)$ is the laser intensity, and $V(t)$ is the signal on the probed conductor. The laser intensity $P(t)$ and hence also $I(t)$ are a series of impulses:

$$I(t) = \frac{I_0}{f_0} \sum_{m=-\infty}^{m=+\infty} \delta(t - m/f_0), \tag{5–12}$$

and thus $i(t)$ is

$$i(t) = I(t) + \frac{\pi}{V_\pi}\frac{I_0}{f_0} \sum_{m=-\infty}^{m=+\infty} V(m/f_0)\delta(t - m/f_0).$$

The probed voltage, $V(t)$, is periodic at frequency $(Nf_0 + \Delta f)$, and hence

$$\begin{aligned} i(t) &= I(t) + \frac{\pi}{V_\pi}\frac{I_0}{f_0} \sum_{m=-\infty}^{m=+\infty} V\left(\frac{m}{f_0} - \frac{mN}{Nf_0 + \Delta f}\right)\delta(t - m/f_0) \\ &= I(t) + \frac{\pi}{V_\pi}\frac{I_0}{f_0} \sum_{m=-\infty}^{m=+\infty} V\left(\frac{m\Delta f}{f_0(Nf_0 + \Delta f)}\right)\delta(t - m/f_0) \\ &= I(t) + \frac{\pi}{V_\pi}\frac{I_0}{f_0} V\left(\frac{t\Delta f}{Nf_0 + \Delta f}\right) \sum_{m=-\infty}^{m=+\infty} \delta(t - m/f_0) \\ &= I(t)\left[1 + \frac{\pi}{V_\pi} V\left(\frac{t\Delta f}{Nf_0 + \Delta f}\right)\right]. \end{aligned} \tag{5–13}$$

Equation 5–13 shows that the photodiode current is a pulse train whose amplitude varies at a rate Δf. The receiver then averages (low-pass filters) the photocurrent over a period much longer than $1/f_0$, eliminating the individual pulses. The averaged photocurrent $i_{out}(t)$ is then continuous and varies with $V(t)$, but at a slow repetition rate Δf:

$$i_{out}(t) = I_0\left[1 + \frac{\pi}{V_\pi} V\left(\frac{t\Delta f}{Nf_0 + \Delta f}\right)\right]. \tag{5–14}$$

Typically $f_0 \cong 82$ MHz, N varies from 1 to 500 for circuit drive frequencies to 40 GHz, and Δf is 10 to 100 Hz. Equivalent-time sampling is similar to pump/probe experiments, where an optical pulse is split into two beams, with one beam exciting a response in a sample and the other beam measuring the sample response. By repeating the measurement for different delays between the excitation beam and the probe beam the time response of the sample is measured. In contrast to pump/probe sampling, which has one probe pulse for every pump signal, in equivalent time sampling the signal repeats N times between probe pulses. Because the pulse repetition rate is harmonically related to the signal repetition rate, Nyquist's sampling theorem (which states that the maximum recoverable signal bandwidth is half the sampling rate) does **not** apply in terms of setting the bandwidth of this measurement. Instead, the bandwidth is determined by three factors: the sampling pulse width, the relative jitter between the sampling pulse and the repetitive signal, and the interaction time (due to the transit time in the electro-optic substrate) between the pulse and the signal. These factors are discussed in detail in Sec. 5.6.1.

5.5.2 Harmonic Mixing

Equivalent-time sampling in the time domain can be described in the frequency domain as harmonic mixing between the spectrum of the laser intensity and the spectrum of the electrical signal being measured. The frequency-domain analysis more fully describes sampling and describes how frequency-domain transfer function measurements are made with the system.

The time-domain signal detected by the photodiode receiver, proportional to the product of the laser signal and the measured signal in the time domain, has a frequency domain representation determined by the frequency spectrum of the laser *convolved* with the frequency spectrum of the measured signal.[31] In the frequency domain,

$$\mathcal{F}(i(t)) \cong \mathcal{F}(I(t)) * \left[\delta(f) + \frac{\pi\mathcal{F}(V(t))}{V_\pi}\right], \tag{5–15}$$

where $\mathcal{F}$ is the Fourier transform operator, $\delta(f)$ is the delta function, and * represents the convolution operation. Figure 5–11 shows a schematic representation of this convolution for a mode-locked laser spectrum and a single microwave frequency signal. Scaled replicas of the signal appear in the laser intensity spectrum as amplitude-modulation sidebands around each laser harmonic. At baseband, the magnitude and phase of the signal is recovered with a vector voltmeter or a synchronous detector.

5.5.3 Electro-Optic Sampling System

The sampling system, shown schematically in Fig. 5–12, can be grouped into three sections; the laser system for optical pulse generation, the microwave instrumentation for driving the IC under test, and the receiver

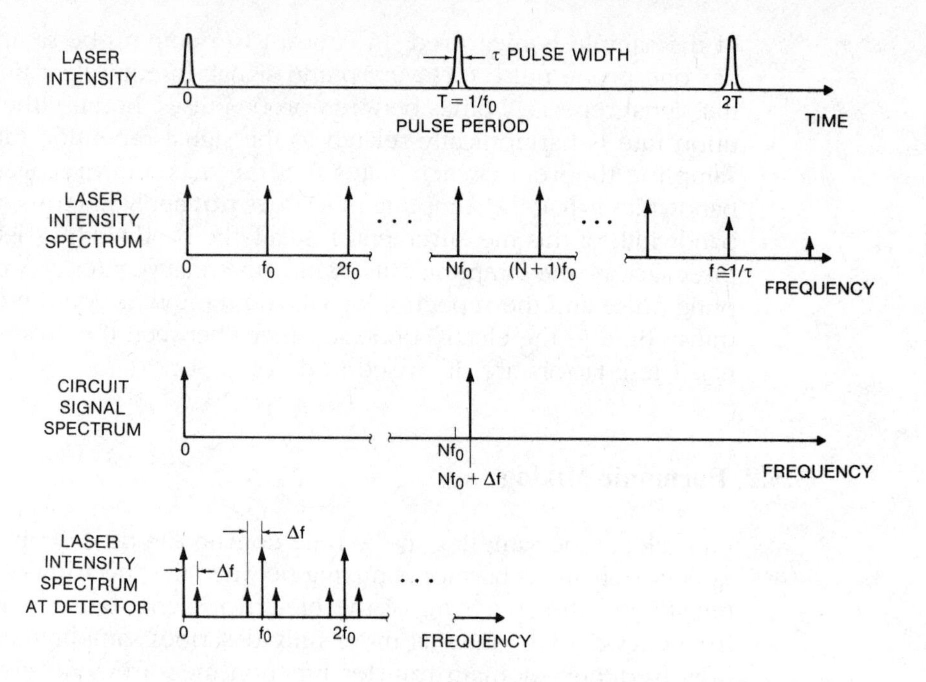

Fig. 5–11. Electro-optic harmonic mixing.

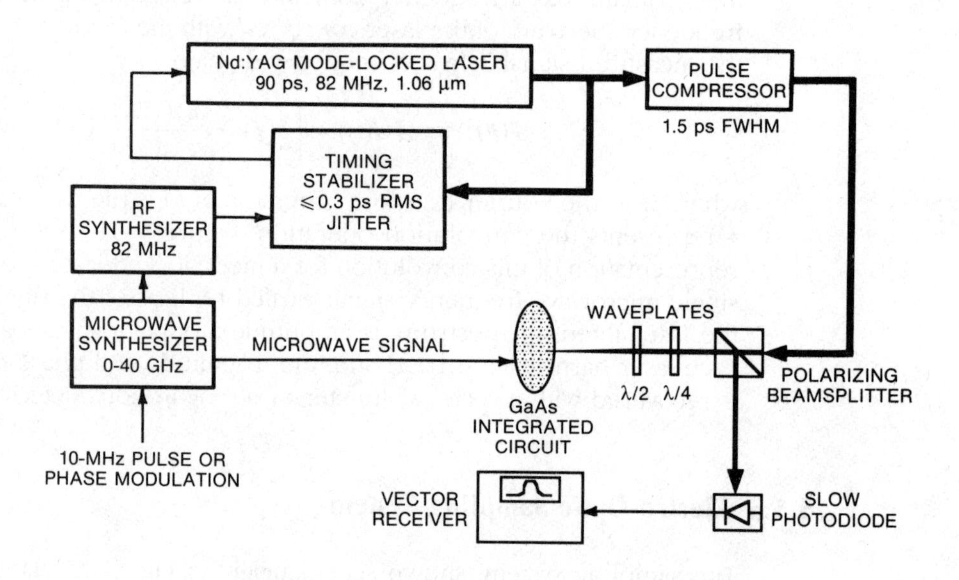

Fig. 5–12. System for direct electro-optic sampling.

system for signal processing and data acquisition. The laser system consists of a mode-locked, Nd:YAG laser, a fiber-grating pulse compressor, and a timing stabilizer feedback system. The Nd:YAG laser, a commercially available system, produces 1.06-μm, 90-ps pulses at an 82-MHz rate. The laser has free-running pulse-to-pulse timing fluctuations of 4 ps rms, reduced to less than 300 fs rms by a phase-locked-loop feedback system[32] which synchronizes and stabilizes the laser pulse timing with respect to the microwave synthesizer. The fiber-grating pulse compressor shortens the pulses to 1.5 ps FWHM (full width at half maximum).[33–35] The beam passes through a polarizing beamsplitter and two waveplates to adjust its polarization, then is focused through the IC substrate with a microscope objective to a 3-μm spot on the probed conductor (backside probing) or a 10-μm spot on the ground plane adjacent to the probed conductor (frontside probing). The reflected light is analyzed by the polarizing beamsplitter; the change in intensity, proportional to the voltage across the GaAs substrate, is detected by a photodiode connected to a vector receiver.

To drive the IC, a microwave synthesizer generates either sinusoidal excitation for microwave circuits, or the clock or data signals for digital circuits. For wafer-level testing of ICs the microwave signal is delivered with a microwave probe station (Cascade Microtech Model 42, Fig. 5–13) modified to allow for backside electro-optic probing. The transmission line probes used with this test station allow for launching a signal on the IC with repeatable, low-reflection connections in a 50-Ω envinronment to 26 GHz.

Equivalent-time sampling (Sec. 5.5.1) is used to view time waveforms: the synthesizer is set to an exact multiple of the laser repetition rate (82 MHz), plus a small frequency offset Δf (1 to 100 Hz). To suppress laser intensity noise (Sec. 5.6.2), the synthesizer is pulse or phase modulated at 1 to 10 MHz; the resulting intensity modulation at the photodiode varies at the slow offset rate Δf in proportion to the measured signal and is detected by a narrow-band synchronous receiver. Harmonic mixing (Sec. 5.5.2) is used for vector voltage measurements; the synthesizer is set to an exact multiple of the laser repetition rate plus a 1- to 10-MHz frequency offset, and the receiver is configured as a vector voltmeter to measure the magnitude and phase of the received signal.

Direct electro-optic sampling has also been demonstrated using a mode-locked InGaAsP injection laser to generate sampling pulses of ~10-ps pulse width.[36,37] This system uses two synthesizers, referenced to a master clock, to drive the laser and to supply a signal to the IC. The longer pulse widths of the injection laser decrease the system's time resolution compared to the 1.5-ps pulse widths generated using the Nd:YAG/pulse compressor; however, the injection laser pulses have almost no timing jitter and the laser is a very compact and reliable optical source. This system

(A) Wafer, wafer stage, microwave wafer probes, and viewing objective.

Fig. 5–13. Cascade Microtech microwave wafer probe station modified for electro-optic sampling. *(After Cascade Microtech, Inc.[6])*

has also been used to measure electrical signals in InP (indium phosphide) ICs.[38]

5.6 System Performance

5.6.1 Bandwidth

The system's bandwidth or time resolution is determined by the optical pulse width, the pulse-to-pulse timing jitter of the laser with respect to the microwave synthesizer driving the circuit, and the optical transit time of the pulse through the GaAs substrate where fields are present. The overall time resolution is the root-mean-square sum of these values:

$$\Delta\tau_{\text{total}} = \sqrt{(\Delta\tau_{\text{pulse}})^2 + (\Delta\tau_{\text{jitter}})^2 + (\Delta\tau_{\text{ott}})^2}\,, \tag{5–16}$$

where $\Delta\tau_{\text{pulse}}$ is the rms optical pulse width, $\Delta\tau_{\text{jitter}}$ is the rms pulse-to-pulse jitter, and $\Delta\tau_{\text{ott}}$ is the rms optical transit time of the pulse through the substrate.

(B) View showing probe beam focusing objective below the wafer stage.

Fig. 5–13 *continued*

The relation between the time resolution and the frequency bandwidth for the optical pulse is given by

$$\Delta\tau_{\mathrm{FWHM}} = \frac{0.312}{f_{3\mathrm{dB}}}, \qquad (5\text{–}17)$$

where, for a Gaussian pulse shape, $\Delta\tau_{FWHM} = 2.35\Delta\tau_{rms}$ is the full width at half-maximum (optical pulse width is typically measured in terms of its FWHM, not its rms value), and Δf_{3dB} is the half-power frequency. This time-bandwidth product is reduced by a factor of $1/\sqrt{2}$ from the more common time-bandwidth product of 0.441, due to the square-law photodiode detector, i.e. the optical intensity of the pulse is converted to a voltage in the receiver with a resulting power spectrum related to the square of this voltage.

The 90-ps pulse width from the Nd:YAG laser, given the above relation, results in a bandwidth of 3.5 GHz and is clearly not suitable for high-bandwidth circuit measurements. To reduce the pulse width, a fiber-grating pulse compressor is used. This system, based on the Kerr effect, or self-phase modulation, in a single-mode optical fiber, generates a linear frequency chirp on the laser pulse as it propagates through the fiber. The light emerging from the fiber has its leading edge slightly red-shifted and the trailing edge slightly blue-shifted with the frequency varying linearly across the pulse duration. These new frequency components are then recombined into a compressed pulse by passing the light through a grating pair. Because the time-of-flight delay through the grating pair is linearly proportional to the light's wavelength, it acts as a dispersive delay line, allowing the blue-shifted light components to "catch up" to the red-shifted light components. The wavelength-dependent delay is adjusted to match the frequency chirp of the light, producing a compressed pulse.

A number of effects limit the amount of pulse compression available with this technique; simulated Raman scattering limits the maximum optical power focused into the fiber core, and deviation from a linear frequency chirp on the pulse due to nonideal pulse shapes can generate long pedestals on an otherwise short pulse. A pulse compression factor of 60 is obtained on a routine, day-to-day basis for mode-locked Nd: YAG using a fiber length of 1 km in the pulse compressor. However, several research groups have used two-stage compression to achieve subpicosecond pulse widths as short as 200 fs.[39,40]

The compressed pulse width, 1.5 ps FWHM, measured with an optical autocorrelator, has spectral content to 200 GHz assuming a Gaussian pulse shape. Since the autocorrelation does not uniquely determine the optical pulse shape, a more quantitative estimate of the spectral content of the pulse is determined by numerically Fourier transforming the autocorrelation, giving the power spectral density of the pulse with no assumption about the pulse shape.[31] This calculation shows the half-power bandwidth reduced to roughly 100 GHz, attributed to the nonideal frequency chirp generated by the Gaussian-shaped input pulse, which results in slight "wings" or pedestals on the pulse.

Timing jitter influences both bandwidth and sensitivity; the impulse response of the sampling system is the convolution of the optical pulse shape with the probability distribution of its arrival time (neglecting optical transit time), while those Fourier components of the jitter lying within the detection bandwidth of the receiver introduce noise proportional to the time derivative of the measured waveform. Stabilization of the laser timing is thus imperative for low-noise measurements of microwave or picosecond signals. To address this issue a timing stabilization feedback system is used to reduce the jitter to a level less than the optical pulse width.[32] Figure 5–14 shows the measured phase noise of one harmonic of the laser with an HP 8662 low-phase noise synthesizer as the reference for the feedback system. From this measurement the time jitter is calculated to be less than 300 fs rms.

In general, the optical transit time of the pulse in the GaAs substrate can be neglected for microwave ICs. Because the optical and microwave dielectric constants in GaAs are nearly equal, microwave transmission lines have a cutoff frequency for higher-order modes roughly equal to the inverse of the optical transit time. Well-designed microwave circuits operate at frequencies well below the multimode cutoff frequency. Only when measuring interconnects near or above the cutoff frequency (where dis-

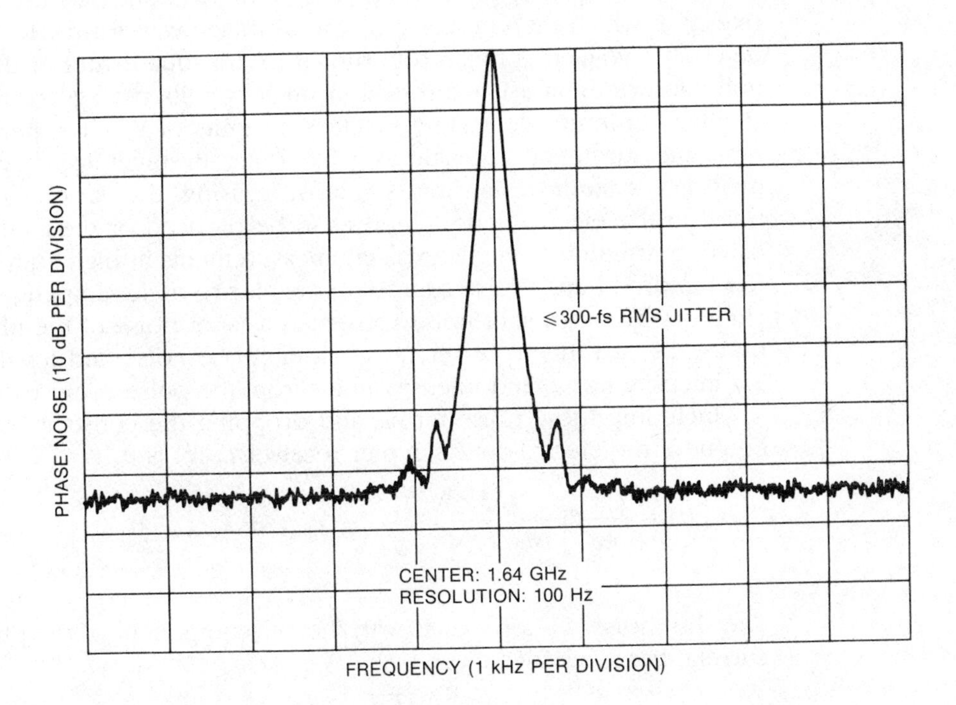

Fig. 5–14. Laser phase noise at the 20th harmonic of the pulse repetition rate, measured with a photodiode and a spectrum analyzer. The noise floor of this measurement is instrumentation limited, giving an upper limit for the timing jitter of 300 fs rms.

persive characteristics are of interest) must the optical transit time be considered. For example, the optical transit time for a 125 μm thick substrate, typical of MMICs operating at frequencies below 40 GHz, is 3 ps, corresponding to a 3-dB response rolloff of greater than 100 GHz.

5.6.2 Sensitivity

If the high measurement bandwidth provided by the electro-optic sampler is to be useful, the instrument must also provide sufficient sensitivity to readily observe the comparatively small voltages typical in high-speed GaAs circuits. As in any system, sensitivity is determined by the signal-to-noise ratio; the instrument's sensitivity, or minimum detectable voltage, is the probed voltage which results in a measured signal equal to the measurement system's noise voltage. Most noise sources have power spectral densities which are independent of frequency ("white" noise), resulting in a noise voltage proportional to the square root of the signal acquisition bandwidth. The acquisition bandwidth is the bandwidth or integration time of the low-frequency photodiode/receiver system and sets the maximum rate Δf at which the sampler can scan a voltage waveform. The minimum detectable voltage is thus proportional to the square root of the acquisition bandwidth, and is expressed in units of volts per square root hertz; smaller minimum detectable voltages, in units of $V/\sqrt{Hz}$, permit more rapid measurement acquisition for fixed measurement accuracy. With appropriate system design and signal processing, the various sources of noise in the electro-optic sampler can be reduced or their effect eliminated, permitting low-noise voltage measurements at high scan rates of a few hundred hertz. Noise sources in the electro-optic sampling system include probe beam shot noise (observed as shot noise of the photodiode quiescent current), receiver noise, laser phase noise and low-frequency $1/f$ intensity noise, and intensity noise from the pulse compressor.

Including these noise terms, and dropping the constant term I_0, the output of the electro-optic sampler receiver $i_{out}(t)$ is

$$i_{out}(t) = \frac{I_0 \pi}{V_\pi} V\left(\frac{t\Delta f}{Nf_0 + \Delta f}\right) + i_{sn} + i_{phase} + i_{receiver} + i_{laser} + i_{compressor} . \tag{5-18}$$

The shot noise, i_{sn}, associated with the DC component of the photodiode current, has a variance given by

$$\overline{(i_{sn})^2} = 2qI_0B ,$$

where q is the electron charge, the horizontal bar denotes the statistical expectation, and B is the measurement acquisition bandwidth. The receiver equivalent input noise current, i_{receiver}, is

$$\overline{(i_{\text{receiver}})^2} = \frac{4kTB}{R_L} + \overline{(i_{\text{amp}})^2} + \frac{1}{{R_L}^2}\overline{(v_{\text{amp}})^2},$$

where k is Boltzmann's constant, T is the absolute temperature, and R_L is the photodiode load resistor. The first term is the Johnson or thermal noise of the load resistor, and $\overline{(i_{\text{amp}})^2}$ and $\overline{(v_{\text{amp}})^2}$ are the equivalent input noise current and equivalent input noise voltage of the amplifier following the photodiode. Noise (i_{phase}) arises in proportion to the product of the laser timing jitter and the time derivative of the measured waveform, and has a variance approximated by

$$\overline{(i_{\text{phase}})^2} \cong \frac{I_0^2\pi^2}{V_\pi^2}\overline{\left(\frac{dV}{dt}\right)^2}\int_0^B \frac{\mathcal{L}(f)}{f_0^2}\,df$$

where $\overline{(dV/dt)^2}$ is the mean-squared slope of the voltage waveform $V(t)$ and $\mathcal{L}(f)$ is the single-sided laser phase noise spectral density relative to the carrier power at the first harmonic of the laser pulse repetition frequency. The spectral densities of i_{laser}, the laser $1/f$ intensity noise, and $i_{\text{compressor}}$, the compressor noise, depend on the design and adjustment of these components.

Of the above noise terms, most can be reduced to a negligible level compared with the shot noise. Receiver noise is reduced to a level below the shot noise limit by appropriate receiver design; R_L is made large in comparison with $2kT/qI_0$ so that the Johnson noise is well below shot noise and the photodiode amplifier is selected for low input noise. At frequencies below ~100 kHz the laser intensity noise is approximately 80 dB above the shot noise level, contributing a $10^4:1$ degradation to the minimum detectable voltage. To translate the signal detection to a frequency where this $1/f$ laser noise is below the shot noise limit, the microwave signal to the IC is modulated at 1 to 10 MHz. The resulting modulation component of the photocurrent, proportional to the sampled voltage, is detected with a 1- to 10-MHz narrow-band vector receiver. For sequential digital circuits which do not operate correctly with chopped excitation, a small-deviation phase modulation is used. In this case the received signal, proportional to the derivative of the sampled waveform, is integrated in software.[41] The pulse compressor is also a source of several types of excess amplitude noise, due to stimulated Raman scattering (SRS) and temperature-induced polarization drift in the non–polarization-preserving fiber.[42] These noise sources are reduced by keeping the optical power in the fiber well below the Raman threshold and by temperature stabilizing the fiber to suppress polarization drift. However, the pulse compressor typically contributes 5 to 10 dB of excess amplitude noise when used with the relatively long 1-km fiber. Finally, the laser timing stabilizer reduces $\mathcal{L}(f)$ by approximately 25 dB at low frequencies; phase noise is not significant to frequencies of 40 GHz.

Given suppression of these excess noise sources, the system sensitivity is set by the signal to shot noise ratio. Setting the signal current V_{min}/V_π equal to the shot noise current, and normalizing to a 1-Hz acquisition bandwidth, $\delta f = 1$ Hz, the minimum detectable voltage is

$$V_{min} = \frac{V_\pi}{\pi}\sqrt{\frac{2q}{I_0}}\left(\frac{\mathrm{V}}{\sqrt{\mathrm{Hz}}}\right). \qquad (5\text{–}19)$$

For the reflection-mode probing geometries, $V_\pi \cong 5$ kV, while the average photocurrent I_0 is typically 1 mA. Then, the minimum detectable voltage is

$$V_{min} = \frac{30\ \mu\mathrm{V}}{\sqrt{\mathrm{Hz}}}.$$

Typically, $V_{min} \cong 70\ \mu\mathrm{V}/\sqrt{\mathrm{Hz}}$ is observed experimentally due to 5 to 10 dB of excess noise from the pulse compressor; this sensitivity is sufficient to acquire measurements at scan rates of 10 to 100 Hz with a noise floor of a few millivolts.

5.6.3 Spatial Resolution

The spatial resolution of the probe is set by diffraction-limited focusing of the 1.06 μm wavelength beam. The minimum achievable spot diameter for diffraction limited optics is

$$\omega_0 \cong \frac{\lambda\sqrt{1 - \mathrm{NA}^2}}{2\mathrm{NA}}, \qquad (5\text{–}20)$$

where λ is the optical wavelength and NA is the numerical aperture of the focusing lens. With a high-NA lens spot sizes approaching the optical wavelength are possible. Standard microscope objectives (focal length of 8 mm and NA of 0.4, for example) routinely achieve spot sizes of 3 μm, suitable for probing most IC interconnects, but impractical for probing very small features such as submicrometer gate lines.

5.6.4 Linearity

Because typical circuit voltages are small compared with the half-wave voltage, V_π, the probe beam intensity modulation is very small and the electro-optic probe is then very nearly linear with respect to the probed voltage. Kolner[24] has analyzed the linearity and dynamic range of the probe due to the sinusoidal dependence of the probe intensity with respect to signal voltage. With the system set at the quarter-wave bias (the linear region of the sinusoidal transmission) as described in Eq. 5–10, the probe is linear to within 1 percent for signal voltages of 200 V and less.

5.6.5 Invasiveness

One important feature of optical probing of ICs is the noninvasive, nondestructive nature of the technique. Compared with conventional electrical probes, the optical probe makes no mechanical contact to the IC, avoiding physical damage to the circuit, it does not require the test point to drive a 50-Ω load impedance, and has no parasitic impedances. The lack of parasitic impedances is an important characteristic for measurement frequencies in the upper microwave and millimeter-wave region, where even the small parasitic impedances of well-designed electrical probes become significant.

The optical probe can perturb the circuit by photoconductively generating carriers, which then change the substrate conductivity and generate photorefractive or photovoltaic potentials. Direct band-to-band absorption of the probe is avoided because the photon energy of the 1.06-μm optical probe, 1.17 eV, is well below the bandgap energy, 1.42 eV, of GaAs. However, the presence of impurities in the GaAs results in deep levels, i.e. the presence of allowed electron states at energies near mid-bandgap. These deep levels, primarily the EL2 level, allow for the residual absorption of the 1.06-μm light in the GaAs and the associated generation of free carriers. Absorption can also occur if the probe beam is too intense: because of the tight beam focusing and short pulse duration, probe beams of average intensities approaching 100 mW have peak pulse intensities sufficient to permit two-photon band-to-band absorption.

With an intense 125-mW probe beam of 1.06-μm wavelength, we have observed small changes, of approximately 0.1 dB, in the forward gain (S_{21}) of microwave distributed amplifiers. If the probe beam is focused *directly* within the active FET channel, significant changes in drain current can be induced; however, for making circuit measurements, the probe is not focused within the device, but on the adjacent metal interconnects. For testing digital circuits probe beam intensities are typically kept below ~50 mW. Probe beam intensities can be reduced by a factor of ten, reducing the associated photogenerated carrier concentration by a similar factor, before the sensitivity degrades (Eq. 5–19) to levels approaching that of competing techniques. Residual invasiveness of the probe can also be reduced by increasing the probe beam wavelength from 1.06 to 1.3 μm, where the absorption due to the EL2 deep level is reduced by a factor of five.

5.7 Circuit Measurements

5.7.1 Realistic Circuit Testing Conditions

Optical probing, providing access to the high-impedance internal nodes of ICs with picosecond time resolution and micrometer spatial resolution, permits direct measurements of the performance of state-of-the-art microwave and digital GaAs circuits. To permit meaningful evaluations of a circuit's performance and to provide meaningful comparisons between competing circuit technologies, these measurements must be made under realistic circuit operating conditions. For example, the propagation delay of switching devices in simple test circuits, measured by either electro-optic sampling or by conventional methods, are used to project the maximum clock frequency of these devices used in digital systems. Unless the test circuit provides representative switching voltages, interface impedances, and fan-outs, the measured delays will not correlate well with the maximum clock frequency of circuits such as shift registers, binary multipliers, and memory. For example, the response of a transistor driven by a low-impedance photoconductor and loaded by a low-impedance, 50-Ω transmission line is in general much faster than the response of a logic gate driven by the normal output impedance of a driving gate and loaded by the normal input capacitances of cascaded gates. In addition, if the test circuit is constructed in hybrid form with wire bonds between the tested device and the transmission lines, the interconnection parasitics may dominate the circuit response.

The simplest representative test structures, ring oscillators and inverter strings are often used as benchmarks of circuit speed; these circuits load the gates with unity fan-out and tend to give optimistically small delay measurements. Inverter strings operate with full logic-level swings, while ring oscillators often operate small-signal, without full logic-level swings. Master-slave flip-flops, connected as binary frequency dividers, operate with realistic signal levels and with each gate loaded by a fan-out of two, and serve as better performance indicators.

For microwave/analog circuits such as distributed amplifiers, appropriate test signals are swept-frequency sinusoids for small-signal transfer function measurements, or single-frequency signals set to larger amplitudes for large-signal and saturation measurements. Signal sources and terminations should have 50-Ω load impedances to provide suitable and realistic microwave loads.

5.7.2 Digital Circuit Measurements

The electro-optic sampler is ideally suited for logic timing and gate propagation delay measurements of GaAs digital circuits, providing general evaluation of new digital circuit technologies. Individual devices within a logic gate can be tested for their rise times and propagation delays. As discussed above, measurement results depend strongly on test conditions and will not adequately predict a gate's performance in a logic system unless representative test structures are used.

Ring oscillators, a standard IC test circuit, provide a measure of a gate delay from the frequency of a free-running signal propagating around an odd-numbered ring of inverters. These free-running circuits are not readily clocked with an external signal, making synchronization to the probe pulses for electro-optic sampling difficult. Inverter chains, however, must be clocked with an external signal, permitting the synchronization of probe pulses for electro-optic measurements of gate propagation delays. Inverter chains, each of which is a series of cascaded inverting logic gates, are used with sampling oscilloscopes to make average gate delay measurements; the propagation delay of the entire chain is measured and divided by the number of inverters to obtain the average delay of an individual inverter. For electro-optic testing the input inverter is switched with a microwave synthesizer, generating a square wave that ripples through the test structure. The first several inverters condition the input signal, sharpening the switching transients until the signal rise times and fall times reach a steady-state value. The time waveform, measured directly at individual gate nodes, allows for direct measurement of the propagation delays and signal rise times at gate nodes and nodes internal to single gates.

Figure 5–15 shows a gate delay measurement on an inverter chain implemented in 1 μm gate length buffered-FET logic MESFETs, and Fig. 5–16 shows an SEM picture of one inverter. The delay between curves A and B of Fig. 5–15 is the propagation delay of the inverting common-source stage, 60 ps, while the delay between curves B and C is the delay of the source-follower buffer and diode level-shifter, 15 ps. The inverter chain from Lawrence Livermore National Labs[43] consisted of 20 gates each with a fan-in and fan-out of unity.

The timing of this inverter chain has also been examined by Zhang *et. al.*,[44] optically triggering an inverter in the chain and using electro-optic sampling to measure the circuit response and gate propagation delays in a pump/probe configuration. A frequency-doubled portion of the probe beam (at λ = 532 nm) focused on the gate region of an FET photoconductively generates carriers, turning on the FET and generating a switching transient that propagates down the test structure. The probe beam, posi-

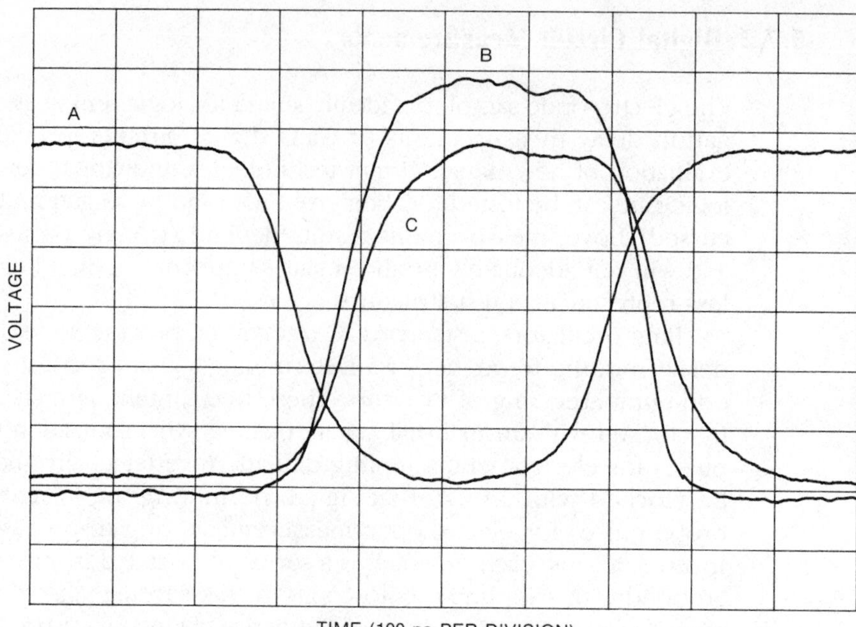

Fig. 5–15. Voltage waveforms at the input (A), source-follower gate (B), and output (C) of GaAs buffered-FET-logic inverter gate. ***(After Swierkowski, Mayeda, and McConaghy[43])***

tioned at a node after the switched gate, is successively delayed with respect to the switching pulse to map the transient waveform. This technique offers an all-optical approach, avoiding microwave connection to the IC, which is suitable for testing of simple IC test structures. For full-scale circuits optical triggering is an unsuitable drive signal.

A higher-level test circuit for IC performance is the static frequency divider, consisting of two flip-flops in a master-slave divide-by-two arrangement.[4] The maximum clock frequency of the divider, set by the propagation delays through the master-slave feedback path, provides an indirect measure of the devices' speed. Testing this circuit is normally accomplished by increasing the clock rate of the divider until its divide-by-two output fails.

The schematic of such a frequency divider is shown in Fig. 5–17. The circuit, from Hughes Research Laboratories, uses 0.2-μm e-beam written gates, molecular-beam epitaxy grown channels, air-bridge interconnects, and optimized feedback to achieve high-frequency clock rates (see Fig. 5–18). The dividers were implemented in two circuit families, buffered-FET logic (BFL) and capacitively enhanced logic (CEL). Conventional testing, using transmission line probes to drive the circuit and monitor its output

Fig. 5–16. A single inverter within the buffered-FET-logic inverter chain. ***(Courtesy Lawrence Livermore National Labs)***

on a spectrum analyzer, indicated correct circuit operation to 18 GHz. However, the spectrum analyzer gives inconclusive evidence of correct divider operation, since it measures only the output frequency and not the time waveforms. By direct waveform measurements using electro-optic sampling, correct divide-by-two operation was verified, gate propagation

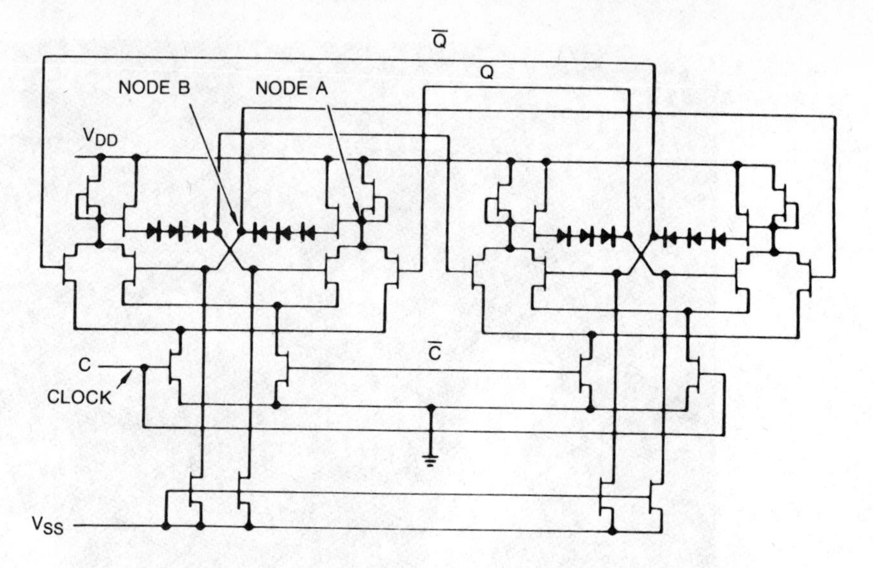

Fig. 5–17. Schematic diagram of a static frequency divider using buffer-FET logic. The labeled points refer to the electro-optically measured data in Fig. 5-19.

delays of 20 to 30 ps were measured and correlated to maximum clock frequencies (Fig. 5–19), and the internal delays through the inverting and source-follower stages of individual BFL gates were identified.[45] Note that while the scaled 0.2 μm gate length FETs had significantly shorter delay through the inverting stage, the delay through the buffer state was comparable to the 1-μm BFL from LLNL. This data suggests the speed limitation through the buffer stage is no longer *transistor* limited but limited by the resistance-capacitance time constant of the level-shifting diode resistance and the input capacitance of the cascaded gates.

The spatial resolution of the electro-optic sampler permits probing of MSI GaAs digital ICs to determine signal rise times and relative timing. Fig. 5–20 shows a serial output waveform probed on a 2-μm conductor internal to the output buffer in a 2.7-GHz 8-bit multiplexer/demultiplexer from TriQuint Semiconductor[46] and Fig. 5–21 shows the eight-phase clock waveforms probed on 4-μm metal interconnects. Similar measurements have recently been made on gigahertz logic flip-flops and counters.[47]

5.7.3 Microwave Circuit Measurements

At microwave and millimeter-wave frequencies, where conductor lengths and circuit element sizes often become large with respect to the electrical wavelength, direct measurements of conductor voltages and currents are difficult, particularly with conventional electrical test instrumentation. Di-

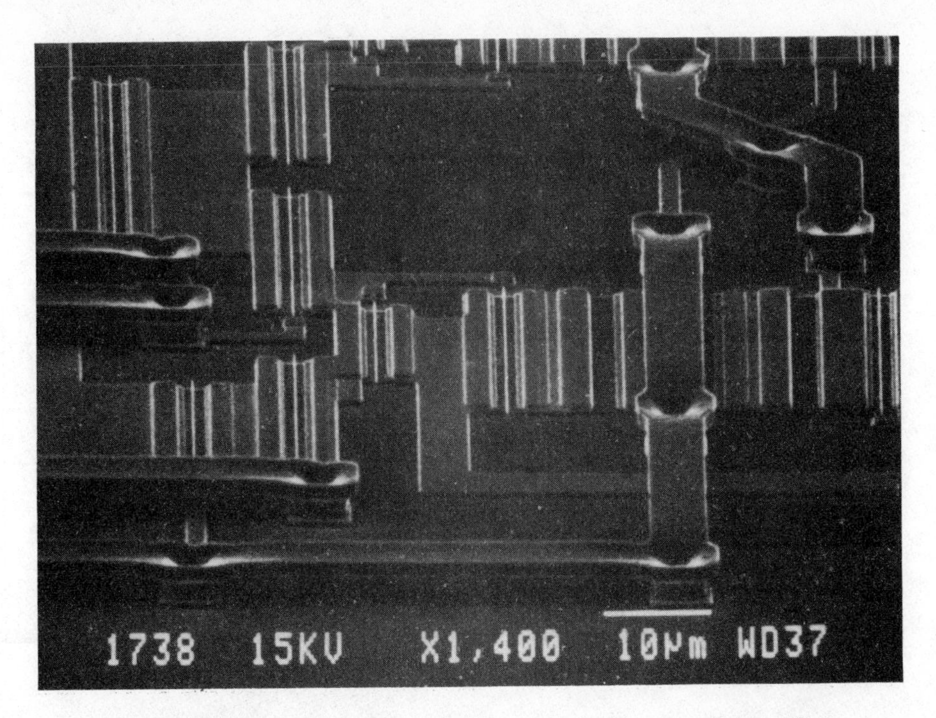

Fig. 5–18. Scanning-electron microscope picture of a section of an 18-GHz static frequency divider.[4] The circuit uses 0.2-μm e-beam written gates, molecular-beam epitaxially-grown channels, air-bridge lines to reduce interconnect capacitance, and optimized feedback to achieve high-frequency clock rates. *(After Jensen* et al.[4] *Courtesy Hughes Research Laboratories)*

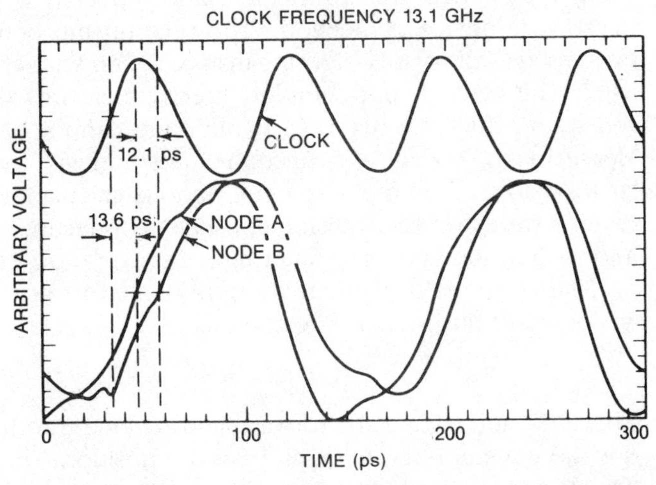

Fig. 5–19. Clock signal and divide-by-two waveforms measured with direct electro-optic sampling. The test points are indicated in Fig. 5-17.

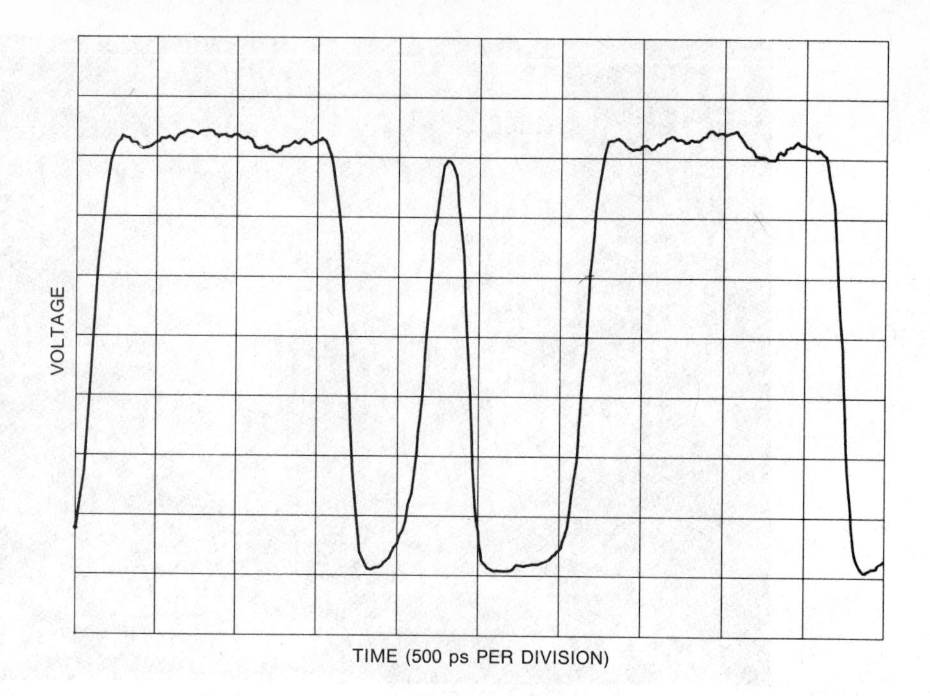

Fig. 5–20. Serial output waveform of a 2.7-GHz, 8-bit multiplexer/demultiplexer[46] measured by electro-optic sampling. The output word is 11110100.

rectional couplers and directional bridges separate the forward and reverse waves on a transmission line; standard microwave test instruments use these to measure the incident and reflected waves at the ports of a microwave device or network. The relationship between these waves is expressed as the wave scattering matrix, **S**, known as the *scattering parameters.*[48] The electro-optic sampler directly measures voltages, but not currents, and thus the two-port parameters cannot be directly measured. Measuring the voltage as a function of position with the optical probe, similar to a slotted-line measurement, permits calculation of the incident and reverse waves on the transmission lines connecting to devices. From this information the network scattering parameters can then be determined.

The vector voltage due to the sum of the forward and reverse traveling waves on a transmission line conductor is

$$V(z) = V^{+}e^{-j\beta z} + V^{-}e^{+j\beta z}, \tag{5–21}$$

where V^{+} and V^{-} are the forward and reverse traveling-wave coefficients, β is the wavenumber $2\pi/\lambda$, and z is the position. The traveling-wave coefficients are calculated by measuring this vector voltage as a function of position along a conductor using the optical probe then solving for these

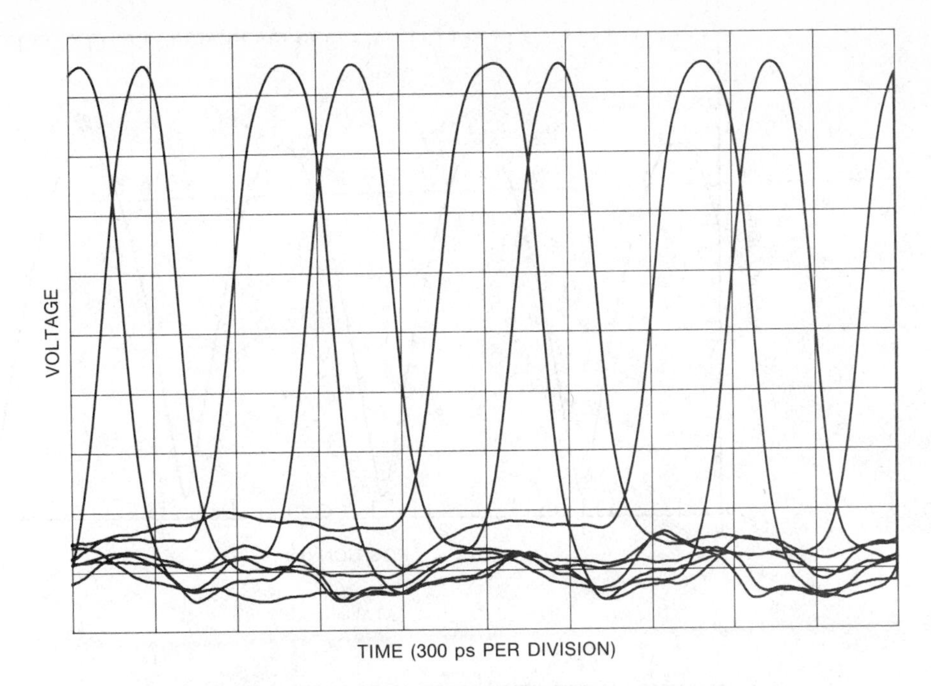

Fig. 5–21. Eight-phase clock waveforms in 2.7-GHz multiplexer. An asymmetrical 2.7-GHz clock input results in timing skew in the eight-phase counter.

coefficients. For a one-port transmission line the ratio of the traveling-wave coefficients V^+ and V^- is the reflection coefficient Γ, or S_{11}, the return loss. An example of such a measured standing wave and the calculated reflection coefficient for a drive frequency of 40 GHz is shown in Fig. 5–22. Extending this technique to calculate the incident and reflected waves on the input and output ports of a network allows for calculation of the *S*-parameters with a reference plane defined *on* the integrated circuit.

On GaAs microwave amplifiers and similar MMICs, the propagation of microwave signals internal to the circuit can be measured. Figure 5–23 shows a monolithic 2- to 18-GHz MESFET distributed amplifier from Varian Research Labs[49,50] with coplanar-waveguide transmission line interconnects. The circuit diagram is shown in Fig. 5–24. In a distributed amplifer, a series of small transistors are connected at regular spacings between two high-impedance transmission lines. The high-impedance lines and the FET input and output capacitances together form synthetic transmission lines, generally of 50-Ω characteristic impedance. Series stubs are used in the drain circuit, equalizing the phase velocities of the two lines and, at high frequencies, providing partial matching between the low impedance of the output line and the higher output impedances of the FETs, thereby

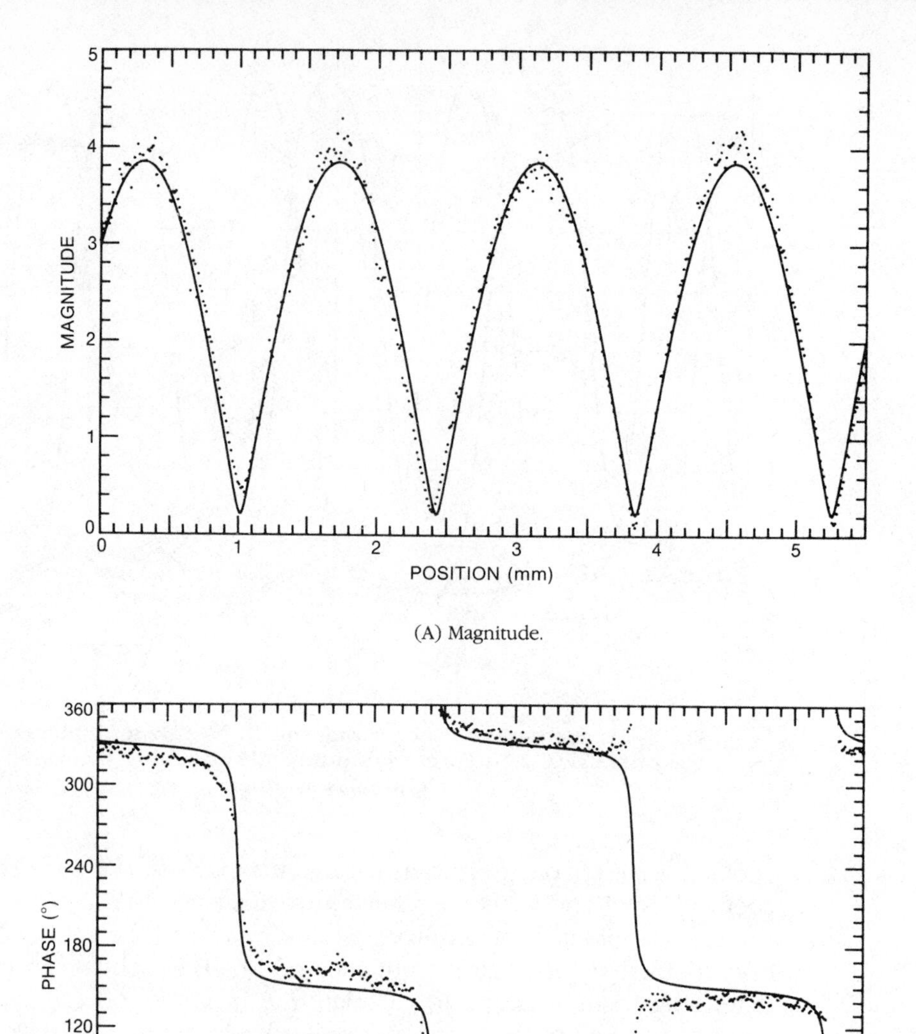

(A) Magnitude.

(B) Phase.

Fig. 5–22. A 40-GHz voltage standing wave on an open-terminated GaAs coplanar waveguide transmission line. The points are the data and the solid line is the fitted curve. From this measurement a reflection coefficient of 0.90 @ −80° is calculated. Each division of 10° in phase corresponds to 0.7 ps in time.

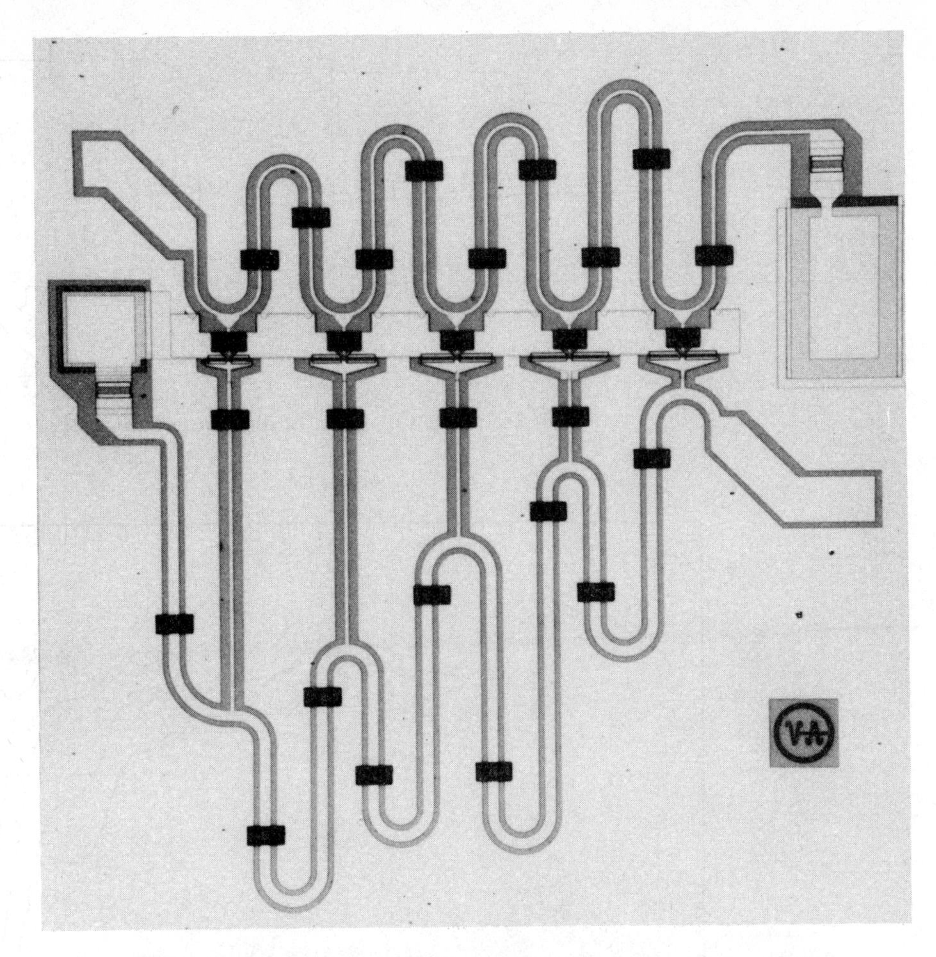

Fig. 5–23. Traveling-wave amplifier using coplanar-waveguide transmission line interconnects. *(After Zdasiuk* et al.[50] *Courtesy Varian Research Center)*

peaking the gain. Measurements of interest include the relative drive levels to the individual FETs as influenced by the loss and cutoff frequency of the synthetic gate line, the small-signal voltage at the drain of each FET, and identification of the saturation mechanisms leading to amplifier gain compression.

Figure 5–25 shows the small-signal gate voltages versus frequency for the amplifier of Fig. 5–23; several features can be noted. The rolloff beyond 18 GHz is the cutoff frequency of the periodically loaded gate line, the slow rolloff with frequency is the gate line attenuation arising from the real part of the FET input admittance, and the ripples are standing waves resulting from mistermination of the gate line (i.e. the load resistance not equal to the synthetic line's characteristic impedance).

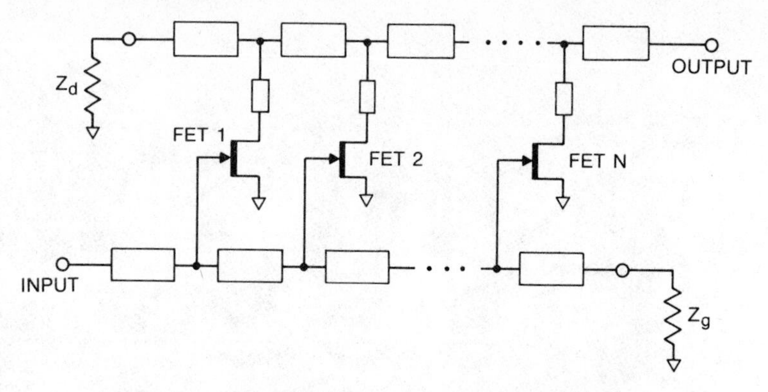

Fig. 5–24. Circuit diagram of distributed amplifier.

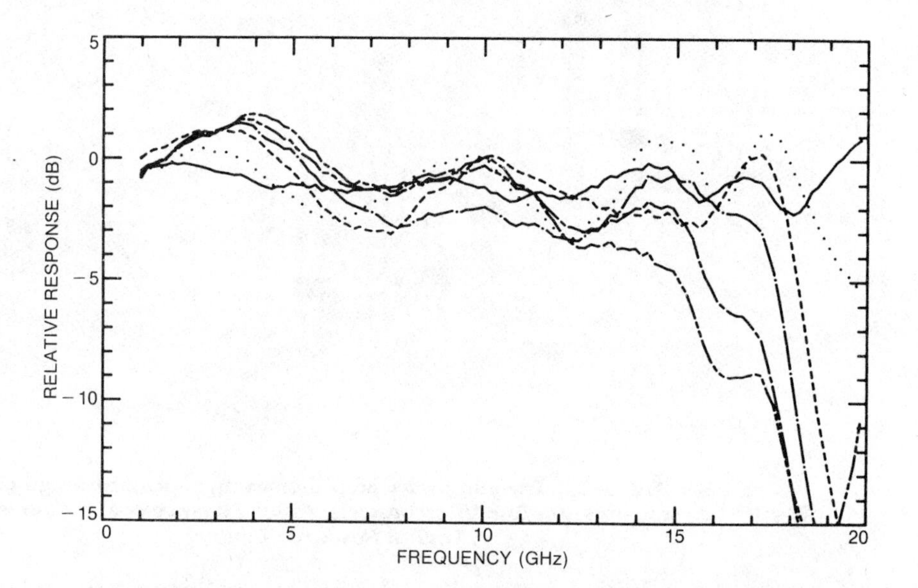

Fig. 5–25. Small-signal voltages at the five gates and at the gateline termination resistor in the coplanar 2- to 18-GHz distributed amplifier of Fig. 5–23.

Similar measurements can be made on MMICs using microstrip transmission lines. Figures 5–26 and 5–27 show the small-signal drain voltages versus frequency for a distributed amplifier with microstrip transmission lines (Fig. 5–28, also from Varian Associates). The strong variation with frequency and position results from interference of the forward and reverse waves on the drain transmission line.

Used in the synchronous sampling mode, the electro-optic sampler can measure the voltage waveforms at internal nodes under conditions of

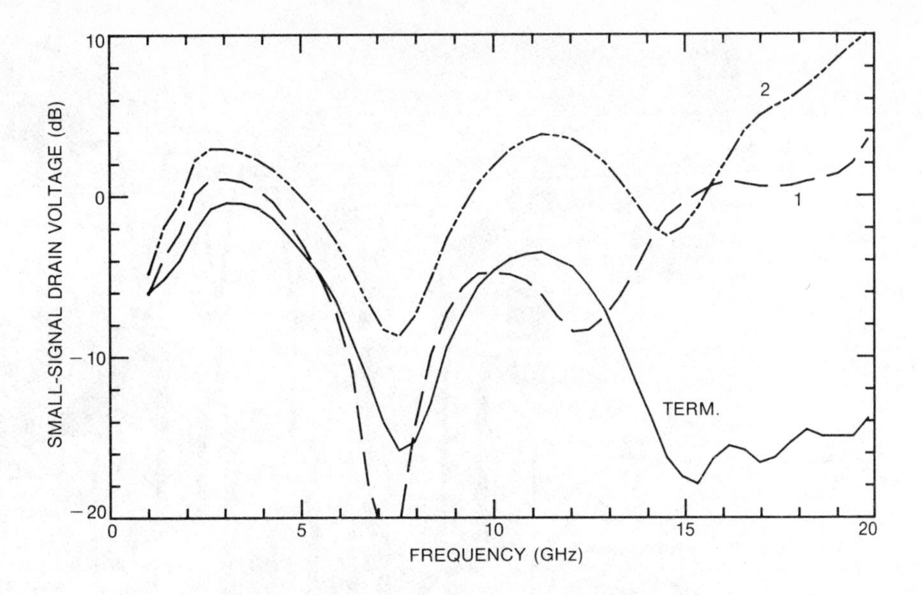

Fig. 5–26. Small-signal voltages at the first and second drains, and at the drainline reverse termination resistor, in a 2- to 18-GHz distributed amplifier using microstrip transmission lines. *(After Rodwell[49])*

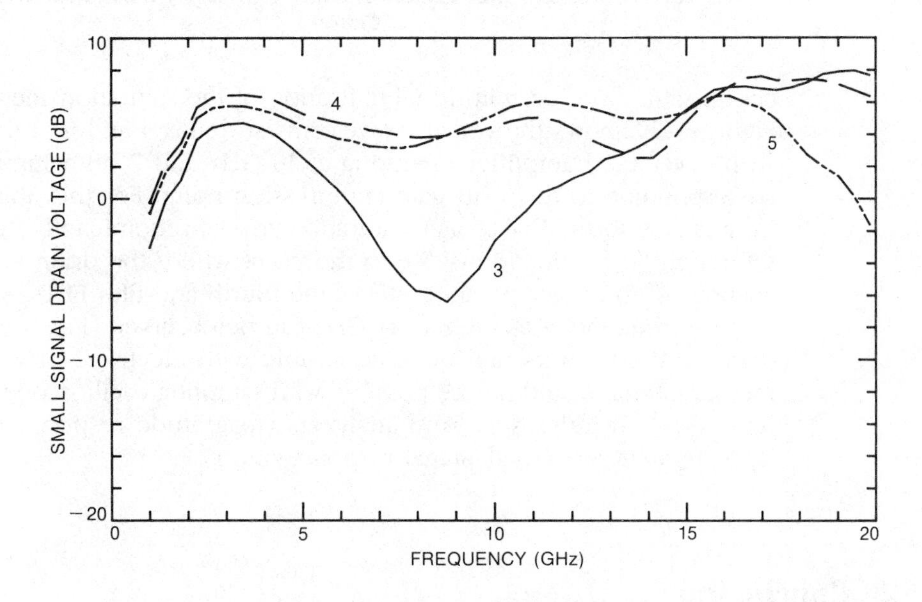

Fig. 5–27. Small-signal voltages at drains 3, 4, and 5 of the microstrip distributed amplifier.

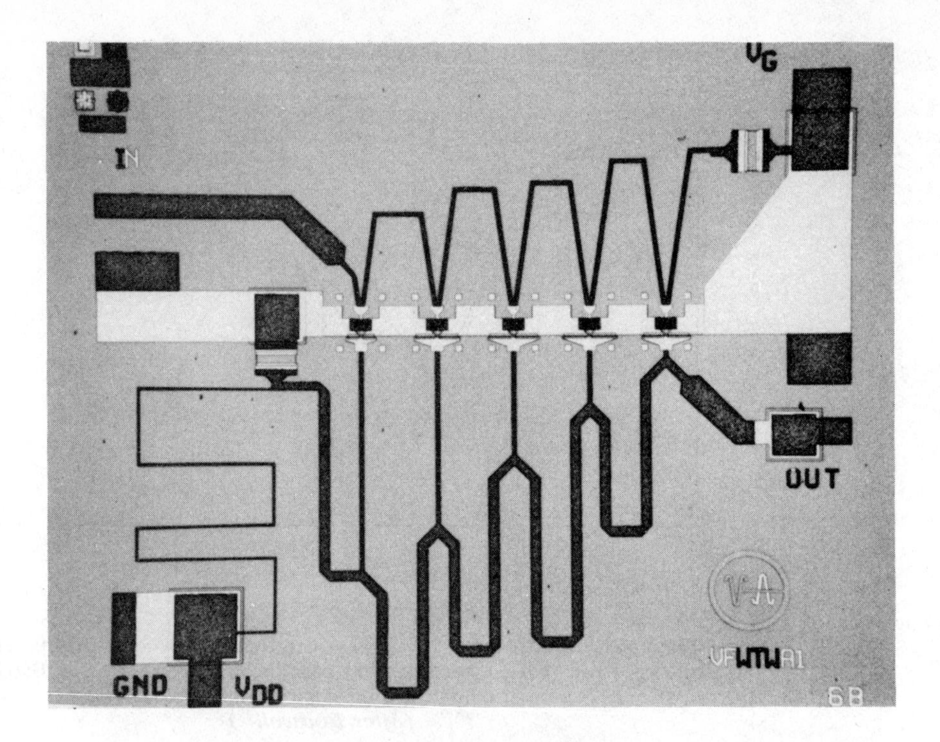

Fig. 5–28. Traveling-wave amplifier using microstrip transmission line interconnects. ***(After Zdasiuk*** **et al.**[50] ***Courtesy Varian Research Center)***

circuit saturation, permitting identification of the saturation mechanisms. Figure 5–29 shows the voltage waveforms at drains 4 and 5 of the microstrip distributed amplifier operating at 10 GHz and 7 dBm input power, corresponding to the 1-dB gain compression point. For this amplifier, at frequencies above 5 GHz, gain saturation arises predominantly from drain saturation (i.e. reduction of V_{dg} to the point where the drain end of the channel is no longer pinched off) of the fourth and fifth FETs. Saturation at drive frequencies as high as 21 GHz can be observed (Fig. 5–30). Even if the probed points had been accessible with electrical probes, these measurements would not be possible with sampling oscilloscopes (due to limited bandwidth), spectrum analyzers (magnitude response only), or network analyzers (small-signal response only).

5.8 Conclusion

A variety of new probing techniques for high-speed integrated circuits have been investigated, with the objective of providing an instrument

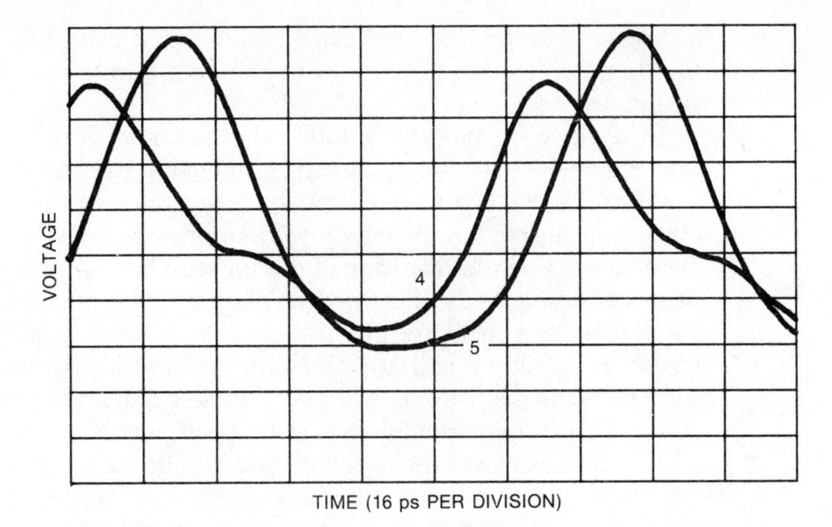

Fig. 5–29. Saturation at drains 4 and 5 of the microstrip distributed amplifier at 10 GHz and 7 dBm input power, the 1-dB gain compression point.

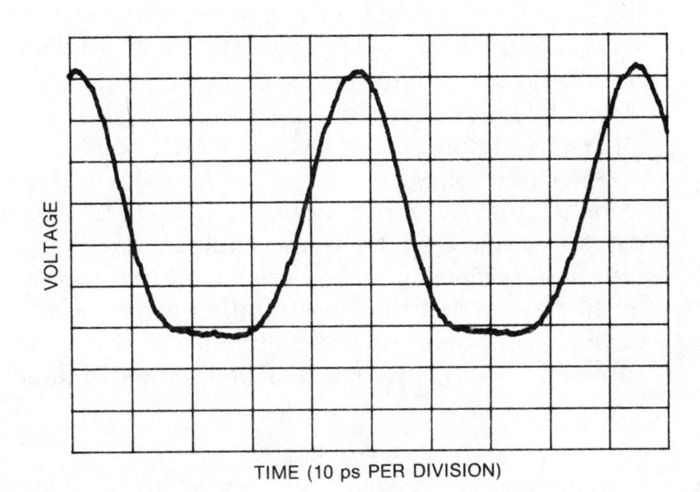

Fig. 5–30. Saturation at drain 3 of the microstrip distributed amplifier at 21-GHz drive frequency.

whose bandwidth is larger than that of the tested device, whose spatial resolution is sufficient to permit access to the most finely spaced conductors within these circuits, and whose circuit perturbation is insufficient to degrade measurement accuracy. Of the techniques reviewed in Sec. 5.3, all show some degree of invasiveness, whether due to electrical interconnections loading the circuit, due to dielectric materials introducing capacitive loading, or due to electron beams and electron extraction fields producing charge concentrations within the semiconductor substrate or at its surface.

In direct electro-optic sampling the substrate of the GaAs IC itself is used as the electro-optic modulator, eliminating external electro-optic elements and their associated invasive aspects, and permitting access to arbitrary points within the circuit without physical contact and with a resolution limited by the diameter of the focused laser probe beam. With the substrate serving as the electro-optic element, the optical properties of the circuit must be considered (polishing of the substrate backside and metallization reflectivity), and any absorbtion of the probe beam in the substrate through deep levels will result in some degree of perturbation to the circuit, placing limits on the probe beam intensity.

In addition to internal node testing, the direct electro-optic sampling system described in this chapter has several special features critical for IC testing. Synchronization of the laser probe pulses to the signal generator driving the IC permits operation of the circuit in its normal fashion, driven by microwave or digital signals from electronic sources, and the integration of the electro-optic sampling system with a microwave wafer probe station permits wafer-level optical probing of high-speed ICs. The sampler has been demonstrated on a variety of GaAs analog microwave and high-speed digital integrated circuits. Digital IC results include gate propagation delay and logic timing measurements of an 8-bit multiplexer/demultiplexer clocked at 2.6 GHz, a 20-gate inverter chain using 1-μm MESFET technology, and 18-GHz static frequency dividers using 0.2-μm MESFET technology. Results with microwave and millimeter-wave ICs include internal signal measurements on 2- to 18-GHz microstrip and coplanar waveguide traveling-wave amplifiers and voltage standing wave and reflection coefficient measurements on transmission lines at frequencies to 40 GHz. Extending the standing-wave measurements to two-port devices will permit on-wafer millimeter-wave scattering parameter measurements with the reference planes located at the device terminals.

References

1. H. Q. Tserng and B. Kim, "110 GHz GaAs FET oscillator," *Electron. Lett.,* 21:178 (1985).
2. T. Henderson *et al.,* "Microwave performance of a quarter-micrometer gate low-noise pseudomorphic InGaAs/AlGaAs modulation-doped field effect transistor," *IEEE Electron. Dev. Lett.,* EDL-7 (1986).

3. T. C. L. G. Sollner, E. R. Brown, W. D. Goodhue, and H. Q. Le, "Observation of millimeter-wave oscillations from resonant tunneling diodes and some theoretical considerations of ultimate frequency limits," *Appl. Phys. Lett.,* 50 (6): 332 (1987).
4. J. F. Jensen, L. G. Salmon, D. S. Deakin, and M. J. Delaney, "Ultra-high speed GaAs static frequency dividers," *Tech. Dig. 1986 Int. Electron Div. Mg.,* 476 (1986).
5. E. W. Strid, K. R. Gleason, and T. M. Reeder, "On-wafer measurement of gigahertz integrated circuits," *VLSI Electronics: Microstructure Science, Vol. 11,* New York: Academic Press (1985).
6. Cascade Microtech, Inc., P.O. Box 2015, Beaverton, OR 97075.
7. Hypres, Inc., 175 Clearbrook Rd., Elmsford, NY 10523.
8. T. E. Everhart, O. C. Wells, and R. K. Matta, "Evaluation of passivated integrated circuits using the scanning electron microscope," *J. Electrochem. Soc.,* 111:929 (1964).
9. C. W. Oatley, "Isolation of potential contrast in scanning electron microscope," *J. Phys. E: Sci. Instrum.,* 2:742 (1969).
10. "Electron-beam testing of VLSI chips gets practical," *Electronics,* 51–54 (March 24, 1986).
11. R. Iscoff, "E-beam probing systems: filling the submicron gap," *Semiconductor International,* 62–68 (September 1985).
12. P. May, J.-M. Halbout, and G. Chiu, "Picosecond photoelectron scanning electron microscope for noncontact testing of integrated circuits," *Appl. Phys. Lett.,* 51:145 (1987).
13. R. B. Marcus, A. M. Weiner, J. H. Abeles, and P. S. D. Lin, "High-speed electrical sampling by fs photoemission," *Appl. Phys. Lett.,* 49:357 (1986).
14. J. Bokor, A. M. Johnson, R. H. Storz, and W. M. Simpson, "High-speed circuit measurements using photoemission sampling," *Appl. Phys. Lett.,* 49:226 (1986).
15. A. M. Weiner, P. S. D. Lin, and R. B. Marcus, "Picosecond temporal resolution photoemissive sampling," *Appl. Phys. Lett.,* 51:358 (1987).
16. H. K. Heinrich, D. M. Bloom, and B. R. Hemenway, "Noninvasive sheet charge density probe for integrated silicon devices," *Appl. Phys. Lett.,* 48:1066 (1986).
17. H. K. Heinrich, B. R. Hemenway, K. A. McGroddy, and D. M. Bloom, "Measurement of real-time digital signals in a silicon bipolar junction transistor using a noninvasive optical probe," *Electron. Lett.,* 22:650 (1986).

18. J. A. Valdmanis, G. A. Mourou, and C. W. Gabel, "Picosecond electro-optic sampling system," *Appl. Phys. Lett.,* 41:211 (1982).

19. B. H. Kolner, D. M. Bloom, and P. S. Cross: "Electo-optic sampling with picosecond resolution," *Electron. Lett.,* 19:574 (1983).

20. J. A. Valdmanis, G. A. Mourou, and C. W. Gabel, "Subpicosecond electrical sampling," *IEEE J. Quantum Electron.,* QE-19: 664 (1983).

21. J. A. Valdmanis and G. Mourou, "Subpicosecond electrooptic sampling: principles and applications," *IEEE J. Quantum Electron.,* QE-22:69 (1986).

22. J. A. Valdmanis and S. S. Pei, "1-THz-bandwidth prober for high-speed devices and integrated circuits, *Electron. Lett.,* 23:1308 (1987).

23. B. H. Kolner and D. M. Bloom, "Direct electrooptic sampling of transmission-line signals propagating on a GaAs substrate," *Electron. Lett.,* 20:818 (1984).

24. B. H. Kolner and D. M. Bloom, "Electrooptic sampling in GaAs integrated circuits," *IEEE J. Quant. Electron.,* QE-22:79 (1986).

25. I. P. Kaminow, *An Introduction to Electro-Optic Devices,* New York: Academic Press (1974).

26. I. P. Kaminow and E. H. Turner, "Electrooptic light modulators," *Proc. IEEE,* 54:1374 (1966).

27. A. Yariv and P. Yeh, *Optical Waves in Crystals,* New York: New York: John Wiley & Sons, 286–287 (1984).

28. *Ibid,* p. 230.

29. J. L. Freeman, S. K. Diamond, H. Fong, and D. M. Bloom, "Electrooptic sampling of planar digital integrated circuits," *Appl. Phys. Lett.,* 47:1083 (1985).

30. W. H. Knox, R. L. Fork, M. C. Downer, R. H. Stolen, and C. V. Shank, "Optical pulse compression to 8 fs at a 5-kHz repetition rate," *Appl. Phys. Lett.,* 46:1120 (1985).

31. R. N. Bracewell, *The Fourier Transform and Its Applications*, New York: McGraw-Hill (1978).

32. M. J. W. Rodwell, K. J. Weingarten, D. M. Bloom, T. Baer, and B. H. Kolner, "Reduction of timing fluctuations in a mode-locked Nd: YAG laser by electronic feedback," *Opt. Lett.,* 11:638 (1986).

33. E. B. Treacy, "Optical pulse compression with diffraction gratings," *IEEE J. Quant. Electron.,* QE-5:454 (1969).

34. D. Grischkowsky and A. C. Balant, "Optical pulse compression based on enhanced frequency chirping," *Appl. Phys. Lett.,* 41:1 (1982).

35. J. D. Kafka, B. H. Kolner, T. Baer, and D. M. Bloom, "Compression of pulses from a continuous-wave mode-locked Nd:YAG laser," *Opt. Lett.,* 9:505 (1984).

36. A. J. Taylor, J. M. Wiesenfeld, G. Eisenstein, R. S. Tucker, J. R. Talman, and U. Koren, "Electrooptic sampling of fast electrical signals using an InGaAsP injection laser," *Electron. Lett.,* 22:61 (1986).

37. A. J. Taylor, J. M. Wiesenfeld, R. S. Tucker, G. Eisenstein, J. R. Talman, and U. Koren, "Measurement of a very high-speed InGaAs photodiode using electrooptic sampling," *Electron. Lett.,* 22:325 (1986).

38. J. M. Wiesenfeld, R. S. Tucker, A. Antreasyan, C. A. Burrus, and A. J. Taylor, "Electrooptic sampling measurements of high-speed, InP integrated circuits," *Appl. Phys. Lett.,* 50:1310 (1987).

39. A. S. L. Gomes, W. Sibbet, and J. R. Taylor, "Generation of subpicosecond pulses from a continuous-wave mode-locked Nd:YAG laser using a two-stage optical compression technique," *Opt. Lett.,* 10: 338 (1985).

40. B. Zysset, W. Hodel, P. Beaud, and H. P. Weber, "200-femtosecond pulses at 1.06 μm generated with a double-stage pulse compressor," *Opt. Lett.,* 11:156 (1986).

41. M. J. W. Rodwell, K. J. Weingarten, J. L. Freeman, and D. M. Bloom, "Gate propagation delay and logic timing of GaAs integrated circuits measured by electro-optic sampling," *Electron. Lett.,* 22:499 (1986).

42. K. J. Weingarten, M. J. W. Rodwell, and D. M. Bloom, "Picosecond sampling of GaAs integrated circuits," *Picosecond Electronics and Optoelectronics,* New York:Springer-Verlag, 18 (1987).

43. S. Swierkowski, K. Mayeda, and C. McGonaghy, "A sub-200 picosecond GaAs sample-and-hold circuit for a multi-gigasample/second integrated circuit," *Tech. Dig. 1985 Int. Electron Dev. Mg.,* 272–275 (1985).

44. X.-C. Zhang and R. K. Jain, "Measurement of on-chip waveforms and pulse propagation in digital GaAs integrated circuits by picosecond electro-optic sampling," *Electron. Lett.,* 22:264 (1986).

45. J. F. Jensen, K. J. Weingarten, and D. M. Bloom, *Picosecond Electronics and Optoelectronics,* New York: Springer-Verlag, 184 (1987).

46. G. D. McCormack, A. G. Rode, and E. W. Strid, "A GaAs MSI 8-bit multiplexer and demultiplexer," *Proc. 1982 GaAs IC Symp.,* 25–28 (1982).

47. X.-C. Zhang, R. K. Jain, and R. M. Hickling, "Electrooptic sampling analysis of timing patterns at critical internal nodes in gigabit GaAs

multiplexers/demultiplexers," *Picosecond Electronics and Optoelectronics,* New York: Springer-Verlag, 29 (1987).

48. R. E. Collins, *Foundations of Microwave Engineering,* New York: McGraw-Hill (1966).
49. M. J. W. Rodwell, M. Riaziat, K. J. Weingarten, B. A. Auld, and D. M. Bloom, "Internal microwave propagation and distortion characteristics of travelling-wave amplifiers studied by electro-optic sampling," *IEEE Trans. Microwave Theory Technol.,* MTT-34:1356 (1986).
50. G. Zdasiuk, M. Riaziat, R. LaRue, C. Yuen, and S. Bandy, "Enhanced performance ultrabroadband distributed amplifiers," *Picosecond Electronics and Optoelectronics,* New York: Springer-Verlag, 160 (1987).

Chapter 6

Growth and Properties of GaAs-based Devices on Si

H. Ünlü, H. Morkoç, and S. Iyer

Advances in the science and technology of heterostructure thin films have revolutionized semiconductor research programs. The scientifically and technologically important semiconductors GaAs and Si can now be used to form heterojunctions despite the 4-percent lattice parameter difference. The ability to synthesize and tailor the structure of thin films has already made it possible to demonstrate GaAs-based high-speed transistors, optical devices and integrated circuits on silicon substrates. These hetero-structure systems require expertise in semiconductor physics, material science and device engineering. The concept, technology and performance of these devices are reviewed. Issues fundamental to the growth and characterization of GaAs/Si heterostructures and GaAs-based devices on Si substrates are discussed in this chapter.

6.1 Introduction

Recent advances in lattice mismatched epitaxial technology are making possible the demonstration of GaAs-based high-speed devices on Si substrates which may have profound implications in the field of microelectronics. Until recently, GaAs-based heterostructures with extremely high speed performance and light emission capability could only be prepared on GaAs substrates which suffer from many imperfections. On the other hand, silicon, the second most abundant material on earth, is extremely attractive as a substrate because of its high mechanical strength, high degree of crystalline perfection, large diameter (6 to 8 inches), and low cost. It is now possible to combine the excellent properties of Si as a substrate material with the unique electronic and optical properties of heterostructures that are made of III-V and II-VI semiconductor compounds,[1] as shown in Fig. 6–1, where energy gaps versus lattice constants for various semiconductors are illustrated. For example, highly developed silicon integrated circuit technology and high-speed GaAs and large-bandgap (Al, Ga)As technology can be integrated for the development of a new class of device systems. Examples of such systems would be the utilization of high current GaAs devices to drive off-chip pins in low-current NMOS and CMOS chips to increase the speed and optical interconnects.

In future microelectronics, the Si integrated circuits may have high-speed GaAs local-area core circuits. This hybrid integration may also involve GaAs light emitters and Si light detectors[2] to replace pins and interconnects for the ultimate speed. The successful preliminary experiments on HgCdTe/GaAs/Si also show the potential development of focal plane array infrared detectors for image processors.[3] The availability of high-quality Si substrates allow the large-area arrays to be fabricated, with HgCdTe being used for detection and the underlying Si circuit for signal processing.

GaAs-on-Si technology, although explored for solar cell applications for its low cost, did not originally look promising because of the antiphase disorder and 4.1-percent lattice mismatch. The antiphase domain is a region of the GaAs crystal where the nearest neighbor bond is not between a Ga and an As atom but between two Ga or two As atoms. With the advent of a highly controllable vacuum process, molecular-beam epitaxy (MBE),

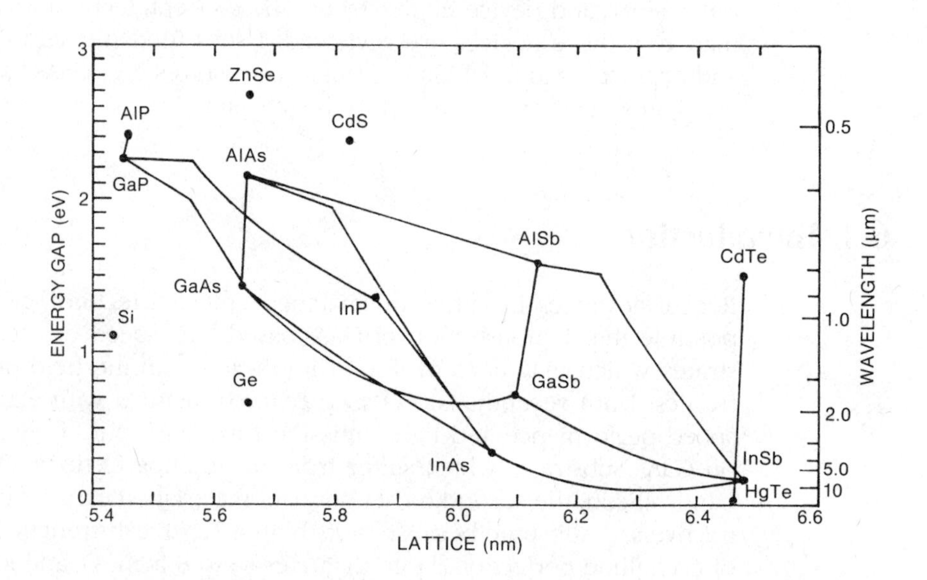

Fig. 6–1. Energy gap versus lattice constant of elemental, III-V and II-VI compound semiconductors.

the challenges have been overcome. Antiphase domains have been entirely suppressed by the proper choice of substrate orientation[4] and a method of producing atomically clean Si surfaces. A Si surface which is cut a few degrees off the (100) direction has predominantly double steps (i.e., the step height is twice a single atomic step). MBE growth of GaAs on such surfaces leads to material free of antiphase domains, a fact that has been confirmed by the combination of Raman scattering experiments and transmission electron microscopy (TEM) studies,[5] X-ray scattering,[6] and selective anisotropic etching techniques.[7]

The 4.1-percent lattice mismatch can lead to imperfections called misfit dislocations at a density of 1×10^{12} cm^{-2} which can destroy any effort to produce high-quality GaAs-based devices in the epilayer on Si substrates. Furthermore, solute concentration gradients imposed by diffusion of large amounts of impurities (e.g. B, P, As) have been known to cause generation of misfit dislocation networks in highly perfect single crystal silicon wafers.[8] Thermal expansion mismatch between GaAs and Si results in strain during the cooldown to room temperature after growth, which may affect the long-term stability of the material. By forcing the dislocations to propagate in a plane parallel to the GaAs/Si interface and thus away from the epilayer surface, GaAs epilayers with dislocation densities of about 1×10^{5} cm^{-2} have been prepared.[4] Strained-layer superlattices and thermal annealing techniques have also been shown to confine the dislocations at the GaAs/Si interface.

To demonstrate the high potential of GaAs on Si, majority and minority carrier high-speed devices have been characterized at both DC and RF levels. GaAs metal gate (0.25 μm gate length) semiconductor field-effect transistors (MESFETs) on Si substrates have been fabricated.[9] The transconductance is about 360 mS/mm, with a cutoff frequency of 55 GHz and a noise figure of 2.8 dB at 18 GHz (1.4 dB higher than a comparable device on GaAs substrates). High performances were also obtained with (Al, Ga)As/GaAs modulation doped FETs (MODFETs) (1 μm gate lengths) which rely on interface conduction between wide-bandgap (Al, Ga)As and narrow-bandgap GaAs.[10] The obtained current gain cutoff frequency is about 15 GHz both on GaAs and Si substrates.

Minority carrier and optical devices, which are more sensitive to defects, have also been demonstrated. The (Al, Ga)As/GaAs heterojunction bipolar transistors (HBTs) with 4- $\times$ 20-μm emitter size showed a current gain cutoff frequency of 30 GHz.[11] The high quality of Si encouraged the exploration of light emitters as well, which are used for light wave communication on Si circuits. The monolithic integration of GaAs MODFETs with Si N-MOS devices without any degradation of Si devices has also shown the compatibility of GaAs and Si device technology.[12]

Furthermore, vertical and lateral integration with the use of two- or three-level heteroepitaxy (fluorides and silicides) is certain to create new opportunities and applications as well as new and exciting device concepts. In this chapter the state of the art of molecular-beam epitaxy of GaAs on Si substrates and phenomena fundamental to the understanding of interfacial properties are discussed. In addition, properties and factors affecting the performances of GaAs-based majority and minority carrier and optical devices grown on Si substrates are also discussed.

6.2 Why Gallium Arsenide on Silicon?

Ever since the invention of transistors in the late 1940s, remarkable advances have been made in the science and technology of semiconductor devices. Over the past ten years or so, device dimensions, in both the vertical and lateral directions, have been greatly reduced to the point where quantum size effects must now be included to describe transistor operations accurately. Short transit times simply imply that the input signal can be transmitted to the output with minimal delay. The reduction in the terminal spacing (lateral dimension) is closely tied to the state of the art of device processing technologies such as pattern generation and transfer. Currently, the minimum dimension obtained reproducibly by optical lithography is about 0.5 μm. For smaller dimensions, X-ray lithography and electron-beam lithography must be used. It is obvious that increased speed by size reduction only is analogous to reducing the distance that must be traveled. It must also be remembered that the transit time can be reduced very effectively by choosing semiconductor structures in which the signal propagates at a faster rate. Since transit times in tens of picoseconds or less are involved, the ability to reach the ultimate velocity is very important as well. This acceleration is termed the *mobility* in semiconductors, which is a measure of how fast the electrons can reach their ultimate velocity in an accelerating electric field.

The electron mobility and velocity are determined by the semiconductor material system. For example, GaAs has very high low-field electron mobility [about six times that of Si for a donor (responsible for free electrons) concentration of 1×10^{17} cm^{-3}]. Second, the peak velocity in GaAs is almost twice that of Si and occurs at much lower (3.5 kV/cm) electric fields, meaning that GaAs devices can operate much faster than Si devices for comparable sizes and at much smaller voltages. This can be seen in Fig. 6–2, where the room temperature electron velocity versus field characteristics of GaAs, Si, InP, and $In_{0.53}Ga_{0.47}As$ semiconductors are compared. Additionally, the larger electron mobility means smaller access resistances and reduced degradation of device performance. This significant difference between GaAs and Si is due to the fact that the lowest energy state in the GaAs conduction band occurs at the point of zero wavevector (commonly denoted by Γ and is nondegenerate). Mobile electrons tend to occupy states near this minimum at thermal equilibrium. The effective mass of these electrons is $0.067m_0$ (m_0 is the mass of a free electron in vacuum) versus $0.19m_0$ in Si, which accounts for the superior electron mobility in GaAs over Si.

The larger direct bandgap coupled with a short minority carrier lifetime gives GaAs a distinct advantage over Si in optical signal generation via electrical stimulation. Gallium arsenide substrates can be grown with very high resistivities which lend themselves to use as a dielectric medium for high-frequency microwave and millimeter-wave integrated circuits. The high-resistivity GaAs substrate also simplifies device isolation for digital circuits. GaAs has other advantages which are of paramount importance as well. For example, the ternary $Al_xGa_{1-x}As$ with its larger bandgap lattice matches GaAs and together they form heterojunctions which are the basis for modern devices. The barrier formed in the conduction band between GaAs and $Al_xGa_{1-x}As$ due to the abrupt change in energy bandgap can be used to confine electrons with properties much more conducive to devices. With sophisticated growth techniques, it is possible to grow III-V heterostructures on GaAs substrates with properties tailored for high-frequency performance. The optical properties of GaAs heterostructures make combinations of GaAs digital, microwave, and optical circuits feasible.

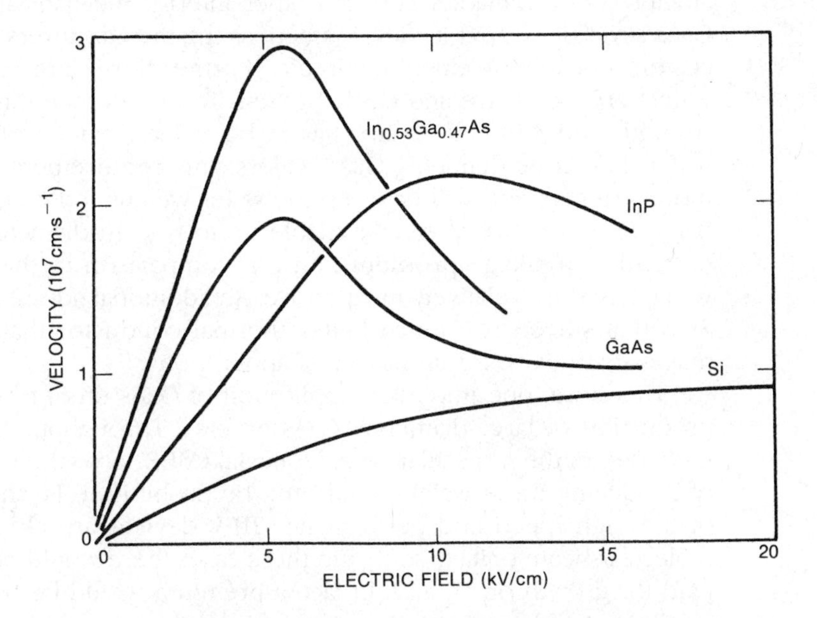

Fig. 6–2. Velocity-field characteristics of Si, GaAs, InP and $In_{0.53}Ga_{0.47}As$ semiconductors at room temperature. The donor concentration in all cases is about 1×10^{15} cm^{-3} range.

Because of the highly developed processing technology that exists for Si, many advantages of GaAs and other III-V compounds are not fully utilized at present. There are certain functions for which GaAs and other III-V compounds are better suited than Si, and, moreover, others which cannot be done at all with Si but can be done with the III-V compounds. GaAs and other III-V compound heterostructures have more drive capability than do Si devices for the same level of power dissipation. Therefore, they are well suited for off-chip communications (communications from one integrated-circuit chip to another) which require high drive capability. Furthermore, the possibility of optical off-chip communication, which would reduce the drive requirements, exists in GaAs whereas it does not in Si.

Despite its great potential, GaAs has shortcomings some of which are associated with the fact that it is a binary semiconductor. In order to prevent dissociation of the surface (As desorption), precise control over temperatures must be maintained during processing. The use of diffusion to achieve the desired doping properties has been more or less unsuccessful in GaAs. Unlike Si, GaAs does not have a stable, easily grown native oxide which has been so important to the development of silicon metal-oxide field-effect transistor (MOSFET) technology. Further, the surface of GaAs is more susceptible to attack by chemicals used in processing.

At present, high-quality GaAs substrate material larger than a 3-in diameter is not available. This and other inferior mechanical properties of GaAs severely restrict its development as the manufacturers must establish custom lines to produce its circuits. Furthermore, large-diameter GaAs wafers are expensive and easily broken. To provide the same mechanical strength, the 3-in GaAs wafer has to be at least twice as thick as the Si wafer. For large-diameter GaAs wafers, this requirement will be much more stringent and will drive up the wafer weight and cost. On the other hand, Si wafers are already available up to 8 in in diameter without any particular breakage problems, and in comparison to the cost of GaAs wafers, they are relatively inexpensive. An additional advantage of GaAs on Si is that silicon is a much better thermal conductor than GaAs, which makes it attractive for a number of applications.

Therefore, one important application of GaAs on Si technology is the production of large-diameter "GaAs wafers." Throughput can thereby be increased in the molecular-beam epitaxial (MBE) growth of GaAs. The cost of producing these wafers would not be prohibitive. Furthermore, many of the high-speed and performance III-V devices are already grown by molecular-beam epitaxy so that in these cases there would be no premium paid for growth on Si, and, in fact, a premium would be paid for growth on GaAs. Another application of this technology would be in monolithic GaAs-over-Si systems. It has already been demonstrated that Si device fabrication is compatible with the GaAs-on-Si process. Therefore, the possibility of using GaAs devices in Si integrated circuits for critical areas where high speed or high drive capability are needed, is also made available. As mentioned above, optical signals cannot be generated in Si. Therefore, if a direct-bandgap semiconductor such as GaAs, which can generate optical signals, is monolithically integrated with Si, then in essence the "Si circuit" could generate optical signals.

The more highly developed processing technology of Si over GaAs allows very high density circuits of high complexity. On the other hand, GaAs has intrinsically higher performance advantages over Si. Therefore, the monolithic integration of GaAs and Si is an extremely attractive possibility for high-speed devices.

6.3 Growth of GaAs on Si

The growth of polar semiconductor on nonpolar semiconductor was first done by Anderson, who fabricated the first Ge/GaAs heterojunction.[13] Later studies focused on applications to heterojunction bipolar transistors (HBTs) in the GaP/Si system,[14] and to GaAs solar cells in the GaAs/Ge system.[15] Although nonpolar on polar epitaxy met with marginal success[13] the reverse case, polar on nonpolar epitaxy, has proved to be much more difficult.

The principal difficulty in growing polar material on a nonpolar substrate is that there is little preference, if any, in bonding sites for cations (e.g. Ga) and anions (e.g. As) in the initial growth stages. Therefore, the growth in some regions may begin with a cation plane, while in some others with an anion plane. This was demonstrated in the growth of GaAs/Ge superlattices, where the GaAs was observed to grow in islands on (100) Ge with antiphase disorder.[16] Studies of vapor phase epitaxial (VPE) GaAs on (100) Ge and GaP (100) Si also showed antiphase disorder.[17] Wright *et al.*[18] have suggested the use of higher index planes such as (211). Since there is a different bond structure for each of the two nonpolar sublattices with this orientation, a preferential bonding takes place which suppresses the antiphase domains.

As mentioned in the introduction, the second difficulty is the 4.1-percent lattice mismatch, the GaAs lattice being larger than that of Si. Therefore, for every 25 atomic planes one dislocation is required, which results in a minimum point-like misfit dislocation density of 1×10^{12} cm^{-2} at the substrate epilayer interface. The large difference between thermal expansion coefficients α of GaAs (6.0×10^{-6} K^{-1}) and Si (4.2×10^{-6} K^{-1}) results in appreciable strain after cool down from growth. Some have taken the approach of growing a Ge interlayer on Si followed by the GaAs/(Al, Ga)As,[19–22] as a large degree of success has been obtained in growing

GaAs on (100) Ge.[15,23,24] The successful growth of GaAs directly on (100) Si has shown that the method of a Ge interlayer is not necessary.[25–26]

6.3.1 Substrate Preparation

In the epitaxy of polar GaAs on nonpolar Si, the first task is to find a surface preparation technique for the Si substrate which will produce an atomically clean surface for the initiation of the epitaxial growth. Standard surface cleaning techniques are used to remove the native oxide, oxygen and carbon contaminants. The surface is then protected (while being transferred from the wet cleaning bench to the MBE) with a chemically grown thin oxide layer. Since the oxides of Si are very stable, very high temperatures are necessary to thermally desorb them. This high temperature requirement to thermally desorb the Si oxide can be reduced somewhat by the use of special preparation techniques that produce very thin and volatile oxide layers that are much more readily desorbed. Several methods have been reported in the literature.[18,27,28] Of these, there are basically two types: those that rely on thermal desorption of a thin volatile oxide layer and those that use a group III beam to reduce the Si oxide, such as Al,[25] Ga,[29] and In.[30] These techniques do not require a very high out-gassing temperature. However, they will most likely cause diffused layers of the reducing element in the substrate, which would cause problems for device fabrication. The second kind of surface cleaning technique relies on thermal desorption of a thin volatile oxide layer. A detailed outline of the method is given in Chart 6–1. The basic method was originally developed by Henderson[31] and consists of several iterations of growing and stripping a chemical oxide. The final stage of this scheme produces a thin volatile oxide layer which can be desorbed thermally at about 800 to 900°C.

Figure 6–3 shows the auger electron spectra of the surface of a Si sample before and after oxide desorption. As shown, no carbon or oxygen is detected. This technique should be able to detect impurities at the level of about (0.01 monolayer) coverage. There, the oxide is completely desorbed and no detectable residual carbon is introduced. It is important that the preparation technique does not leave carbon contamination, because carbon will react with Si at these temperatures to form SiC, which is impossible to remove thermally.

6.3.2 Growth Procedure

After a clean substrate surface is obtained, the growth can be initiated. The problem of antiphase domain suppression can be overcome if the proper growth conditions are used to start the polar on nonpolar epitaxy. For a

Chart 6–1. Processing Steps for Si Substrate Preparation (Revised from Fischer[45])

The first step in the process is to cut the wafers into ~1-in × 1-in squares for mounting in special holders designed for direct heating in a Riber model 1000 MBE system. The wafers are scribed along ⟨011⟩ directions using a Tempress automatic diamond stylus scriber. Once lines are scribed, it is possible to break the Si wafers along the scribe lines to get accurate control of size.

The wafers are then subjected to a heavy metal removal process.

1. Place wafer in 1:1 (HNO_3:H_2SO_4) solution at 70°C for 5 min.
2. Rinse well in deionized H_2O.

Once the organic contaminants have been removed by the above procedures, the wafers are subjected to an oxide/oxide removal treatment:

1. Place wafer in oxidizing solution consisting of 5:3:3 (HCl:H_2O:H_2O_2) at 80°C for 5 min.
2. Rinse wafer in deionized H_2O.
3. Place wafer in oxide stripping solution consisting of 1:10(HF:H_2O) for 20 s.
4. Rinse wafer and place in 5:3:3 oxidizing solution at 80°C for 2 min.
5. Rinse in H_2O (deionized).
6. Place wafer in 1:10 oxide stripping solution for 20 s.
7. Rinse in H_2O and repeat steps 4 through 7 until the wafer has seen four 2-min oxidations.
8. Place wafer in 5:3:3 oxidizing solution at 80°C for 5 min. This is the final oxide to be removed by thermal desorption in the MBE system.
9. Rinse thoroughly in deionized H_2O and blow dry with filtered N_2 gas.

After step 9, the wafer is ready to be mounted and loaded into the MBE system.

Two points should be noted about the preparation steps above. First, when stripping the oxide with HF, the wafer should not be allowed to remain in the HF longer than the 20 s. If it is, the wafer may become stained. Second, when mixing the 5:3:3 solution, the HCl and H_2O should be mixed first and heated before adding the H_2O_2. Otherwise, most of the H_2O_2 will boil off before the solution can reach 80°C.

---

zinc-blende lattice oriented in the (100) direction, the atomic planes alternate between cation and anion. Therefore, on an atomically smooth surface of Si, if one ensures that the entire plane is covered with cation or anion, then antiphase domains will be suppressed.[32]

Another technique to suppress the APDs is through the use of (211) substrate orientation.[18] On the (211) surface, a bonding site in one sublattice has only one dangling bond. Therefore, there is a distinction between the sublattices such that there could be a preference as to bonding sites between cation and anion, which may suppress the APDs. However, this

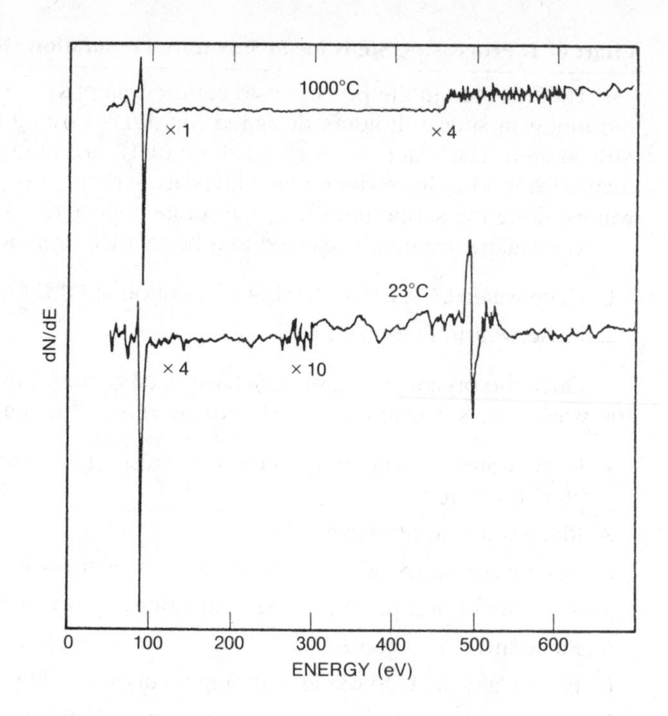

Fig. 6–3. Auger electron spectra from a Si substrate prepared using the method described in the text, before and after heating to 1000°C.

higher-index surface causes large dislocation densities which are necessarily unacceptable.

Although prelayers will suppress APDs on atomically clean substrate surfaces, the presence of steps will modify the situation. It is important that the correct growth conditions must be used for formation of a particular prelayer. For an As prelayer, a low growth temperature as low as 450°C[25,32] or less than 500°C compared to the typical 580°C used for GaAs growth is used since As is a volatile species. Since low growth temperatures are required for an As prelayer coverage, the growth rate should be reduced to about 0.1 μm/h (as compared with the typical 1 μm/h) to maintain crystal quality. For an As pre-exposure, the As sticking coefficient on the As-covered Si surface is low[33] and a perfect control of the pre-exposure duration is not critical. However, the situation for Ga pre-exposure is different because of its nonvolatile nature. Although a higher starting temperature can be used, a low growth rate is needed for it would be difficult to control a one monolayer deposition with a 1-μm/h growth rate.[32] A higher growth initiation temperature can also be obtained by the use of an AlAs/GaAs superlattice at the substrate/epilayer interface. Since the AlAs bond is tighter than that of GaAs, the initial growth temperature can be increased.

6.4 Defects at the GaAs/Si Heterointerface

When a heterojunction is formed between a polar and nonpolar semiconductor, such as GaAs and Si, so that the heterointerface plane is normal to the cubic body diagonal in both semiconductors, a piezoelectrically active heterojunction is formed. In its simplest form, a stress applied normal to the heterointerface induces a parallel polarization in the zinc-blende polar material (e.g. GaAs). Although a piezoelectric effect is negligible for heterojunctions grown on (100) surfaces, it will modify the interface properties of heterojunctions grown on (110), (111), or other surface orientations. As discussed above, there may be antiphase disorder resulting from the growth of a polar crystal on a nonpolar crystal and a high density of threading dislocations formed. Furthermore, the misfit between the covalent atomic radii of impurities and host material can generate disorder networks of defects at the GaAs/Si interface. These defects can degrade the properties of GaAs-based high-speed devices that are grown on Si substrates unless they are minimized. In the following, we will discuss these problems in detail.

6.4.1 Antiphase Disorder

Silicon has a diamond crystal structure which consists of two interpenetrating face-centered cubic sublattices. The atoms in the two sublattices differ from one another only in the spatial orientation of the four tetrahedral bonds that connect each atom to its four nearest neighbor atoms on the other sublattice. However, GaAs has a zinc-blende structure in which the two sublattices are now occupied by different atoms, one by Ga and the other by As atoms. The antiphase disorder occurs when there is a disorder in bonding the cation and anion atoms (e.g. Ga-Ga, As-As) inside the crystal.

Figure 6–4 illustrates the problem of antiphase disorder (APD) at the polar/nonpolar epitaxial interface. GaAs on Ge is chosen for simplicity since they are nearly lattice matched but the idea is the same for GaAs on Si. The first atomic layer bonding to the Si substrate can be either Ga or As which will induce APDs as shown. Since these bonds are electrically charged defects, the resulting excess charge will act like a scattering center which will lower the mobility. The antiphase domains are always expected to form when GaAs is grown directly on (100)-oriented Si substrates.

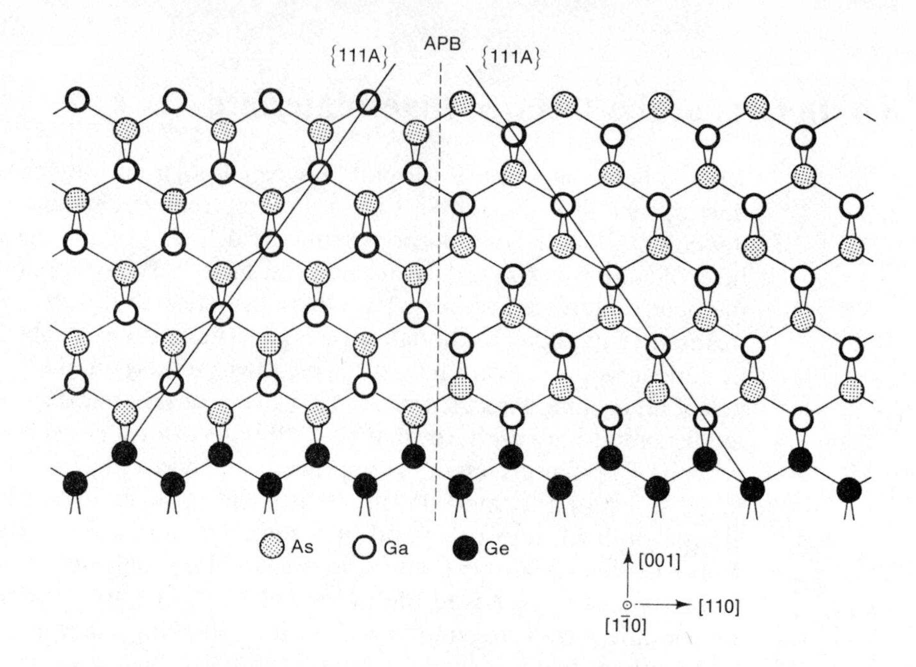

Fig. 6–4. Plane view of a polar semiconductor GaAs on a nonpolar substrate, e.g. GaAs on Ge. This representation shows the formation of an APB resulting from Ga and As atoms occupying the same sublattice in various parts of the surface. In addition the reversal of (111)A (Ga planes) across the APB is shown.

6.4.2 Lattice and Thermal Mismatch

The difference between the lattice constants and linear thermal expansion coefficients of the two materials is of great importance in the epitaxial growth of one on the other. Consider the lattice structures of GaAs and Si situation shown in Fig. 6–5, where lattices differ slightly in size. The epitaxial growth of GaAs on Si substrate can occur either as unstrained epitaxy or as strained epitaxy. In the former (see Fig. 6–5A) both GaAs and Si retain their individual structure and at the interface of GaAs and Si, every 25th row of Si atoms will have only three bonding neighbors. Rows of these improperly bonded atoms form so-called misfit dislocations which can seriously degrade the properties of the interface. In the strained epitaxy, on the other hand (see Fig. 6–5B), the fourfold bonding at the interface is preserved by compression of the larger GaAs lattice along the interface and elongation perpendicular to it. Such strained-layer epitaxy is possible as long as the stored strained energy in the film is lower than the reduction in the dangling bond energies. After 5-nm thickness of GaAs, the interface returns to the former case.

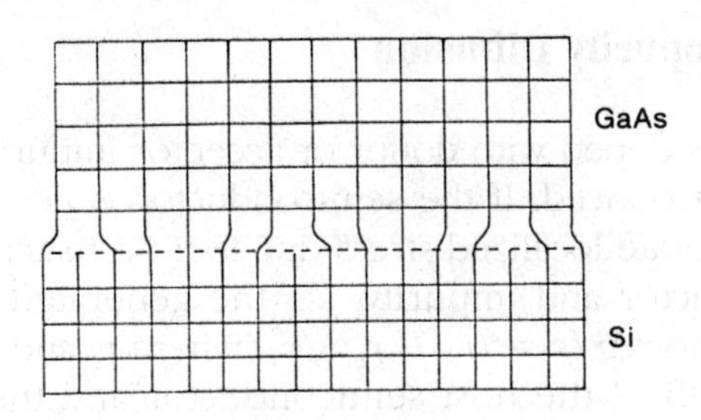

(A) The case of large lattice mismatch, GaAs on Si, where the strain must be accommodated by the generation of misfit dislocations. The dislocation shown is a perfect edge dislocation running perpendicular to the paper.

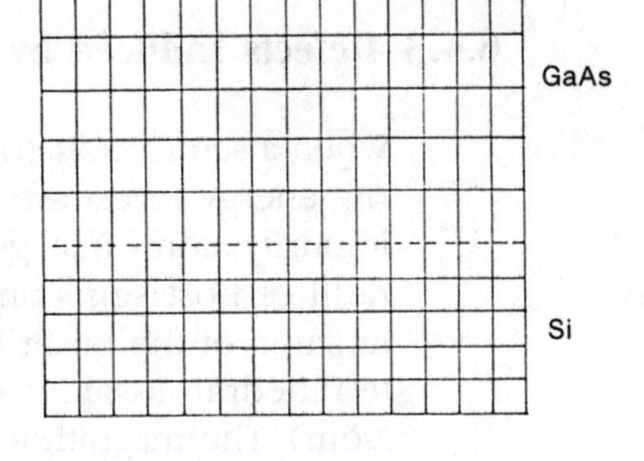

(B) A small lattice mismatch which is accommodated by a coherent strain in the film. This particular example is for a film ($In_{0.15}Ga_{0.85}As/GaAs$) under compressive strain because of its large lattice constant.

Fig. 6–5. Lattice arrangement of an epitaxial layer on a substrate.

Besides the lattice constant mismatch, there is also a mismatch in the thermal expansion coefficients of GaAs and Si. The thermal expansion coefficients of GaAs and Si are $6.0 \times 10^{-6}\,K^{-1}$ and $4.2 \times 10^{-6}\,K^{-1}$, respectively. As the growth temperature is raised, the lattice atoms vibrate with increasing amplitude about their equilibrium position. Since the potential energy of the atoms is not symmetric about the equilibrium position, the mean separation increases linearly with temperature, that is, the lattice constant of the crystal increases with temperature as $a = a_0(1 + \alpha\Delta T)$. [Here a_0 (a) is the room (growth) temperature value of the semiconductor lattice constant.] Since GaAs is polar semiconductor, the position of its molecules can be disturbed by the thermal motion. Any electric field would tend to orient the polar molecules. This orientation is temperature dependent, since heating intensifies the thermal motion of molecules and interferes with their orientation. At higher temperatures, if the orientation is in the wrong direction, thermal mismatch will change sign therefore putting the GaAs film under tensile strain.

After the growth, observations during cool down show that mostly the elastic deformation takes place. In any case, however, the structure becomes fixed below a certain temperature where only elastic formation is possible. Therefore, strain always occurs in the case of thermal mismatch. Strain resulting from lattice mismatch and different thermal expansion coefficients of the substrate and overlayer in semiconductor heteroepitaxy leads to wafer warpage or even cracks (in sufficiently thick films) in the device quality active layers.

6.4.3 Defects Induced by Impurity Diffusion

When a semiconductor is doped with donor or acceptor impurities, impurity energy levels are introduced. If the semiconductor is heavily doped, impurity atoms may generate localized stress due to a mismatch of atomic radii of host semiconductor and impurity, i.e. the generated stress is a function of the misfit factor $\Delta r/r = (r_s - r_{im})/r_s$ (where r_s and r_{im} are the tetrahedral covalent radii of the host semiconductor and the impurity atom). The magnitude of the stress is given by[8]

$$\sigma_{im} = [1 - (1 - \Delta r/r)^3]\frac{C_s}{C_L}Y_L/(1 - \nu) \qquad (6\text{–}1)$$

This assumes that the impurity atoms enter the crystal lattice substitutionally. In the case of interstitial impurity migration the generated stress is a function of the ratio of the radius of the impurity atom to the radius of the voids between the lattice atoms. In Eq. 6–1, C_s, C_L, and Y_L are the impurity concentration at the semiconductor surface, density of semiconductor lattice atom sites, and Young's modulus, respectively. Finally, ν is the Poisson's ratio. From Eq. 6–1 it is easy to see that when $r_s = r_{im}$ there is no stress generated, i.e. $\sigma_{im} = 0$. If the semiconductor is doped with impurities with a larger atomic radii than the host ($r_{im} > r_s$), the semiconductor will be under localized compressional forces ($\sigma_{im} < 0$). In the opposite case, where $r_{im} < r_s$, the tensile type localized forces will be generated. For example, if GaAs and Si are doped with Si and As, respectively, the mismatches between the impurity and host atoms radii are 4.2 and 4.43 percent for GaAs (Si) and Si(As). That is, GaAs (Si) will be under localized tensile (compressive) forces when the doping concentrations are close to the solubility limit.

The stress generated within the host semiconductor lattice due to the presence of impurity atoms may have an effect on the motion of a dislocation line present in the lattice. Misfit dislocation networks may occur or be prevented from propagating in semiconductors as a result of high impurity concentration (near the maximum solid solubility). This may occur internally in the case of GaAs/Si heterojunctions. For example, during the growth of heavily Si-doped GaAs on Si (diffusion may automatically cause this to happen), the first few monolayers of GaAs closest to the interface can be Si doped to the solid solubility limit. Further Si atoms may move to the GaAs/Si interface region from the bulk GaAs and bond with Si atoms. Therefore, the GaAs lattice in the vicinity of the interface will be under localized tensile type forces. As long as these defects remain confined to within a few atomic monolayers of the GaAs/Si interface, the device quality layers grown on GaAs buffer are protected from such dislocations.

6.5 Characterization and Reduction of Defects

The effects of defects discussed in the previous section are the decrease of the performance and lifetime of GaAs-based high-speed devices, especially those grown on Si substrates. They may be severe for devices whose operation depends on the minority carriers, such as bipolar transistors and optical devices (lasers, light-emitting diodes, etc.). In order to improve the device performance and reliability, it is essential that these defects be minimized or eliminated when possible. Only then may the low current density lasers on Si substrates be possible which would have a great impact, for example, making optical off-chip communication for Si integrated circuits a reality. The introduction of strained-layer superlattices (SLSs),[34] such as (In, Ga)As/GaAs SLSs,[32,35] before the active layer is grown, may be capable of bending the dislocations. Strain relief might occur via plastic deformation caused by the movement of dislocations through the SLS's material or along the GaAs/Si interface. The low elastic stiffness coefficients of the SLS material may also help concentrate remaining stress in the SLSs film. Annealing is also often employed to reduce dislocations.[36–41]

6.5.1 Suppression of Antiphase Disorder

As mentioned previously, since the sublattice structure is uniform throughout the entire Si (100) single crystal, there will not be any antiphase disorder. However, in GaAs these planes alternate between being composed entirely of cation Ga and anion As atoms. If this ordering is not maintained somewhere in the GaAs layers grown on Si substrates, Ga to Ga and As to As bonding will occur, leading to massive antisite defects leading to antiphase domains, APDs. Since these bonds are electrically charged defects, the resulting excess charge will act like a scattering center lowering the mobility. A schematic representaion of an APD is shown in Fig. 6–4 for GaAs on a nonpolar Ge substrate. To the left of the vertical line, the growth is started with As atoms, whereas to the right, the growth is started with Ga atoms. Between these two regions, there exists an antiphase domain boundary (APB) with As-to-As and Ga-to-Ga bonds. The APB is arbitrarily chosen to propagate along the ⟨100⟩ direction. The As-to-As bonds provide an excess of negative charge, and Ga-to-Ga bonds provide an excess of positive charge. This leads to an oscillating microscopic dipole field structure with intolerable local perturbation. If a different surface orientation, e.g. (211), is used, electrically neutral interface can be obtained as pointed out by Wright *et al.*[18,42]

Until recently work on the sticking probability of As on Si was not well known and in some cases was believed to be nearly zero. Molecules of As_2 (dimeric As), for example, are known to have a larger sticking coefficient than molecues of As_4 (tetrameric As) and may thus be preferred. Since it may not always be possible to deposit only a monolayer of As atoms very precisely, a substrate temperature for the initiation of As deposition must be chosen such that the sticking probability of As to Si is much larger than that of As to As. For certain device applications Ga and/or Al prelayers can be used to maintain atomic ordering in GaAs and its ternary alloy (Al, Ga)As.

So far we have assumed that the surface of Si is atomically flat and the growth is to nucleate in a two-dimensional form uniformly everywhere, and that the first sublattice can be deposited followed by the more conventional MBE growth. However, such an ideally flat surface cannot always be obtained. In such case the growth of GaAs on Si is by island formation followed by the merger of these nucleation centers (see Fig. 6–6). Single atomic steps on the Si surface can also give rise to APDs as shown in Fig. 6–7 even when the monolayer of As or Ga or Al is successfully deposited. On the other hand, if the surface has double atomic steps only, the prelayer concept can be employed to prevent the formation of APDs, as shown in Fig. 6–8. It is unlikely that the surface of Ge or Si has only double steps (regardless of how it is prepared), particularly those with perfect (100) alignment. Henzel and Clabes[43] indicated that on a misoriented Si (100) surface, there is a tendency towards step doubling with an increasing annealing temperature. This was then confirmed by Kaplan,[44] who reported that on Si surfaces tilted by a few degrees from the (100) plane towards the (011) planes (with the appropriate thermal treatment performed), most steps are two atoms high. (Excellent reviews of the APD problem are given by Fischer[45] and Kroemer.[46]) Calculations by Aspnes and Ihm[47] predict that the bonds parallel to the steps are favored on Si surfaces. Fischer *et al.*[4] has achieved such APD-free material grown on tilted surfaces. However, it is very unlikely that every step on the surface is an integral multiple of double steps, although it is believed that the surface with double steps has a minimum total energy.

Antiphase disorders can be forced by deliberately starting with neighboring Ga and As prelayers. This can be accomplished by inducing a temperature gradient on the Si surface before depositing an As prelayer, since the sticking coefficient of As is apparently high in cool regions and negligible in hotter regions (unless Ga is also present). The temperature of the cool part of the substrate should be sufficiently low so that the As sticking probability is substantial (Fig. 6–9). On the other hand, the temperature of the hot part must be such that As does not stick at all. Under these experimental conditions when the growth is started with As prelayer, the hot part will actually start with the Ga prelayer as depicted in Fig. 6–10A. On

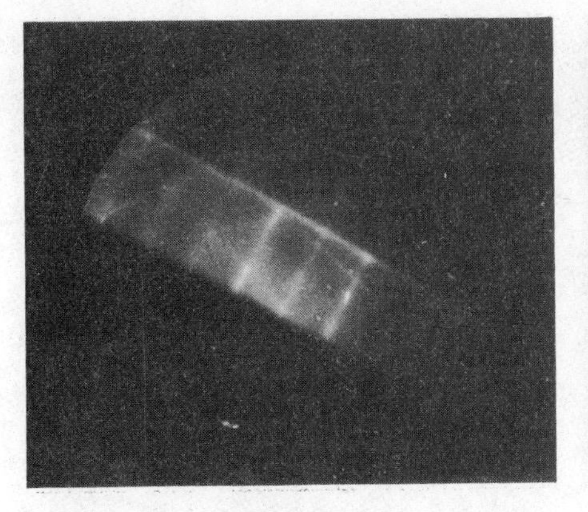

(A) Reflection high-energy electron diffraction pattern of a clean Si surface with a classical 2 × 1 reconstruction.

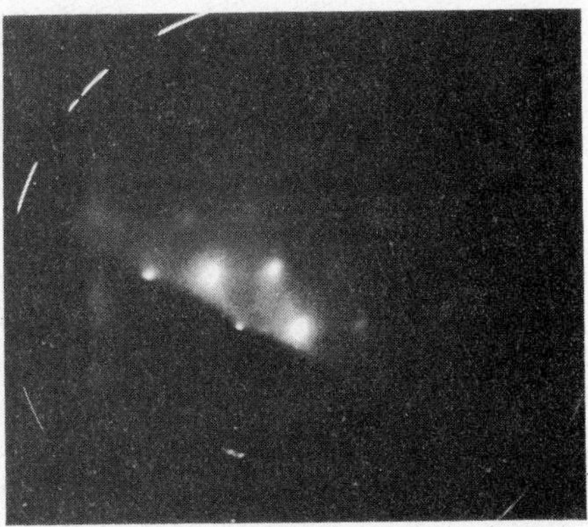

(B) The same as (A) after about 20 nm of GaAs growth. The dotted nature of the pattern indicates that the growth is taking place in island.

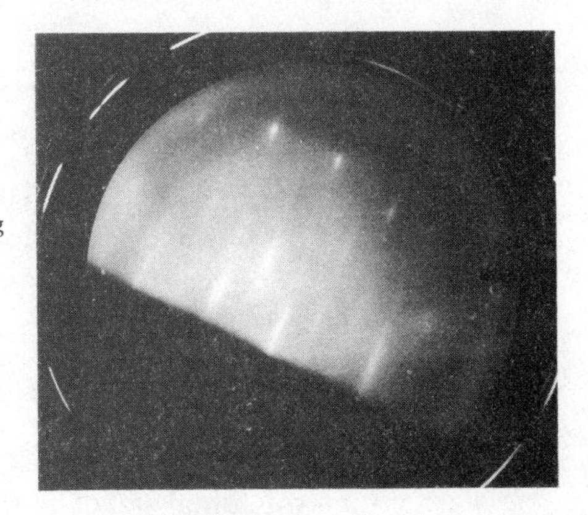

(C) The pattern after 1.5 μm of GaAs growth by which time the surface is very planar. Depending on the conditions used, the planar growth is reached even in layers as thin as 20 nm.

Fig. 6–6. The growth of GaAs on Si.

the left side the growth is started with the Ga prelayer, whereas the right side shows the As prelayer region. The substrate orientation is 4° off (100) towards (001). Figures 6–10B, 6–10C, and 6–10D show the magnified views of these three regions. Where the Ga and As prelayer regions meet, APDs occur. Near the As prelayer side of the transition region, the size of As prelayer domains are larger than the Ga prelayer islands. Near the Ga prelayer transition region, the situation is reversed. In the middle of the mixed domain region, the island sizes are expected to be comparable.

Experiments performed in the authors' laboratory have shown that even in growth on "off the shelf" Si substrates of (100) orientation, APDs

Fig. 6–7. Stick model of GaAs started with an As prelayer on a surface with single atomic steps. The GaAs is not APD free because of single steps on the surface.

do not form. This may be attributed to possible unintentional tilting present on the substrate surface. This observation can also be explained by two other phenomena. One model assumes the massive mass transfer during the initial stages of the growth as discussed above. The other concerns the way in which the APBs propagate. If one assumes that the APBs are inclined, as opposed to being orthogonal to the plane of the growth, and that the slope depends on the direction of the single step on the substrate which produces the APB, then the APBs can annihilate one another some distance away from the surface as depicted in Fig. 6–11. The etching experiments assume that the bonds are parallel to the steps as supported by the predictions of Aspnes and Ihm.[47]

Various techniques can be used to determine whether or not a given GaAs layer on Si contains APDs. These include plane view transmission electron microscopy (TEM),[5] X-ray scattering,[6,48] and anisotropic chemical etching.[7] None of these techniques has shown any evidence for antiphase disorder. One of the simplest techniques is the use of selective etching if the size of the APDs is reasonably large ($\geq 0.2\ \mu m$). Certain chemical etches for GaAs such as $NH_4OH:H_2O_2:H_2O$ attack (111) B (As) planes but

Fig. 6–8. Deliberate misorientation of the substrate off of (100) by several degrees is believed to produce double atomic steps which avoid the formation of APDs when the growth is initiated with an As prelayer.

not the (111) A (Ga) planes.[45] The principle of the experiment is shown in Fig. 6–12. For example, if a bar of photoresist is defined on the GaAs surface along the ⟨110⟩ direction, the slope of the side walls after etching (See Fig. 6–13) will change, from being inward to outward as one crosses an APB. Figure 6–14 shows the scanning electron micrographs of etched GaAs on Si using an As prelayer (Fig. 6–14A) and Ga prelayer (Fig. 6–14C). Figure 6–14B shows a sample where antiphase disorder was intentionally generated to demonstrate the ability of this technique to determine the antiphase disorder.

The etching technique is capable of revealing antiphase disorder if the domain sizes are larger than about 0.2 μm. To detect antiphase domains of smaller sizes, X-ray scattering can be used[6,48] by looking at the fundamental and superstructure peaks with scans parallel and perpendicular to the scattering vector. Neumann *et al.*[6,48] scanned both fundamental [(004), (008), etc.] and superstructure [(002), (006), etc.] reflections. If antiphase domains are present, the superstructure peaks, since they depend on the

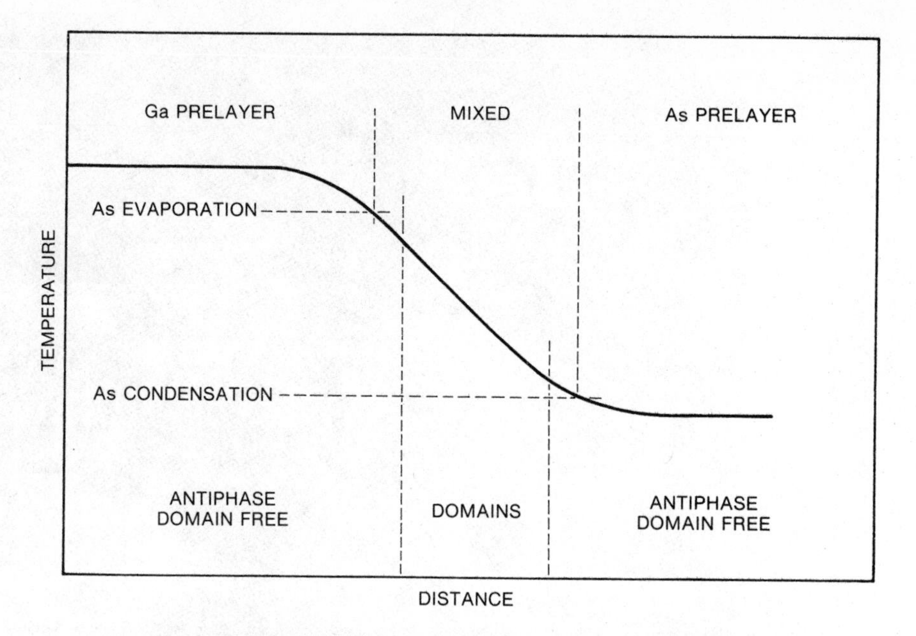

Fig. 6–9. Phenomenological description of domains formed by As and Ga prelayers on either side of a temperature gradient introduced during the initiation of the growth. In the regions where the domains merge, APDs are formed.

structure factor, will be broadened with respect to the fundamental peaks. A typical example is shown in Fig. 6–15, where the widths of (H00) reflections (parallel to film surface) and (00L) reflections (perpendicular to the film surface) are plotted. As can be seen, the superstructure reflections (002), (006), and (200) exhibit the same widths as the fundamental reflections (004), (008), (400) and (800). This demonstrates that if APBs exist, their separation must be larger than the instrumental resolution, i.e., larger than 400 nm. Furthermore, the same measurements indicate that GaAs is under tensile strain.

6.5.2 Dislocations: Characterization and Reduction

High-resolution electron microscope (HREM) studies[49] have shown that there are two predominant types of dislocations, one which is contained at the GaAs/Si interface (type I), while the other inclines (type II). The type I misfit dislocations are the pure edge type: their Burgers vectors and dislocation lines are perpendicular to each other. Type II misfit dislocations are of the mixed type: their Burgers vectors are inclined to dislocation lines by 60° and at 45° to the plane of the substrate.

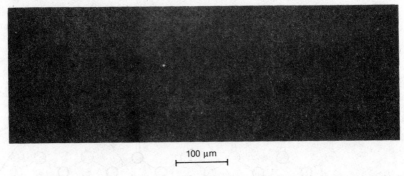

(A) All three regions: Ga prelayer on the left, As prelayer on the right and mixed domains in the center.

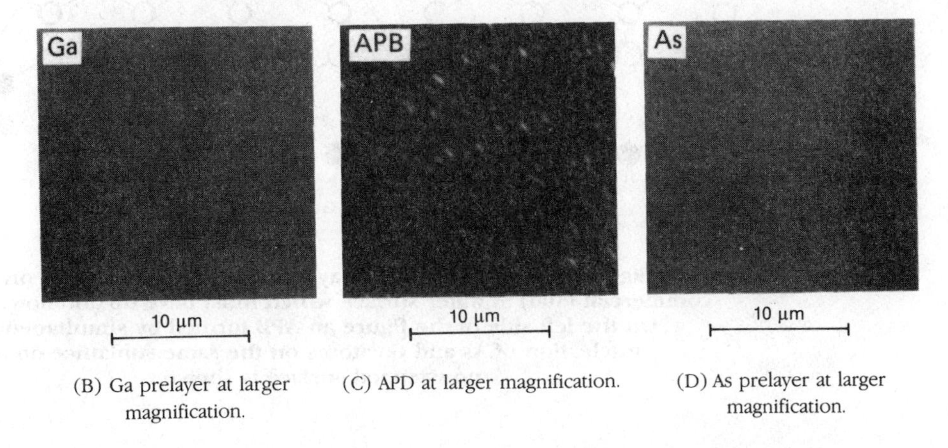

(B) Ga prelayer at larger magnification.

(C) APD at larger magnification.

(D) As prelayer at larger magnification.

Fig. 6–10. Microphotograph of a surface of GaAs film (2 µm thick) on Si near the temperature gradient region described in Fig. 6–9.

Shown in Figs. 6–16A and 6–16B are the magnified images of these type I and type II misfit dislocations, respectively, taken from the HREM studies.[49] As can be seen in the micrographs, two misfit dislocations have different core structures. The type I dislocations (Fig. 6–16A) have extra lattice planes along both (111) and $(1\bar{1}\bar{1})$ orientations as indicated by arrows. The type II dislocations (Fig. 6–16B), on the other hand, have an extra lattice plane only along the (111) orientation. The type I dislocation has Burgers vectors of ½[011], which is parallel to the heterointerface, and the type II ones has Burgers vectors of ½[101], which is inclined from the interface by 45°. Shown in Figs. 6–17A and 6–17B are the projections of the atomic arrangements in the misfit dislocations with these Burgers vectors. The presence of extra lattice planes in these projections are in agreement with that in the observed images. As the images show, the lattice mismatch at the GaAs/Si heterointerface is accommodated by misfit dislocations rather than by uniform elastic strains.

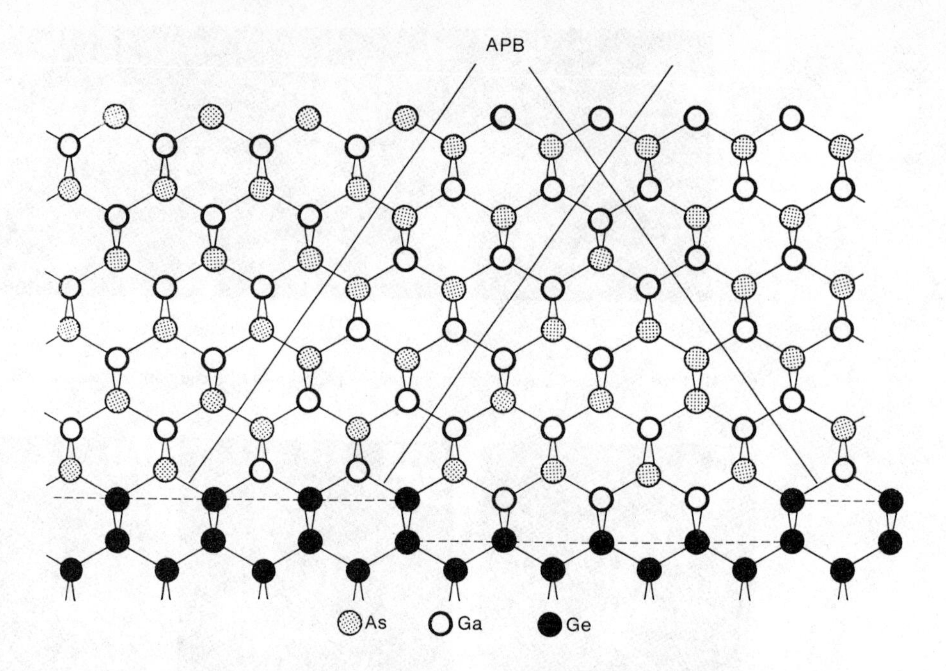

Fig. 6–11. A model which may explain the lack of APDs on a commercial (100) Si wafer surface which must have up and down steps. On the left side of the figure an APB formed by simultaneous nucleation of As and Ga atoms on the same sublattice on a nonstepped surface is shown.

Since the misfit accommodated by a dislocation is the projection of its Burgers vector onto the substrate plane, a type I dislocation is more effective in accommodating misfit than a type II dislocation and, as a result, fewer of them are required. A type I dislocation is desirable from the standpoint of minimizing the number of interfacial dislocations. A second point making the type II dislocation undesirable is that these dislocations can only move by slip along crystallographic planes which contain both Burgers vectors and dislocation lines, the (100) plane for type I dislocations and (111) planes; it is unlikely that type I misfit dislocations will propagate towards the film surface by slip. They will stay in the GaAs/Si interface. In contrast, type II misfit dislocations can easily move through the GaAs film from the interface to the film surface. Therefore, this type of misfit dislocation can become a highly active source for the generation of threading dislocations.

Several techniques have been used to minimize these dislocation densities: the use of tilted substrates,[4,32,35] strained-layer superlattices,[32,35] and thermal annealing.[37,39] In the following subsections, these techniques for reducing the threading dislocations will be discussed in detail.

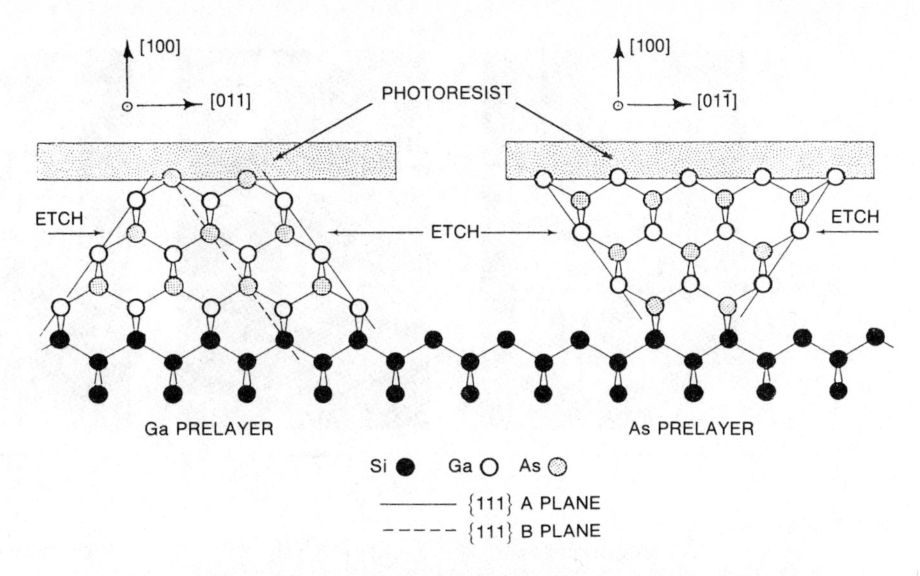

Fig. 6–12. Diagram showing the zinc-blende lattice of the GaAs along a ⟨011⟩ type cross-section. In the plane of the figure for different prelayer types the direction parallel to the interface is [011] or [$0\bar{1}1$]. As shown, the etch reveals the (111)A planes. The slope of the mesa walls are different along the [011] and ⟨$0\bar{1}1$⟩ directions.

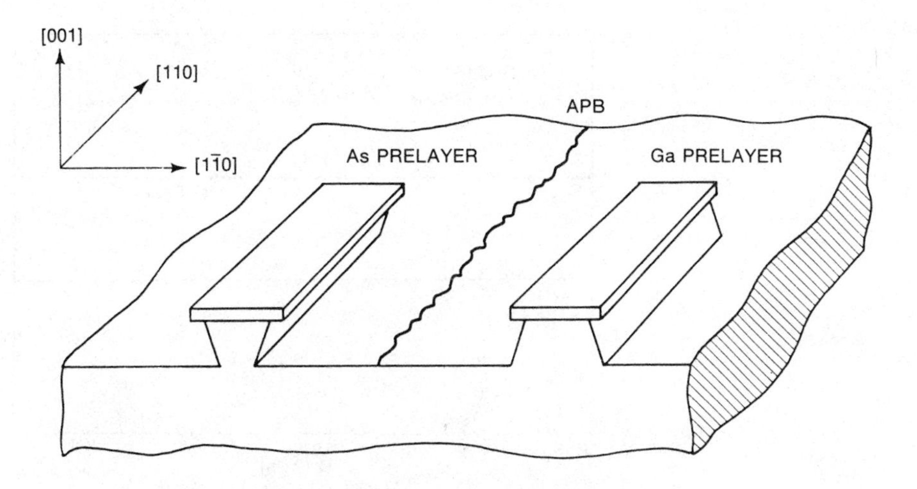

Fig. 6–13. Schematic diagram of the etch pattern produced under photoresist patterns on either side of an APB. The wall shape across the boundary changes from being inward to being outward delineating the (111)A planes because the preferential etch used does not attack the (111)A planes.

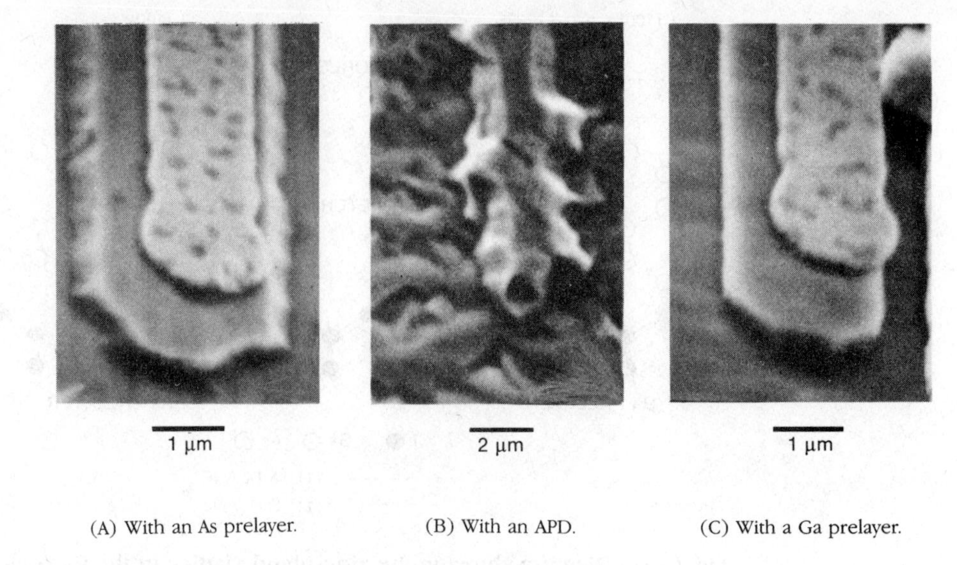

(A) With an As prelayer. (B) With an APD. (C) With a Ga prelayer.

Fig. 6–14. Etching experiment in order to reveal APD. Sidewall shapes of GaAs epitaxial layers on Si. (The tilt of the substrate has the same orientation in A and C.)

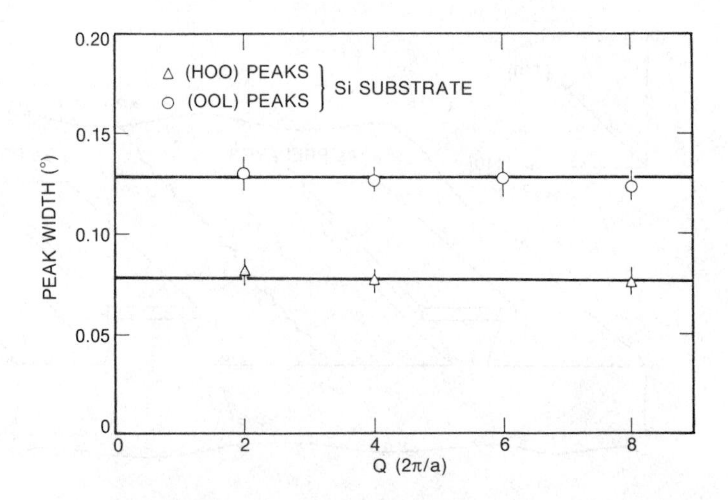

Fig. 6–15. Peak widths as a function of the normalized scattering vector $Q(2\pi/a)$. The fact that the superstructure reflections do not exhibit an appreciable broadening compared to the fundamental peaks, indicates that if APDs exist, they must be larger than 400 nm. (*After Neumann* et al.[6])

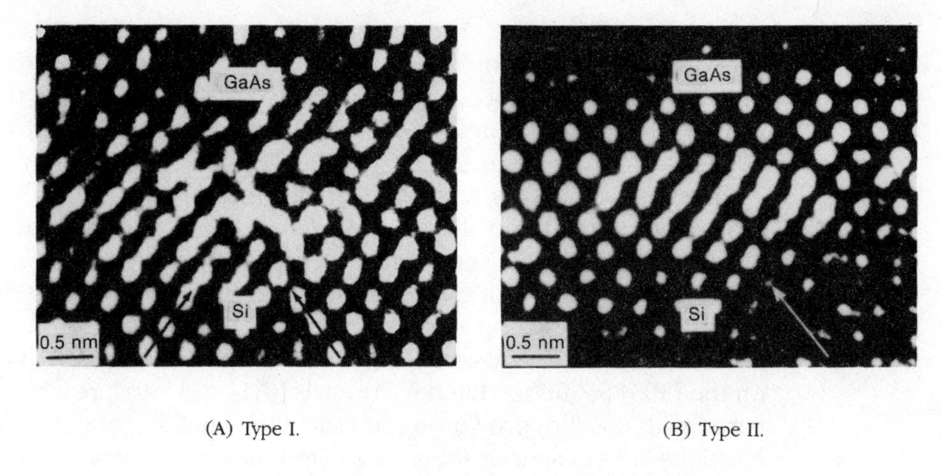

(A) Type I. (B) Type II.

Fig. 6–16. Magnified images of misfit dislocations.

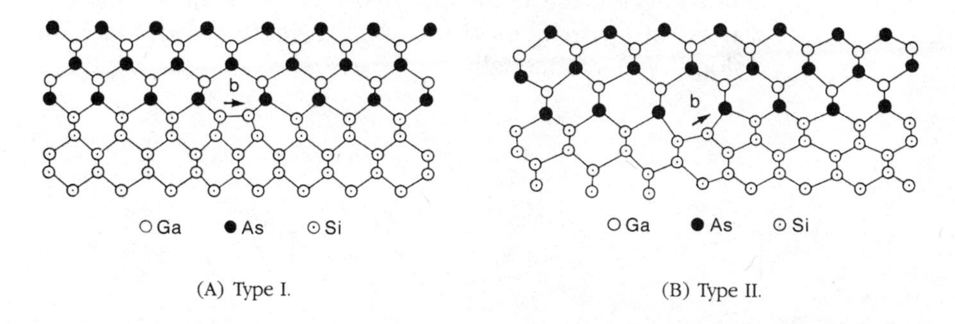

(A) Type I. (B) Type II.

Fig. 6–17. Projections of atomic arrangements of misfit dislocations. Directions of Burgers vectors are shown by arrows in figures. (*After N. Otsuka, Purdue University*)

6.5.2.1 Substrate Tilt

It has been shown[44] that the steps at the Si (100) surface provide edges where the type I dislocations can be initiated. Therefore, the use of a surface which intentionally causes steps, e.g., the tilt of the substrate surface (as shown in Fig. 6–18A), can preferentially induce type I dislocations; therefore, reducing the density of the active sources (e.g., type II dislocations) for the generation of threading dislocations.[4,32,35] One type I dislocation is necessary for every 25 atomic planes in accommodating the lattice mismatch. Therefore, a substrate tilt provides one step for every 25 atomic planes. The steps in the substrate surface tilted toward (011) run along (⟨011⟩) directions preferentially,[49] which results in type I dislocations to be nucleated along one direction only, as shown in Fig. 6–17B. However, in order to further reduce the dislocation density steps running in both the [011] and [0$\bar{1}$1] directions would be desirable. This can be obtained with the tilt of the (100) surface plane of the substrate toward [001] direction since steps occur only along a [011] direction. This may further reduce the dislocation density, but complicates the double step arguments.

Figure 6–19 shows a cross-sectional bright field image view of GaAs film grown directly on the (100) surface taken by transmission electron microscopy (TEM).[35] The image view shows massive dislocations and stacking faults. The dislocation density is as high as 10^8 cm^{-2}. Figure 6–19 shows the cross section of GaAs on straight (100) Si, whereas Figs 6–20 and 6–21 show the cross-sectional TEM image view of GaAs layer grown on the (100) Si surface tilted 4° towards [011] and [001], respectively. Compared with the film grown on the exact (100) surface, fewer threading dislocations are present in these films, and most are confined in the region approximately 1 mm in thickness at the GaAs/Si interface. The dislocation density near the GaAs surface is in the range of 1×10^5 cm^{-2}. Further examination of these figures demonstrates that the substrate surface tilt towards [001] reduces the defect density (both stacking faults and dislocations) substantially.

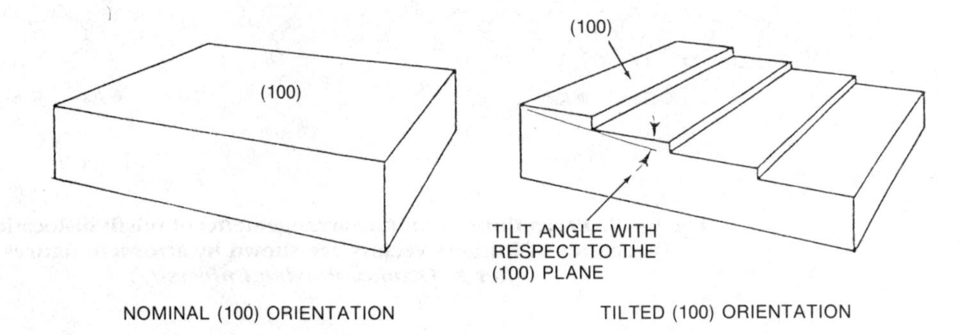

(A) Orientation tilted off (100) toward (011).

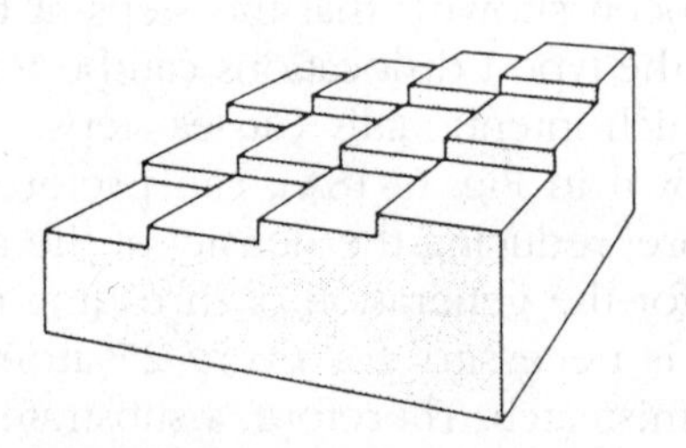

(B) Orientation tilted off (100) toward (001).

Fig. 6–18. Diagrams showing the effect of tilting on the substrate surface.

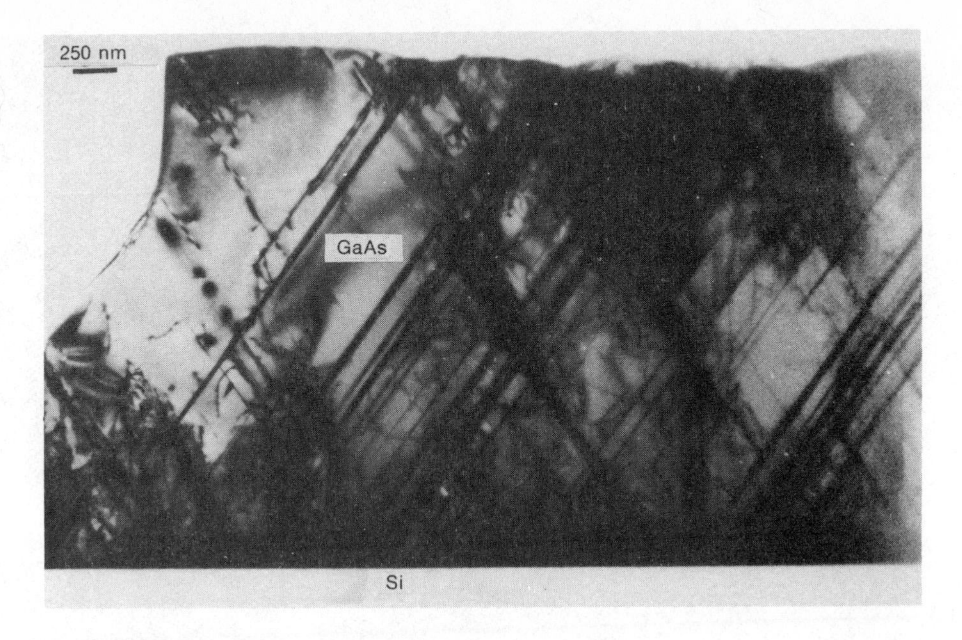

Fig. 6–19. TEM cross-sectional bright field image of a GaAs layer grown on an exact (100) Si surface.

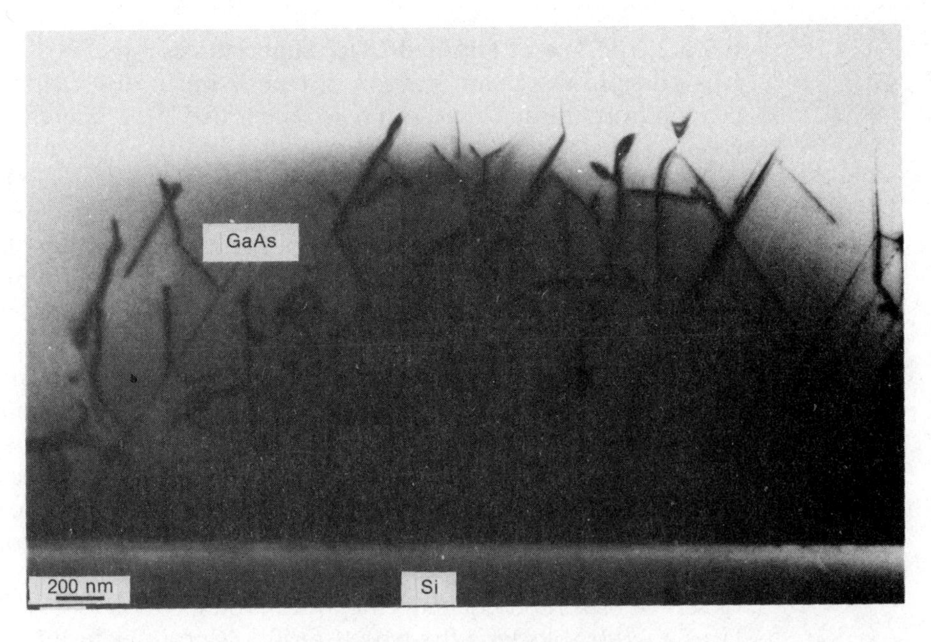

Fig. 6–20. TEM cross-sectional bright field image of a GaAs layer grown on (100) Si tilted 4° off toward [011].

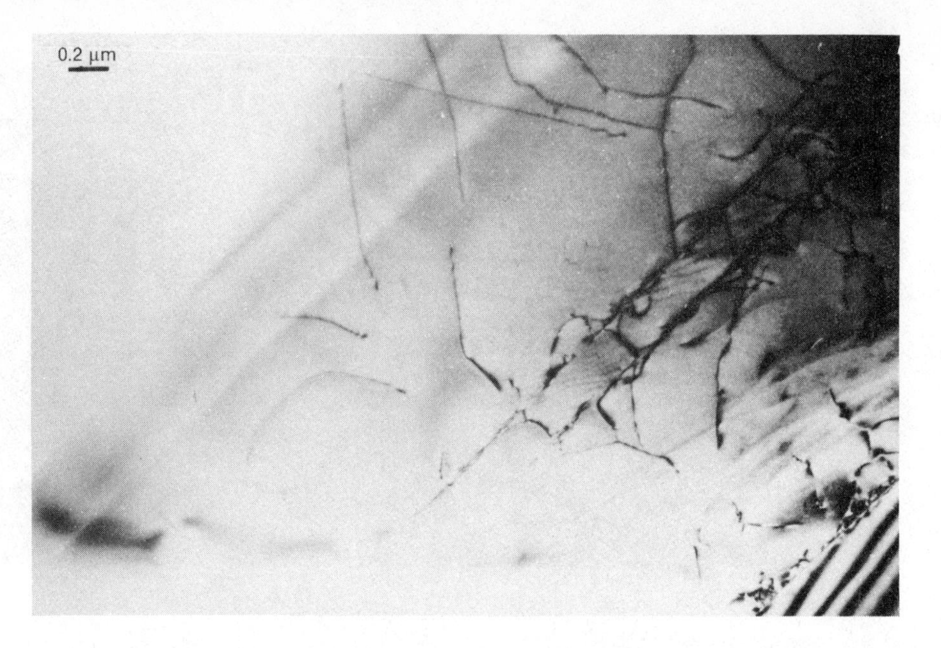

Fig. 6–21. TEM cross-sectional bright field image of a GaAs layer grown on (100) Si tilted 4° toward [001].

6.5.2.2 The Use of Strained-Layer Superlattices

It is possible that some fraction of type II misfit dislocations can propagate up from the substrate and reach the active layer. Therefore, it is desirable to reduce these type of dislocations as much as possible. This can be done by the use of a strain field in a strained-layer superlattice[32,35] [e.g. (In, Ga)As/GaAs SLS] to turn aside the type II dislocations propagating up to the active layer. The principle has been predicted by Matthews and Blakeslee.[34] The dislocations that encounter a strain field are bent over and forced to move laterally towards the edge of the substrate. In the authors' studies, a compressive strain in the GaAs tends to prevent the dislocation from propagating in the direction of the growth. Shown in Fig. 6–22 is a TEM cross-sectional image of a GaAs on Si epilayer which incorporates 10 periods of 10-nm GaAs/10-nm $In_{0.15}Ga_{0.85}As$ strained-layer superlattices. As can be seen in Fig. 6–22, the superlattice bends or terminates (by looping) many of the type II threading dislocations and therefore reduces the density above it by about one order of magnitude. The incorporation of intermediate thick layers of GaAs will allow the extension of the number of sets of SLSs. Consequently, this will introduce more strained-layer interfaces, thereby blocking the type II misfit dislocations further.

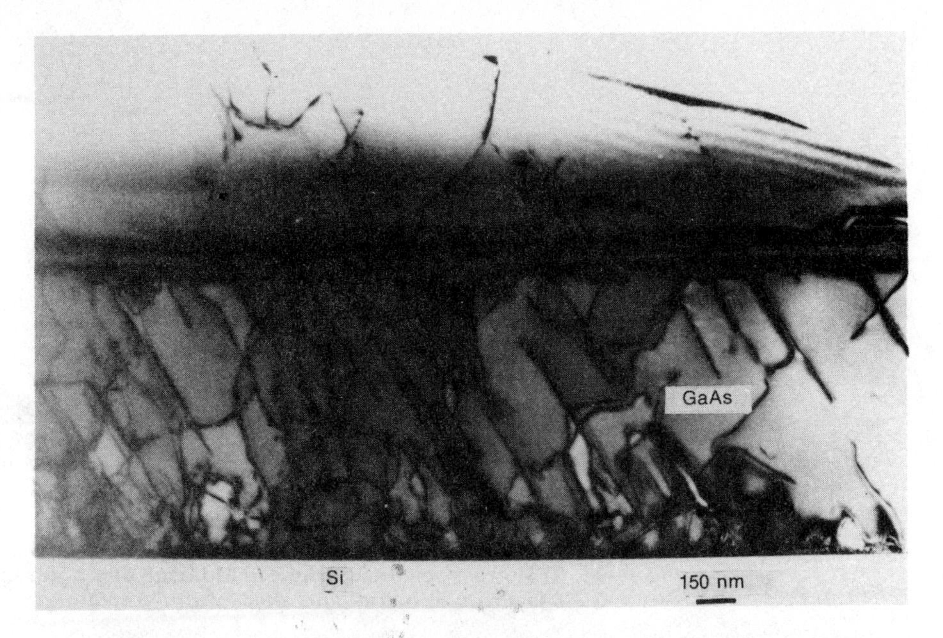

Fig. 6–22. TEM cross-sectional bright field image of a GaAs layer grown on a tilted (100) Si surface with a 10-period 10-nm $In_{0.15}Ga_{0.85}As$ SLS.

6.5.2.3 The Use of Thermal Annealing

Recent improvements of GaAs epitaxial layer in terms of both dislocation density[37,39] and optical properties[36,38,40] have shown that the dislocation density could be reduced even further. The annealing temperature ranges from 650 to 850°C. Different types of annealing can be used including *ex situ* and *in situ*. The *ex situ* annealing can be done by either (*a*) annealing in a furnace under As pressure with a typical duration of 1/2 h[37] or (*b*) rapid thermal annealing (RTA), where the temperature is pulsed at a higher temperature (900 to 950°C) for a few seconds.[36] The *in situ* annealing which is done in the MBE system during the growth can be done either (*a*) statically, where the growth is stopped and substrate temperature is raised under As overpressure, or (*b*) dynamically, where instead of stopping the growth an (Al, Ga)As layer is grown at high temperature (~700°C).[41] This annealing must be done shortly after the thickness of the film has reached the critical dislocation thickness.

Shown in Fig. 6–23[41] are the bright field images of a 2-μm GaAs epilayer grown on a (100) substrate tilted 4° toward [001] before (Fig 6–23A) and after (Fig. 6–23B) *ex situ* annealing at 850°C. As can be seen in Fig. 6–23A, massive dislocations appear in a region within 0.7 μm of the GaAs/Si interface, and many of them, together with stacking faults, appear to have threaded to the free film surface. After *ex situ* annealing at 850°C for one

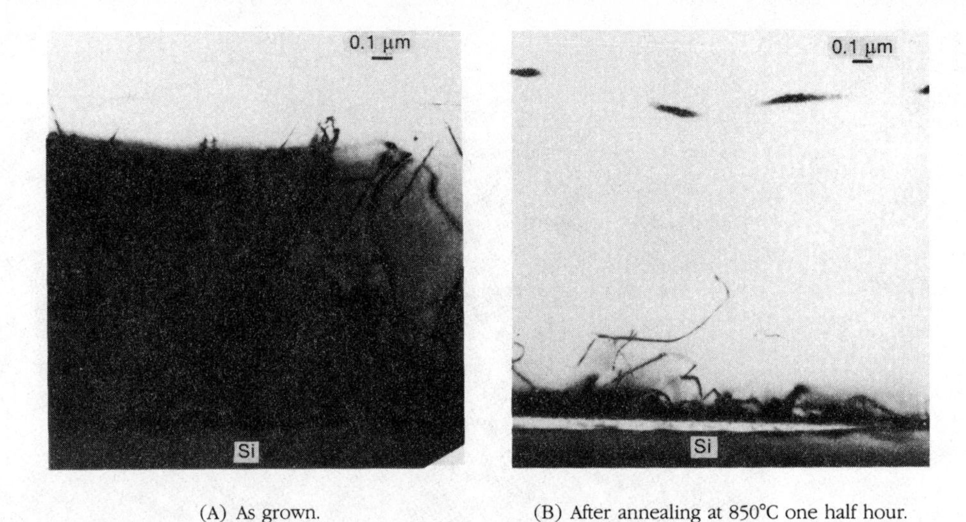

(A) As grown. (B) After annealing at 850°C one half hour.

Fig. 6–23. TEM cross-sectional bright field image of a 2 μm thick GaAs layer on Si before and after *ex situ* annealing.

half hour, a dislocation-free material is achieved near the film surface, with majority of the dislocations being confined at the interface. Figure 6–24 shows the identical study on a layer that includes two sets of strained-layer superlattices. The sample consists of a 4-μm GaAs layer interlaid with two sets of 5-period (10 nm/10 nm) $In_{0.15}Ga_{0.85}As$/GaAs SLSs, separated by 0.3 μm of GaAs. Comparison of Fig. 6–24A and Fig. 6–24B shows that the dislocation density is greatly reduced and the SLS is shown to be still effective in confining the dislocations after high-temperature annealing.

The relatively high annealing temperatures may not be compatible with some heterostructure and modulation doping applications. This objection can be taken care of by annealing the sample within the MBE system, prior to the growth of the critical layers. Although in most present MBE system temperatures of 850°C may not be possible, this technique has been proved to be effective even at lower temperatures. As shown in Fig. 6–25, excellent results are possible with a static anneal at 700°C.[37]

6.6 Deformation Potential Theory to Investigate GaAs/Si Interface Properties

The lattice, impurity, and thermal mismatch effects on the GaAs/Si interface properties can be formulated by the deformation potential theory in terms of local changes in the potential energy across the interface experi-

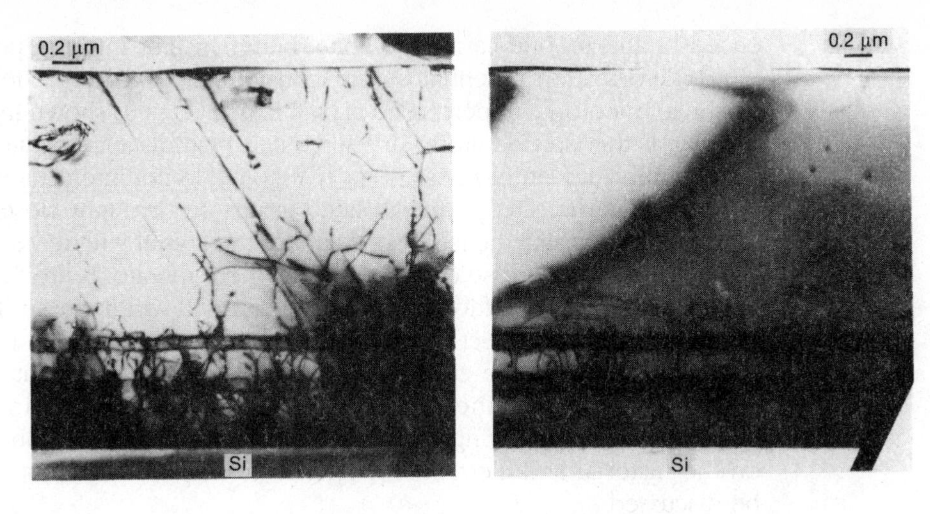

(A) As grown. (B) After annealing at 850°C one half hour.

Fig. 6–24. TEM cross-sectional images of a GaAs layer on Si incorporating two SLSs.

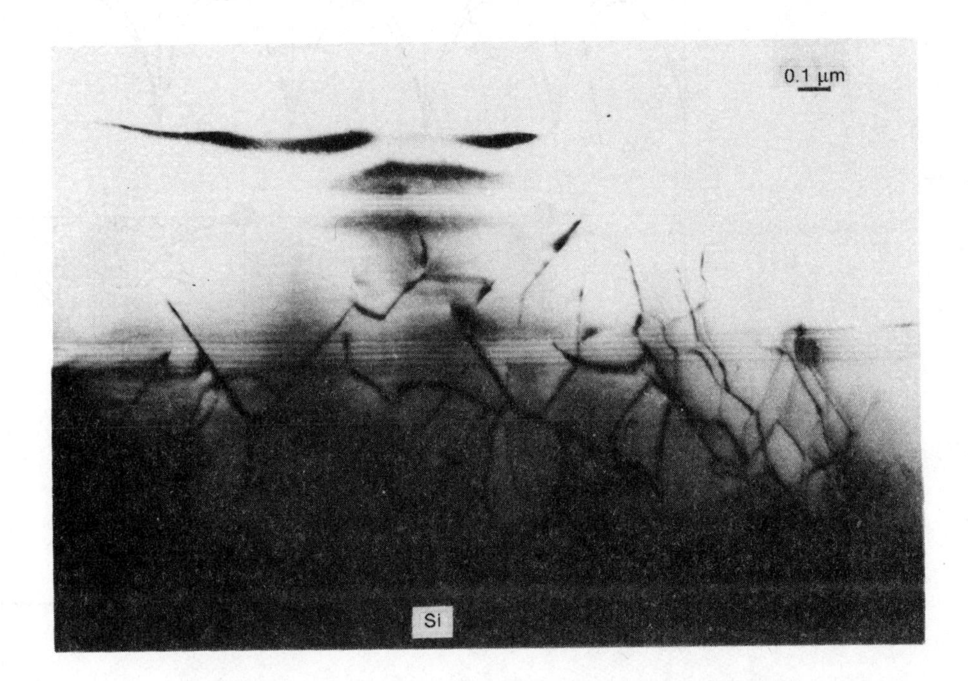

Fig. 6–25. TEM cross-sectional bright field image of GaAs layer on Si incorporating two SLSs with an *in situ* annealing at 700°C, one half hour, prior to the growth of the superlattices.

enced by the mobile carriers.[50] For a better insight into the peculiarities of the deformation potential theory, consider the one-dimensional potentials in a monotonic unperturbed crystal lattice ($\epsilon = 0$) shown in Fig. 6–26. As long as the isolated atoms are separated by a distance greater than the true equilibrium lattice constant a_0, the atoms do not interact and their energy remains discrete. As the isolated atoms are brought closer together (compressive strain) the lattice constant of the crystal will decrease (Fig. 6–26). In the opposite case (tensile strain) where the atoms are farther apart from each other, the lattice constant of the crystal will increase (Fig. 6–27). In this respect the effect of compression and extension can be substantially different from the effect of the external electric field that always displaces the edges of the conduction and valence bands in the same direction. In the following paragraphs the cause of the strain and its effects on the material and electronic properties of GaAs/Si heterostructure will be discussed.

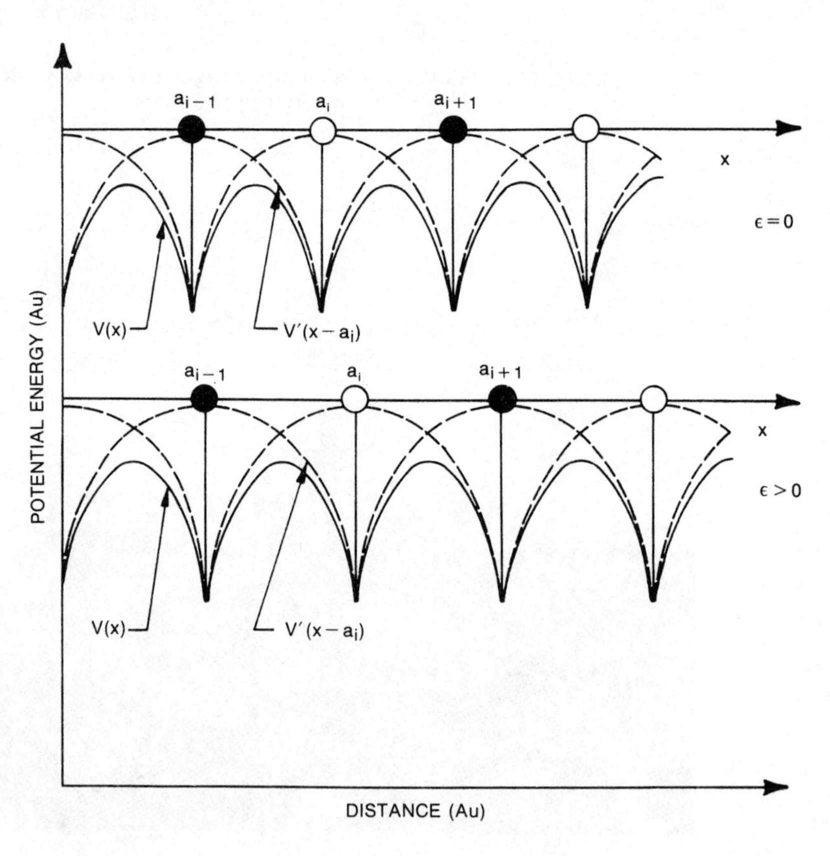

Fig. 6–26. Schematic representation of one-dimensional potentials in a monatomic lattice for an unperturbed lattice ($\epsilon = 0$) and for the effect of tensile strain ($\epsilon > 0$).

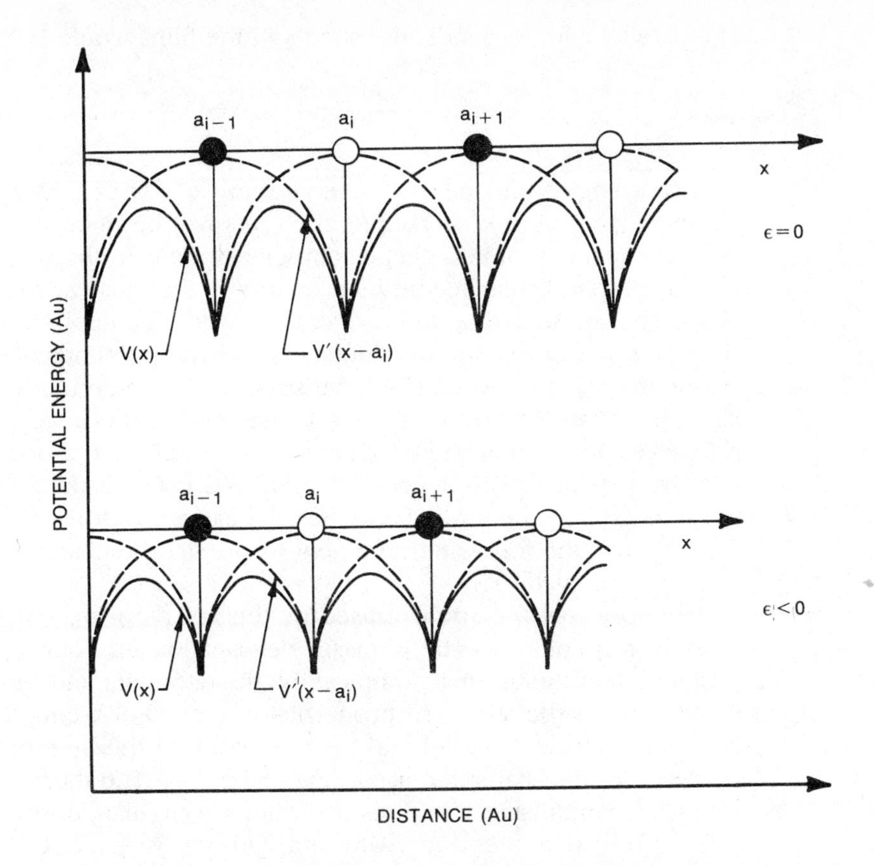

Fig. 6–27. One-dimensional potential in a monatomic lattice for an unperturbed lattice ($\epsilon = 0$), and for the effect of compressive strain ($\epsilon < 0$).

Matthews and Blakeslee[34] have shown that when two sufficiently thin materials form a mismatched heterostructure, the lattice mismatch will be accommodated by uniform elastic strain. The lattice constants of strained layers in the planes parallel to the heterointerface is given by

$$a_{\|} = \frac{a_1 G_1 t_1 + a_2 G_2 t_2}{G_1 t_1 + G_2 t_2}, \tag{6–2}$$

where a_i, G_i, and t_i ($i = 1, 2$) are the lattice constant, shear modulus and layer thickness of materials, respectively. The lattice constants in the planes perpendicular to the heterointerface are given by

$$a_{\perp}^{i} = a_i - 2\frac{C_{12}^{i}}{C_{\|}^{i}}(a_{\|} - a_i). \tag{6–3}$$

The parallel and perpendicular strains in the films would be given by

$$\epsilon_{\parallel}^{i} = \frac{a_{\parallel} - a_i}{a_i} \quad \text{and} \quad \epsilon_{\perp}^{i} = \frac{a_{\perp}^{i} - a_i}{a_i}. \tag{6–4}$$

According to this theory,[34] when a thin GaAs layer is deposited on an infinitely large Si (100) substrate, the GaAs will be under strain. To compensate for this strain, the lattice constant of GaAs in the directions parallel and perpendicular to the heterointerface will change to minimize its elastic energy. According to Eq. 6–4, GaAs would be under large compressive and tensile strains in directions parallel and perpendicular to the heterointerface, respectively. The large strain will shorten the lifetime of the laser diode fabricated on an Si (100) substrate, and change various material properties such as the bandgap energy, valence band degeneracy, and effective mass, etc. This suggests that Eq. 6–2 may not describe accurately the strained-layer heteroepitaxy when the lattice mismatch is large. Therefore, we use the following argument to predict the change in the lattice constants for such cases.

Suppose that the strain caused by the lattice mismatch is largely relaxed by forming dislocations, plastic deformation, etc., during the film deposition, then the thermal strain can be the dominant and responsible for the change in the structure properties of films. For example, the lattice constant in planes parallel and perpendicular to the heterointerface will change with thermal strain according to Eq. 6–4. The thermal strains and stresses for multilayer thin films deposited on circular substrates were calculated by Yamada *et al.*,[51] with and without warpage. Considering an isotropic strain field $\epsilon_{xx} = \epsilon_{yy} = \epsilon_{\parallel}$ and $\epsilon_{\perp} = \epsilon_{zz} = -2\nu\epsilon_{\parallel}/(1 - 2\nu)$ such that the multilayered film is constrained and cannot warp, the thermal expansion takes place in both parallel and perpendicular directions. The strains are given by[51]

$$\epsilon_{\parallel}^{i} = \Delta T(\alpha_1 - \alpha_i) + \Delta T \sum_{i=2}^{n} \frac{t_i E_i (1 - \nu_1)}{t_1 E_1 (1 - \nu_i)} (\alpha_i - \alpha_1) \tag{6–5A}$$

and

$$\epsilon_{\perp}^{i} = -\frac{2\nu_i}{1 - \nu_i} \epsilon_{\parallel}^{i}, \tag{6–5B}$$

where α_i, E_i, ν_i, and t_i are the thermal expansion coefficient, elastic modulus, Poisson's ratio, and thickness of the substrate ($i = 1$) and films ($i \leq 2$), respectively.

When Eqs. 6–5A and 6–5B are substituted into Eq. 6–4 to calculate the lattice constants for GaAs/Si and GaAs/Ge heterostructures, good agreement is found with the X-ray measurements of Neumann *et al.*[48] This indicates that the strain caused by the lattice mismatch is indeed relaxed by forming dislocations or plastic deformation. Judging from the residual strain caused by the thermal mismatch, one can surmise that the film is relaxed at growth temperature.

Recent X-ray experiments by Lucas *et al.*[52] have also shown that the tetragonal distortion of the GaAs lattice at the interface is caused by the thermal strain. The in-plane (solid line) and out-of-plane (dashed line) lattice constants of both GaAs and Si are plotted as a function of the annealing temperature in Fig. 6–28. The in-plane thermal expansion of GaAs lattice follows that of Si lattice, smaller than the bulk value of GaAs. However the out-of-plane thermal expansion of GaAs lattice is larger than its bulk value.

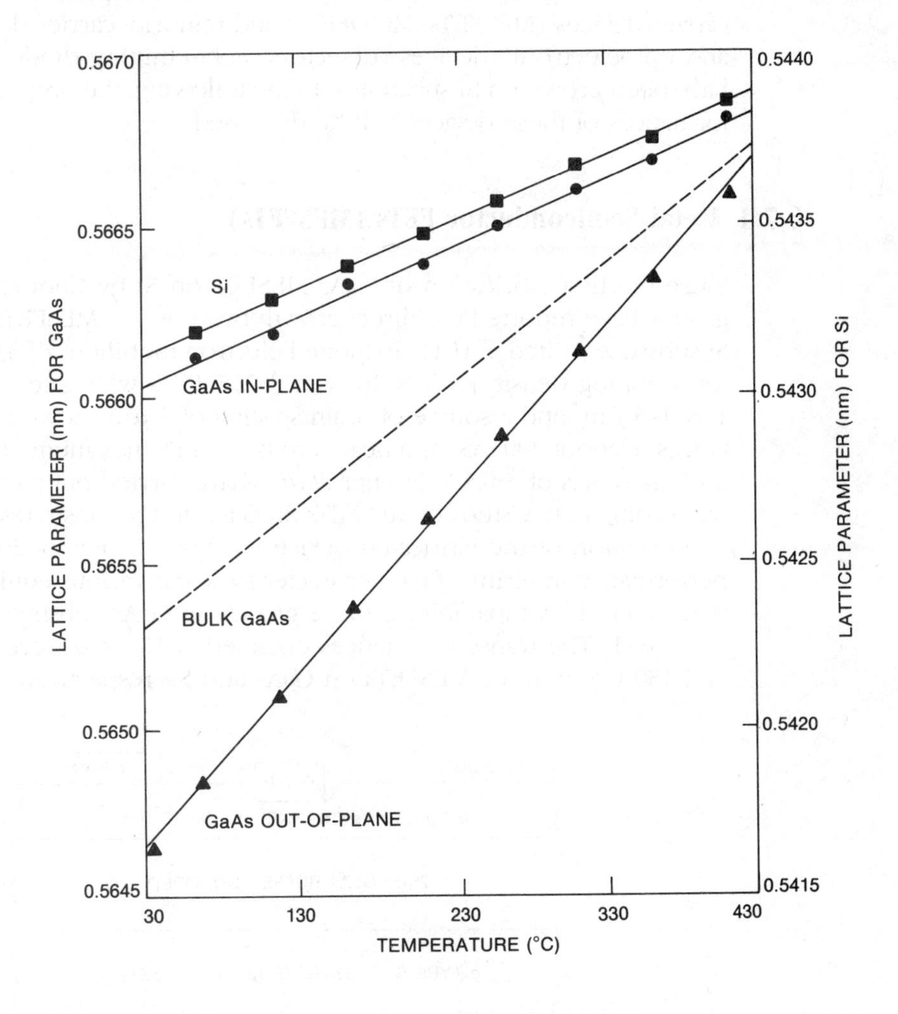

Fig. 6–28. Lattice parameter as a function of temperature of a GaAs on Si sample. The different lines are labeled in the figure.

Naturally, in- and out-of-plane lattice constants of GaAs converge at about 490°C annealing temperature, which is in good agreement with the growth temperature of the first few hundred monolayers.

6.7 High-Speed GaAs-based Devices on Si Substrates

Recent studies have already demonstrated that remarkably good device performance at both dc and microwave frequencies can be achieved from GaAs-based high-speed devices grown on Si substrates. All the majority carrier devices (MESFETs, MODFETs) and minority carrier devices (HBTs) and optoelectronic devices (detectors, light-emitting diodes, and lasers) have been grown on Si substrates. In the following, the properties and performances of these devices will be discussed.

6.7.1 Metal-Semiconductor FETs (MESFETs)

Since the first fabrication of GaAs MESFET on Si by Choi *et al.*[53] several groups have reported the direct growth by MBE.[54–57] MESFETs grown on a Si substrate[55] tilted 4° (011) exhibited electron mobilities of 3350 $cm^2/V \cdot s$ for a doping density of 3×10^{17} cm^{-3}. MESFETs with gate dimensions of 1×145 μm^2 and a source of drain spacing of 3 μm showed transconductances of about 200 mS/mm near zero bias and a maximum of 225 mS/mm for a gate bias of +0.5 V. Fischer *et al.*[56] have carried out a DC and microwave comparative study of MESFETs on GaAs and Si substrates. A schematic cross section of the fabricated device is shown in Fig. 6–29. The device performance in terms of gain characteristics and various equivalent circuit parameters is comparable to those grown on GaAs substrates, as seen in Table 6–1. The transconductances obtained at $V_{GS} = 0$ were 170 mS/mm and 180 mS/mm for MESFETs on GaAs and Si, respectively. Moreover, as

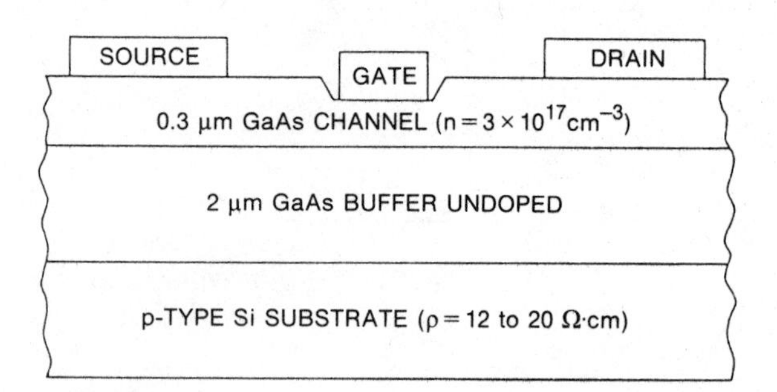

Fig. 6–29. Schematic cross-section of the MESFET structure on Si.

Table 6–1. Equivalent-Circuit Values of GaAs-Based and Si Substrate-Based GaAs MESFETs (V_{ds} = 3 V, V_{gs} = 0 V)

Parameters	On Si	On GaAs	Units
I_{ds}	131.8	83.8	mA
R_{in}	9.4	7.5	Ω
C_{gs}	0.711	0.747	pF
C_{dg}	35	64	fF
g_{mo}	52	59	mS
r_d	205	321	Ω
C_{ds}	160	85	fF
τ_t	3.75	4.41	ps
f_t	11.1	11.58	GHz

shown in Fig. 6–30 (where the carrier saturation velocity profiles for devices on GaAs and Si is compared), the carrier velocity remains constant to within about 10 to 15 nm of the interface for both types of devices.

For ultimate performance, GaAs MESFETs with 0.25 μm gate lengths have also been fabricated using 3-in (100) Si substrates tilted by 4° off toward (011) having a resistivity of 10 to 20 Ω · cm.[9] The output characteristics of a 150 μm wide device are shown in Fig. 6–31, giving good characteristics and a transconductance of 360 mS/mm at V_{GS} = 0.5 V. The short-circuit current gain calculated from microwave scattering parameters is shown in Fig. 6–32 as a function of frequency up to 20 GHz. The extrapolation of the gain gives a cutoff frequency of 55 GHz, which is comparable to the best GaAs MESFET grown on GaAs with identical device

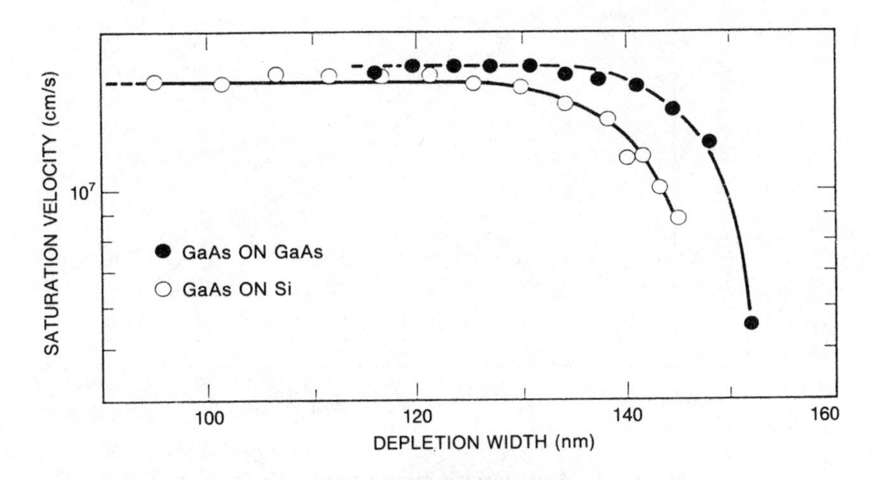

Fig. 6–30. Carrier saturation velocity versus depletion width for MESFETs on GaAs and Si substrates.

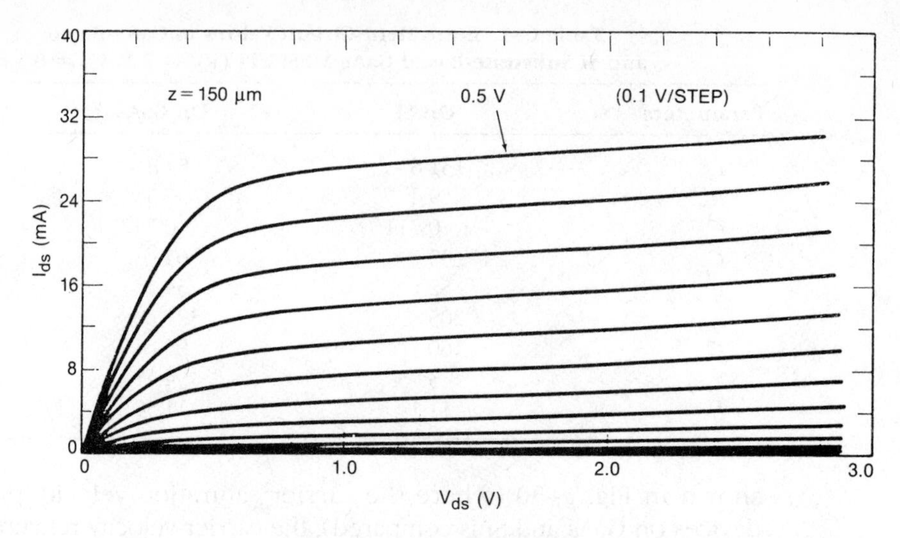

Fig. 6–31. Current-voltage characteristics for 0.25- × 150-μm GaAs MESFET on Si.

geometry and determined by the same extrapolation method. The noise figure of 2.8 dB measured at 18 GHz is about 1.4 dB larger than that grown on GaAs substrates. These results demonstrate that any parasitics are not of any specific problem with respect to the use of Si as substrates.

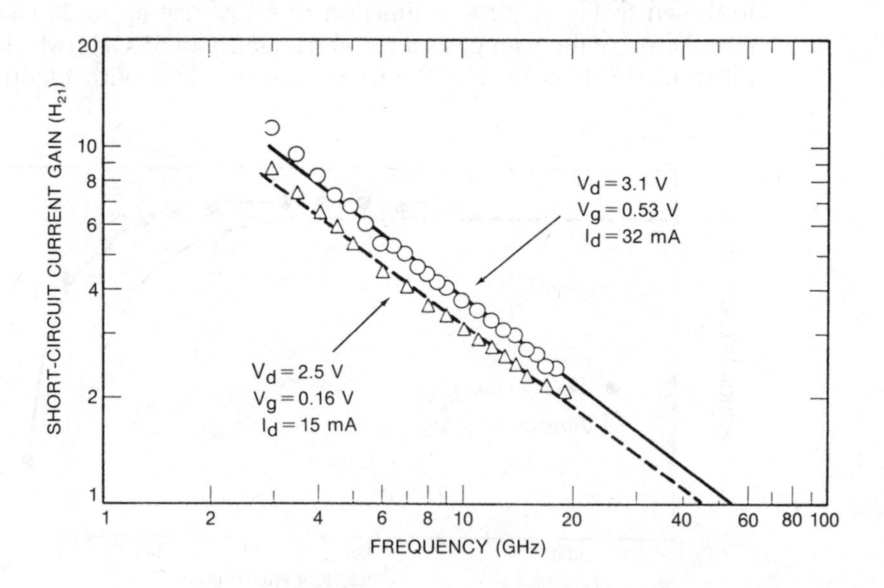

Fig. 6–32. Short-circuit current gain for 0.25- × 150-μm GaAs MESFET on Si under the bias conditions of V_{ds} = 3.1 V, I_{ds} = 32 mA, and V_{ds} = 2.5 V, I_{ds} = 15 mA.

6.7.2 Modulation Doped FETs (MODFETs)

Although both the MODFETs and MESFETs are majority carrier devices, there are important differences between the two types of devices. Unlike in MESFETs, electrons in n-MODFETs are confined to an interface where two-dimensional electron gas (2DEG) is formed (in MESFETs they are confined to a much broader area). Furthermore, the channel layers of MODFETs are of very high purity GaAs [or others, e.g., (In, Ga) As] while in MESFETs the doping levels are in the range of 1×10^{17} cm^{-3} to 1×10^{18} cm^{-3}.

Since electrons in an n-MODFET are closely confined to an interface [e.g., (Al, Ga)As/GaAs, (AlGa)As/(InGa)As] any roughness at the GaAs/Si interface is capable of degrading the performance of these structures. Therefore, the problem of achieving a high-performance MODFET on Si is a more difficult task than that of MESFET on Si. Because the tilting of the Si substrate and the three-dimensional character of GaAs/Si initial growth can induce dislocations, these dislocations (type II) can propagate up to the channel and can degrade the interface quality and hence high-performance character of MODFETs.

In any n-MODFET structure the carrier concentration in the ith sub-band ($i = 0, 1$) will be modified to[50]

$$n_s = \sum_{i=0}^{2} N_i = \sum_{i=0}^{2} \frac{m^* kT}{\pi \hbar^2} \ln\left[1 + \exp\left(\frac{E_F - E_i}{kT}\right) \exp\left(\frac{\delta(E_F - E_i)}{kT}\right)\right], \tag{6–6}$$

where m^* is the strain-dependent electron effective mass, and $\delta(E_F - E_i)$ is the shift in the energy difference between the Fermi level and ith sub-band level due to strain. Therefore, any change of 2DEG concentration with internal (mismatch strain) or external (bias strain) will influence the I-V and C-V characteristics and thus the high performance of modulation doped FETs.

The feasibility of MODFETs on Si have already been demonstrated[10] giving performances comparable to those on GaAs. After the first demonstration of MODFET on Ge,[24] several results have been reported on MODFETs grown on Si substrates.[58–60] Shown in Fig. 6–33 is the schematic cross section of the fabricated GaAs/(Al, Ga)As MODFET studied by Fischer *et al.*[10] The device consists of a 2 μm thick GaAs buffer layer and a 3-nm undoped $Al_{0.28}Ga_{0.72}As$ setback layer, a 35-nm doped ($N_D = 5 \times 10^{18}$ cm^{-3}) $Al_{0.28}Ga_{0.72}As$ and a 20-nm doped ($N_D = 5 \times 10^{18}$) GaAs cap layer. Fischer *et al.*[10] measured and compared MODFETs on both GaAs and Si substrates at microwave frequencies (see Fig. 6–34). Table 6–2 compares the equivalent-circuit parameters of MODFETs on Si and GaAs. As can be seen, the agreement is remarkable considering the drastically different technologies involved which also indicates the degree of reproducibility that can be obtained in GaAs on Si.

Recently developed $In_{0.15}Ga_{0.85}As/Al_{0.15}Ga_{0.85}As$ pseudomorphic MODFETs on Si substrates were also investigated at both DC and microwave frequencies. Although these pseudomorphic MODFETs on GaAs substrates outperformed GaAs/(Al, Ga)As MODFETs by a large margin, the ones on Si substrates have performances comparable to GaAs/(Al, Ga)As MODFETs on GaAs substrates. Figure 6–35 shows the output *I-V* characteristics at 77 K in dark and ambient light (solid lines) where the *I-V* collapse is absent. It is believed that they should perform better if the defects and

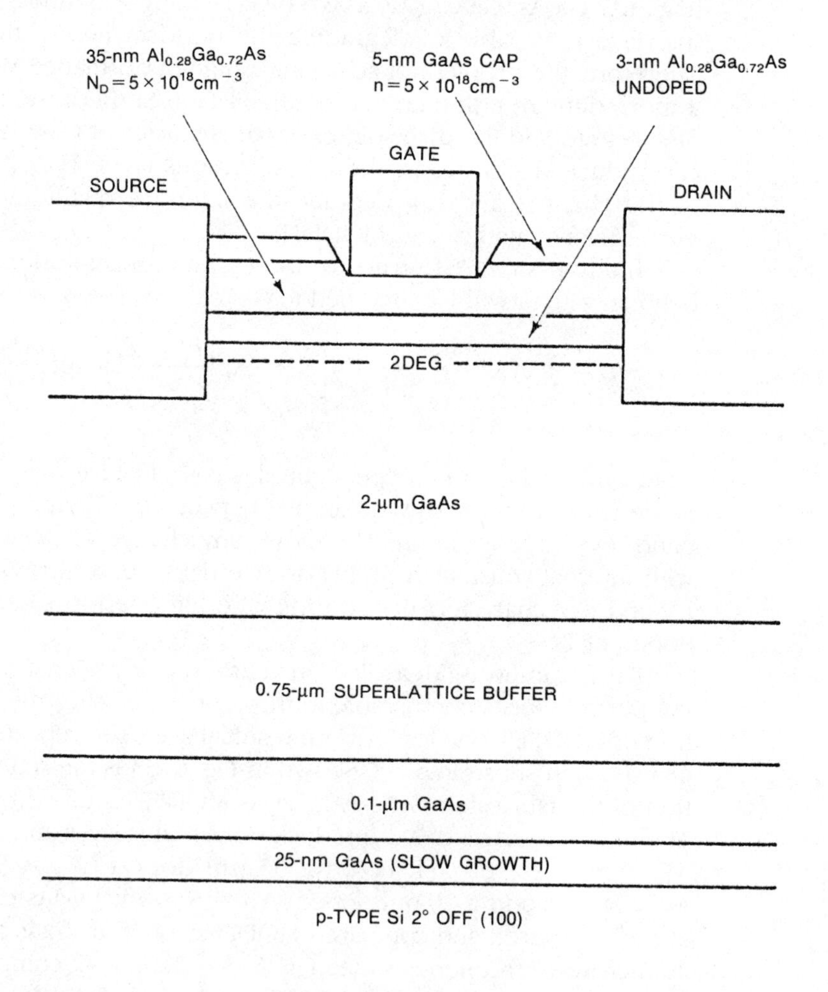

Fig. 6–33. Cross-sectional view of MODFET grown on Si substrate.

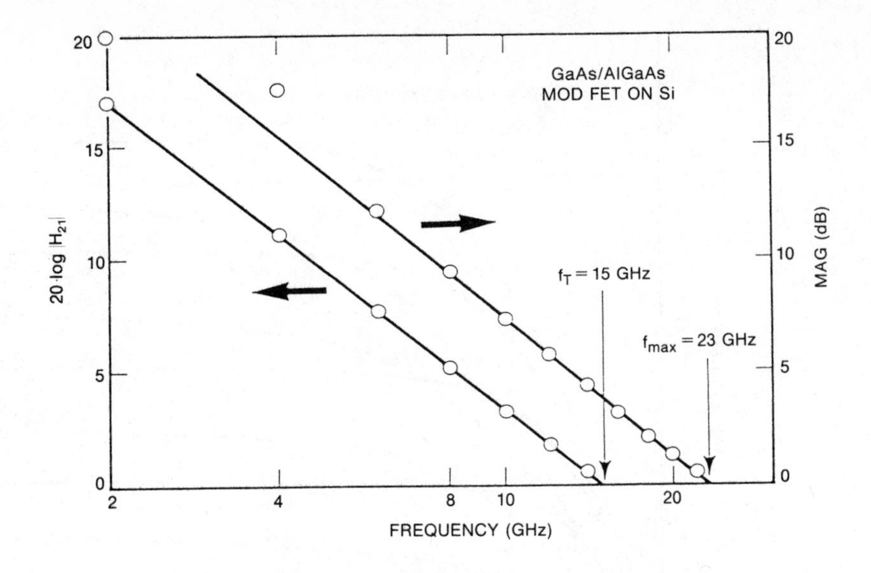

Fig. 6–34. Short-circuit gain and maximum available gain as a function of frequency for MODFETs grown on Si.

Table 6–2. Equivalent-Circuit Parameters for MODFETs on Si and GaAs Substrates (L_g = 1 μm)

Parameter	On Si	On GaAs	Units
g_m	82	118	mS
C_{gs}	0.87	1.25	pF
τ	1.98	2.37	ps
C_{dg}	14.3	12.4	fF
C_{ds}	114	117	fF
G_{ds}	6.3	5.09	mS
R_i	4.0	4.61	Ω
R_g	6.64	7.77	Ω
R_d	2.54	2.53	Ω
R_s	5.62	5.27	Ω
L_g	0.29	0.38	nH
L_d	0.35	0.33	nH
L_s	0.036	0.084	nH
f_T	15	15	GHz
f_{max}	24.5	25	GHz

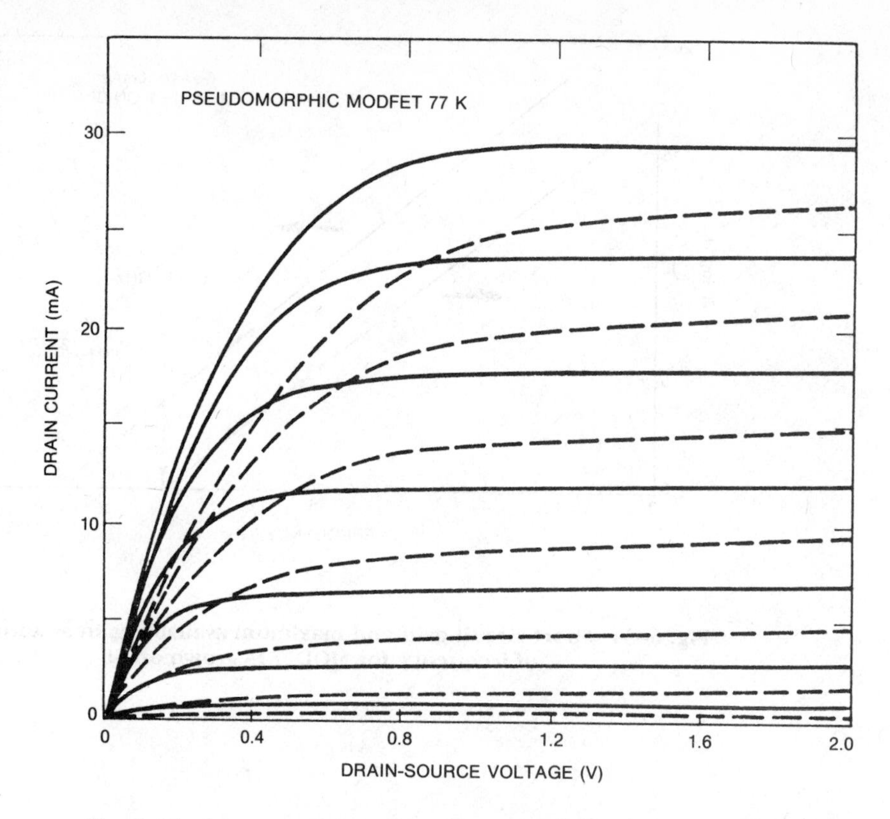

Fig. 6–35. Current-voltage characteristics of a pseudomorphic MODFET grown on Si substrate.

dislocation propagation from the GaAs/Si interface are terminated before the (In, Ga)As quantum well is grown.

6.7.3 Heterojunction Bipolar Transistors (HBTs)

Since the field-effect transistors (FETs) are majority carrier devices, their properties are not as sensitive to the GaAs/Si heterointerface quality as the minority carrier bipolar transistors. Any imperfection at the GaAs/Si heterointerface can propagate to the base-collector and base-emitter junctions and will degrade the effectiveness of the minority electrons or holes. It will create interfacial recombination of electrons and holes at the emitter-base and collector-base junctions. Any large tensile strain across the collector-base junction will reduce the collector current at voltages up to about 0.6 V. This will reduce the common-emitter current gain, β, increase the base resistance, cause short circuits, and soft breakdown voltage of the collector-base junction. For example, neglecting the interfacial recombination component of the current at moderate current levels, the common-emitter current gain of an npn HBT can be written as[50]

$$\beta(\epsilon) = \beta_0 \frac{\alpha_B(\epsilon)}{\alpha_E(\epsilon)}, \tag{6–7}$$

where β_0 is the current gain without strain and $\alpha_B(\epsilon)$ and $\alpha_E(\epsilon)$ represent the strain effects in the base and emitter regions of the emitter-base junction. Shown in Fig. 6–36 is the cross section of the heterojunction GaAs bipolar transistor grown on Si (100) studied by Fischer *et al.*[11] The Si (100) substrate is misoriented toward [011] direction by 4°. Figure 37 shows the current gain and the maximum oscillation frequency of the devices used in their study where they used 4 × 20 μm^2 emitter areas. Although the studies on the gain of (Al, Ga)As/GaAs on Si have shown low current gain, Fischer *et al.*[11] found good microwave performance for a relatively large geometry of HBT on Si (4 × 20 μm^2 emitter area). The current gain cutoff frequency of f_t is 30 GHz, and the maximum oscillation frequency of f_{max} = 11.3 GHz has been obtained, as shown in Fig. 6–37. These results are comparable with values of f_T = 40 GHz and f_{max} = 26 GHz for an HBT structure on GaAs with an emitter width of about 1.5 μm.[62]

Npn bipolar transistors, both homojunction[61] and heterojunction,[11,63] have already been demonstrated on Si substrates, with satisfactory perfor-

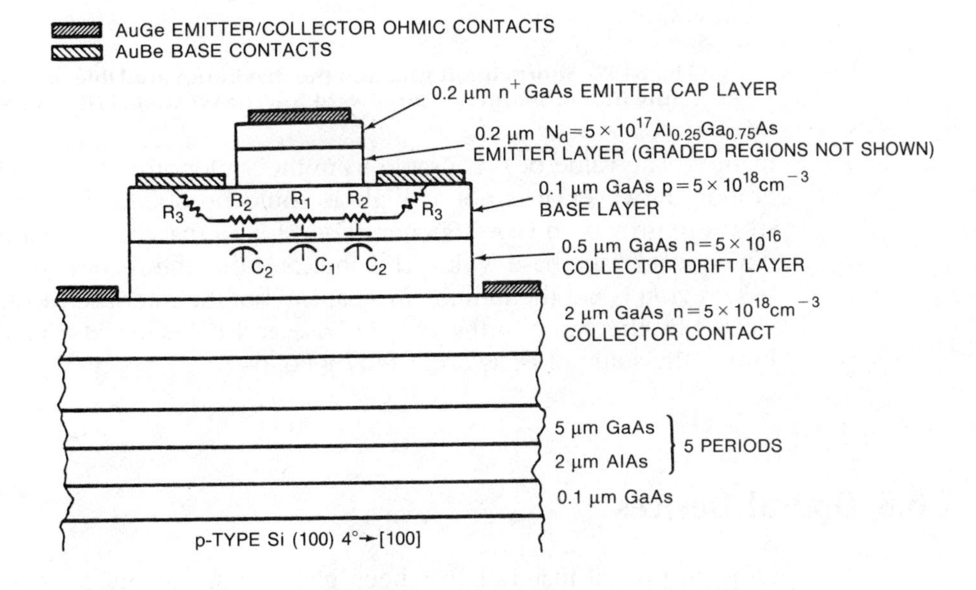

Fig. 6–36. Cross-sectional view of the GaAs/AlGaAs heterojunction bipolar transistor grown on Si.

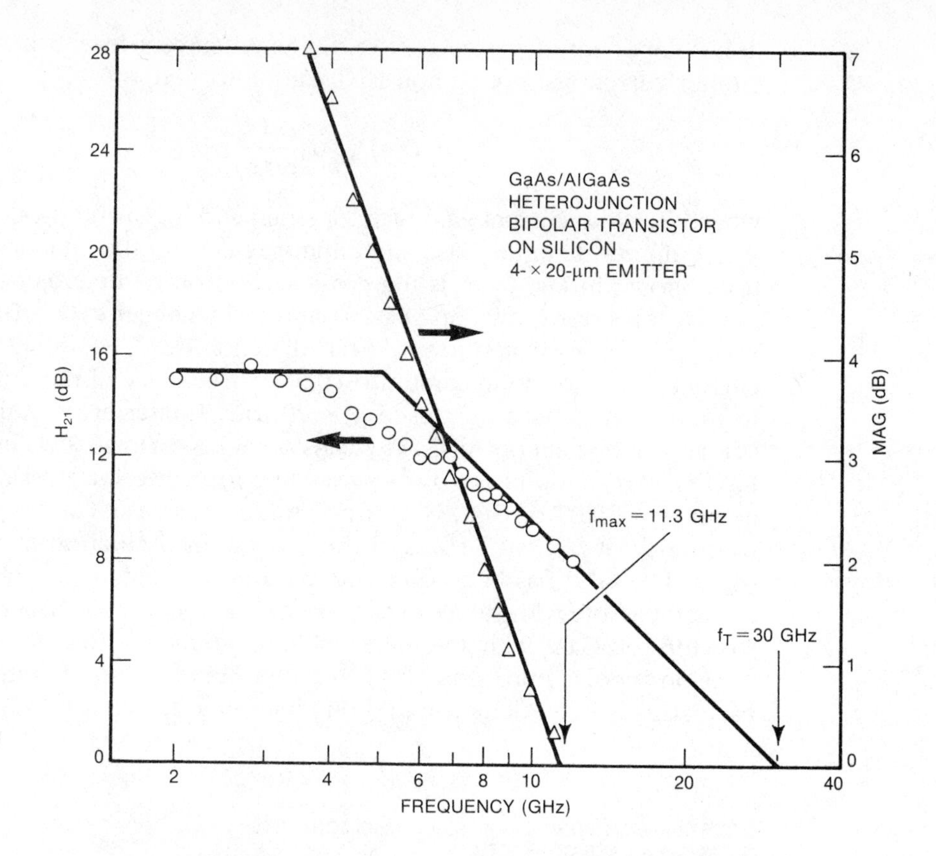

Fig. 6–37. Short-circuit gain and the maximum available gain as a function of frequency for 4 × 20 μm^2 GaAs/AlGaAs HBT on Si.

mances. The value of β is usable from the standpoint of some circuit operation; however, it is not as high as could be obtained from the same structure grown on GaAs. Fischer *et al.*[61] found that the current gain is independent of the base width. This indicates that the recombination in the base region is not the limiting mechanism but the interface recombination caused by the strain at the emitter-base and collector-base junction that lowers the value of β, as predicted by Eq. 6–7.

6.8 Optical Devices

Effort in optical materials has been given to semiconductors because of their importance in light-emitting devices (lasers and LEDs), detectors, and microelectronics, as well as the hope of combining both electronic and optical functions on the same chip. The operation of these devices depend on the carrier confinement in the layer. As shown in Sec. 6.6, the carrier confinement is degraded by the anisotropy of the conduction band structure (such as that of silicon). Therefore, it has been necessary to use the semiconductors that have direct-bandgap structure, such as GaAs or some III-V compounds. Since the mobile carriers are very sensitive to the misfit induced strain, the growth of GaAs lasers on Si is a good opportunity to test their performance in an effort to combine the electronic and optical functions on the same chip (Si substrate). Misfit dislocation, if allowed to propagate from the substrate into the active layer, can act as nonradiative recombination centers because of the strain. The use of tilted substrate and strained-layer superlattices can be used to reduce these dislocations.

Windhorn *et al.* achieved the first successful growth of laser diodes by using a Ge-coated Si[20] and direct growth on Si.[64] Sakai *et al.*[65,66] have achieved the first successful growth of lasers on Si by using metal-organic chemical-vapor epitaxy. The heterostructure they used was the Si/GaP/GaAs system. Improved lasers grown directly on Si were demonstrated by Fischer *et al.*[67] in 1986. These lasers had pulsed current threshold densities of about 6.9 kA/cm^2. The tilted substrates and SLSs used to reduce the dislocation density in this case are speculated to be responsible for the low current threshold. Figure 6–38 is a power versus current plot for these GaAs lasers on Si. The lasing wavelength was 810 nm. Typical slope efficiencies, per facet, of 0.132 W/A with a best value of 0.15 W/A were obtained. These efficiencies compare with a typical value of about

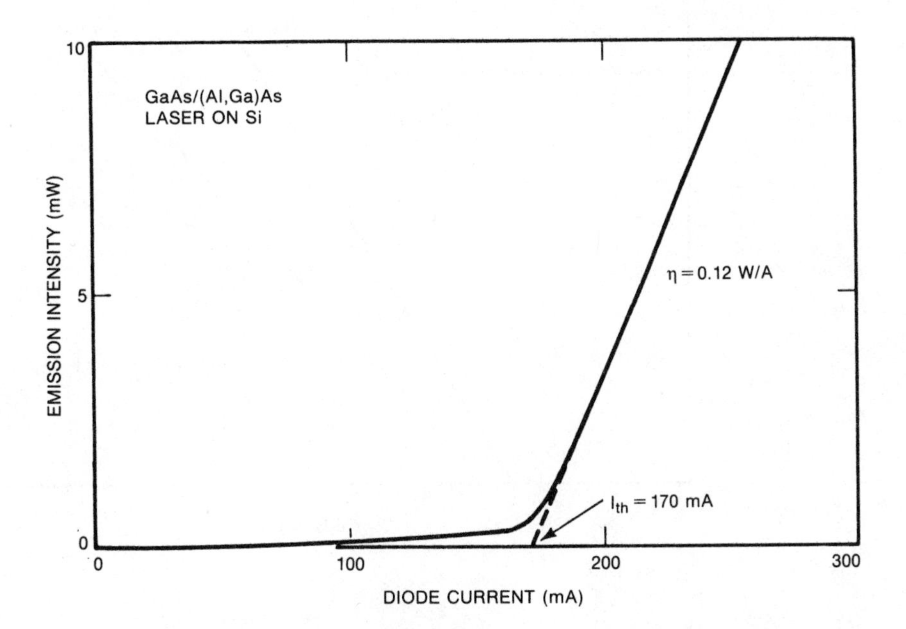

Fig. 6–38. Power output (from one facet) versus pulsed drive current for the 10-μm GaAs/AlGaAs laser on Si.

0.3 W/A for this type of structure grown on GaAs substrates. From measurements over the temperature range of 10 to 50°C, T_0 is 165 K, which compares well to that obtained for similar structures on GaAs substrates.

To date the stripe geometry lasers (about 10 × 500 μm) have room temperature threshold currents of 50 mA.[68] Threshold current densities of 300 A/cm and 214 A/cm have already been obtained.[69,70] These threshold current densities are comparable to those obtained on GaAs substrates in most laboratories and their low values have allowed for the operation of broad-area devices at room temperature for at least 5 min. Figure 6–39 shows the light vs current characteristics of GaAs lasers on GaAs and Si substrates. More work, however, remains to be done to enhance the properties of these devices for possible uses in optical interconnects and hybrid signal processing and radar applications. High-speed modulation experiments show a corner frequency of 2.5 GHz in 10- × 600-μm ridge lasers on Si substrates which is comparable to those obtained on GaAs substrates.[71]

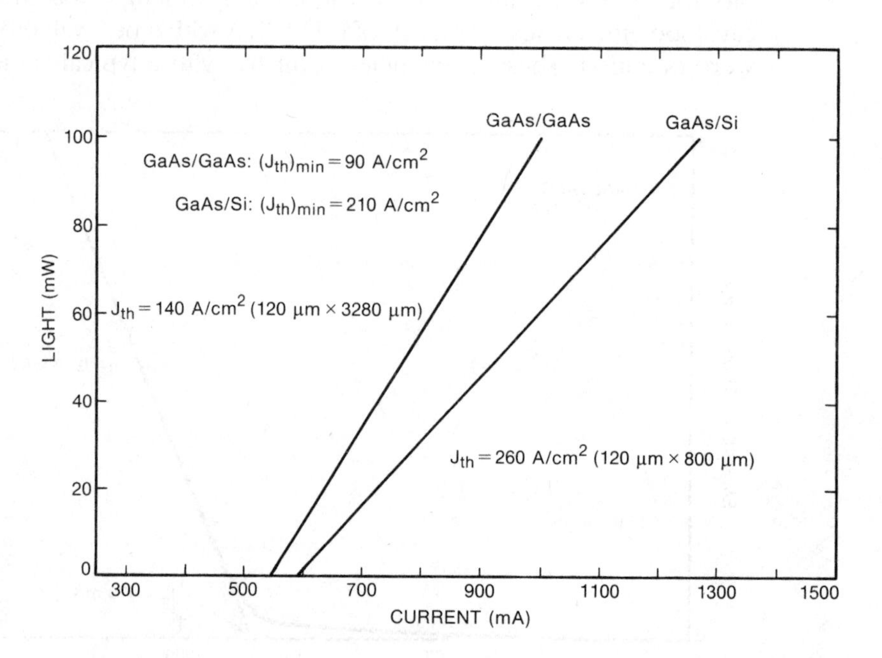

Fig. 6–39. Comparison of the threshold current densities of GaAs on Si and GaAs on GaAs broad-area lasers.

6.9 Hybrid Integration

One of the greatest potentials of GaAs/Si is in partitioning functions between GaAs and Si devices on the same chip for the most optimum operation of the overall circuit. For example, clocks and shift registers and the low-density, low-power fast-cache memories can be made of GaAs, whereas the LSI type circuits can be built in the mature Si technology.

To this end individual components of Si and GaAs devices have already been demonstrated in GaAs/Si. Among them are the Si MOSFET and GaAs LEDs,[2] and Si MOSFETs and GaAs MODFETs.[12] GaAs LEDs were in fact driven by the Si MOSFETs at up to 100 MHz, which was limited by the speed of the large size of the MOSFETs driver.

Multifunction multimaterial integration is also under study with great anxiety. For example, current focal-plane array-detector technology operating at wavelengths greater than 3 μm utilizes HgCdTe photovoltaic detectors which are In bumped to Si CCDs for signal processing. The In bumping for which the alignment is at room temperature can cause alignment problems at the operating temperature of 80 K for even the medium-size arrays. Utilization of GaAs as an interlayer on the Si CCD circuitry for the subsequent deposition of HgCdTe detectors can potentially produce large-area focal-plane arrays at a much lower cost. Significant progress has already been made in demonstrating each components of this approach without degrading the others. For example, HgCdTe material has already been demonstrated on Si substrates following a process that is friendly to Si CCDs.[3] Very recently HgCdTe photovoltaic detectors have been operated on GaAs/Si substrates, which represents a significant milestone toward large-area focal-plane arrays.[72]

6.19 Summary and Conclusions

Recent progress made in the GaAs/Si technology brings forward the serious questions of what needs to be done so that this potentially fruitful technology can be put to use. It appears that applications such as large-area focal-plane arrays, where a significant bottleneck can be overcome with GaAs/Si, may be the first serious application of this technology. Another area which is the forefather of GaAs/Si technology is the lightweight solar cells for space applications. This is a result of the combined factors of lightweight of Si for a given thickness as compared to GaAs and the superior strength of Si which allows the use of thinner substrate. Radiation-hard solar cells have already been demonstrated in GaAs/Si technology with great potential also because of the availability of large-area substrates. This composite technology can find applications in GaAs ICs and MMICs, in light emitters for Si ICs, and integration of GaAs fast circuits with high-density Si circuits.

References

1. H. Morkoç, "High speed GaAs based devices on Si substrates," *Mat. Res. Soc. Symp. Proc.,* 67:149–155 (1986).
2. H.K. Choi, J.P. Mattia, G.W. Turner, and B.-Y. Tsaur, "Monolithic integration of GaAs/AlGaAs LED and Si driver Circuit," *Electron Dev. Lett.,* EDL-9:512 (1988).
3. K. Zanio, R. Bean, K. Hay, R. Fischer, and H. Morkoç, "A HgCdTe/CdTe/GaAs/Si epitaxial structure," *Mat. Res. Soc. Symp. Proc.,* 67:141–146 (1986).
4. R. Fischer, W.T. Masselink, J. Klem, T. Henderson, T.C. McGlinn, M.V. Klein, H. Morkoç, J. Mazur, and J. Washburn, "The growth and properties of GaAs/AlGaAs on nonpolar substrates using molecular beam epitaxy," *J. Appl. Phys.,* 58:374–381 (1985).
5. R.P. Gale, J.C.C. Fan, B.-Y. Tsaur, G.W. Turner, and F.M. Davis, "GaAs shallow homojunction solar cells on Ge-coated Si substrates," *IEEE Electron Dev. Lett.,* EDL-2:169 (1981).
6. D.A. Neumann, Z. Zhu, H. Zabel, T. Henderson, R. Fischer, W.T. Masselink, J. Klem, C.K. Peng, and H. Morkoç, "Structural properties of GaAs on Si and Ge Substrates," *Proc. 6th MBE Workshop,* August 14–16, 1985 (University of Minnesota) and *J. Vacuum Sci. and Technol.,* B-4:642 (1986).
7. R. Fischer, N. Chand, W. Kopp, H. Morkoç, L.P. Erickson, and R. Youngman, "GaAs bipolar transistors grown on (100) Si substrates by molecular beam epitaxy," *Appl. Phys. Lett.,* 47:397 (1985).
8. S. Dash and M.L. Joshi, "Silicon defect structure induced by arsenic diffusion and subsequent steam oxidation," *IBM J. Res. Develop.,* 14:453 (1970).
9. M.I. Aksun, H. Morkoç, P.C. Chao, M. Longerbone, and L.P. Erickson, "Performance of quarter-micron GaAs metal-semiconductor field-effect transistors on Si substrates," *Appl. Phys. Lett.,* 49:1654 (1986).
10. R. Fischer, W.F. Kopp, J.S. Gedymin, and H. Morkoç, "Properties of MODFETs grown on Si substrates at DC and microwave frequencies," *IEEE Trans. Electron Dev.,* ED-33:1407 (1986).

11. R. Fischer, J. Klem, C.K. Peng, J.S. Gedymin, and H. Morkoç, "Microwave properties of self-aligned GaAs/AlGaAs heterojunction bipolar transistors on silicon substrates," *IEEE Electron Dev. Lett.,* EDL-7:112 (1986).
12. R. Fischer, T. Henderson, J. Klem, W. Kopp, C.K. Peng, H. Morkoç, J. Detry, and S.C. Blackstone, "Monolithic integration of GaAs/AlGaAs modulation doped field effect transistors and n-metal oxide semiconductor silicon circuits," *Appl. Phys. Lett.,* 47:983 (1985).
13. R.L. Anderson, "Experiments on Ge-GaAs heterojunctions," *Solid State Electronics,* 5:341 (1962).
14. A.G. Milnes and D.L. Feucht, *Heterojunctions and Metal-Semiconductor Junctions,* New York: Academic Press, 58–93 (1972).
15. D.L. Miller and J.S. Harris, "Molecular beam epitaxial GaAs Heteroface solar cell grown on Ge," *Appl. Phys. Lett.,* 37:1104 (1980).
16. P.M. Petroff, A.C. Gossard, A. Savage, and W. Wiegmann, "Molecular beam epitaxy of Ge and $Ga_{1-x}Al_xAs$ ultra thin film superlattices," *J. Crys. Growth,* 46:172 (1979).
17. K. Morizane, "Antiphase domain structures in GaP and GaAs epitaxial layers grown in Si and Ge," *J. Cryst. Growth,* 38:249 (1977).
18. S.L. Wright, H. Kroemer, and M. Inada, "Molecular beam epitaxial growth of GaP on Si," *J. Appl. Phys.,* 55:2916 (1984).
19. H.K. Choi, B.-Y. Tsaur, G.M. Metze, G.W. Turner, and J.C.C. Fan, "GaAs MESFETs fabricated on monolithic GaAs/Si substrates," *IEEE Dev. Lett.,* EDL-5:207 (1984).
20. T.H. Windhorn, G.M. Metze, B.-Y. Tsaur, and J.C.C. Fan, "AlGaAs double heterostructure diode lasers fabricated on a monolithic GaAs/Si substrate," *Appl. Phys. Lett.,* 45:309 (1984).
21. P. Sheldon, K.M. Jones, R.E. Hayes, B.-Y. Tsaur and J.C.C. Fan, "Growth and patterning of GaAs/Ge single crystal layers on Si substrates by molecular beam epitaxy," *Appl. Phys. Lett.,* 45:274 (1984).
22. R.M. Fletcher, D.K. Wagner, and J.M. Ballantyne, "GaAs light-emitting diodes fabricated on Ge-coated Si substrates," *Appl. Phys. Lett.,* 44:967 (1984).
23. W.T. Masselink, R. Fischer, J. Klem, T. Henderson, P. Pearah, and H. Morkoç, "Polar semiconductor quantum wells on non-polar substrates: (Al, Ga)As/GaAs on (100) Ge," *Appl. Phys. Lett.,* 45:457 (1984).
24. R. Fischer, J. Klem, T. Henderson, W.T. Masselink, W. Kopp, and H. Morkoç, "GaAs/AlGaAs MODFETs grown on (100) Ge," *IEEE Electron Device Lett.,* EDL-5:456 (1984).
25. W.I. Wang, "Molecular beam epitaxial growth and materials properties of GaAs and AlGaAs on Si (100)," *Appl. Phys. Lett.,* 44:1149 (1984).

26. W. T. Masselink, T. Henderson, J. Klem, R. Fischer, P. Pearah, H. Morkoç, M. Hafich, P. D. Wang, and G. Y. Robinson, "Optical properties of GaAs on (100) Si using molecular beam epitaxy," *Appl. Phys. Lett.,* 45:1309: (1984).

27. H. Christou, E. D. Richmond, B. R. Wilkins, and A. R. Knudson, "Surface treatment of ($1\bar{1}0$) sapphire and (100) Si for molecular beam epitaxial growth," *Appl. Phys. Lett.,* 44:796 (1984).

28. D. E. Aspnes and A. A. Studna, "Chemical etching and cleaning procedure of Si," *Appl. Phys. Lett.,* 39:316 (1981).

29. S. Wright and H. Kroemer, "Reduction of oxides on silicon by heating in a gallium molecular beam at 800°C," *Appl. Phys. Lett.,* 36:210 (1980).

30. H. T. Yang and P. M. Mooney, "The characterization of indium desorbed Si surfaces for low temperature surface cleaning in Si molecular beam epitaxy," *J. Appl. Phys.,* 58:1854 (1985).

31. R. C. Henderson, "Silicon cleaning with hydrogen peroxide solutions: a high energy electron diffraction and Auger electron spectroscopy study," *J. Electrochem. Soc.,* 119:772 (1972).

32. R. Fischer, H. Morkoç, D. A. Neumann, H. Zabel, C. Choi, N. Otsuka, M. Longerbone, and L. P. Erickson, "Material properties of high quality GaAs epitaxial layers grown on Si substrates," *J. Appl. Phys.,* 60:1640 (1986).

33. R. I. G. Uhrberg, R. O. Bringhans, R. Z. Bachrach, and J. E. Northrup, "Symmetric arsenic dimers on the Si (100) surface," *Phys. Rev. Lett.,* 56:520 (1986).

34. J. W. Matthews and A. E. Blakeslee, "Defects in epitaxial multilayers, II, dislocation pile-ups, threading dislocations, slip lines and cracks," *J. Crystal Growth,* 29:273 (1975).

35. R. Fischer, D. A. Neumann, H. Zabel, H. Morkoç, C. Choi, and N. Otsuka, "Dislocation reduction in epitaxial GaAs on Si (100)," *Appl. Phys. Lett.,* 48:1223 (1986).

36. N. Chand, R. People, F. A. Baiocchi, K. W. Wecht, and A. Y. Cho, "Significant improvement in the crystalline quality of molecular beam epitaxially grown GaAs on Si (100) by rapid thermal annealing," *Appl. Phys. Lett.,* 49:815 (1986).

37. C. Choi, N. Otsuka, G. Munns, R. Houdré, H. Morkoç, S. L. Zhang, D. Levi, and M. V. Klein, "Effect of in-situ and ex-situ annealing on dislocations in GaAs on Si substrates," *Appl. Phys. Lett.,* 50:992 (1987).

38. T. C. Chong and C. G. Fonstad, "Growth of high quality GaAs layers on Si substrates by molecular beam epitaxy," *J. Vac. Sci. Technol.,* B-5:815 (1987).

39. J.W. Lee, H. Schichijo, H.L. Tsai, and R.J. Matyi, "Defect reduction by thermal annealing of GaAs layers grown by molecular beam epitaxy on si substrates," *Appl. Phys. Lett.*, 50:31 (1987).
40. P.L. Gourley, G. Munns, R. Houdré, and H. Morkoç, "Photoluminescence microscopy of epitaxial GaAs on Si," *Appl. Phys. Lett.*, 51:599 (1987).
41. R. Houdré, G. Munns, H. Morkoç, C. Choi, N. Otsuka, S. L. Zhang, D. Levi, and M.V. Klein, "Dislocation reduction via annealing of GaAs grown on Si substrates," presented at the SPIE meeting, Bay Point, Florida (March 1987).
42. S.L. Wright, M. Inada, and H. Kroemer, "Polar-on-nonpolar epitaxy: sublattice ordering in the nucleation and growth of GaP on Si (211) surfaces," *J. Vac. Sci. Technol.*, 21:534 (1982).
43. M. Henzler and J. Clabes, "Structural and electronic properties of stepped semiconductor surfaces," *Japan J. Appl. Phys.*, Suppl. 2, Part 2: 389 (1974).
44. R. Kaplan, "LEED study of the stepped surface of vicinal Si (100)," *Surf. Sci.*, 93:145 (1980).
45. R.J. Fischer, An Investigation of Gallium Arsenide Materials and Devices Grown on Silicon Substrates, Ph.D. thesis, University of Illinois at Urbana-Champaign (1986).
46. H. Kroemer, "Polar-on-nonpolar epitaxy," *J. Cryst. Growth*, 81:193 (1987).
47. D.E. Aspnes and J. Ihm, "Biatomic steps on (001) Si surfaces," *Phys. Rev. Lett.*, 57:3054 (1986).
48. D.A. Neumann, H. Zabel, R. Fischer, and H. Morkoç, "Structural properties of GaAs on (001) oriented Si and Ge substrates," *J. Appl. Phys.*, 61:1023 (1987).
49. N. Otsuka, C. Choi, Y. Nakamura, S. Nagakura, R. Fischer, C.K. Peng, and H. Morkoç, "High resolution electron microscopy of misfit dislocations in the GaAs/Si epitaxial interface," *Appl. Phys. Lett.*, 49:277 (1986).
50. H. Ünlü and H. Morkoç, "Modulation doped FETs and other III-V high-speed transistors," presented at the SPIE conference, Bay Point, Florida (March 1987).
51. K.H. Yamada, T. Ogawa, and K. Wada, "Elastic calculations of the thermal strains and stresses of multilayered plate," *J. Appl. Phys.*, 62:62 (1987).
52. N. Lucas, H. Zabel, H. Morkoç, and H. Ünlü, "Anisotropy of thermal expansion of GaAs on Si (001)," *Appl. Phys. Lett.*, 52:2117 (1988).

53. H. K. Choi, B.-Y. Tsaur, G. M. Metze, G. W. Turner, and J. C. C. Fan, "MESFETs fabricated on monolithic GaAs/Si substrates," *IEEE Electron Dev. Lett.,* EDL-5:207 (1984).

54. G. M. Metze, H. K. Choi, and B.-Y. Tsaur, "Metal-semiconductor field-effect transistors fabricated in GaAs layers grown directly on Si substrates by molecular beam epitaxy," *Appl. Phys. Lett.,* 45:1107 (1984).

55. H. Morkoç, C. K. Peng, T. Henderson, W. Kopp, R. Fischer, L. P. Erickson, M. D. Longerbone, and R. C. Youngman, "High-quality GaAs MESFETs grown on silicon substrates by molecular-beam epitaxy," *IEEE Electron Dev. Lett.,* EDL-6:381 (1985).

56. R. J. Fischer, N. Chand, W. F. Kopp, C. K. Peng, H. Morkoç, K. R. Gleason, and D. Scheitlin, "A dc and microwave comparison of GaAs MESFETs on GaAs and Si substrates," *IEEE Trans. Electron Dev.,* ED-33:206 (1986).

57. H. Shichijo, J. W. Lee, W. V. McLevige, and A. Taddiken, "Performance of digital GaAs E/D MESFET circuits fabricated in GaAs-on-Si substrate," in *Proc. IEDM Meeting,* Los Angeles, Calif. (December 1986).

58. R. Fischer, J. Klem, T. Henderson, W. T. Masselink, W. Kopp, and H. Morkoç, "GaAs/AlGaAs MODFETs grown on (100) Ge," *IEEE Electron Dev. Lett.,* EDL-5:456 (1984).

59. R. Fischer, T. Henderson, J. Klem, W. T. Masselink, W. Kopp, H. Morkoç, and C. W. Litton, "Characteristics of GaAs/AlGaAs MODFETs grown directly on (100) silicon," *Electron. Lett.,* 20:945 (1984).

60. D. K. Arch, H. Morkoç, P. J. Vold, M. Longerbone, "High performance self-aligned gate (Al, Ga)As/GaAs MODFETs on MBE layers grown on (100) silicon substrates," *IEEE Electron Dev. Lett.,* EDL-7:635 (1986).

61. R. Fischer, C. K. Peng, J. Klem, T. Henderson, and H. Morkoç, "III-V semiconductors on Si substrates: new directions for heterojunction electronics," *Solid State Electronics,* 29:269 (1986).

62. P. M. Asbeck, A. K. Gapta, F. J. Ryan, D. L. Miller, R. J. Anderson, C. A. Liechti, and F. H. Eisen, "Microwave performance of GaAs/AlGaAs heterojunction bipolar transistors," *IEDM Tech. Dig.,* 864 (1984).

63. T. Won, C. W. Litton, H. Morkoç, and A. Yariv, "A high-gain GaAs/AlGaAs n-p-n heterojunction bipolar transistor on (100) Si grown by molecular beam epitaxy," *IEEE Electron Dev. Lett.,* EDL-9:405 (1988).

64. T. H. Windhorn and G. M. Metze, "Room temperature operation of GaAs/AlGaAs diode lasers fabricated on a monolithic GaAs/Si substrates," *Appl. Phys. Lett.,* 47:1031 (1985).

65. S. Sakai, T. Soga, M. Takeyasu, and M. Umeno, "AlGaAs/GaAs DH Lasers on Si substrates grown using super lattice buffer layers by MOCVD," *Jap. J. Appl. Phys.,* 24:L666 (1985).

66. S. Sakai, T. Soga, M. Takeyasu, and M. Umeno, "Room-temperature laser operation of AlGaAs/GaAs double heterostructure fabricated on Si substrates by metalorganic chemical vapor deposition," *Appl. Phys. Lett.,* 48:413 (1986).

67. R. Fischer, W. Kopp, H. Morkoç, M. Pion, A. Specht, G. Burkhart, H. Appelman, D. McGougan, and R. Rice, "Low threshold laser operation at room temperature in GaAs/AlGaAs structure grown directly on (100) Si," *Appl. Phys. Lett.,* 48:1360 (1986).

68. H. Z. Chen, J. Paslaski, A. Ghaffari, H. Wang, H. Morkoç, and A. Yariv, "High speed modulation and CW operation of AlGaAs/GaAs lasers on Si," *Proc. IEDM,* 238 (1987).

69. H. K. Choi, J. W. Lee, J. P. Salerno, M. K. Connors, B.-Y. Tsaur, and J. C. C. Fan, "Low threshold GaAs/AlGaAs lasers grown directly on Si by organometallic vapor phase epitaxy," *Appl. Phys. Lett.,* 52:1114 (1988).

70. H. Z. Chen, A. Ghaffari, H. Wang, H. Morkoç, and A. Yariv, "Continuous wave operation of extremely low threshold GaAs/AlGaAs broad area lasers, injection lasers on Si(100) substrate at room temperature," *Opt. Lett.,* 12:812 (1987).

71. H. Z. Chen, J. Paslaski, A. Yariv, and H. Morkoç, "High-frequency modulation of AlGaAs/GaAs lasers grown on Si substrates by molecular beam epitaxy," *Appl. Phys. Lett.,* 52:605 (1988).

72. R. Kay, R. Bean, K. Zanio, C. Ito, and D. McIntyre, "HgCdTe photovoltaic detectors on Si substrate," *Appl. Phys. Lett.,* 51:2211 (1987).

Chapter 7

Recent Developments in Diffusion in GaAs and $Ga_{1-x}Al_xAs$

R. J. Roedel

7.1. Introduction

Over the last 20 years, GaAs and related compounds have emerged as extremely useful materials for the fabrication of both high-speed and optoelectronic devices and circuits. Associated with the improvements and increasing sophistication of GaAs electronic devices have been the remarkable advances in the quality of the material and the processing procedures used in GaAs device fabrication. Large-diameter (3 in or more), high-resistivity ($10^8\ \Omega \cdot$ cm or higher), thermally stable GaAs substrates for fully ion-implanted MESFET or JFET circuit technologies are readily available; large-area, uniformly doped (n- or p-type), stable, low-dislocation (1000 cm^{-2} or less) GaAs substrates for optoelectronic device applications are also plentiful. Epitaxial technologies, which are required for the generation of quantum-well, superlattice, heterojunction, and virtually all optoelectronic devices, have never been more reliable, more nearly perfect, or more routine than at the present time. Tremendous advances in metal-organic chemical-vapor deposition (MOCVD), molecular-beam epitaxy (MBE), and even liquid-phase epitaxy (LPE) have made this possible. Ion implantation into substrate and epitaxial layers has become more controllable and more reproducible, especially with the advent of coimplantation procedures and novel annealing schemes. The judicious application of all of these technologies in combination has resulted in the recent development of a 2.6-W phase-locked array of $Ga_{1-x}Al_x$ As lasers,[1] 1-ns GaAs HEMT 5- × 5-bit parallel multipliers,[2] and 40-GHz $Ga_{1-x}Al_xAs$ heterojunction bipolar transistors,[3] to name a few recent dramatic accomplishments.

And yet, the relatively simple and uncomplicated procedure of diffusion of impurities such as zinc into GaAs and its alloys has remained a common processing technique in III-V device fabrication. How is it that such a prosaic procedure has remained viable even in the high-powered III-V processing laboratories of today? The answer is as uncluttered as the procedure itself—it has remained significant because it is the *only technique available* for certain specialized device structures. It is evident that diffusion cannot be employed in most modern GaAs devices and circuits: the three structures mentioned above were fabricated without the use of a diffusion procedure, and probably would not have performed as well if diffusion had indeed been used. But one of the purposes of this chapter is to describe those areas where diffusion is the processing method of choice.

There are several examples, both in optoelectronics and high-speed circuits, where diffusion would be required. Although most current GaAs laser structures are fashioned with elaborate epitaxial growth procedures, such as buried heterostructure devices, the transverse junction stripe (TJS) laser can be made only with diffusion.[4] The TJS laser, which is a promising candidate for integrated optoelectronics, is shown in cross section in Fig. 7–1. Zinc diffusion is carried out to convert the N-n-N heterostructure to P-p-P in a portion of the device and a transverse active region is created in the process. The device is planar, both contacts are topside, and the structure is compatible with semi-insulating substrates. Diffusion is used instead of implantation because of the *depth* requirement. Typical TJS lasers will employ epitaxial layers of total thickness 2.0 μm or more; ion implantation is incapable of achieving ion penetration to this depth, and diffusion must be employed.

Solar cells, fabricated from GaAs, can be made more efficient by texturing the surface to help reduce photon reflections. Random texturing can be employed, but uniform grooving is likely to be more efficient and reliable.[5] In GaAs, ⟨110⟩-oriented grooves may have cross sections shaped either like a "V" or a "dovetail." Grooved n-type substrates or epitaxial layers could be made into pn junction photovoltaic structures by a shallow zinc diffusion.[6] As shown in Fig. 7–2, the diffusion can produce remarkably

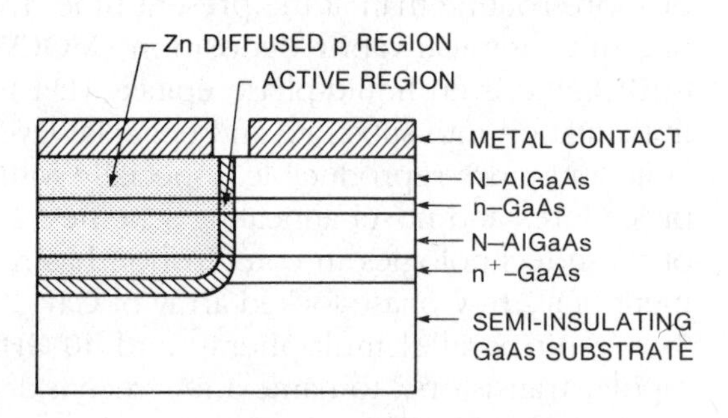

Fig. 7–1. Cross-sectional view of a transverse junction stripe (TJS) laser.

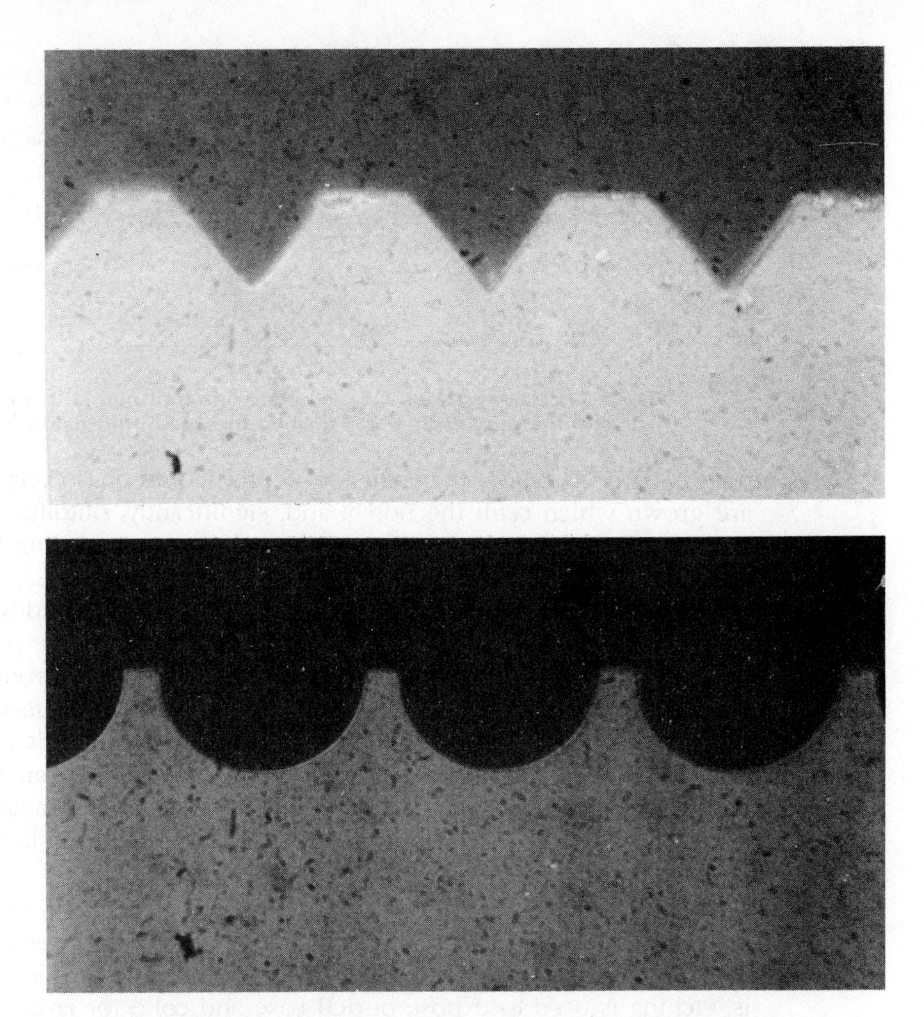

Fig. 7–2. Photomicrographs of two diffused, groove geometry solar cells. The center-to-center spacing of the grooves is 20 μm; the junction depth is 200 nm.

uniform junctions in etched structures; in fact, a junction with constant depth can even be produced in material with reentrant corners. This is a consequence of the vapor-phase character of most diffusion procedures. Although the shallow junction depth requirements for GaAs photovoltaics would allow the use of implantation, implanted ions follow straight line paths toward the target, and could not possibly produce pn junctions that uniformly follow an irregular surface.

Another example of the use of diffusion in optoelectronic structures is in the fabrication of planar, pn junction confined high-radiance light-emitting diodes. One possible embodiment is shown in Fig. 7–3. P-type diffusion is performed into an n-type GaAs substrate, after which a circular

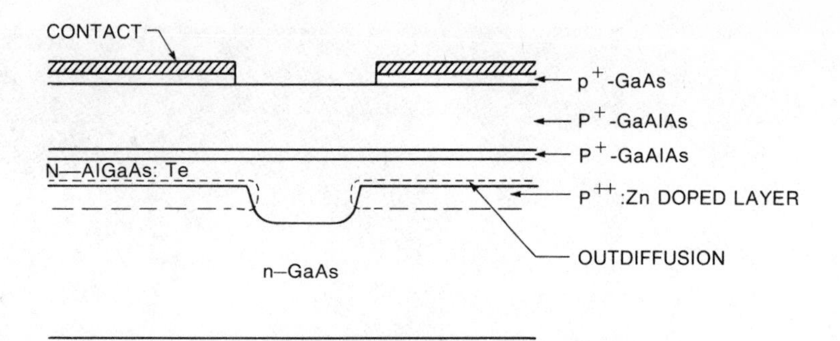

Fig. 7–3. Cross-sectional view of a current-confined LED, whose confinement region was formed by zinc outdiffusion.

notch is etched below the diffusion depth. Liquid-phase epitaxial layers are grown which refill the notch, and, significantly, outdiffusion of zinc from the diffused layer into the N-$Ga_{1-x}Al_xAs$ occurs during the epitaxy step. This converts the N-type confining layer into P-type $Ga_{1-x}Al_xAs$. The p-N-p layers that surround the active portion of the LED, and which could act as a parasitic phototransistor in the device, thus become p-N-P-p. This very effectively spoils the gain of the phototransistor, and produces a very efficient planar LED. Although ion implantation, or even epitaxy, could be employed to make the p-type outdiffusing layer, diffusion is most convenient because of the enormous density of impurities that can be incorporated in the layer during the procedure. It consequently makes a better source for outdiffusion than either a lower doped epitaxial layer or a thin implanted layer.

Heterojunction bipolar transistors (HBTs) are currently receiving considerable attention for their use in high-speed or high-power applications. Most HBT embodiments are fabricated with a nonplanar technology; that is, etching is used to expose buried base and collector layers for contact metallization. As a result, many HBTs resemble “wedding cakes” at completion. Although this may be suitable for discrete devices, it could be problematic in dense integrated circuits (ICs), with interconnections passing over and around the etched terraces. A planar HBT technology would clearly be preferable, but local ion implantation, used to reach buried layers, may be unsuitable because of the thickness of contact and emitter layers. Diffusion of zinc to reach a deep p-type base is an alternative processing scheme.[7] A diffused HBT is depicted, in cross section, in Fig. 7–4.

In addition, there are several areas in device engineering where a quick and simple diffusion can be very cost-effective and desirable. One example is the use of a very shallow diffusion of zinc into an existing p-type layer to produce a degenerately doped surface. Such a surface would make the subsequent fabrication of ohmic contacts easier and more reliable. Another example could be in the generation of differentially

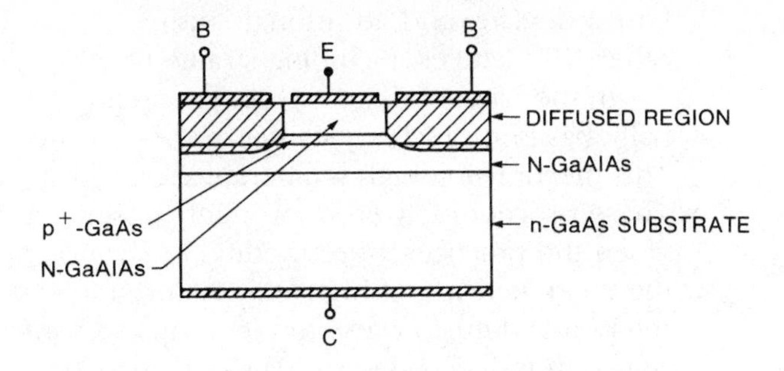

Fig. 7–4. Cross-sectional view of a planar heterojunction bipolar transistor (HBT), whose base is contacted through a zinc diffused region.

doped photocathodes. All GaAs photocathodes are p-type and the uniform doping level in the material is a compromise value. A light doping would produce a long electron diffusion length and aid carrier collection at the surface; on the other hand, high acceptor concentration would be useful in generating an enhanced surface field and improved escape probability.[8] A novel GaAs photocathode could be envisioned that has very light doping in its bulk followed by degenerate doping at the surface. The high doping at the surface could, of course, be produced by either implantation or epitaxy, but the uniformity and quickness of diffusion makes this the most attractive procedure. Furthermore, the maximum acceptor concentration in an implanted layer occurs not at the surface (where it would be most beneficial in this structure) but slightly inward, and the epitaxial growth of degenerate p-type layers is occasionally troublesome.

Recently, it has been demonstrated that impurities diffusing through a GaAs-$Ga_{1-x}Al_xAs$ superlattice can disorder the lattice and convert the superlattice into a uniform ternary layer with aluminum fraction the same as the spatial average of the original structure.[9] This phenomenon, known as diffusion-induced disordering (DID), has once again stimulated a great deal of interest in the physics and technology of diffusion in GaAs and $Ga_{1-x}Al_xAs$. DID can be carried out with acceptor diffusion, such as zinc,[9,10] with donor diffusion, such as sulfur[11] or silicon,[12] or even lattice defect (vacancy) diffusion.[13] Utilizing DID has permitted the fabrication of special buried quantum-well–laser structures: the diffusion produces material of higher average aluminum fraction than the laser's active layers, and the embedded device has superlative carrier confinement and index guiding.[13–15] In fact, the procedure has been employed to make a novel version of a phase-locked laser array.[16] In addition, it has been postulated that the mechanics of DID are remarkably complex, requiring an 11-step procedure to interchange vacancy positions on the sublattice.[17] It is evident that DID will continue to be the source of new and unique GaAs-based devices and additional studies of its mechanism of operation in other III-V lattices, including strained-layer superlattices.

In the nascent days of GaAs processing, diffusion procedures were initially based on techniques that had been successful in silicon processing. This meant using high temperatures, gas sources, and open-tube furnaces. These procedures were thoroughly unsuitable for volatile GaAs, and even when the practices were modified so as to "protect" the GaAs substrates, the continued use of high-temperature (1000°C, or 85 percent of the melting point) diffusion resulted in samples characterized by a vast amount of diffusion-induced defects. These defects typically included both precipitates[18–21] and dislocations,[21,22] and these defects produced irregularities and instabilities in the devices made with these materials. Diffusion in GaAs rapidly developed a reputation as a processing procedure that was to be avoided if possible.

However, the advent of improved diffusion procedures has changed this picture significantly. Modern GaAs diffusion practices can achieve high throughput, and are economical and reliable techniques. This has stemmed from an understanding that low-temperature procedures, coupled with better contamination control, can yield reproducible and consistent results, highlighted by substantially reduced defect content. In sum, the need for simple and economical processing procedures and the steady search for novel device structures has led to a reexamination of diffusion, and the results from the last five years have been quite remarkable.

In this chapter, recent advances in the theory, technique, and application of diffusion in GaAs and $Ga_{1-x}Al_xAs$ will be presented. Specifically, the use of "semisealed" or "leaky-tube" furnaces will be especially stressed, but other viable, modern practices will be discussed briefly as well, starting in the next section. The current results in p-type diffusion will be described in the third section, and a similar treatment for n-type diffusion will be presented in the fourth section. The overall results and trends, directions, and remaining problems will be covered in the last portion of this chapter. This chapter will only deal with GaAs and $Ga_{1-x}Al_xAs$. There is insufficient material to discuss the status of diffusion in GaSb and related compounds, and the case of GaP and $GaAs_{1-x}P_x$ has been covered elsewhere.[23] Although the practice of diffusion in InP and related materials is widespread and certainly deserves consideration, it is beyond the scope of this work and will be omitted.

7.2. Current Diffusion Technologies

As mentioned in the introduction, original diffusion technologies that intended to mimic the style of conventional silicon processing were doomed to failure.[24,25] All III-V compounds have extremely volatile components, and high-temperature or long-duration exposures to environments with flowing gases will invariably result in severe surface erosion. The most common type of surface deterioration, in the case of GaAs,[26] is thermal dissociation of the gallium and the arsenic during the diffusion. The situation is aggravated by thermal gradients, flowing gases, and related effects in an "open-tube" furnace. Furthermore, deterioration may also occur when the diffusing species directly interacts with the GaAs surface.[26] At high temperatures and high source pressures, zinc (a column II element and potential acceptor) may directly alloy with the wafer surface.[27] In addition, sulfur, selenium, and tellurium (column VI elements and potential donors) have high chemical activities, may attack the GaAs surface and form glassy coatings on the surface.[28]

Consequently, workers have traditionally resorted to a "closed-tube" type of diffusion system, originated in the 1960s[27,29,30] and still performed in research laboratories to this day.[31,32] In the closed-tube system, the GaAs wafer is sealed in an evacuated ampoule with a source for the dopant, and usually a source of aresnic to provide an arsenic overpressure during the diffusion. The diffusion is carried out by placing the ampoule into a furnace, allowing the diffusion to proceed, and then terminating the process by quenching the ampoule. The quenching stops the diffusion abruptly and thus prevents any outdiffusion of the impurity; it also helps to reduce condensation of the source onto the wafer surface. This procedure has been employed for both p-type and n-type diffusions. Dopants that can be used for acceptor diffusion include the column II elements, such as beryllium, magnesium, calcium, zinc, cadmium, and mercury (substituting for the column III gallium), and the column IV elements, such as carbon, silicon, germanium, tin, and lead (substituting for the column V arsenic), at least in principle. In practice, for reasons to be described in the next section, this list has been pared to just two elements, *zinc* and *cadmium*. For n-type diffusion, there is also an extensive group of potential donors, including the same column IV elements listed above (now substituting on the gallium sites), and the column VI elements, such as oxygen, sulfur, selenium, and tellurium (substituting on the arsenic sites). This list also has yielded only two practical candidates, *silicon* and *tin*.

The closed-tube approach provides, as Tuck points out,[33] a closed thermodynamic system which is ideally suited for a study of the diffusion mechanisms, measurements of solid solubility, and other related phenomena. Consequently, in the case of zinc diffusion into GaAs with the ampoule approach, a wide variety of possible zinc sources have been investigated including elemental zinc,[27,34] gallium-zinc alloys,[26] zinc diarsenide,[35] and gallium-zinc-arsenic alloys.[36] Casey and Panish have studied the role of the Zn-Ga-As phase diagram in the selection of sources for zinc diffusion into GaAs and have concluded that an alloy of 5, 50, and

45 atomic percent Ga, As, and Zn, respectively, is a near ideal diffusion source that would yield reproducible diffusions without surface deterioration in a closed-tube system in the 700 to 800°C range.[36] They further conclude that a zinc diffusion from any other source composition (for temperatures from 700 to 1000°C) in the ampoule system would probably result in nonuniform diffusion fronts, surface erosion, or a lack of repeatibility. This conclusion is based on the thermodynamic fact that in a closed system the actual diffusion source may be dramatically different in composition from the starting source—the actual source will try to approach an equilibrium ternary mixture, the composition of which is determined by the amount of each element present, the temperature and the volume of the ampoule. It is thus imperative that tight control of all of these factors be maintained to achieve acceptable, repeatable results. The 5/50/45 alloy is selected because it is not particularly sensitive to partial pressure variations; and other sources are far more sensitive. Because such painstaking efforts are required for the closed-tube approach, it has been termed "cumbersome" by most who have used it. Furthermore, it is an approach that is suitable only for small quantities of material. For a technique that is to be economical and easily employed, one must look elsewhere.

Two major alternative techniques have emerged in the last few years, namely, an open-tube approach that utilizes doped-glass sources that cover the wafers, and a semisealed or "leaky-tube" technique that resembles a compromise between the ampoule and the flowing-gas, open-tube approaches. Both techniques have significant advantages when compared with the closed-ampoule technique. In the first place, the efforts needed to produce repeatable partial pressures are relaxed. There is no longer a pressing need to weigh the source components with extreme precision or to establish an exact ampoule volume, because the source component quantities are determined by sample layer thicknesses or fixed furnace geometries. Secondly, the new techniques are typically easy to operate. There is no longer a need to carry out a difficult ampoule sealing procedure because the new procedures involve simple furnace loading and unloading without vacuum seals. And, the new procedures are economical and give high throughput by design. The closed-tube system requires the (nonrecoverable) use of expensive, high-quality quartz ampoules and can usually accommodate only a small quantity of GaAs material. The open- or leaky-tube approaches are intended to be used with large numbers of samples, and the economy of scale becomes applicable.

7.2.1 Doped Film Sources

Diffusion into GaAs from a doped-glass source covering the wafer surface was first proposed and carried out by von Muench 20 years ago.[37] At that time, the processes of depositing the SiO_x ($1 < x < 2$) films by pyrolitic decomposition and simultaneous doping were relatively complicated and expensive. The doping was achieved with the use of metal alkyls, such as diethylzinc (DEZ) for p-doping and tetraethyltin (TET) for n-type sources. Although initial tests were promising, the overall results were not encouraging because of inconsistencies in dopant incorporation and layer composition. The procedure was improved significantly by workers at Rensselaer Polytechnic Institute who developed a procedure for the simultaneous oxidation of tetramethyltin (TMT) and silane[38] for n-sources and DEZ and silane for p-sources.[39] These modified procedures were shown to have vastly improved reliability, precise control (especially for shallow diffusions), and genuine convenience of operation.[40,41] However, only a few researchers employed this method of diffusion, possibly because of reluctance to use the metal-organic growth procedures. Major attention for solid source diffusion arose with the development of "dopant emulsions" or "spin-on" dopant sources. Dopant emulsions are commercially available, blended solutions whose exact composition is generally proprietary. They typically include a film former (silicic acid ester usually), reaction components (acids, water), dopants (usually salts), and a solvent such as ethanol. The solution is applied to the semiconductor surface by dipping, spraying, or most often by spinning, using equipment suitable for photoresist application. After application, the emulsion requires a curing cycle to permit the SiO_2-forming reaction to take place. Normally this involves several exposures to an inert ambient at elevated temperature. There may be a low-temperature (200°C) annealing step to densify the film, and to drive off some of the volatile organic components, and there may also be a high-temperature step (600°C or more) to complete the SiO_2-forming reaction and solidification. This final, high-temperature step may also serve as the diffusion step for the wafer itself.

Doped spin-on films have been used for p-type diffusions, but they have found much more extensive use in n-type diffusion experiments and applications. The primary reason for this is that donor-doped films represent perhaps the easiest approach to n-type diffusion. As we shall see, there are very few vapor sources for donors (needed for the leaky-tube process) that have suitable vapor pressures or manageable toxicology. Tin-doped spin-on films can supply an adequate amount of donors for diffusion with a hazard-free procedure. Researchers at Lincoln Labs have used tin-doped emulsions to produce shallow n-type layers in GaAs photovoltaic structures,[42] and several groups have fabricated GaAs MESFETs with n-type active layers generated by a spin-on diffusion process.[43,44] The use of tin-doped spin-on sources has recently been reviewed by Arnold *et al.*[44]

P-type diffusions from spin-on sources usually involve zinc-doped films, and the main application of the doped films has been in the fabrication of highly doped contact layers.[45] There has been very little published work in the use of acceptor-doped spin-on films for GaAs, when compared with that for donor-doped films. It is evident that most p-type diffusions are being carried out by other methods.

The composition of the dopant emulsion and the subsequent annealing procedure can affect the final dopant diffusion profile. It will be demonstrated in later sections that the diffusion coefficient of any impurity is strongly dependent upon both the gallium and the arsenic vacancy concentrations in the material. These concentrations in turn can depend upon surface conditions. It has been established that gallium atoms can outdiffuse into overlying SiO_2 films, thus upsetting the normal vacancy densities near the surface.[46,47] In spin-on SiO_2 films, there are several factors that can affect this Ga outdiffusion. These include the porosity of the film, the degree of film hydroscopicity, and the presence of background gallium in the film.[44] Ma and Miyauchi[48] and Hayashi *et al.*[49] have examined the Ga outdiffusion effect in spin-on films that were undoped, arsenic-doped (4 atomic percent), or gallium-doped (5 atomic percent). In brief, Auger analysis verified that the Ga outdiffusion was moderate with the undoped films, quite extensive with the As-doped films, and undetectable in the Ga-doped layers. In the use of tin-doped emulsions, one relies upon the availability of gallium vacancies to provide substitutional locations for the tin atoms. Consequently, moderate Ga outdiffusion is probably a desirable occurrence. One might conclude that in order for this spin-on procedure to be repeatable, control of the outdiffusion would be critical. However, Hiyashi's work has shown that the Ga outdiffusion is a slow process compared with the indiffusion of Sn, so that only a shallow surface layer is affected by this phenomenon.[49]

The tin incorporation from Sn-doped emulsions also depends upon the doping source, i.e., the use of $SnCl_4$ produces different results from SnO_2 in the spin-on film.[44] Nissim *et al.*[43,50] investigated tin diffusion from spin-on films that contained tin in the oxide form, and found that there could often be surface deterioration and reduced electrical activity near the surface after diffusion. These effects have been attributed to the high oxygen content in the film (specifically, the SnO_2/SiO_2 ratio), which can accelerate the formation of Ga_2O_3 (eroding the surface) and Sn_3As_2 (diminishing the tin's efficacy as a dopant). These effects have not been observed when $SnCl_4$ is used as the tin source.[44] In fact, the SIMS profile and the electron concentration profile generally are identical in form and value for such diffusion sources. Characteristic results for tin diffusion from spin-on sources include:

a. surface concentration exceeding 1.0×10^{18} cm^{-3},

b. an atomic profile characterized by a region of nearly constant tin concentration and a very steep gradient at the diffusion front, and

c. an activity coefficient very close to unity.

Of all the possible donor elements available for GaAs doping, only tin has been used in dopant emulsions. Other potential donors, such as the column VI elements (sulfur, selenium, and tellurium) have not been investigated as spin-on film dopants. As pointed out previously, the tendency of the chalcogens to react with the GaAs surface and to form complex glassy layers would probably make them unsuitable emulsion dopants. Furthermore, even though "silicon-doped" SiO_2 films could probably be prepared, silicon diffusion into GaAs is usually done by another method.

One last diffusion approach that uses solid films as sources has been described in the literature. The element chosen to act as the diffusing species is deposited onto the sample surface by evaporation or sputtering. It may be anodized to form its own doped-oxide source, or it may act as its own elemental film source. Sakai *et al.*[51] have described a method in which zinc metal is evaporated onto the GaAs (typically 50 nm) and anodization is carried out until the zinc film is completely oxidized. Diffusion of the zinc takes place when the sample is placed into an open-tube furnace, in the presence of flowing nitrogen gas, at 600 to 800°C. The authors maintain that their approach is reliable and reproducible, but care must be taken to ensure that the anodization is stopped before the GaAs itself is oxidized, or difficulties with the diffusion will ensue.

Silicon can be utilized as an n-type diffusant in GaAs and very often it is diffused into GaAs wafers from films of the pure element itself. This has been carried out by Greiner and Gibbons using rapid thermal processing[52] and with longer term thermal treatments by Meehan *et al.*;[53] the Si film is normally 10 nm thick. The rapid thermal processing approach has been carried out at temperatures ranging from 850 to 1050°C, for durations from 3 to 300 s; a representative diffusion depth for the rapid processing is 0.25 μm for a 3-s, 1050°C treatment.[52] With near-equilibrium thermal processing, a similar junction depth would be achieved at 850°C in approximately 20 min. Because silicon diffusion is relatively novel, very little work has been performed to determine whether crystallographic defects are being produced by either the rapid or the near-equilibrium heat treatments.

7.2.2 "Leaky-Tube" or "Box" Diffusion Methods

The second alternative to ampoule diffusion is the so-called leaky tube or box diffusion method. In essence, diffusion takes place within a special chamber that is placed inside a normal open-tube furnace. This special chamber may be a compact capsule, such as that described by SpringThorpe and Svilans,[54] a modified liquid-phase epitaxial boat as used by Yuan *et al.*[55] and Prince *et al.*[56] or a long tubular insert as described by Roedel *et al.*[57] The capsule of SpringThorpe and Svilans is actually a pair of matched, ground-silica, cone joints, as shown in Fig. 7–5. The male cone is rigidly fixed in place, and the female cone, which holds the samples and diffusion source, can be moved into the furnace along the tube axis to mate with the male portion.[54] The cone can accommodate several GaAs slices at once and has a subcompartment that can contain polycrystalline GaAs to provide a small partial pressure of arsenic. This system has been used to perform zinc diffusions in GaAs and $Ga_{1-x}Al_xAs$ (and other III-Vs, as well), and the source is usually a Zn/Ga alloy. By varying the composition of the alloy source, it is possible to alter the zinc diffusion depth for fixed temperature and times. This will be explained in more detail in the next section.

The modified LPE boats or graphite holders, as described in Refs. 55 and 56, have been used for zinc diffusion in GaAs and $Ga_{1-x}Al_xAs$ and sulfur diffusion in GaAs. In both cases, the graphite boat contains a slot for the GaAs wafer, a compartment that holds the diffusion charge, and machined passageways to permit the dopant, in vapor form, to reach and interact with the substrate. These boats can be dissembled, cleaned, and loaded outside of the furnace itself. Yuan *et al.*[55] used a diffusion charge consisting of elemental Ga, elemental Zn, and polycrystalline GaAs for p-type diffusions in GaAs; Prince *et al.*[56] had a source consisting of GaAs and Ga_2S_3.

The leaky-tube diffusion method of Roedel *et al.*[6,57,58,59] has been used for zinc diffusion into GaAs and $Ga_{1-x}Al_xAs$, and tin diffusion into GaAs. The leaky-tube system consists of a conventional open-tube furnace equipped with a quartz liner sealed at one end. It is shown schematically in Fig. 7–6. Ultrapure helium can be passed through the outer open tube, or through the liner by means of a small purge tube. The helium flow can be halted in the liner to produce a stagnant, inert ambient within the liner. A quartz boat is used to hold the wafers (more than four 2-in wafers can be accommodated horizontally, more than twenty vertically) and a crucible for the source as well. The source for p-type diffusions is usually elemen-

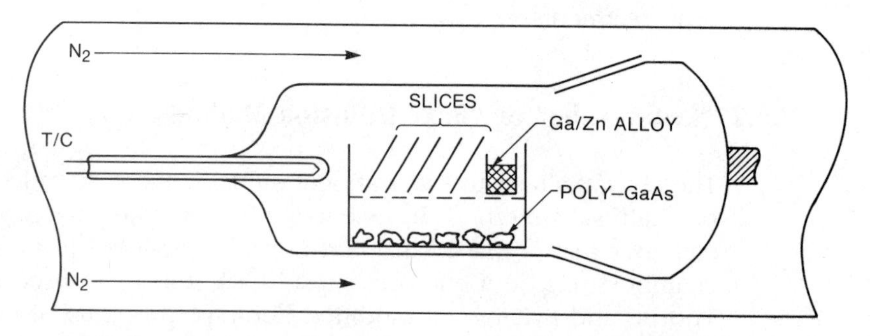

Fig. 7–5. The diffusion capsule used by SpringThorpe *et al.* for semisealed zinc diffusion. (*After SpringThorpe* et al.[54])

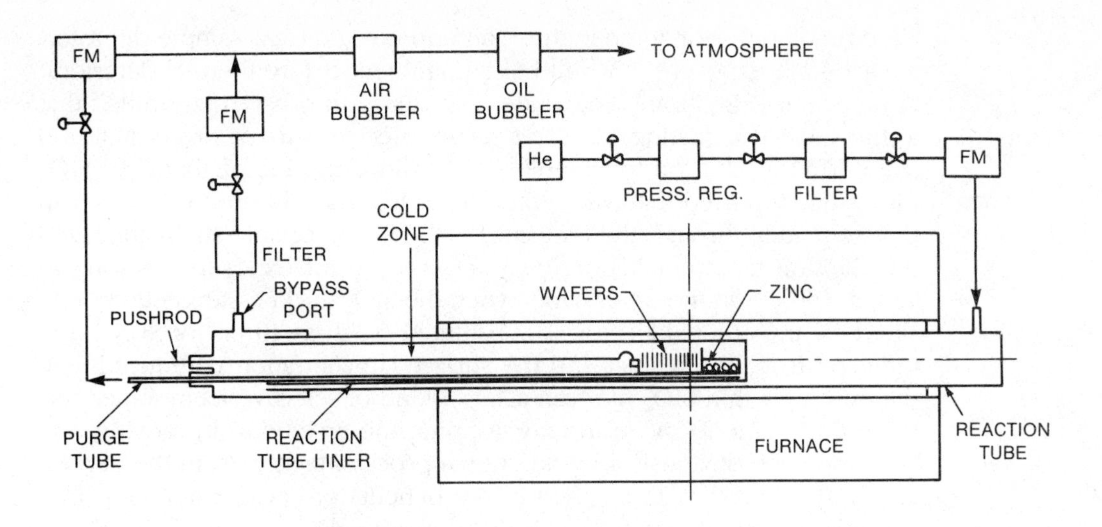

Fig. 7–6. A schematic of the leaky-tube diffusion system described by Roedel *et al.* (*After Roedel* et al.[57])

tal zinc or cadmium; for n-type experiments, SnS has been used.[59] The source crucible can hold in excess of 40 g of material; this large charge essentially provides "infinite-source" conditions.

The leaky-tube and other box diffusion systems possess several significant advantages. In the first place, they are ideal for the manufacturing environment. They are convenient and easy to use and they have a high throughput in general. In addition, the results of the diffusion are normally consistent, reliable, and reproducible. Although it has been pointed out that using elemental zinc or zinc/gallium alloys should result in poorly controlled diffusions,[36,54] most workers who have used box-type diffusion systems have found just the opposite results in practice.[54–59] This can probably be attributed to the constancy in experimental conditions achieved in run after run. Certainly, in the case of the leaky-tube furnace, the large quantity of source material and the stagnant ambient in the diffusion zone lead to a stable and consistent zinc vapor environment (infinite-source conditions) in every run. This should be contrasted with a closed-tube method in which the zinc vapor pressure can be drastically altered by a small variation in ampoule volume or source mass.

In several of the box diffusion systems, workers typically add a small amount of GaAs powder (or some other arsenic source) to generate a slight As overpressure to help reduce GaAs sample surface deterioration. Roedel and coworkers have utilized sources without arsenic bearing compounds and have found that precautions may be required to avoid surface erosion.[60] In the first place, they demonstrated that at 600°C (a common diffusion temperature), an unprotected GaAs sample can tolerate a 15-min exposure in a still helium ambient before thermal deterioration becomes significant. For longer exposures, they have determined that a thin dielectric coating can serve as an effective surface protectant, and still permit indiffusion to take place. In an investigation of sputtered SiO_2 films, they reported that, when the film is less than 10 nm thick, zinc can diffuse readily through the dielectric with no appreciable attenuation and the GaAs surface remains perfectly intact, even for exposures as long as 16 h at 600°C. On the other hand, when the SiO_2 thickness exceeds 15 nm, or when any other dielectric is substituted (such as 10 nm of Si_3N_4), the diffusion front is irregular and the surface concentration is reduced. Evidently, only a thin SiO_2 film can act as a kind of selective membrane, permitting the zinc to move in and stopping the arsenic from moving out. This approach obviously adds additional processing steps with the deposition and removal of the glass, but the benefits of having an arsenic-free source and furnace are not to be dismissed lightly.

In the leaky-tube furnace, the temperature profile is usually adjusted so that the source crucible is 3 or 4°C cooler than the wafers to be diffused. This is done so that the dopant vapor will not condense on the wafers themselves. In spite of this small temperature gradient, and a more significant gradient at the mouth of the tube, there is no evidence that any "cold-finger" effects are taking place. That is, at least for sources with low vapor pressures, there is no significant transport of the dopant vapor toward the front of the furnace by convection, but the vapor moves across to the wafers solely by diffusion.

Characteristic results for zinc diffusion from semisealed diffusion furnaces include:

(*a*) surface concentration as high as $1.0 \times 10^{20}\,cm^{-3}$,

(*b*) an atomic profile characterized by a plateau, followed by a very steep gradient at the diffusion front, and

(*c*) an activity coefficient close to unity.

To summarize, two diffusion techniques different from the closed-ampoule approach have emerged in the last few years. One approach, using doped glasses (deposited both thermally and in emulsion form), has become quite common for n-type diffusion involving tin. The other, using semisealed containers, is routinely employed for p-type diffusion using zinc. In the author's laboratory, leaky-tube diffusion has been used almost exclusively in recent years; in the next two sections, the current understanding of the mechanics of p-type and n-type impurity diffusion with the leaky-tube system will be described.

7.3 P-type Diffusion

In the previous section, it was mentioned that there are a large number of potential acceptor atoms for GaAs. The list includes column IV elements such as carbon, silicon, and germanium, and column II elements such as beryllium, calcium, magnesium, zinc, cadmium, and mercury. Many of these atoms are unsuitable for diffusion purposes because of *(a)* low solubility, vapor pressure, or diffusivity (Mg, Ca, C), *(b)* undesirable amphoteric or n-type activity (Si, Ge) or (c) unmanageable toxicity (Hg, Be). The remaining two elements, zinc and cadmium, are the most convenient species for p-type diffusion in GaAs. Because of their relatively high solid solubility, high diffusivity, high vapor pressure, and relatively low potential for hazardous operation, they are the elements that have been used the most for p-type diffusions in all III-V materials. This section will concentrate on the mechanics of p-type diffusion as carried out with leaky-tube diffusion apparatus, and with a special emphasis on zinc diffusion.

7.3.1 Zinc in n-type GaAs

When zinc is to be diffused into n-GaAs with the leaky-tube furnace, the following procedure is employed. The GaAs samples (substrates or epitaxial material) are thoroughly cleaned by degreasing in boiling 1,1,1-trichloroethane, boiling acetone, and boiling methanol, followed by etching in dilute HCl in water to remove native oxides, and finally degreased again. In general, a thin (7.5 to 10 nm) protective SiO_2 film is then deposited onto the samples by conventional RF sputtering. The sputter-coated samples are degreased one last time before being loaded into the diffusion furnace.

Some of the diffusion furnace quartzware is cleaned before each run as well. The purge tube, push rod, quartz boat, and quartz crucible are etched in dilute HCl and dilute HF, and rinsed in deionized water and methanol. The leaky-tube inner liner requires etching and cleaning infrequently, perhaps after 50 h of use, to remove the small amount of zinc that condenses at the cold end of the liner. Of course, the outer quartz tube, which is protected by the inner reaction liner, never requires any cleaning maintenance.

Although high-purity Zn pellets (6 9s purity or higher) are used in the first run, the pellets melt and reform into a zinc bar upon cooling. The bar can be used repeatedly, and is cleaned prior to each diffusion by etching in dilute nitric acid. For each diffusion operation, a few pellets are added to the crucible so that the total charge (bar plus pellets) weighs 40 g. Normally, a fine quartz powder is added to the bottom of the crucible to prevent the bar from adhering to the crucible.

After these preparatory steps, the furnace can be loaded. The wafers and zinc source are placed on the boat, which is put into the cold end of the furnace. The end cap is secured and the purging sequence commences. Ultrapure helium enters the back of the furnace and purges both the outer tube and the inner liner. The helium enters the liner at the front of the furnace, flows to the back of the liner, and ultimately exits the furnace by means of the purge tube. It takes approximately 1 h for the helium to displace all of the air within the furnace, but the purging is generally allowed to proceed for twice that length of time. At the completion of the helium flushing, the flow rate through the purge tube is reduced to zero, creating a nearly stagnant, inert environment within the liner. The helium flow into the furnace is reduced to a small value simply to maintain positive pressure on the end cap and to reduce the risk of any backstreaming through the outlet ports. The boat can then be pushed into the hot zone of the furnace so that diffusion can begin. At the conclusion of the diffusion, the boat is pulled back to the cold zone and is cooled with the aid of an external fan. The boat is usually allowed to remain in the cold zone for 30 min before being removed from the furnace.

Leaky-tube diffusion of zinc into GaAs is typically carried out in the temperature range 500 to 700°C, and for times that vary from 5 min to 24 h. At these temperatures, the elemental zinc source is molten (Zn melting point: 419°C) and the vapor pressure of the zinc varies from 1.3 to 60.3 torr. This is sufficient to achieve "infinite-source" conditions at the GaAs wafer surfaces.

Samples that have been zinc diffused with the leaky-tube approach have been examined with a variety of characterization techniques. The overall uniformity of the diffusion can be examined with cathodoluminescence (CL) in a scanning electron microscope (SEM); the junction depth can be determined with cleaving and staining, with CL, and with electrochemical profiling (ECP); the carrier profile can also be determined with ECP, and the atomic profile can be measured with secondary ion mass spectrometry (SIMS); and the transmission electron microscopy (TEM) can be used to evaluate the diffused layers from a physical point of view and to search for diffusion-induced defects.

Figure 7–7 shows (in plan view) the CL intensity from a GaAs epitaxial layer that has been diffused with Zn for 30 min at 600°C in the leaky-tube diffusion apparatus. The undulations in the surface are normal liquid-phase epitaxial terraces. It is evident that the CL emission is bright and even, and is indicative of the uniformity of the diffusion process. In Fig. 7–8, the same sample is shown in cross section, and the CL has been used to reveal the location of the pn junction. Since electrons and holes are swept rap-

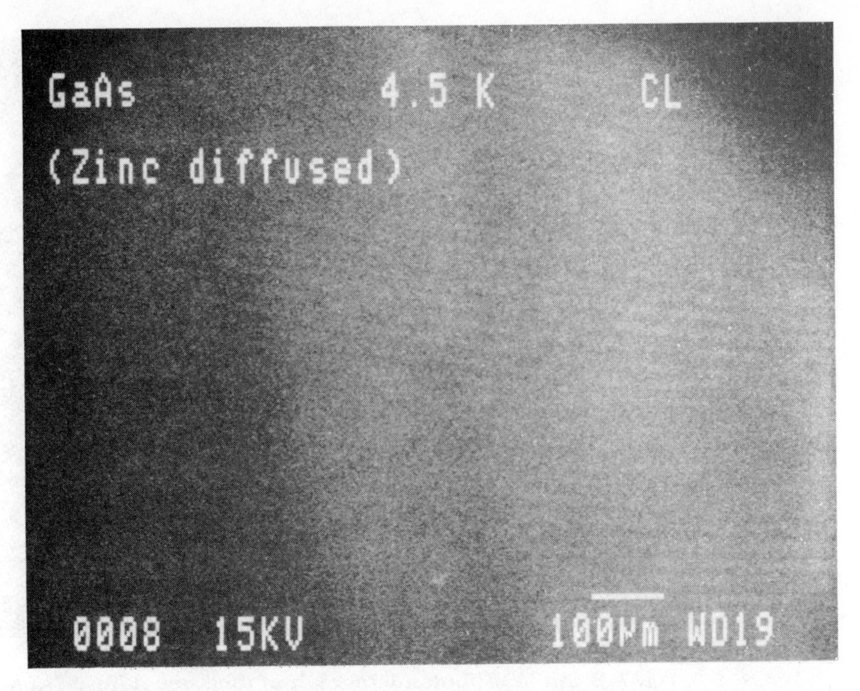

Fig. 7–7. An SEM photomicrograph of a zinc-diffused GaAs epitaxial layer, as viewed from the top, using cathodoluminescence imaging.

idly through the depletion region of a pn junction, the CL intensity of the space charge region is usually very low. Thus in Fig. 7–8 the dark band running parallel to the surface is roughly centered over the metallurgical junction. Again, it is evident that the junction is quite planar, has no evidence of defects being punched out in its vicinity, and has extremely uniform depth. These results are indeed typical for the leaky-tube diffusion process, and should be contrasted with the physical irregularities that have been shown for zinc diffusions carried out at 1000°C.[18–22]

It is customary in diffusion experiments to plot the junction depth versus the square root of diffusion time, as most simple theories of diffusion predict that

$$x_j \propto \sqrt{Dt}\,, \qquad (7\text{–}1)$$

where x_j is the junction depth, D the diffusion coefficient, and t the diffusion time. It will be demonstrated that this can be expected even for the relatively complicated case of zinc diffusing in GaAs, and in Fig. 7–9, the junction depth versus $t^{1/2}$ is plotted for the case of 600°C diffusions into GaAs substrates with silicon doping ($N = 1.0 \times 10^{18}$ cm^{-3}). It is apparent that the junction depth is indeed proportional to the square root of the diffusion time, and this holds for all temperatures in the 500 to 700°C range, for substrate and epitaxial material alike, and for all n-doping in the

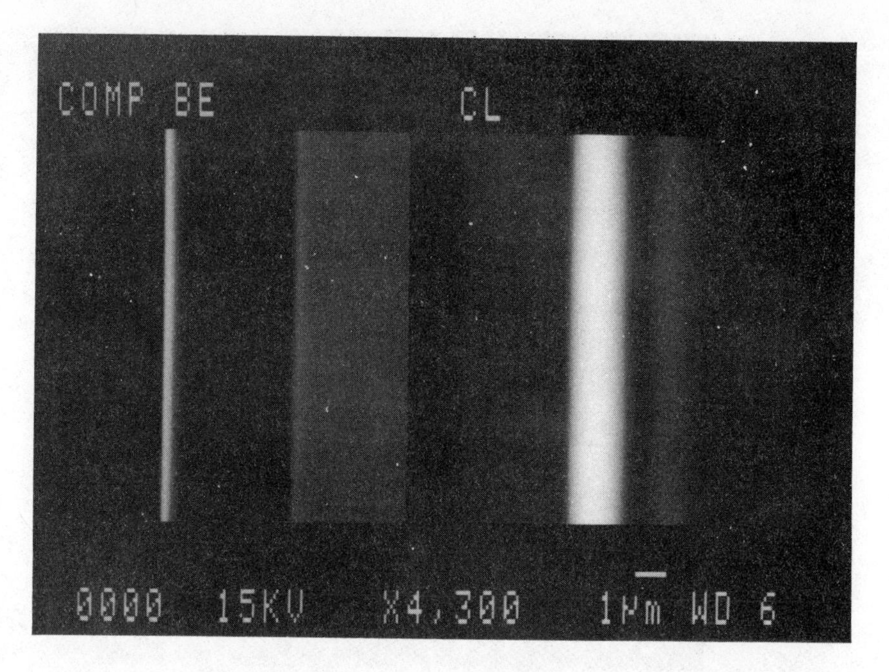

Fig. 7–8. An SEM photomicrograph of the same diffused sample as in Fig. 7–7, now shown in cross section. The pn junction is vertical in these images. The figure on the left is made with backscattered electrons, and shows a bright band at the surface, and the epi-substrate interface. The figure on the right is made with CL imaging and shows the high CL efficiency of the diffused region, a dark band at the junction, a dimmer image in the remaining portion of the epi-layer, and no signal coming from the substrate.

10^{16} cm^{-3} to 10^{19} cm^{-3} range. It is interesting to note that in Fig. 7–9 the line does not pass through the origin. Figure 7–10, which shows the short diffusion time data more clearly, reveals that an extrapolation to zero junction depth occurs at a time of approximately 1 min. Beyond this so-called incubation time, the junction depth versus $t^{1/2}$ is linear, and for $t_{diff} < t_{inc}$ there is no measurable junction depth. The carrier profiles corroborate this picture. For diffusion times less than t_{inc}, the sample remains n-type, but with some reduction in electron density, presumably due to some initial compensation. The incubation time depends upon the furnace temperature, the distance the zinc vapor must diffuse to reach the samples, and the time required to establish near steady-state conditions at the sample surface, but is typically 3 min or less. To generate Figs. 7–9 and 7–10, the diffusion depths were measured with both optical microscopy and electrochemical profiling.

As the diffusion temperature increases, the diffusion depth also increases since the diffusion coefficient typically obeys this type of relation:

$$D = D_0 e^{-\Delta E/kT}, \tag{7–2}$$

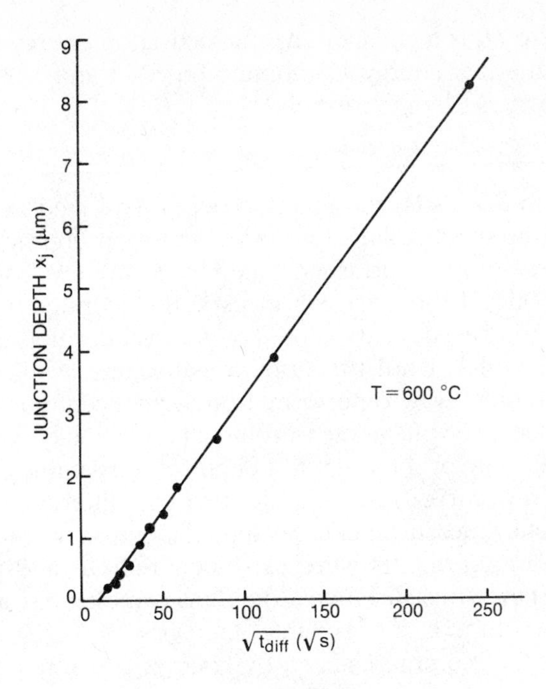

Fig. 7–9. Plot of junction depth versus square-root diffusion time from junction measurements.

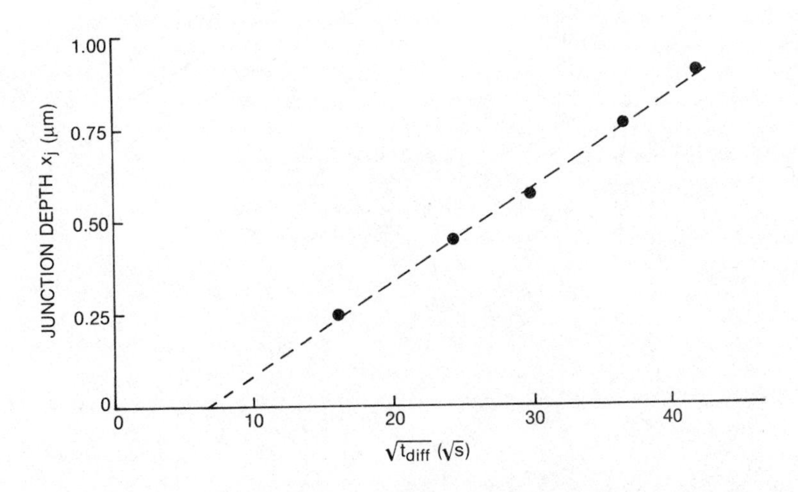

Fig. 7–10. A replotting of the data from Fig. 7–9 to show the short diffusion time data in more detail.

where D_0 is a constant, ΔE the activation energy for the diffusion, and kT the thermal energy. Combining Eqs. 7–1 and 7–2 produces an expression known as Chang's formula:[61]

$$x_j = A\sqrt{t}e^{-\Delta E/2kT}. \tag{7–3}$$

In Fig. 7–11, the junction depth is plotted versus $1/T$ for 30- and 60-min leaky-tube diffusions. The activation energy, ΔE, does not vary with diffusion time, and is estimated to be 1.20 eV from this figure.

One of the great advances in semiconductor characterization equipment in recent years is the commercial development of the electrochemical profiler. Until this apparatus became available, there really was no convenient way to perform a carrier concentration profile. One could use capacitance-voltage measurements on Schottky-contacted samples, but this could only probe a limited depth of the sample (until reverse-bias breakdown occurred). One could also use differential Hall measurements or Schottky measurements on angle lapped samples, but the sample preparation requirements were extremely tedious and troublesome. Spreading resistance measurements on GaAs samples has always been difficult because of the lack of suitable standards. However, the electrochemical profiler, first commercialized by Polaron Instruments, based on a design by

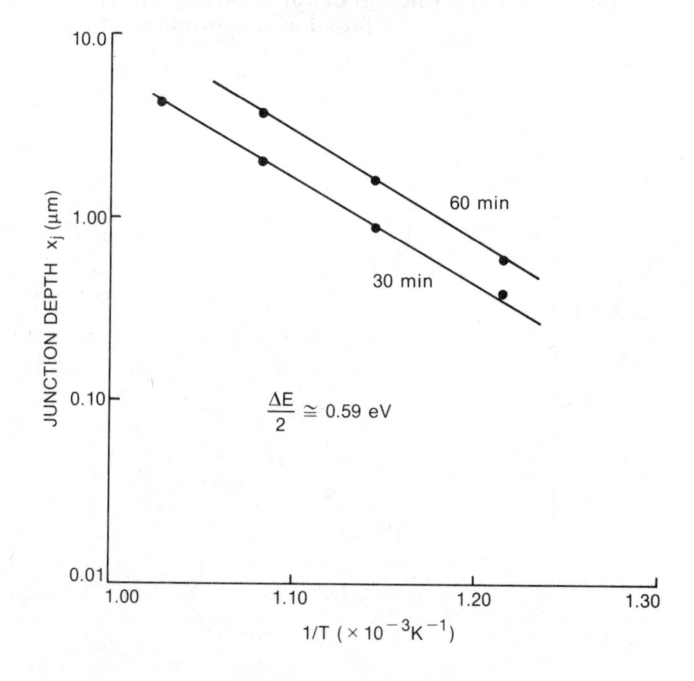

Fig. 7–11. An Arrhenius-type plot of junction depth versus reciprocal time for 30- and 60-min diffusions.

the British Post Office, has revolutionized this extremely important analytical technique. In brief, the profiler holds the sample in contact with an electrolytic solution forming an electolyte-semiconductor contact. This contact is very similar electrically to a metal-semiconductor contact,[62,63] and capacitance-voltage measurements can be carried out to extract the effective carrier concentration at the edge of the depletion region. One can, however, adjust the potentials within this electrochemical cell, comprised of the semiconductor, the electrolyte, and a platinum electrode, so that the semiconductor can actually be etched. The etching can be timed and the etch depth can be monitored by the use of Faraday's law and corroborated at a later time by a surface profilometry measurement. The etching is confined to a small circular region defined geometrically by a special sealing ring in the cell. After etching, the *C-V* measurement is carried out again, the carrier concentration is again determined, and this procedure is repeated until the desired profile is completed.[64] What has made this procedure so extremely useful are the following considerations: (1) the depth resolution (minimum depth step) is less than 3.5 nm, (2) it can be used for n-type or p-type material, and in fact can continuously profile through a pn junction, (3) it can be used to profile heterojunction samples, and (4) it can accurately measure carrier concentrations in the range 1.0×10^{13} cm^{-3} to 1.0×10^{21} cm^{-3}! In the next several figures, typical ECP profiles found in leaky-tube zinc-diffused GaAs will be depicted.

In Fig. 7–12, the carrier concentration profiles are shown in GaAs substrates zinc diffused at 600°C for 5, 10 and 15 mins; the background substrate concentrations are shown as well. Figures 7–13 and 7–14 show additional profiles, with the substrate profiles deleted, for diffusions as long as 60 min and as long as 16 h, respectively. As mentioned previously, 600°C diffusions as short as 1 and 2.5 min have been attempted. In the 1-min case, a slight reduction in the electron concentration was detected; in the 2.5-min case, the sample had converted to *p*-type close to the surface, but the hole concentration was small (less than 10^{17} cm^{-3}) and rather shallow (less than 100 nm). In every diffusion longer than 5 min, the zinc diffusion profile has the following characteristics. The concentration at the surface is high, but the maximum occurs approximately 100 nm below the sample surface. This maximum value can be as high as 1.5×10^{20} cm^{-3}. This subsurface peak disappears within 250 nm of the surface, and a region of relatively uniform concentration begins. This plateau region eventually gives way to a very steep gradient at the pn junction itself. Samples diffused only for 5 min show the subsurface peak, but the plateau region has not yet developed. Similar profiles are obtained for 500, 550, and 650°C diffusion runs. It was first believed that the curious subsurface peak was somehow associated with the leaky-tube procedure itself, but this was found not to be the case. Figures 7–15 and 7–16 are comparison ECP profiles for 600°C zinc diffusion into GaAs substrate and liquid-phase epitaxial

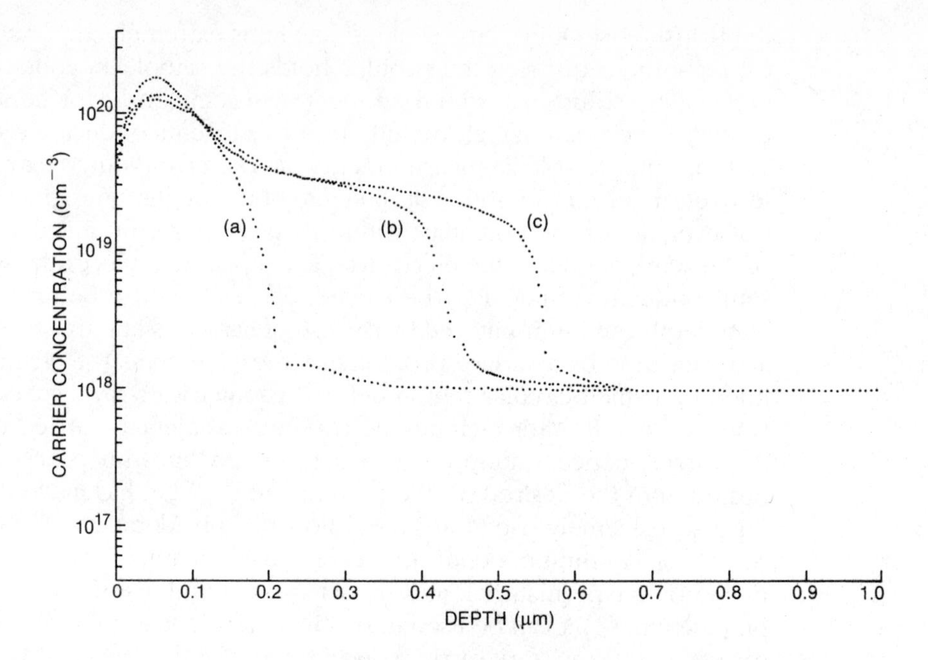

Fig. 7–12. Carrier concentration profiles for 5-, 10-, and 15-min zinc diffusions.

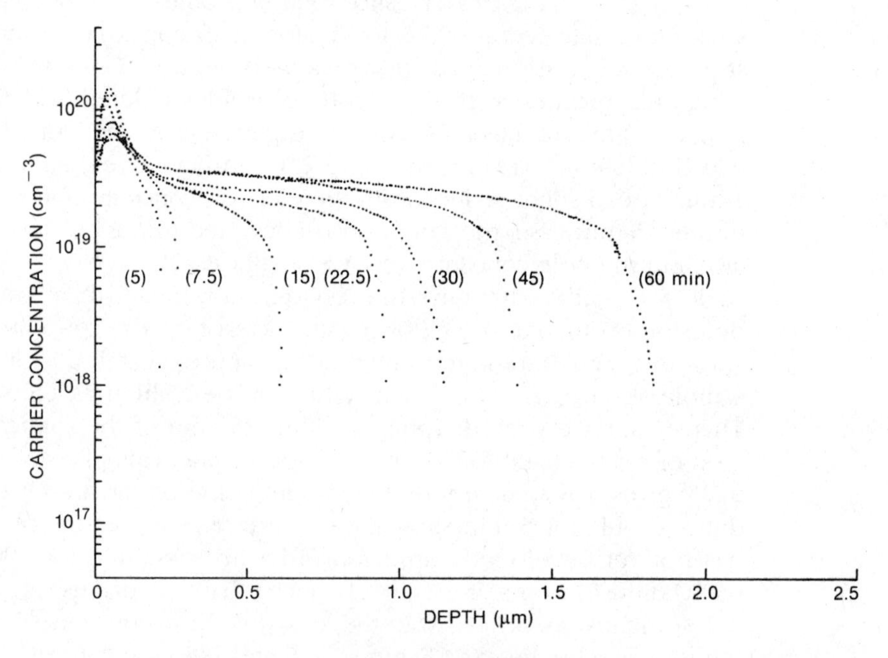

Fig. 7–13. Carrier concentration profiles for 5-, 7.5-, 15-, 22.5-, 30-, 45-, and 60-min zinc diffusions.

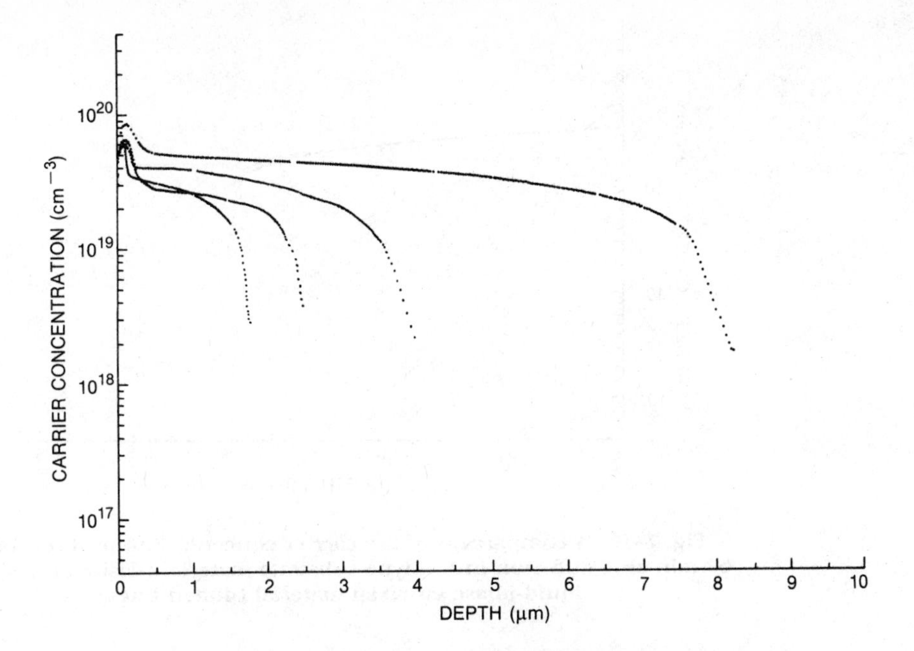

Fig. 7–14. Carrier concentration profiles for 1-, 2-, 4-, and 16-h zinc diffusions.

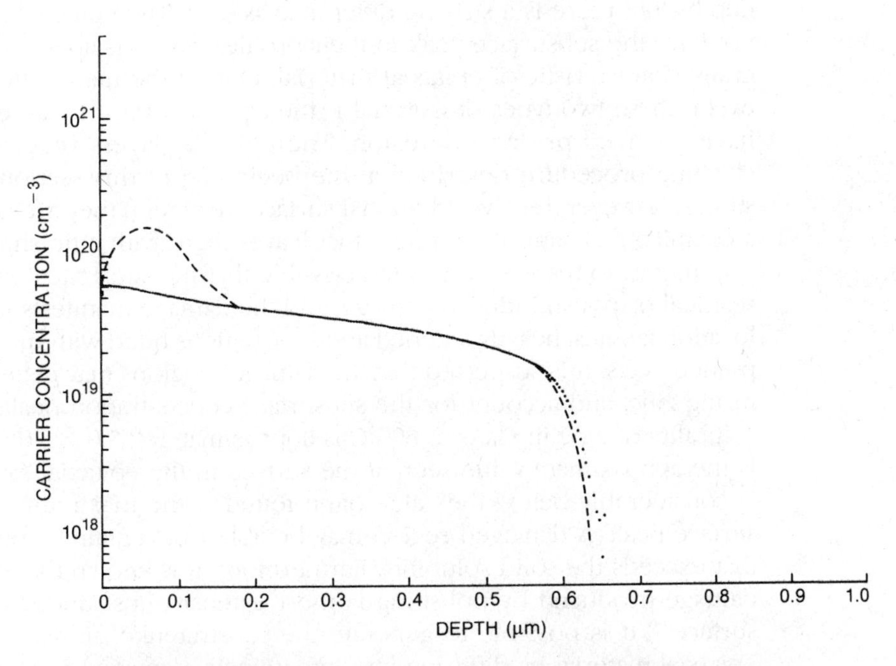

Fig. 7–15. A comparison of the carrier concentration profiles for a 15-min zinc diffusion into n-type substrate material (dashed lines) and liquid-phase epitaxial material (dotted lines).

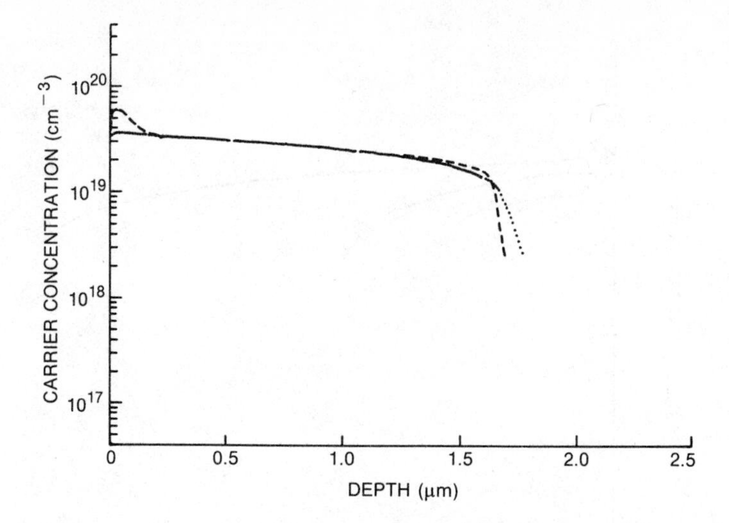

Fig. 7–16. A comparison of the carrier concentration profiles for a 60-min zinc diffusion into n-type substrate material (dashed lines) and liquid-phase epitaxial material (dotted lines).

material, for 15 and 60 min, respectively. The two profiles in each figure have virtually identical plateau regions, diffusion fronts, and junction depths, but there is a striking difference as well. The epitaxial samples do not have the subsurface peak in their profiles, and this appears to be generally characteristic of epitaxial material. One of the main differences between these two types of material is the type of surface treatment the two have received prior to diffusion. The epitaxial layers receive only the cleaning procedure described at the beginning of this section. The substrates, however, receive additional surface treatment: they are polished by a chemico-mechanical treatment that leaves them with a finish that is mirror smooth to the eye. It is quite possible that the substrates actually have residual or polish induced damage, and the damage manifests itself as dislocation tangles, heavily strained areas, or regions filled with miscellaneous point defects. It is suspected that the damaged regions may getter the indiffusing zinc, and account for the subsurface concentration peak. The solid solubility of zinc in GaAs at 600°C is approximately[65] 7.0×10^{19} cm^{-3}; this is the approximate value seen at the surface in the epitaxial samples, and is considerably below the value often found at the maximum of the subsurface peak. A damaged region may be able to maintain a concentration that exceeds the solid solubility. Furthermore, it is known that mechanical damage produced by polishing is most extensive just under the sample surface.[66] It is possible to generate the substratelike diffusion profile in epitaxial material by diffusing into intentionally damaged material. The intentional damage can be introduced by ion bombardment, for example. It appears that the development of a subsurface peak during a zinc diffusion experiment can be used as an indicator of subsurface damage.

Zinc diffused samples have also been examined with secondary ion mass spectrometry (SIMS), so that zinc atomic profiles can be generated. A direct comparison of atomic and carrier profiles can be used to determine the activation of the zinc, the amount of compensation, and whether some anomalous conditions have developed. In principle, apart from a slight rearrangement of free carriers due to ambipolar diffusion, the two profiles should be nearly identical. Figures 7–17, 7–18, and 7–19 are atomic profiles for samples whose carrier profiles have previously been displayed in parts of Figs. 7–12, 7–13, and 7–14. The SIMS analysis was performed on a state-of-the-art Cameca IMS-3f instrument. In general, the SIMS profiles and the ECP profiles have very similar form. The plateau regions, the steep gradients at the diffusion front, and even the subsurface peaks are clearly evident in the SIMS profiles. At the surfaces themselves, however, the carrier concentration drops abruptly from the subsurface peak value, while the SIMS profile shows the highest concentration right there. There are two possible explanations for this discrepancy. If the atomic profile is cor-

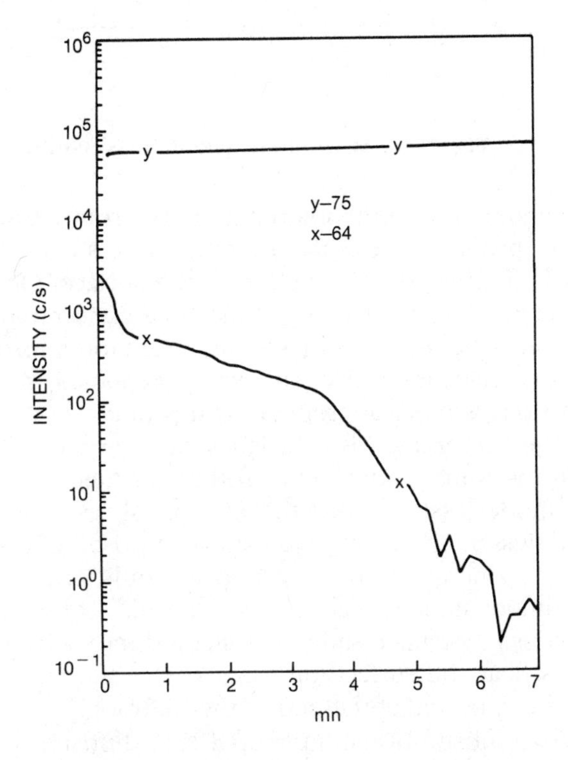

Fig. 7–17. SIMS atomic profile of a 5-min zinc diffusion. The zinc curve is marked by the x's; the reference arsenic profile by the y's.

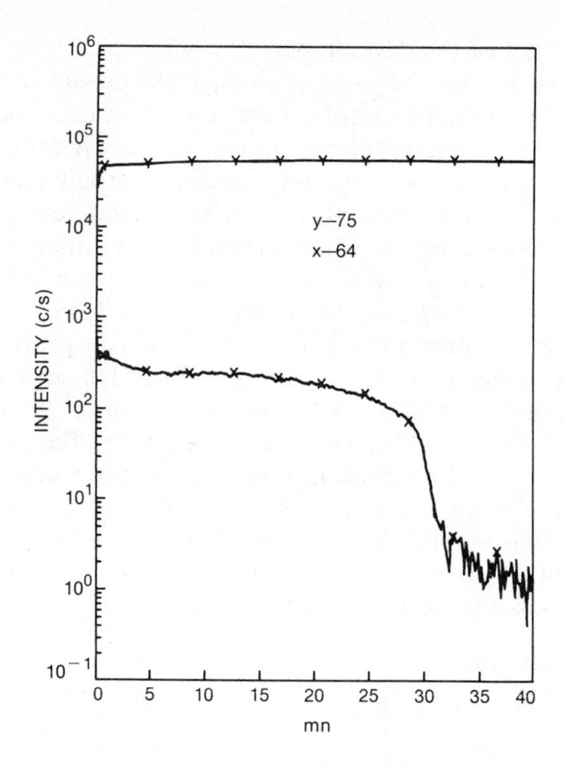

Fig. 7–18. SIMS atomic profile of a 60-min zinc diffusion.

rect, and there is a monotonic decrease from the surface inward, then the carrier profile suggests that there must be a considerable amount of electrically inactive zinc very close to the surface. On the other hand, if the carrier profile is correct, and the hole concentration peaks at a location somewhat below the surface, then the atomic profile could be in error due to surface ion mixing artifacts. It is not known which of these two explanations is more appropriate at this time.

The author and his colleagues have carried out some modeling to explain the unusual diffusion profiles for zinc in GaAs substrate material. The model assumes that the ECP profile is essentially correct and that the diffusion, therefore, is actually comprised of three components. These are: (1) zinc atoms moving into the bulk of the semiconductor by the substitutional-interstitial (SI) mechanism,[33,67,68] (2) zinc atoms diffusing by the conventional substitutional method into subsurface sites, and (3) zinc atoms near the surface diffusing out of the material during the 30-min cooldown period at the end of the diffusion.

The substitutional-interstitial (SI) diffusion process, or dissociative mechanism, has been examined by several researchers, and only a brief review, following Tuck's treatment,[33] is presented here. In the SI process,

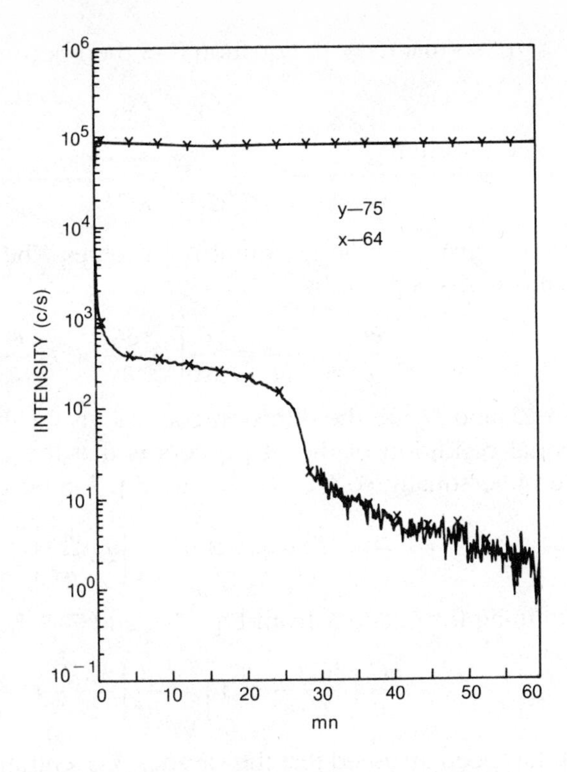

Fig. 7–19. SIMS atomic profile of a 120-min zinc diffusion.

zinc atoms are found in two forms, namely, as a rapidly moving, positively charged interstitial ion, Zn_i^+, and a slow moving, negatively charged, substitutional ion, Zn_s^-. The interstitial ions move quickly through the lattice and become immobilized when they encounter gallium vacancies, becoming substitutional ions in the process. The following quasi-chemical reaction describes this conversion:

$$Zn_i^+ + V_{Ga} \rightleftharpoons Zn_s^- + 2h^+ , \qquad (7\text{–}4)$$

where V_{Ga} represents an uncharged gallium vacancy, and h stands for a hole. There has been considerable discussion about the magnitude of the positive charge of the interstitial zinc,[69,70] but the singly positive state appears to provide the best fit to the available data. Equation 7–4 has a mass action law given by

$$C_i^+ C_V = KC_s^-(\gamma p)^2 , \qquad (7\text{–}5)$$

where C_i^+ is the interstitial concentration, C_V the vacancy concentration, K a constant, C_s^- the substitutional concentration, γ the hole activity coefficient, and p the hole density. For simplicity, we will assume that γ is unity and all forms of zinc are fully ionized; in addition, we will assume that

$C_s >> n_i$, so that $C_s \cong p$. Equation 7–5 then becomes

$$C_i C_V = K C_s^3 , \quad (7\text{–}6)$$

and at equilibrium

$$C_i' C_V' = K C_s'^3 , \quad (7\text{–}7)$$

where the primes indicate equilibrium values. The diffusion equation for the zinc atoms is given by

$$\frac{\partial C_i}{\partial t} + \frac{\partial C_s}{\partial t} = \frac{\partial}{\partial x}\left[D_i \frac{\partial C_i}{\partial x} + D_s \frac{\partial C_s}{\partial x}\right], \quad (7\text{–}8)$$

where D_i and D_s are the diffusion coefficients for the two zinc species. An essential condition of the SI process is that the zinc is primarily transported interstitially, so $D_i >> D_s$, and Eq. 7–8 becomes

$$\frac{\partial C_i}{\partial t} + \frac{\partial C_s}{\partial t} = \frac{\partial}{\partial x}\left[D_i \frac{\partial C_i}{\partial x}\right]. \quad (7\text{–}9)$$

Substituting for C_i and K from Eqs. 7–6 and 7–7, Eq. 7–9 becomes

$$\frac{\partial C_s}{\partial t}\left[\frac{3C_s^2 C_i'}{C_s'^3} + 1\right] = \frac{\partial}{\partial x}\left[\frac{3C_i' D_i}{C_s'^3} C_s^2 \frac{\partial C_s}{\partial x}\right]. \quad (7\text{–}10)$$

It has been assumed that the vacancy concentration is near equilibrium at all times. Another essential condition in the SI mechanism is that the interstitial concentration is a small component of the total amount, so

$$C_{\text{Zn}} = C_i + C_s \cong C_s \quad (7\text{–}11)$$

and, since $C_s < C_s'$, we then have

$$\frac{3C_s^2 C_i'}{C_s'^3} << 1 . \quad (7\text{–}12)$$

The effective zinc diffusion equation becomes

$$\frac{\partial C_s}{\partial t} = \frac{\partial}{\partial x}\left[D(C_s)\frac{\partial C_s}{\partial x}\right], \quad (7\text{–}13)$$

where

$$D(C_s) = \frac{3C_i' D_i}{C_s'^3} C_s^2 = A_s C_s^2 . \quad (7\text{–}14)$$

In sum, the effective zinc diffusion coefficient, D_{eff}, becomes dependent upon the square of the concentration. This type of behavior produces a zinc profile that does not follow classical Fickian diffusion results. In fact, it leads to profiles that have flat initial regions followed by very abrupt diffusion fronts. As the concentration begins to drop, D_{eff} drops as well, which leads to a steeper drop in concentration, and so on. A numerical solution of Eq. 7–13, subject to the initial condition, $C_s(x,0) = 0$, and the boundary conditions, $C_s(\infty,t) = 0$ and $C_s(0,t) = C_0$, constitutes the first portion of the zinc diffusion model. Representative solutions are shown in Fig. 7–20.

The second portion of the model is based on the supposition that subsurface damage produces a vast number of vacancies into which zinc atoms can readily diffuse. Because the damage sites are close to the surface, there is little opportunity for the zinc atoms to become interstitial before being immobilized by the vacancies. We therefore model the diffusion that produces the subsurface peak by a conventional substitutional mechanism:

$$\frac{\partial C_s}{\partial t} = D_s \frac{\partial^2 C_s}{\partial x^2}, \tag{7–15}$$

where D_s is a constant. The solution of Eq. 7–15, subject to the initial condition $C_s(x,0) = 0$, and the boundary conditions $C_s(\infty,t) = 0$ and $C_s(0,t) = C_1$ is given in terms of the well-known complementary error function:

$$C_s(x,t) = C_1 \operatorname{erfc}\left(\frac{x}{2\sqrt{D_s t}}\right). \tag{7–16}$$

However, this second portion of the diffusion model is actually an intermediate step. The third portion of the model is an outdiffusion step that takes place during the cool-down period of the leaky-tube process. The outdiffusion may take place because the concentration of zinc achieved in

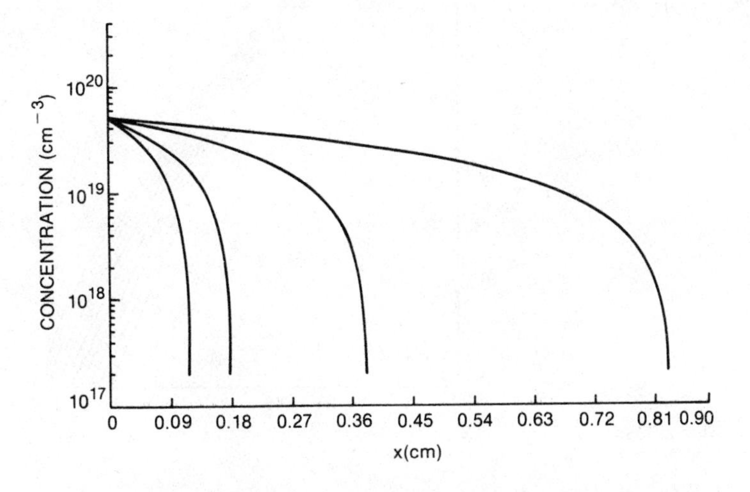

Fig. 7–20. Simulation of substitutional-interstitial zinc diffusion.

the second part of this model probably exceeds the solid solubility. Hence, the third portion is modeled as again substitutional diffusion, but the initial condition is now

$$C_s(x,0) = C_1 \operatorname{erfc}\left(\frac{x}{2\sqrt{D_s T}}\right)$$

and the boundary conditions are

$$C_s(\infty,t) = 0, \quad \left.\frac{\partial C_s}{\partial x}\right|_{x=0} = \text{constant}.$$

The last condition is based on the experimental evidence that the slope at $x = 0$ always appears to be the same; physically, it represents a constant outward flux of zinc atoms. Representative solutions of Eq. 7–15 with these conditions are shown in Fig. 7–21. The proposed final model then consists of the superposition of solutions of the first and third portions described above. in Fig. 7–22, typical solutions are presented, and it is evident that the solutions qualitatively match the experimental results.

However, the match among the carrier (ECP) profiles, atomic (SIMS) profiles, and predicted zinc diffusion profiles is rather good. Comparisons of these three modes are shown in Figs. 7–23, 7–24, and 7–25, for 5-, 60-, and 120-min diffusions, at 600°C, respectively. The carrier profiles are re-

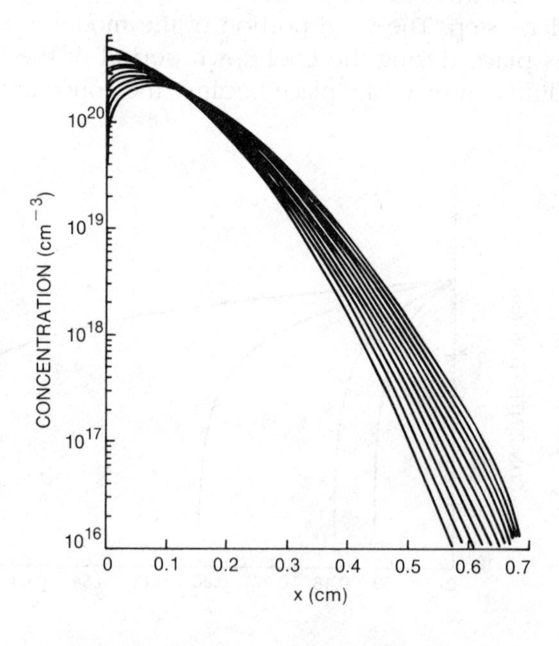

Fig. 7–21. Simulation of zinc outdiffusion from a shallow surface layer, by means of the substitutional mechanism.

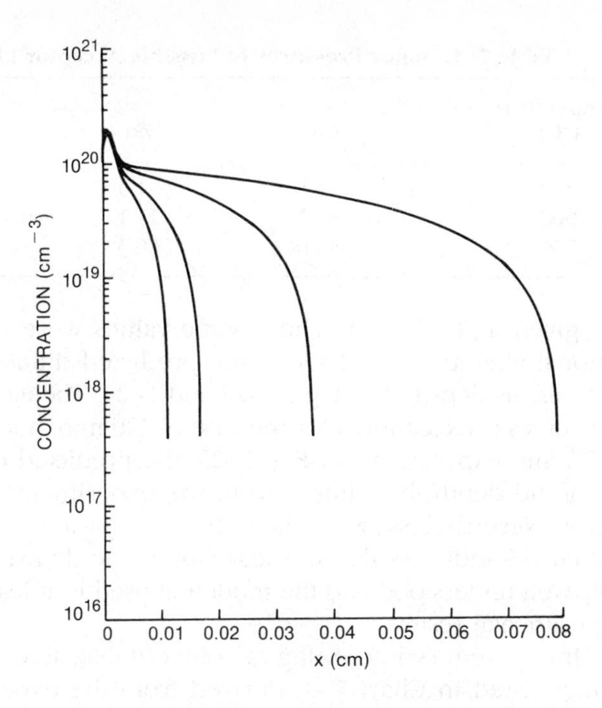

Fig. 7–22. Simulation of zinc diffusion into GaAs substrates combining the mechanisms shown in Figs. 7–20 and 7–21.

plotted from Figs. 7–12, 7–13, and 7–14; the SIMS profiles have been extracted from Figs. 7–17, 7–18, and 7–19. In these comparison plots, the SIMS data, which had not been presented in standard density units previously, have been adjusted to fit the ECP data in the center of the plateau regions; the distance scale for the SIMS profiles was found from a calibration of the ion mill rate, and was found to be very close to the ECP distance scale. When this SIMS work was carried out, independent calibration of the doping density, through the use of ion implanted standards, was not available. Obviously, the adjustment of the atomic density to agree with the ECP profile presupposes a zinc activity coefficient of unity, but the closeness of the atomic and carrier profiles in form, value and depth indicates that this may be a very good approximation. Some curve fitting was employed to bring the predicted diffusion profile in line with the ECP carrier profile, but this was done only in Fig. 7–24, the 60-min diffusion experiment. The values that were adjusted to produce a reasonable fit were (1) C_0, the surface concentration needed for the SI process, (2) A_s, the multiplier in the expression for D_{eff}, (3) C_1, the surface concentration for the substitutional indiffusion and outdiffusion processes, (4) D_s, the substitutional diffusion coefficient, and (5) $\partial C/\partial x\,|_{x=0}$, the concentration gradient at the surface. The values that produced an adequate fit in Fig. 7–24

Table 7–1. Vapor Pressures of Possible Acceptor Elements in Torrs

Temperature (°C)	Cd	Zn	Mg	Ca
500	13.5	1.3	0.1	0.0004
600	85.1	11.1	1.0	0.01
700	355.0	60.3	6.6	0.1

are given in Table 7–1. These same values were used without any additional alteration to generate the predicted diffusion profiles at 5 and 120 min, as depicted in Figs. 7–23 and 7–25. The agreement among all of the curves is exceptional for the case of 120 min, but not quite as good for the 5-min experiment. In Fig. 7–23, the predicted curve has the correct form and depth, but differs from the experimental curves by a scaling factor. Nevertheless, we believe that the basic agreement among all of the curves indicates that the leaky-tube zinc diffusion process is reasonably well understood and the modeling used is at least consistent with the experimental results.

It is extremely interesting to point out that, according to the parameter values listed in Chart 7–1, derived from the experimental figures, the

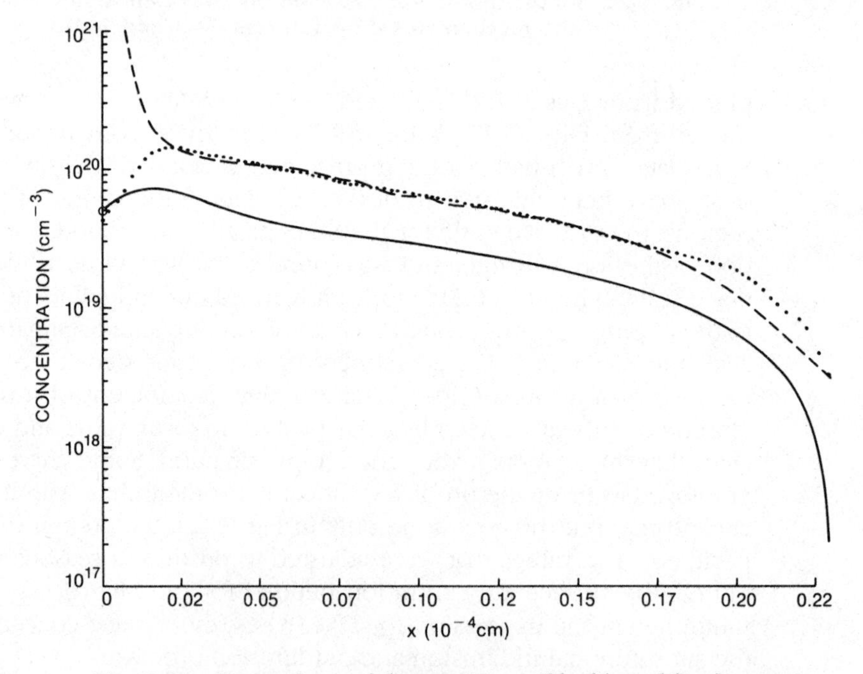

Fig. 7–23. Comparison of the carrier profile (dotted line), atomic profile (dashed line), and theoretical prediction (solid line) for a GaAs substrate diffused for 5 min.

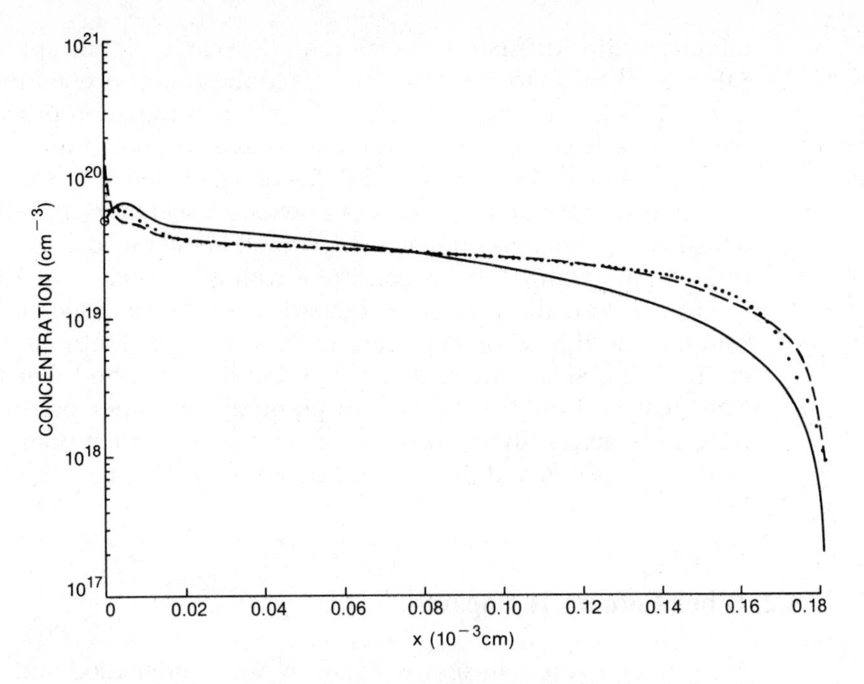

Fig. 7–24. The same comparison as in Fig. 7–23, but for a 60-min diffusion.

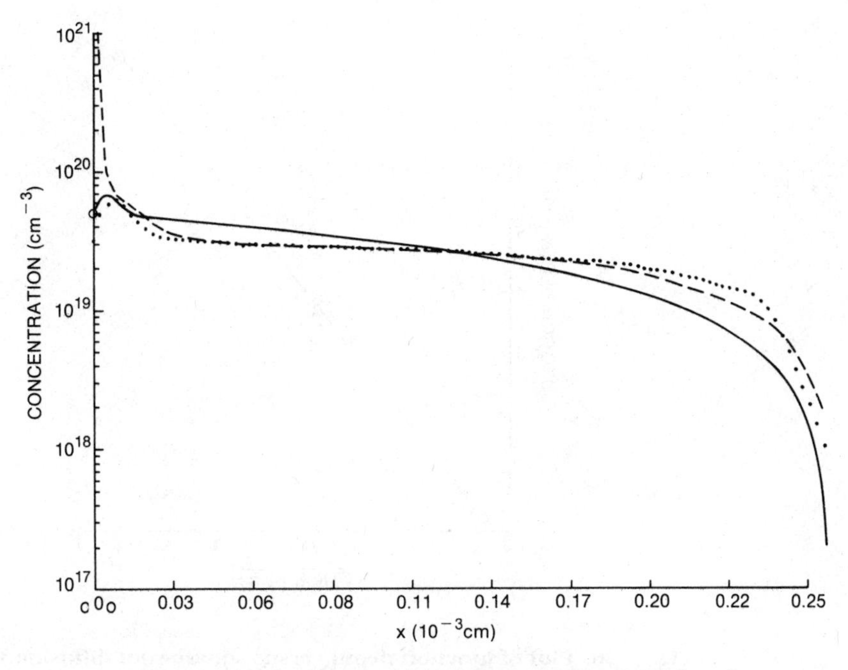

Fig. 7–25. The same comparison as in Fig. 7–23, but for a 120-min diffusion.

effective zinc diffusion coefficient D_{eff} varies from approximately 5.0×10^{-14} cm^2/s to 5.0×10^{-12} cm^2/s as the zinc concentration increases from 5×10^{18} cm^{-3} to 5×10^{19} cm^{-3}, which is the appropriate range for our figures. This effective value is *two* to *three* orders of magnitude larger than the estimated substitutional diffusion coefficient! By way of comparison, by extrapolating the results of previous researchers, the effective zinc diffusion coefficient is estimated[30,71,72] to be in the range 4.4×10^{-13} cm^2/s to 2.0×10^{-10} cm^2/s; this is consistent with our results.

To conclude this portion, a comparison of the expected junction depth with measured junction depth for 600°C leaky-tube diffusions is presented in Fig. 7–26. Again, the experimental and the predicted diffusion depths were matched only at the 60-min point; all the other predicted depths were generated without correction from the one match point. The agreement is excellent, and the junction depth is clearly predicted to be linear in $t^{1/2}$.

7.3.2 Other Column II Dopants

Even though zinc diffusion is relatively well understood and technologically mature, the search for alternative p-diffusants continues for several

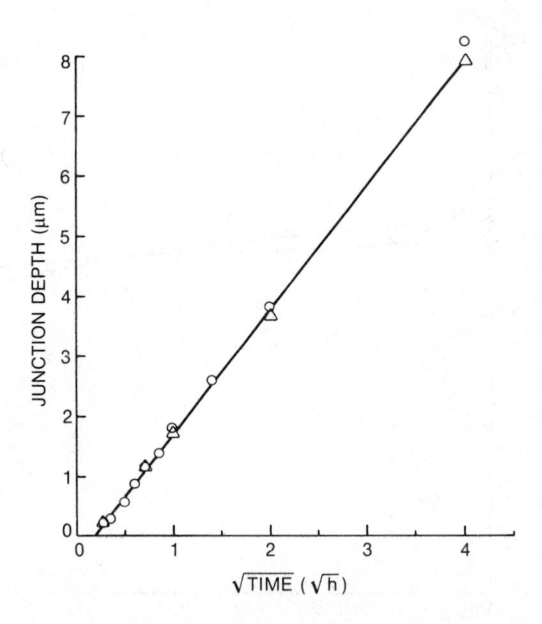

Fig. 7–26. Plot of junction depth versus square-root diffusion time; the circles are the experimental values (same data as in Fig. 7–9), and the triangles are the values found from the simulation.

Chart 7–1. Factors Used in Fitting Predicted and Measured Diffusion Profiles

$$C_0 = 4.0 \times 10^{19}\ \text{cm}^{-3}$$

$$A_s = 2.0 \times 10^{-51}\ \text{cm}^8/\text{s}$$

$$C_1 = 8.0 \times 10^{19}\ \text{cm}^{-3}$$

$$D_s = 5.0 \times 10^{-15}\ \text{cm}^2/\text{s}$$

$$\left.\frac{\partial C_s}{\partial x}\right|_{x=0} = 5.0 \times 10^{24}\ \text{cm}^{-4}$$

reasons. It may be possible to find and utilize dopants with higher or lower solid solubility, or higher or lower diffusion coefficients, than zinc. A lower diffusion coefficient may be especially beneficial for producing shallow junctions, a lower solid solubility for producing lightly doped layers, and a higher solubility for fabricating degenerately doped layers. It was mentioned previously that of all of the potential acceptor dopants in GaAs, only two or three have emerged as suitable candidates for diffusion processes, especially in the leaky-tube process. A suitable diffusant must have a reasonably high vapor pressure at temperatures of 700°C or less, reasonably high diffusion coefficient in the same temperature range, and manageable toxicology. These conditions stem from the basic characteristics of the leaky-tube approach. Because elemental sources have been employed, the vapor pressure of the source must be adequate to provide a copious number of atoms for the diffusion to take place.

A rough empirical study of the proper conditions for diffusion in a leaky-tube system has shown that a source vapor pressure of at least 1 torr is required. In Table 7–1, the equilibrium vapor pressures of possible diffusant elements are listed. Other elements not listed, such as Ge, Si, and so on have negligible vapor pressures in this temperature range; mercury, on the other hand, has a vapor pressure of 250 torr at 300°C, and at typical diffusion temperatures, above 1 atm, and is therefore unsuitable. Of course, the diffusion properties of zinc have been extensively examined, and in this section we will discuss the possibilities of cadmium, magnesium, and calcium diffusion in GaAs.

Cadmium has emerged as the p-type diffusant of choice in InP technology and the main reason for this appears to be its relatively high diffusivity in that material at low processing temperatures.[73–75] InP is even more sensitive to thermal degradation than GaAs. Surprisingly, there has been very little recent work in the diffusion of Cd into GaAs or $Ga_{1-x}Al_xAs$. Two reasons may be cited for this state of affairs. In the first place the effective diffusion coefficient of Cd appears to be at least three orders of magnitude smaller than the coefficient for zinc diffusion. Estimates of the Cd diffusion coefficient in GaAs carried out 20 years ago[30,71,72] put its value in the range

2.3×10^{-18} cm^3/s to 3.3×10^{-16} cm^2/s at 600°C. Our recent examination of cadmium diffusion is consistent with these data. A leaky-tube diffusion of Cd (from an elemental Cd source) into n-GaAs at 600°C for 1 h produced a pn junction with a junction depth of approximately 10 nm. This depth was too small to be detected either with standard microscopy or electrochemical profiling, but SIMS analysis revealed the presence of Cd atoms to that depth. This depth is roughly two orders of magnitude smaller than an equivalent zinc diffusion depth, indicating a diffusion coefficient reduction of four orders of magnitude. This has the unfortunate technological effect of forcing diffusion times to be unmanageably long; for example, a 1-μm Cd diffusion at 600°C would require a diffusion for slightly longer than one year.

The second problem with the use of cadmium, at least in the leaky-tube diffusion system, is its relatively high vapor pressure. In the cold end of the leaky-tube liner, the excess vapor can condense, and for long diffusions with a volatile source, this can often produce a mass of loosely condensed material that nearly fills the mouth of the liner. Because cadmium is believed to have hazardous toxicology, this condensation at the furnace opening presents a genuine concern for safety.

The author and colleagues have attempted to alter the standard leaky-tube procedure to circumvent these problems. In brief, the modified procedure consisted of using a two-zone furnace arrangement. The cadmium source was placed in a low-temperature region (~450°C) and the wafers in a high-temperature zone (~650°C). In this fashion, the Cd vapor pressure would be reduced, but the sample would provide a high-temperature environment for more rapid diffusion. In practice, the surface concentration of cadmium turned out to be vanishingly low, presumably because of the "uphill" diffusion that the cadmium vapor underwent, and cadmium diffusion with a leaky-tube process for GaAs does not appear to be a viable processing procedure.

Two other possible diffusion candidates that may provide an alternative to zinc diffusion by the leaky-tube procedure are magnesium and calcium. As shown in Table 7–1, the vapor pressure of Mg at 650°C exceeds 1 torr. However, the value of the Mg diffusion coefficient is a matter of some controversy. Moore *et al.*[76] have measured the diffusion coefficient at elevated temperatures (in excess of 1000°C) and from their data an estimate of the coefficient at 650°C is 4.0×10^{-17} cm^2/s. On the other hand, Shaw reports that the coefficient may depend upon the purity of the Mg source and may be several orders of magnitude greater than Moore's value. Recently, Small and coworkers[77] have demonstrated that the diffusion coefficient of Mg at 835°C may be as large as 5×10^{-11} cm^2/s. Using the activation energy reported in Refs. 71 and 76, $\Delta E \cong 2.5$ eV, the pre-exponential factor from Small's experiment is estimated to be 12.0 cm^2/s; this yields a Mg diffusion coefficient at 650°C of roughly 3.0×10^{-13} cm^2/s.

This is large enough to permit a 1-μm–deep diffusion to take place in less than 10 h. With these facts in mind, the author's team attempted to perform Mg diffusion into GaAs and $Ga_{1-x}Al_xAs$ with the leaky-tube system at 650°C. The results, however, have not been encouraging. In the first place, the magnesium reacts with the quartz components in the system, especially the quartz source crucible. As opposed to the case of leaky-tube zinc diffusion, in which both the crucible and the source can be used repeatedly, both the Mg source and crucible are rendered useless by this reaction. Secondly, there was no evidence that a pn junction was actually formed during diffusions at 650°C for 60 to 180 min. This procedure does not appear to have much merit at this time.

Calcium is the last column II element that has a nonnegligible vapor pressure in the 500 to 700°C temperature range, as shown in Table 7–1. There is, however, no information in the literature concerning the solid solubility or diffusion coefficient of Ca and GaAs. Although a Ca leaky-tube diffusion has never been attempted, it is worthy of attention, if only to complete the column II diffusion picture.

7.3.3 Zinc in $Ga_{1-x}Al_x$ As

Zinc diffusion into the ternary $Ga_{1-x}Al_xAs$ is often carried out in the fabrication of transverse junction stripe lasers, high-performance photovoltaic cells, or planar heterojunction bipolar transistors. However, the behavior of zinc diffusing into $Ga_{1-x}Al_xAs$ has not been intensely examined, and in fact the published literature on this topic is somewhat inconsistent. Boltaks *et al.*[78] reported that both the diffusion coefficient and the solubility of zinc decreased monotonically in the range $0.0 < x < 0.40$, while the experiments of Lee *et al.*[79] showed that the apparent zinc diffusion coefficient increased monotonically in this same range of aluminum fraction. The work of Yuan *et al.*[80] is in approximate agreement with that of Lee, but the experiments of Blum *et al.*[81] seem to reveal that the diffusion coefficient may first decrease and then increase with increasing aluminum fraction. Because of the inconsistencies in the literature and the relative paucity of the data (most researchers examined only a few $Ga_{1-x}Al_xAs$ samples with large gaps in the aluminum fraction sequence), the author and his colleagues have examined the diffusion of zinc into Ga_{1-x} Al_xAs with more detail employing the leaky-tube system.[58] What has emerged from this analysis is the realization that zinc diffusion into the ternary semiconductor is a most complicated matter. It was found that the zinc diffusion coefficient is not a simple monotonic function of the aluminum fraction and that anomalies in the zinc behavior appear that still do not have a complete explanation.

The ten samples used in this study were N-$Ga_{1-x}Al_xAs$ epitaxial layers grown with liquid-phase epitaxy on n-type GaAs substrates. The ten LPE layers were doped with tellurium ($N_0 \cong 4$ to 6×10^{17} cm^{-3}), nominally 8 μm thick, and with aluminum fraction in the range $0.0 < x_{Al} < 0.43$ in roughly equal increments. The diffusion was performed at 550, 600, and 650°C at both 30 and 60 min, producing 60 separate samples for analysis and characterization. The carrier profile was measured for each sample with the electrochemical profiler, and nearly every sample was examined on the SEM for uniformity and for independent confirmation of the junction depth. Typical ECP doping profiles for the diffused $Ga_{1-x}Al_xAs$ samples are shown in Figs. 7–27 and 7–28. In Fig. 7–27, the carrier concentration profiles for three 30-min diffusions (at 550, 600, and 650°C) for a sample with $x_{Al} \cong 0.17$ are depicted. It is obvious that both the junction depth and the average carrier concentration increase as the temperature of the diffusion is increased. These effects are expected because both the diffusion coefficient and the solid solubility of zinc increase with increasing temperature. It should also be noted that as the temperature increases, a small concentration peak begins to develop near the sample surface. These samples were not covered with the thin protective SiO_2 film and the emergence of the peaks is an indication of the start of surface deterioration. Figure 7–28 shows the results of another pair of $Ga_{1-x}Al_xAs$ samples ($x_{Al} \cong 0.29$) diffused at 600°C for two temperatures. The junction depth increases as expected, and the surface concentration remains constant, indicating that the limit of solid solubility has been achieved. One should note that the profiles have a characteristic plateau followed by a very steep diffusion front. This "signature" profile indicates that the interstitial-substitutional model is probably valid for zinc diffusing in the ternary material.

In spite of the regular appearance of the diffusion profiles shown in Figs. 7–27 and 7–28, a global view of the junction depth and average carrier concentration reveals the presence of some complicated physical processes. Figures 7–29 and 7–30 show the measured junction depths versus aluminum fraction for 30- and 60-min diffusions, respectively. Because of the proportionality between x_j and D_{eff}, these figures also represent a plot of the zinc diffusion coefficient versus x_{Al}.

The lines connecting the data points have no significance other than to accentuate trends in the data. Clearly, as the temperature increases, the junction depth increases for fixed diffusion time and aluminum fraction. At any given temperature, however, as the aluminum fraction increases, the junction depth first decreases, reaching a relative minimum at $x_{Al} \cong 0.07$, then increases, reaching a relative maximum at approximately $x_{Al} \cong 0.18$, then decreases sharply again, going through a second minimum at $x_{Al} \cong 0.21$, and then finally has a monotonic increase through $x_{Al} \cong 0.43$. This rather irregular pattern for the diffusion coefficient cannot be explained by inconsistencies with the diffusion procedure as samples

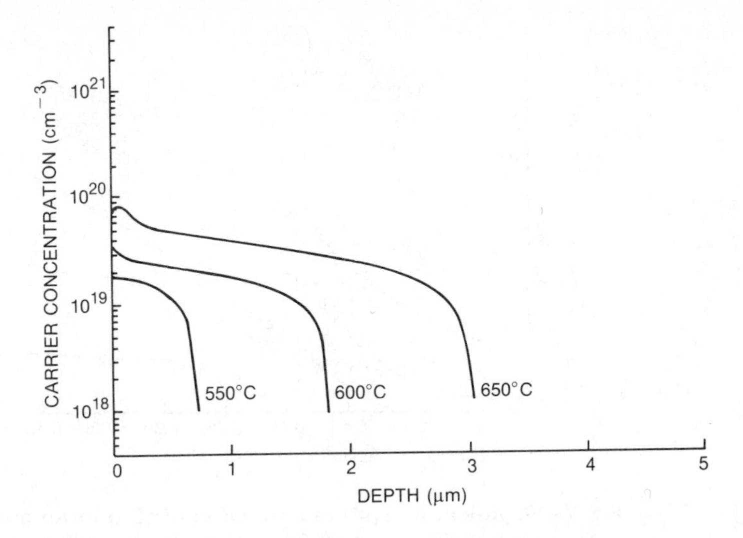

Fig. 7–27. Carrier concentration profiles for $Ga_{0.83}Al_{0.17}As$ samples after 30-min zinc diffusions at 550, 600, and 650°C.

were picked at random for diffusion; that is, one diffusion may have involved the samples with x_{Al} = 0.07, 0.20, and 0.43, while the next may have included x_{Al} = 0.07, 0.16, and 0.00. In addition, the diffusions were carried out in two different furnaces to help demonstrate uniformity with the leaky-tube approach.

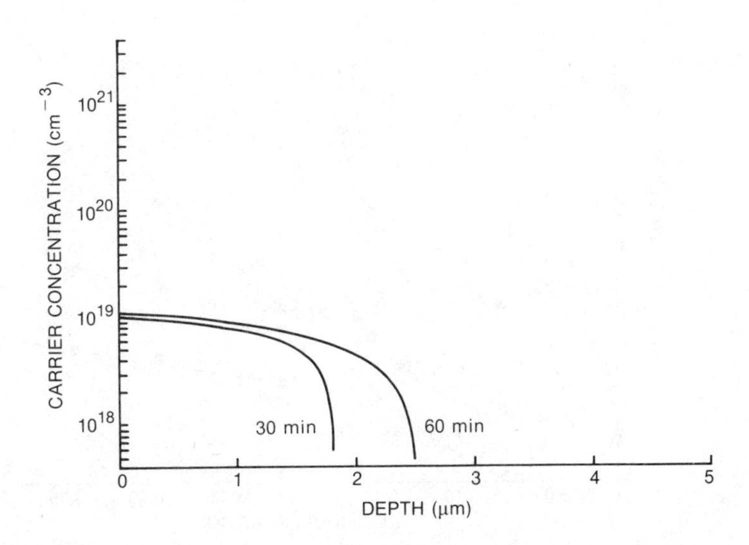

Fig. 7–28. Carrier concentration profiles for $Ga_{0.71}Al_{0.29}As$ samples after 30- and 60-min zinc diffusions at 600°C.

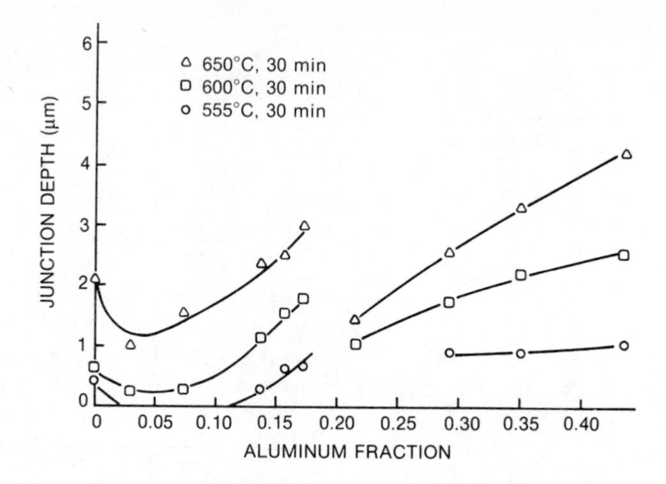

Fig. 7–29. Junction depth as a function of aluminum fraction for 30-min zinc diffusions in $Ga_{1-x}Al_xAs$ at 550, 600, and 650°C.

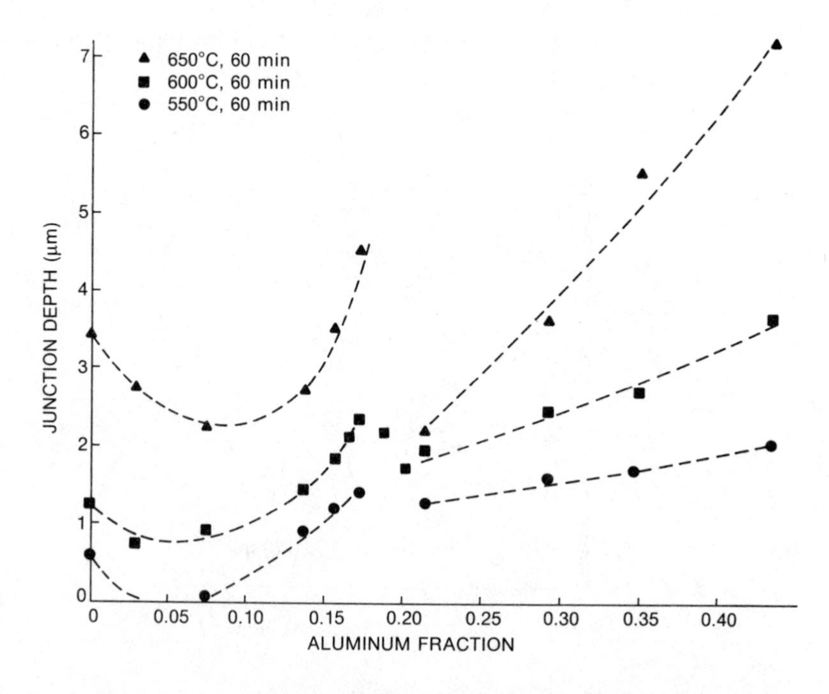

Fig. 7–30. Junction depth as a function of aluminum fraction for 60-min zinc diffusions in $Ga_{1-x}Al_xAs$ at 550, 600, and 650°C.

Previous workers have suggested that the diffusion coefficient for zinc should increase monotonically with increasing x_{Al} for the following reason.[80] The melting point of AlAs is greater than that for GaAs, so the binding energy of the solid and group III vacancy formation energy should increase with increasing x_{Al}. Consequently, the number of group III vacancies at a given temperature should decrease as x_{Al} increases. Since zinc apparently diffuses in $Ga_{1-x}Al_xAs$ also by the interstitial-substitutional mechanism, it follows that the zinc atoms must diffuse further in $Ga_{1-x}Al_xAs$ on the average before encountering vacancies that will stop their motion. In effect, then, the zinc diffusion coefficient should increase uniformly with increasing aluminum fraction. (In addition, the average zinc solubility should decrease because of the reduced number of group III vacancies.) It is interesting to note that with the exception of the 650°C data, the points at $x_{Al} \cong 0.0$, 0.20, and 0.40 alone form nearly straight lines, which would be reminiscent of the results in prior reports. However, it is evident that additional factors must be at work, because the data in Figs. 7–29 and 7–30 cannot be explained by the above model.

The junction depth versus aluminum fraction is replotted on an expanded scale in Fig. 7–31 for the range $0.0 < x_{Al} < 0.18$, and in Fig. 7–32 the doping profiles for the 30-min, 600°C diffusions in this aluminum frac-

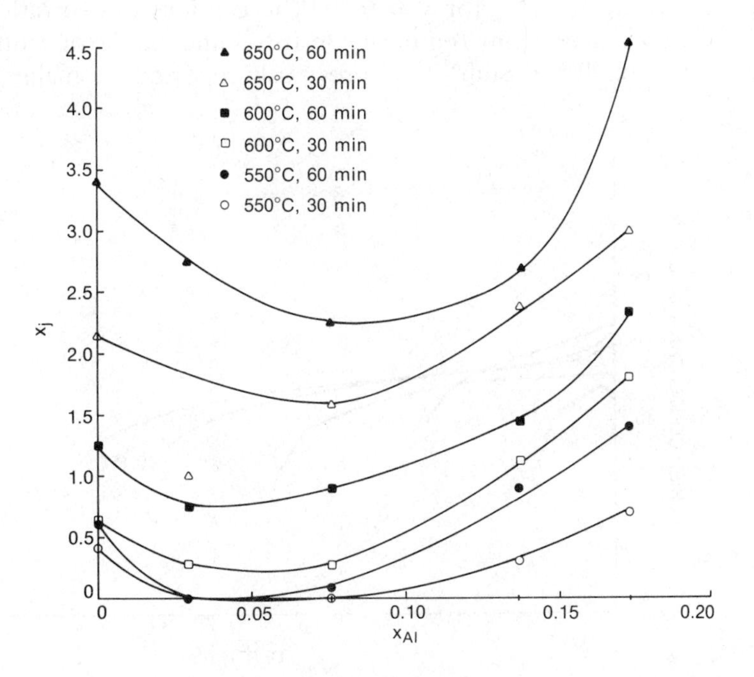

Fig. 7–31. Junction depth as a function of aluminum fraction in the range $0.0 < x_{Al} < 0.17$.

tion range are depicted. A simple examination of all of the samples in this aluminum fraction range has shown that the diffusions were uniform, without surface decomposition, without any obvious anomalies to explain the trends shown in Figs. 7–31 or 7–32. However, it is apparent from Fig. 7–32 that the maximum zinc concentration and the junction depth are directly related. That is, just as the junction depth reaches a minimum near $x_{Al} \cong 0.07$, the zinc concentration also has a local minimum at that same aluminum fraction.

Figure 7–33 includes the doping profiles for the 30-min, 600°C diffusions for the aluminum fraction range, $0.21 < x_{Al} < 0.43$, and a different trend has emerged. As the aluminum fraction increases, the diffusion coefficient increases, but the carrier concentration now decreases monotonically. In Fig. 7–34, the average carrier concentration for the 60-min diffusions is plotted versus x_{Al}. There is clearly a general diminution in the concentration as x_{Al} increases, but a careful look at the data shows a clear local minimum at $x \cong 0.07$.

To summarize then, in accordance with the simple model stated previously, these data indicate that the diffusion coefficient in general increases with increasing aluminum fraction and the carrier concentration decreases in general with increasing x_{Al}. However, for small values of x_{Al}, the diffusion coefficient shows two special features, a dip in value for $x \cong 0.07$ and a peak in value for $x \cong 0.20$. The carrier concentration also displays departures from regularity in this same range of aluminum fraction. Campbell and Shih[82] have previously reported anomalies in the solubility

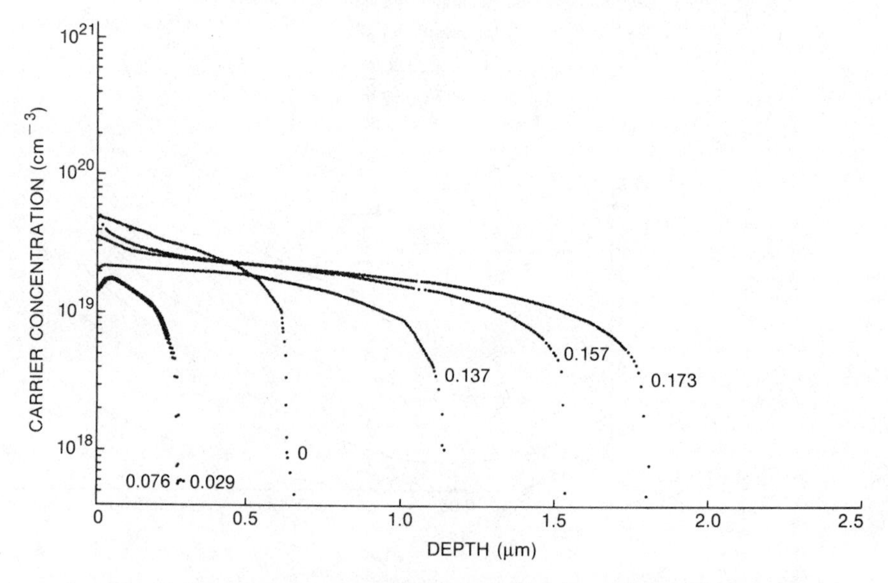

Fig. 7–32. Carrier concentration profiles for the low aluminum fraction $Ga_{1-x}Al_xAs$ samples diffused with zinc for 30 min at 600°C.

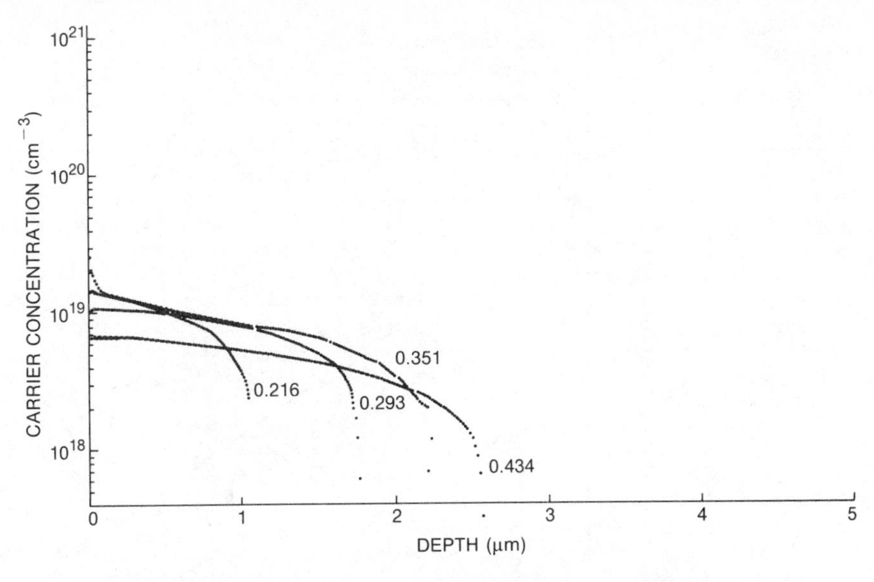

Fig. 7–33. Carrier concentration profiles for the high aluminum fraction $Ga_{1-x}Al_xAs$ samples diffused with zinc for 30 min at 600°C.

for zinc in $Ga_{1-x}Al_xAs$. Their experiments showed that the zinc concentration solubility drops dramatically as x_{Al} increases from 0.0 to 0.20 and then has a steady increase as x_{Al} increases from 0.20 to 0.40. This is vaguely consistent with the data presented in this work, as shown in Fig. 7–34, except that the minimum occurs at $x \cong 0.07$, not at 0.20. Campbell and Shih have suggested that the initial lowering of the zinc concentration is due to the lattice binding considerations mentioned previously and the increase then comes about because of the introduction of some presently unidentified defects. These extra defects may enhance the solubility of zinc by providing additional locations for substitutional zinc. They further suggest that if this picture is valid, there may be another turning point in zinc concentration versus x_{Al} where the mixing effects of Al and Ga on the same lattice become even more significant.[82] The author's data seems to suggest the presence of two turning points, at $x \cong 0.07$ and 0.20, but these values are considerably lower than those described in Ref. 82. It is entirely possible that some other mechanism may be responsible for the unusual dip in these data.

The picture of the diffusion coefficient also requires further explanation. It is possible that for small values of the aluminum fraction ($x_{Al} < 0.07$) the diffusion coefficient drops because the carrier concentration is dropping and it has already been shown that $D_{eff} \propto C_s^2$. For increasing values of x_{Al}, one might expect that D_{eff} would increase uniformly because of the lattice bonding model requirements. This is approximately true except for the peak in the vicinity of $x \cong 0.18$. This resonancelike ef-

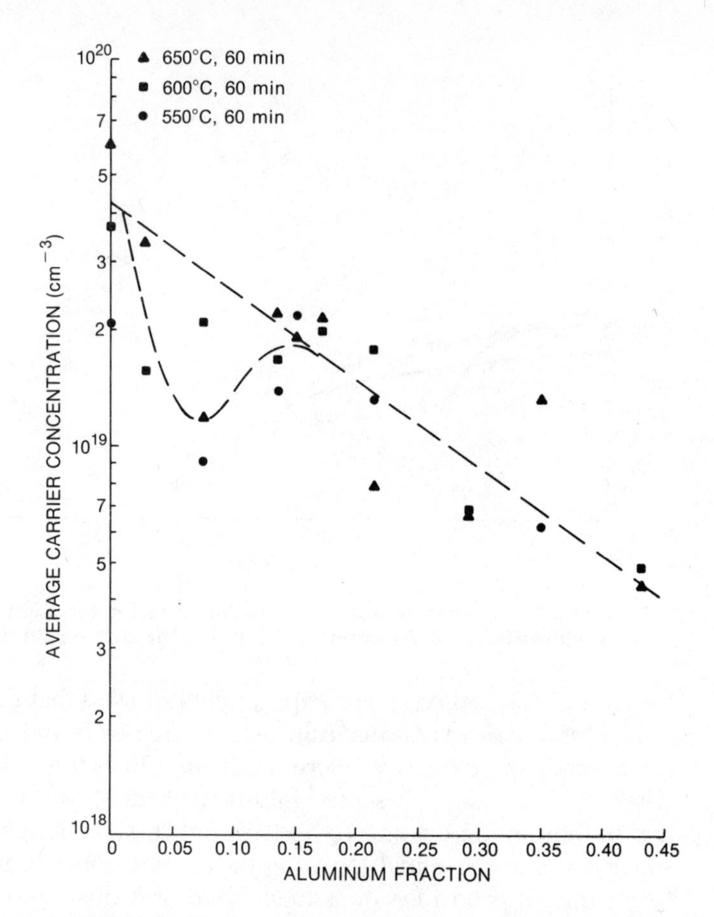

Fig. 7–34. The average carrier concentration versus aluminum fraction for the zinc diffused samples. The straight dashed line shows a simple monotonic drop in concentration with x_{Al}; the line with the dip more accurately reflects the trend in this data.

fect may possibly be associated with the defect mentioned by Campbell and Shih, and the author proposes that the DX center is a potential candidate. This defect, whose exact structure is still a matter of some conjecture, is responsible for persistent photoconductivity, laser pulsations, and other related phenomena. It typically shows an insignificant DLTS signal for $x_{Al} < 0.20$, but for greater values of aluminum concentration, its DLTS signal rises abruptly and its effects on transport and optical properties become manifest.[83] This center, possibly through electrostatic attraction or irregular strain fields, may suddenly augment the zinc diffusion producing the peak in Figs. 7–29 and 7–30. Unhappily, deep-level transient spectroscopy (DLTS) examination of the diffused samples has not yet been carried out, so this potentially clarifying information is not available. For the present the rather unusual behavior of zinc diffusing into $Ga_{1-x}Al_xAs$ grown by LPE must remain somewhat mysterious. It would be of extreme interest to extend this work to include an examination of the effects of different n-type background dopant (Si, Sn, S, etc.) and different growth procedure (MOCVD, MBE), and to examine the samples further with cryogenic photoluminescence, high-resolution electron microscopy, and of course, DLTS. A more complete story might then begin to emerge.

Recently, Quintana *et al.* have reported the results of another examination of zinc diffusion into $Ga_{1-x}Al_xAs$.[84] In this work, zinc was diffused into $Ga_{1-x}Al_xAs$ epitaxial layers of varying composition by means of a sealed ampoule technique in which Zn_2As_3 was the zinc source. The epitaxial layers were grown by MOCVD and were nominally undoped. Figure 7–35 shows the junction depth versus aluminum fraction reported by these researchers for 1 h, 650°C diffusions, with the results of Ageno *et al.*[58] displayed for reference. There are some significant differences as well as similarities in the two sets of data. In the first place, the overall trend for the closed-ampoule diffusions is that of reduced junction depth and a very

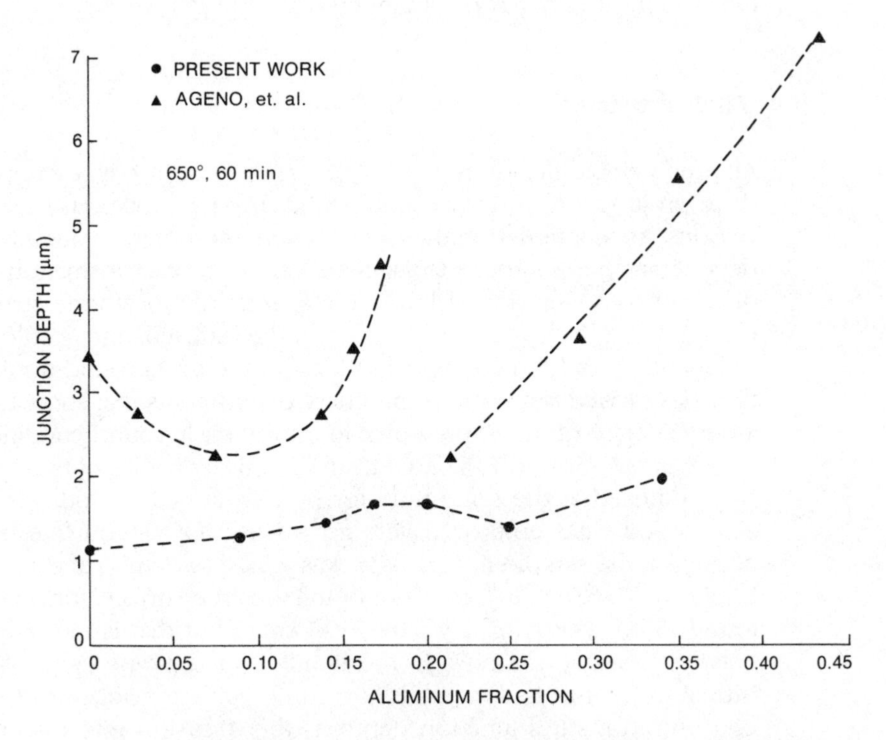

Fig. 7–35. A comparison of the junction depth versus aluminum fraction with two separate diffusion procedures. The triangles are from the work of Agena *et al.*[58] with leaky-tube diffusion; the circles are from the work of Quintana *et al.*[84] with sealed ampoule diffusion.

weak dependence of depth on aluminum fraction. Secondly, there is no local minimum in the junction depth for small x_{Al}, as with the leaky-tube diffusion data. But there is still a measurable, if reduced, peak in the junction depth for $0.15 < x_{Al} < 0.25$. The trend of reduced junction depth in the closed-tube ampoule data is consistent with the thermodynamic predictions of Casey and Panish.[36] An arsenic overpressure is developed in the closed system, which in turn reduces the zinc incorporation rate, and thus the diffusion coefficient; this is not the case for the leaky-tube system so that the diffusion rate will be considerably larger. Although the lack of a minimum in the data for $x_{Al} \cong 0.07$ has no explanation at the present time, the development of a peak in the junction depth for $x_{Al} \cong 0.18$ is quite significant. It should be emphasized that the samples of Ageno *et al.* were grown by LPE and those of Quintana *et al.* by MOCVD, and yet even with two different growth procedures (and presumably different residual defect densities and types) and two different diffusion procedures, the resonancelike behavior of the zinc diffusion manifests itself. This corroborating work of Quintana *et al.* domonstrates again that knowledge of the native defect character in $Ga_{1-x}Al_xAs$ is required for an understanding of the complicated behavior of zinc in this ternary material.

7.3.4 Zinc in p-GaAs

Although zinc diffusion is most often carried out in n-type samples to produce pn junctions, it is also performed in p-type material for a variety of reasons. As mentioned in the introduction, these may include producing a degenerate p++ surface to facilitate ohmic contact formation or to produce differentially doped layers for special photocathode structures. Although it may be expected initially that the diffused zinc profile should be independent of both the type and level of the background doping, the author has carried out some preliminary experiments that show that there is a dependence of the doping profile on the background conditions.

Figures 7–36 and 7–37 are electrochemical profiles of two samples that were diffused at the same time for 30 min at 600°C. Both samples were zinc doped GaAs epitaxial layers grown by MOCVD, but the background doping of the sample in Fig. 7–36 was 1.5×10^{19} cm^{-3}, and in the sample in Fig. 7–37, 6.0×10^{16} cm^{-3}. In both cases the surface concentration was found to be approximately 5.0×10^{19} cm^{-3}, but that is the only similarity between the two samples. In the lightly doped p-type epitaxial layer, the diffusion profile had virtually the same surface concentration, plateau concentration, and junction depth as the diffusion into n-type GaAs for 30 min at 600°C, and we can conclude that for lightly doped material the diffusion results are independent of doping type. On the other hand, the results for the heavily doped sample are dramatically different. In fact, the profile in the first 200 nm of this sample closely resembles a complementary error function. This suggests that for the heavily doped p-type sample, the diffusion conditions appear to be approximately that of an isoconcentration experiment. In a true isoconcentration experiment, a sample is doped to solid solubility with, say, nonradioactive zinc, and then placed in a diffusion apparatus and exposed to a radioactive zinc source. The nonradioactive zinc diffuses out and the radioactive zinc diffuses in, but the total zinc concentration remains approximately constant during the experiment. Chang and Pearson have utilized this technique extensively to determine the diffusion coefficient of zinc for fixed zinc concentrations.[85] The claim that the diffusion conditions resemble an isoconcentration experiment is based on the idea that the indiffusing atoms are encountering a lattice already doped nearly to the solid solubility with the same specie. The lattice sites needed for incorporation are virtually exhausted, and are generated only by simultaneous indiffusion of gallium vacancies from the free surface. The conditions would presumably be completely different if the sample were doped p-type with germanium, for example, which occupies a different lattice site. To conclude, then, the type of background dopant can have a significant effect on the diffusion when its density is large enough to alter the normal (and required) vacancy densities.

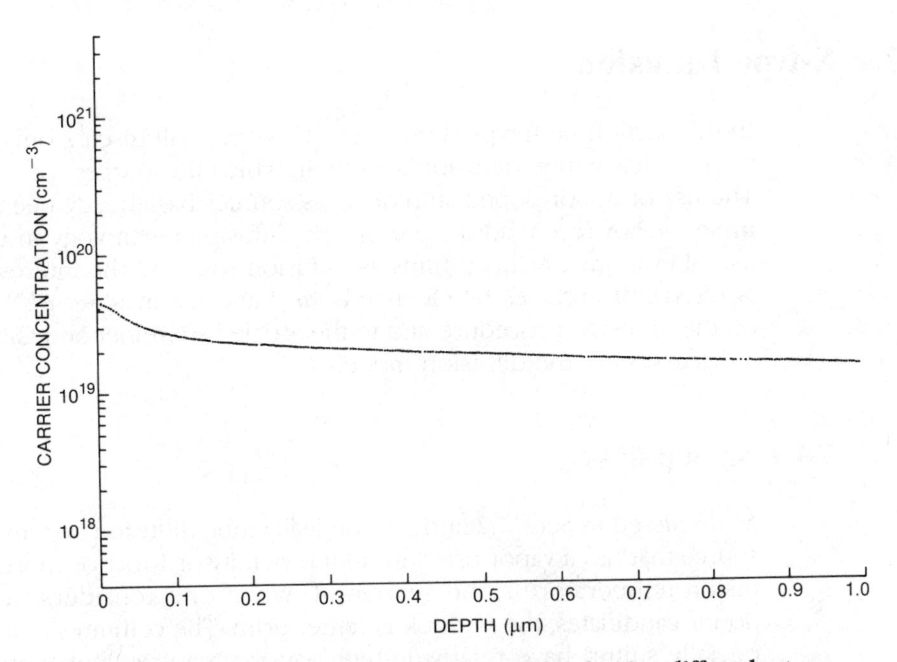

Fig. 7–36. Carrier concentration profile of a zinc diffused p-type epitaxial sample with high background concentration (30 min, 600°C).

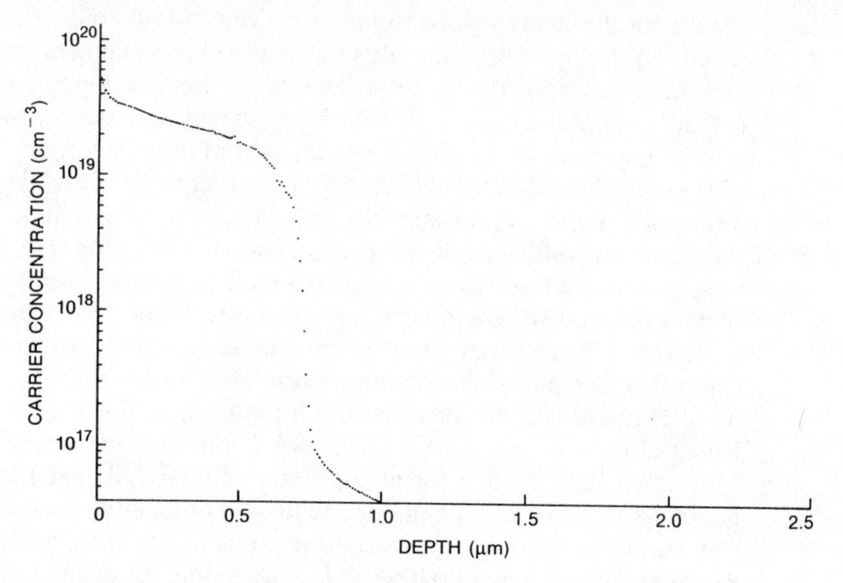

Fig. 7–37. Carrier concentration profile of a zinc diffused p-type epitaxial sample with low background concentration (30 min, 600°C).

7.4 N-type Diffusion

In this section on n-type diffusion in GaAs, we will discuss only the efforts with the leaky-tube diffusion system, in which tin sources were employed. The use of donor doped spin-on glass sources has already been discussed in Sec. 7.2. A recent advance in n-type diffusion technology in GaAs is the use of evaporated silicon films as diffusion sources. The interested reader is referred to articles by Meehan *et al.*[53] and Kavanagh *et al.*[86] for details on the diffusion procedure and to the article by Greiner and Gibbons[52] for a discussion of the diffusion model.

7.4.1 Sn in p-GaAs

As discussed in Secs. 7.2 and 7.3, the leaky-tube diffusion system requires a source that has a vapor pressure in the vicinity of 1 torr or more at the diffusion temperature to be successful. When one considers the possible donor candidates, the outlook is rather grim. The column VI elements, especially sulfur, have relatively high vapor pressures, but their reactivity and toxicology make them poor choices; the column IV elements have typically low vapor pressures. Tin, for example, has a relatively low melting point, but its vapor pressure is miniscule at temperatures below 1600°C. This of course rules out the use of elemental tin as a leaky-tube diffusion source. There are, however, a variety of tin compounds with suitable vapor pressures at temperatures in the 500 to 700°C range which have been used as sources for tin implantation.[87] Tin halides possess substantial vapor pressure at low temperatures, but because of the possibility of halogen etching of the samples, they were not considered. The author has used solid stannous sulfide, SnS, because it has adequate vapor pressure (approximately 1 torr at 750°C) and manageable toxicology.[88]

The samples used in these experiments were p-type GaAs substrates, doped with zinc to a concentration of 2.0×10^{17} cm^{-3}, with (100) orientation, and polished with conventional bromine-methanol procedures. Diffusions were carried out in the temperature range of 650 to 750°C in the leaky-tube system; the only modification from the procedure described previously was the substitution of the SnS for elemental zinc. It was quickly established that the thin SiO_2 film was inadequate for GaAs surface protection when SnS was to be the source. There was drastic etching of the surface, probably due to the formation of volatile gallium and arsenic sulfides. However, the etching problem was circumvented by the use of relatively thick SiO_2 films (100 to 350 mn). It was later determined that the tin concentration reaching the semiconductor surface was quite independent of the dielectric film thickness.

Samples with thick SiO_2 films were diffused for durations from 2 to 24 h. A representative sample, designated as LT097-B, was examined in some detail by optical and electron microscopy, SIMS, and Rutherford backscattering spectrometry (RBS). Figure 7–38 is a photomicrograph of the cleaved and stained edge of sample LT097-B after a 12-h diffusion at 750°C. The line parallel to the surface resembles a standard diffusion front as control samples placed in the furnace for the same period without the tin source do not possess this feature. However, the carrier profile for this diffused sample, as depicted in Fig. 7–39, shows the unfortunate result that the net carrier concentration has not been altered by the diffusion procedure. In fact, approximately 30 samples were diffused in this manner, and each and every one remained p-type! Each sample shows a diffusion front, but none displays a type conversion.

However, there is no question about the addition of tin to the lattice. Figure 7–40 is the SIMS profile for sample LT097-B. It is obvious that a significant amount of tin has been introduced into the lattice. It also appears that the background zinc has redistributed itself during the heat treatment, but it also appears that the zinc concentration is not sufficient to compensate the tin as completely as the ECP profile requires. The scale for the tin concentration was determined by means of RBS calibration. As shown in Fig. 7–41, there is a small but statistically significant peak in the RBS spectrum which sets a lower limit on the tin concentration at the surface of

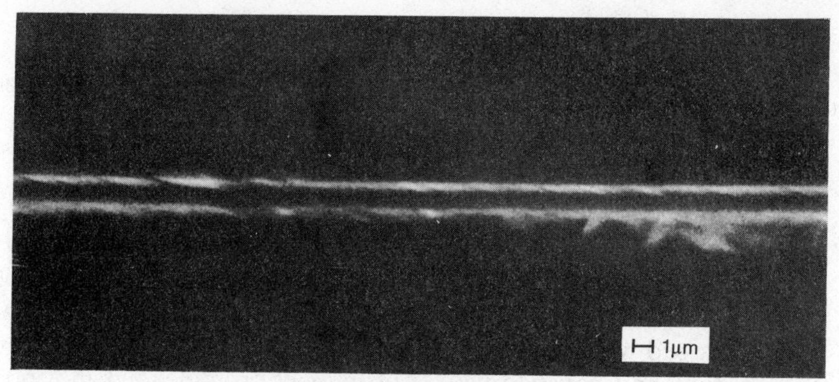

Fig. 7–38. A photomicrograph of a tin diffused GaAs sample shown in cross section after cleaving and staining (12 h, 750°C).

$N_{Sn} \cong 3.3 \times 10^{19}$ cm^{-3}. It is also interesting to point out that the rolloff of the tin SIMS signal corresponds to the same depth where the "diffusion front" appears in the photomicrograph.

Electron microscopy was employed to help shed some light on the mechanism responsible for the close compensation of the indiffusing tin. Two interesting facts emerged in this study by Graham *et al.*[89] Within 100 nm of the surface, small (< 10 nm) precipitates were often observed.

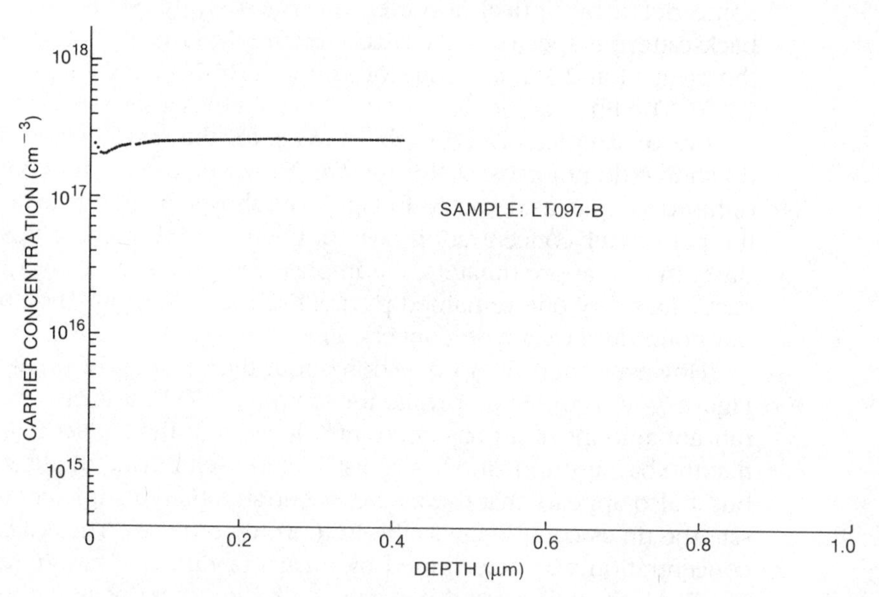

Fig. 7–39. The carrier concentration profile for the sample shown in Fig. 7–38.

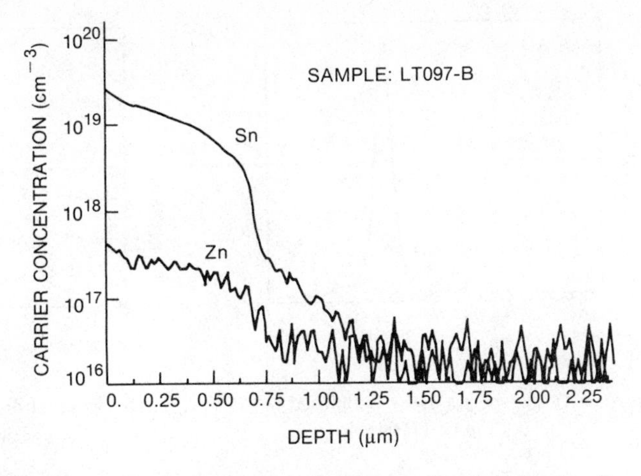

Fig. 7–40. The SIMS atomic concentration profiles for the sample shown in Fig. 7–38.

They were found to be tin rich, but their small size and density precluded exact analysis by convergent-beam electron diffraction. They may be composed of Sn-Ga alloys, which exist over a wide range of compositions, and Sn-As phases, of which Sn_3As_2 and SnAs are the most likely. Even more significantly, cathodoluminescence, carried out in the TEM at 23 K, revealed that the emissions spectra in the diffused and in the undiffused regions were quite different. In Fig. 7–42, the CL spectra are depicted. The spectrum shown in Fig. 7–42A is for the undiffused protion of the sample, and it consists of three transitions. At 1.51 eV, the transition is due to donor to valence band transitions from residual donor impurities; at 1.49 eV, the

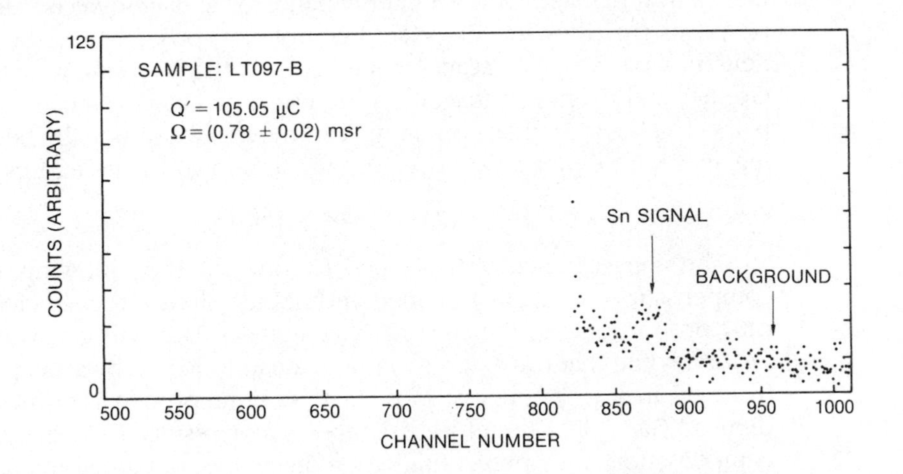

Fig. 7–41. Rutherford backscattering (RBS) data for the sample shown in Fig. 7–38.

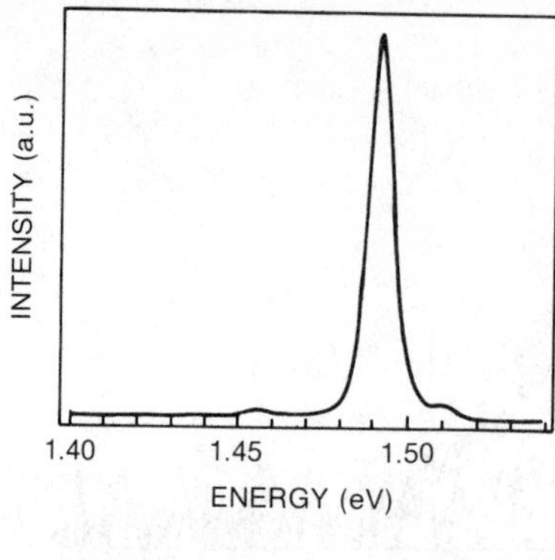

(A) The spectra in the undiffused portion of the sample.

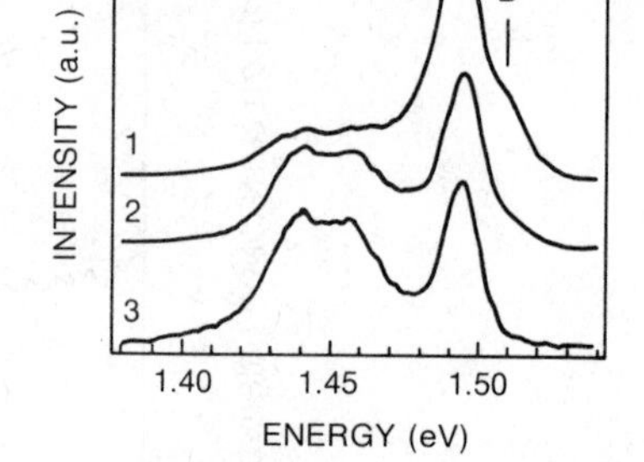

(B) Spectra from three different parts of the diffused portion.

Fig. 7–42. Cathodoluminescence spectra for a tin diffused sample. The sample was held at 23 K during the measurements.

dominant free-to-bound transition through the zinc (and possibly carbon) acceptors; at 1.46 eV, an LO phonon replica of the 1.49 EV transition. Figure 7–42B shows the CL spectra taken at three different points in the diffused portion of the sample. The 1.49 eV free-to-bound acceptor transition is still the dominant radiative path, but an expected peak or shoulder for free-to-bound transitions through a tin donor is weak or nonexistent. Instead, two lower energy transitions, at 1.455 eV and at 1.44 eV, have emerged. In fact, the more dominant these new transitions become, the more reduced the donor related shoulder becomes. Because these transitions differ from other deep transitions reported in the literature involving mechanical damage induced energy bands or antisite defects,[89] the author proposes that they are tin-related complexes, possibly Sn_{Ga}-Sn_{As} nearest neighbor pairs or Sn-vacancy complexes. It has been shown recently by Greiner and Gibbons[52] that silicon diffuses into GaAs by means of coupled pairs. That is, paired silicon atoms can move substitutionally be exchanging sites with either a Ga or As vacancy, as shown by the expression

$$(Si_{Ga}^{+} - Si_{As}^{-}) + V_{Ga} \rightleftarrows V_{Ga} + (Si_{As}^{-} - Si_{Ga}^{+})\,.$$

The consequence of this type of motion is the following: (*a*) close compensation is usually obtained and (*b*) the diffusion coefficient is typically proportional to the silicon concentration. This will give rise to a diffusion profile that has a gently tilting plateau followed by a steep diffusion front. By analogy we may propose that the same process is occurring in tin diffused GaAs. This therefore has three major results: (*a*) extremely close compensation, thus preventing type conversion, (*b*) an atomic profile, as shown in the SIMS profile, Fig. 7–40, that is "intermediate" in form to a complementary error function ($D_{\text{eff}} \cong$ constant) and to the zinc diffusion profile ($D_{eff} \propto C_s^2$), and (*c*) the generation of new optical transition paths at low temperature.

It is well known that various workers have successfully diffused tin into GaAs to produce n-type layers. However, virtually all of these successful diffusions were carried out at temperatures close to 1000°C. In fact, Nissim *et al.*[90] demonstrated at least with spin-on tin sources that tin can be introduced into the lattice at temperatures as low as 700°C, but it remains electrically inactive until a high temperature ($T > 900°C$) annealing step is performed. This is just the kind of processing step that one wants to avoid with the leaky-tube system, which has been designed to provide only low-temperature processing. It is evident that n-type diffusion into GaAs must unavoidably remain a relatively high-temperature processing step, and spin-on or evaporated diffusion sources are better suited to protect the surface and yield high-quality n-type layers.

7.5 Discussion and Summary

It should be fairly evident that a viable diffusion technology is critical for any semiconductor technology to remain prosperous. To recapitulate, diffusion into GaAs and $Ga_{1-x}Al_xAs$ has remained a useful processing procedure for a variety of reasons. In the first place, it can be employed to generate unique electronic and photonic devices that could not possibly be fabricated by ion implantation or epitaxial procedures alone. Some of these structures were listed in the introduction to this chapter. In the second place, as diffusion procedures have matured (in the form of leaky-tube and spin-on source techniques) and as epitaxial growth by sophisticated means has become more commonplace (in the form of MBE and MOCVD procedures), the blend of the two has added another dimension to the concept of bandgap engineering. The use of diffusion to produce controllable disorder and atomic rearrangement has been described previously. We can expect the continued use of the diffusion induced disordering procedure, especially when deep disordering or three dimensionally disordered structures are required. And finally, the more one examines the physics of the diffusion process in detail, the more it becomes evident that the complexities of the process are as intellectually intriguing as they are technologically significant. As described in the section on zinc diffusion into $Ga_{1-x}Al_xAs$, the zinc may be serving as an unusual atomic probe of the native defects in the material.

There are several remaining problem areas in current GaAs diffusion technology. These include the inability to produce degenerate n-type diffused layers, the difficulty in producing anything but degenerate p-type layers, and the lingering problem of masking for limited-area diffusions.

The problems of n-type and p-type doping values may find solutions, but if not, they are problems that can be circumvented by clever device design. On the other hand, masked diffusion has remained a complicated and perplexing problem for many years, and only recently have a few potential solutions begun to emerge.

The heart of the limited-area diffusion problem in GaAs technology lies in the inability to find a suitable dielectric film that can serve as a diffusion mask. Any diffusion mask must be able to block the motion of the diffusing species, must not interact with the semiconductor surface, and must be easily deposited, patterned, and removed. In silicon technology, all of the requirements are readily met with the native thermal oxide, SiO_2. The native oxide of GaAs, which can be grown anodically and is a complicated mixture of As_2O_3 and Ga_2O_3, is not a viable diffusion mask; it acts as an attenuator, but does not halt the flow of diffusing species. Most researchers have employed deposited dielectric films for diffusion masks,

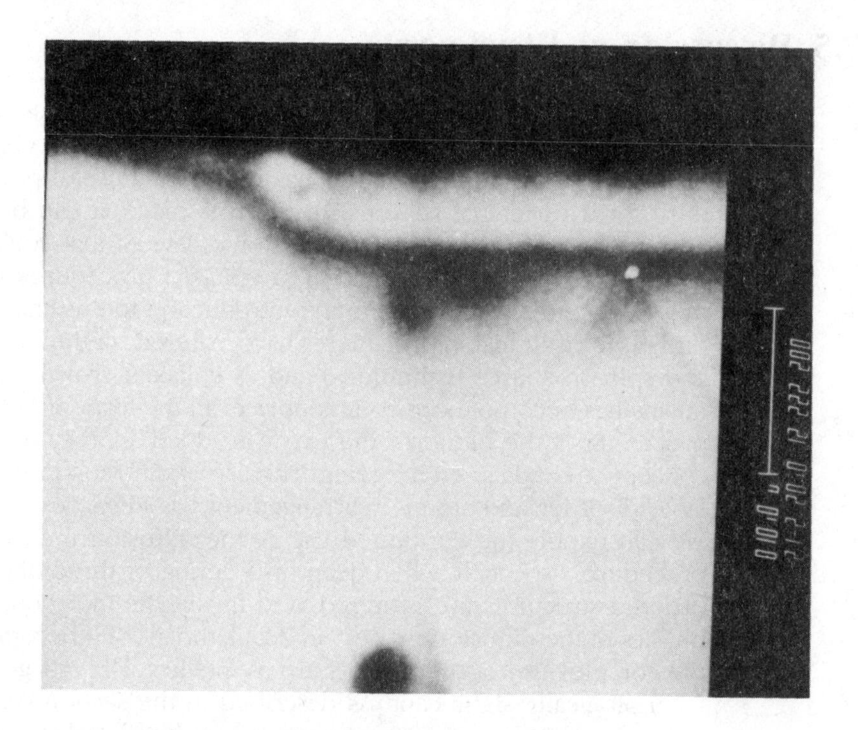

(A) This figure has been made with cathodoluminescence.

Fig. 7–43. SEM photomicrographs of a zinc diffused sample that has substantial laterial diffusion. The sample has been diffused for 60 min at 600°C through a Si_3N_4 mask.

the most common of which are SiO_2 and Si_3N_4, deposited by sputtering, chemical vapor deposition, or plasma enhanced CVD.[91–96] The results of the various research groups have not been consistent, but several general trends have appeared. It has become clear that the mechanical properties of the film are crucial, and those films with a significant difference in thermal expansion coefficient from the GaAs substrate will often produce a dramatic lateral diffusion problem. A typical example of this phenomenon is shown in Fig. 7–43. A cleaved and stained GaAs sample that has been Zn diffused in the leaky-tube furnace for 1 h at 600°C through a plasma deposited Si_3N_4 film is shown in this figure. The mask has interacted with the diffusing zinc so that enhanced diffusion along the substrate-mask interface has taken place. Recent work by Roedel *et al.*[97] has shown that there is a distinct correlation between the extent of the lateral diffusion and the magnitude of the compressive stress in the dielectric film, and the origin of this stress is due to thermal expansion coefficient mismatch and certain deposition parameters. It now appears that most, if not all, dielectric films may in fact be unsuitable for diffusion masks. Silicon dioxide has too large an expansion coefficient mismatch (in excess of 900 percent), and is

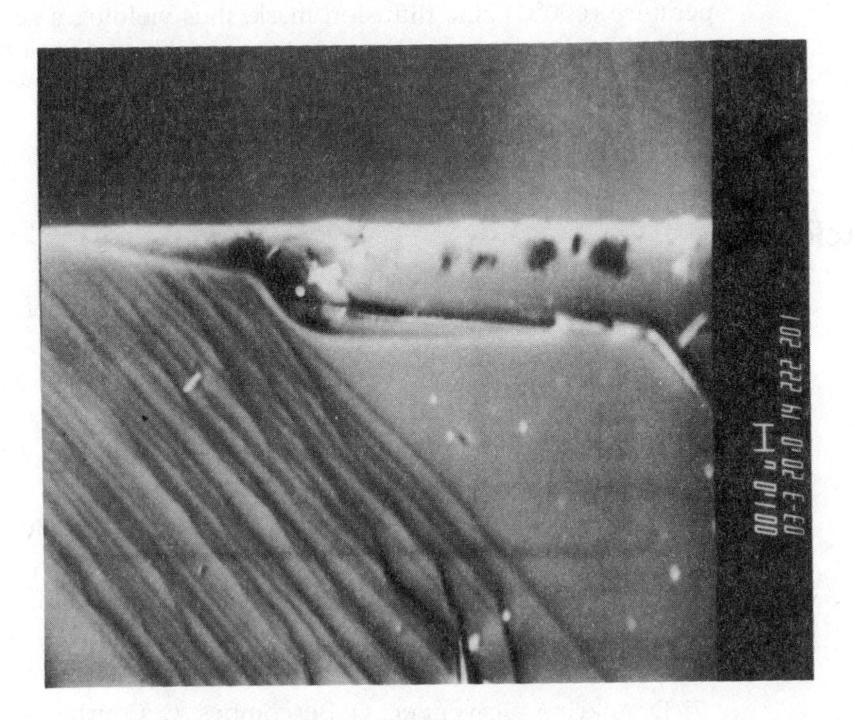

(B) This figure has been made with secondary electron imaging.

Fig. 7–43. *continued*

known to be a very effective sink for gallium outdiffusion. Aluminum oxide has the same expansion coefficient as GaAs, but unfortunately does not halt the flow of zinc atoms.[97] Silicon nitride has a moderate expansion coefficient mismatch (roughly 50 percent), is a good mask even as a very thin film (15 nm), but generally yields uneven results because of variability in its deposition.

A new direction in diffusion masking has developed recently, and this consists of using thin *semiconductor* films as the mask. Chin *et al.*[98] have employed e-beam evaporated silicon films for masked diffusion into GaAs and InP, for both materials, the junctions were devoid of excess lateral diffusion. A similar report by Oe *et al.*[99] described the use of epitaxial InP films as masks for zinc diffusion into GaInAsP laser structures. Although the stress content in these films has not been measured, it is expected that the compressive stress (which can be enormous in thin dielectric films) is probably very low. The use of deposited semiconductor films also leads to another intriguing processing scheme, as reported by Merz.[100] In that work, Si is first deposited as a thin film and then patterned. It is then used as a source for silicon diffusion at 850°C. Not all of the film is consumed during the high-temperature diffusion, and it can then serve as a low-temperature (600°C) zinc diffusion mask, thus yielding a self-aligned JFET technology, for example. Because of the viability of diffusion in GaAs technology and the resourcefulness of the people utilizing it, we can expect to continue seeing additional novel applications.

References

1. D. R. Scifres, C. Lindstrom, R. D. Burnham, W. Streifer, and T. L. Paoli, *Electron. Lett.,* 19:169 (1983).
2. D. K. Arch, B. K. Betz, P. J. Void, J. K. Abrokwah, and N. C. Cirillo, Jr., *IEEE Electron Dev. Lett.,* EDL-7:700 (1986).
3. Y. Yamaguchi and T. Ishibashi, *IEEE Electron Dev. Lett.,* EDL-7:655 (1986).
4. H. Kumabe, T. Tanaka, H. Namizaki, M. Ishii, and N. Susaki, *Appl. Phys. Lett.,* 33:38 (1978).
5. R. J. Roedel and P. M. Holm, *Solar Cells,* 11:221 (1984).
6. P. M. Holm and R. J. Roedel, *18th IEEE Photovoltaics Specialists Conf. Proc.,* Orlando, Florida, 921 (1984).
7. D. Ankri, A. Scavennec, C. Bescombes, C. Courbet, F. Heliot, and J. Riou, *Appl. Phys. Lett.,* 40:816 (1982).
8. J. S. Escher and H. Schade, *J. Appl. Phys.,* 44:5309 (1973).

9. W. D. Laidig, N. Holonyak, Jr., M. D. Camras, K. Hess, J. J. Coleman, P. D. Dapkus, and J. Bardeen, *Appl. Phys. Lett.,* 38:776 (1981).
10. K. Ishida, T. Ohta, S. Semura, and H. Nakashima, *Jpn. J. Appl. Phys., 24,* L620 (1985).
11. E. V. K. Rao, H. Thibierge, F. Brillouet, F. Alexandre, and R. Azoulay, *Appl. Phys. Lett.,* 46:867 (1985).
12. K. Meehan, N. Holonyak, Jr., J. M. Brown, M. A. Nixon, P. Gavrilovic, and R. D. Burnham, *Appl. Phys. Lett.,* 45:549 (1984).
13. D. G. Deppe, L. J. Guido, N. Holonyak, Jr., K. C. Hsieh, R. D. Burnham, R. L. Thornton, and T. L. Paoli, *Appl. Phys. Lett.,* 49:510 (1986).
14. K. Meehan, P. Gavrilovic, N. Holonyak, R. D. Burnham, and R. L. Thornton, *Appl. Phys. Lett.,* 46:75 (1985).
15. T. Fukuzawa, S. Semura, H. Saito, T. Ohta, Y. Uchida, and H. Nakashima, *Appl. Phys. Lett.,* 45:1 (1984).
16. P. Gavrilovic, K. Meehan, J. E. Epler, N. Holonyak, Jr., R. D. Burnham, R. L. Thornton, and W. Streifer, *Appl. Phys. Lett.,* 46:857 (1985).
17. J. A. van Vechten, *J. Appl. Phys.,* 53:7082 (1982).
18. J. F. Black and E. D. Jungbluth, *J. Electrochem. Soc.,* 114:181 (1967).
19. J. F. Black, *J. Electrochem. Soc.,* 114:1293 (1967).
20. W. D. Rhines and D. A. Stevenson, *J. Electron. Mat.,* 2:341 (1973).
21. M. Maruyama, *Jpn. J. Appl. Phys.,* 7:476 (1968).
22. J. F. Black and E. D. Jungbluth, *J. Electrochem. Soc.,* 114:188 (1967).
23. See, for example, A. A. Bergh and P. J. Dean, *Light Emitting Diodes,* London: Oxford University Press, p. 391 (1976).
24. J. van den Boomgaard and K. Schol, *Philips Res. Rep.,* 12:127 (1957).
25. D. P. Miller, J. H. Harper, and T. R. Perry, *J. Electrochem. Soc.,* 105:1123 (1961).
26. S. R. Shortes, J. A. Kanz, and E. C. Wurst, Jr., *TMS-AIME,* 230:300 (1964).
27. F. A. Cunnell and C. H. Gooch, *Phys. Chem. Solids,* 15:127 (1960).
28. B. Goldstein, in *Compound Semiconductors, Vol. 1,* eds. R. K. Willardson and H. L. Goering, New York: Van Nostrand Reinhold, p. 345 (1962).
29. J. W. Allen and F. A. Cunnell, *Nature,* 182:1158 (1958).
30. B. Goldstein, *Phys. Rev.,* 118:1024 (1960).
31. O. Hasegawa and R. Namazu, *Appl. Phys. Lett.,* 36:203 (1980).
32. Y. J. Yang, Y. C. Lo, G. S. Lee, K. Y. Hsieh, and R. M. Kolbas, *Appl. Phys. Lett.,* 49:835 (1986).
33. B. Tuck, *Introduction to Diffusion in Semiconductors,* 171, Stevenage, England: Peter Peregrinus (1974).

34. H. C. Casey, Jr., M. B. Panish, and L. L. Chang, *Phys. Rev.,* 162:660 (1967).
35. A. Juque, J. Martin, and G. L. Araujo, *J. Electrochem. Soc.,* 123:249 (1976).
36. H. C. Casey, Jr., and M. B. Panish, *TMS-AIME,* 242:406 (1968).
37. W. von Muench, *Solid State Electron.,* 9:619 (1966).
38. B. J. Baliga and S. K. Ghandhi, *J. Electrochem. Soc.,* 126:135 (1979).
39. J. R. Shealy, B. J. Baliga, and S. K. Ghandhi, *IEEE Electron Dev. Lett.,* EDL-1:119 (1980).
40. S. K. Ghandhi and R. J. Field, *Appl. Phys. Lett.,* 38:267 (1981).
41. R. J. Field and S. K. Ghandhi, *J. Electrochem. Soc.,* 129:1567 (1982).
42. G. W. Turner, B-Y. Tsaur, J. C. C. Fan, F. M. Davis, R. P. Gale, and M. K. Connors, *Quarterly Tech. Sum. Rep.,* MIT-Lincoln Labs, Air Force Contract No. F19628-80-C-0002 (February 11, 1983).
43. Y. I. Nissim, J. F. Gibbons, C. A. Evans, V. R. Define, and J. C. Norberg, *Appl. Phys. Lett.,* 37:89 (1980).
44. N. Arnold, R. Schmidt, and K. Heime, *J. Phys. D,* 17:443 (1984).
45. P. Kupper and J. Freyer, *Intl. J. Electron.,* 47:469 (1979).
46. K. V. Vaidyanathan, M. J. Helix, D. J. Walford, B. G. Streetman, R. J. Blattner, and C. A. Evans, *J. Electrochem. Soc.,* 124:1781 (1977).
47. S. Mizutani, I. Ohdomari, T. Miyazawa, T. Iwamori, I. Kimura, and K. Yoneda, *J. Appl. Phys.,* 53:1470 (1982).
48. T. P. Ma and K. Miyauchi, *Appl. Phys. Lett.,* 34:88 (1979).
49. H. Hiyashi, K. Kikuchi, T. Yamaguchi, and T. Nakahara, *Gallium Arsenide and Related Compounds, 1980,* (Inst. Phys. Conf. Ser. No. 56, 1981), p. 56.
50. Y. I. Nissim, J. F. Gibbons, F. J. Magee, and R. Ormond, *J. Appl. Phys.,* 52:227 (1981).
51. T. Sakai, T. Suzuki, and H. Hasegawa, *Electron. Lett.,* 14:249 (1978).
52. M. E. Greiner and J. F. Gibbons, *Appl. Phys. Lett.,* 44:750 (1984).
53. K. Meehan, N. Holonyak, Jr., J. M. Brown, M. A. Nixon, P. Gavrilovic, and R. D. Burnham, *Appl. Phys. Lett.,* 45:549 (1984).
54. A. J. SpringThorpe and M. N. Svilans, *Gallium Arsenide and Related Compounds, 1982* (Inst. Phys. Conf. Ser. No. 65, 1983) p. 589.
55. Y-R. Yuan, K. Eda, G. A. Vawter, and J. L. Merz, *J. Appl. Phys.,* 54:6044 (1983).
56. F. C. Prince, M. Oren, and M. Lam, *Appl. Phys. Lett.,* 48:546 (1986).
57. R. J. Roedel, J. L. Edwards, A. Righter, P. Holm, and H. Erkaya, *J. Electrochem. Soc.,* 131:1726 (1984).
58. S. K. Ageno, R. J. Roedel, N. Mellen, and J. S. Escher, *Appl. Phys. Lett.,* 47:1193 (1985).

59. H. H. Erkaya, R. J. Roedel, R. Lareau, P. Williams, J. Leavitt, and A. von Neida, *J. Electrochem. Soc.,* 132:2214 (1985).
60. J. L. Edwards and R. J. Roedel, *Electron. Lett.,* 19:962 (1983).
61. L. L. Chang, *Solid State Electron.,* 7:853 (1964).
62. T. Ambridge, C. R. Elliot, and M. M. Faktor, *J. Appl. Electrochem.,* 3:1 (1973).
63. T. Ambridge and M. M. Faktor, *J. Appl. Electrochem.,* 4:135 (1974).
64. T. Ambridge and M. M. Faktor, *Inst. Phys. Conf. Ser. No. 24,* Bristol (Avon): The Institute of Physics, 320 (1975).
65. D. R. Campbell and K. K. Shih, *Appl. Phys. Lett.,* 19:330 (1971).
66. B. V. Dutt, S. Mahajan, R. J. Roedel, G. P. Schwartz, D. C. Miller, and L. Derick, *J. Electrochem. Soc.,* 128:1573 (1981).
67. R. L. Longini, *Solid State Electron.,* 5:127 (1962).
68. L. R. Weisberg and J. Blanc, *Phys. Rev.,* 131:1548 (1963).
69. C. Chiaretti and C. Cognetti, *J. Electrochem. Soc.,* 128:2199 (1981).
70. A. Luque, J. Martin, and G. L. Araujo, *J. Electrochem. Soc.,* 123:249 (1976).
71. D. L. Kendall, "Diffusion," in *Semiconductors and Semimetals,* ed. R. K. Willardson and A. C. Beer, New York: Academic Press, 4:163–259 (1968).
72. L. M. Kogan, S. S. Meskin, and A. Ya. Goikhman, *Sov. Phys.—Solid State,* 6:882 (1964).
73. B. V. Dutt, A. K. Chin, and W. A. Bonner, *J. Electrochem. Soc.,* 128:2014 (1981).
74. W. Kuebart, O. Hildebrand, H. W. Marten, and N. Arnold, *Inst. Phys. Conf. Ser. No. 65,* 597 (1983).
75. N. Chand and P. A. Houston, *J. Electron. Mat.,* 11:37 (1982).
76. R. G. Moore, Jr., M. Belasco, and H. Strack, *Bull. Am. Phys. Soc.,* 10:731 (1965).
77. M. B. Small, R. M. Potemski, W. Reuter, and R. Ghez, *Appl. Phys. Lett.,* 41:1068 (1982).
78. B. I. Boltaks, T. D. Dzhafarov, Yu. P. Demakov, and I. E. Maronchuk, *Sov. Phys. Semicond.,* 9:545 (1975).
79. C. P. Lee, S. Margalit, and A. Yariv, *Solid State Electron.,* 21:905 (1978).
80. Y-R. Yuan, K. Eda, G. A. Vawter, and J. L. Merz, *J. Appl. Phys.,* 54:6044 (1983).
81. S. E. Blum, M. B. Small, and D. Gupta, *Appl. Phys. Lett.,* 42:108 (1983).
82. D. R. Campbell and K. K. Shih, *Appl. Phys. Lett.,* 19:330 (1971).
83. D. V. Lang, R. A. Logan, and M. Jaros, *Phys. Rev. B,* 19:1015 (1979).

84. V. Quintana, J.J. Clemencon, and A.K. Chin, paper presented at SOTAPOCS VI, Electrochemical Society Meeting, Philadelphia (May 13, 1987).

85. L.L. Chang and G.L. Pearson, *J. Appl. Phys.*, 35:1960 (1964).

86. K.L. Kavanagh, J.W. Mayer, C.W. Magee, J. Sheets, J. Tong, and J.M. Woodall, *Appl. Phys. Lett.*, 47:1208 (1985).

87. D.R. Stull, *Ind. Eng. Chem.*, 39:4 (1947).

88. C.M. Hsiao and A.W. Schlechten, *J. Met.*, 4:65 (1952).

89. R.J. Graham, J.C.H. Spence, and R.J. Roedel, *J. Appl. Phys.*, 59:164 (1986).

90. Y.I. Nissim, J.F. Gibbons, C.A. Evans, V.R. Deline, and J.C. Norberg, *Appl. Phys. Lett.*, 37:89 (1980).

91. K. Oe, S. Ando, and K. Sugiyama, *J. Appl. Phys.*, 51:43 (1980).

92. H.K. Choi and S. Wang, *Appl. Phys. Lett.*, 43:230 (1983).

93. B.J. Baliga and S.K. Ghandhi, *IEEE Trans. Electron Dev.*, ED-21:410 (1974).

94. C. Blaauw, A.J. SpringThorpe, S. Dzioba, and B. Emmerstorfer, *J. Electron. Mat.*, 13:251 (1984).

95. E.A. Rezek, P.D. Wright, and N. Holonyak, *Solid State Electron.*, 21:325 (1978).

96. C.S. Hong, J.J. Coleman, P.D. Dapkus, and Y.Z. Liu, *Appl. Phys. Lett.*, 40:208 (1982).

97. R.J. Roedel, H.H. Erkaya, and J.L. Edwards, *J. Electron. Mat.*, 17:243 (1988).

98. A.K. Chin, I Camlibel, L. Marchut, S. Singh, L.G. Van Uitert, and G.J. Zydzik, *J. Appl. Phys.*, 58:3630 (1985).

99. K. Oe, S. Ando, and K. Sugiyama, *J. Appl. Phys.*, 51:3541 (1980).

100. J. Merz, paper presented at the Device Research Conference, Amherst, Mass. (June 1986).

Chapter 8

Oxidation of GaAs

C. W. Wilmsen

8.1 Introduction

There have been several motivating factors which have encouraged investigation of the GaAs oxides. The first was the desire to emulate the silicon MOSFET technology, which had been developed in the mid-1960s. It has been shown, however, that the thermal oxide of GaAs provides a poor gate oxide since it tends to be polycrystalline, has a rough surface and interface, and collects elemental As at the interface. In 1975, Hartnagel and coworkers[1] reported very encouraging electrical characteristics of MOS capacitors using an anodic oxide insulator. This work stimulated considerable research on the anodic oxide insulator. Unfortunately, this oxide was found to create a high density of surface states and was unstable[2] due to the absorption of water and a change in chemical composition when heated.[3] On the other hand, GaAs oxides have been successfully used to passivate laser diodes, provide a cap for ion implant anneals, and form a thin surface seal which can be evaporated off in a MBE chamber. It has also been recognized that thin oxides grow as a result of chemical etching, exposure to air, and in a deposition chamber containing a reactive oxidizer. This has motivated the study of thin oxides in order either to prevent or to utilize them. Recent research on water-grown oxides has shown that the GaAs surface can be unpinned by photo-oxidation in DI water or air.[4,5] These results will undoubtedly lead to additional investigation of the oxidation process and possible device applications.

The first reported analyses of oxides grown on a III-V compound semiconductor were performed for InSb. Dewald[6] reported results on anodic oxides in 1957, and Rosenberg and Levine[7] published results for the thermal oxidation of InSb in 1960. The first paper on an oxide of GaAs appeared in 1962; in it, Minden[8] reported on the thermal oxidation of GaAs in both dry and wet O_2 over the temperature range 600 to 900°C. He found that the oxide layer was polycrystalline Ga_2O_3. From these early investigations has grown a large body of literature on the thermal, anodic, plasma and chemically grown oxides on GaAs. At the present, there are 200 to 300 published works on the GaAs oxides. While this is a small number compared with that for Si, it is sufficient to provide a basic understanding of oxide growth and composition. Primary to our understanding is the equilibrium Ga-As-O ternary phase diagram and the kinetic factors of diffusion, evaporation and dissolution. As will be evident later, our knowledge of the kinetic factors is not complete, but sufficient to allow formulation of qualitative models.

This chapter serves as a primer on GaAs oxides. It builds upon previous reviews by Schwartz,[9] Croydon and Parker,[10] and the present author.[11–13] While some overlap with these previous works is inevitable, it is hoped that the present work brings a new and more fundamental understanding of the subject.

8.2 Thermodynamics

The composition of an oxide layer is determined by the growth conditions, the thermodynamic stability of the oxidation products, and the growth kinetics. In order to illustrate how these factors depend upon each other, the thermodynamics and kinetics of the thermal oxidation of Si, GaAs, and InP are compared.

The thermal oxidation of Si is the simplest of these three oxide/semiconductor systems, since there is only one substrate element and only one stable bulk oxidation product,[14] i.e. SiO_2. In addition, the oxidizing specie, O_2 or H_2O, diffuses much faster than does Si and therefore all the oxidation takes place at the interface.[15] Thus, there is little possibility for any variation in the resulting thermal oxidation, and near stoichiometric SiO_2 always forms on the Si substrate. At the interface there is a structural problem of fitting the amorphous SiO_2 to the crystalline Si substrate. This causes defects and mixed chemical states at the interface,[16] but the salient chemical features of the bulk oxide and interface do not change with growth conditions or annealing. Increasing the growth temperature or oxidant pressure accelerates the growth kinetics but does not change the final oxidation products.

Growth is more complex and variable on InP since there are two substrate elements which have different diffusion coefficients, oxidation potentials, and vapor pressures. In addition there are a number of different stable oxide compounds, including In_2O_3, P_2O_5, $InPO_4$ and $In(PO_3)_3$. The equilibrium ternary phase diagram[17] of Fig. 8–1A indicates that the reaction of O_2 with InP should yield $InPO_4$ and indeed some very thin oxide layers

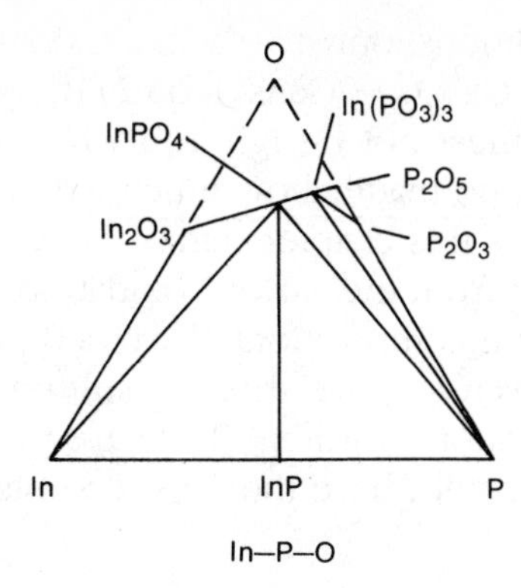

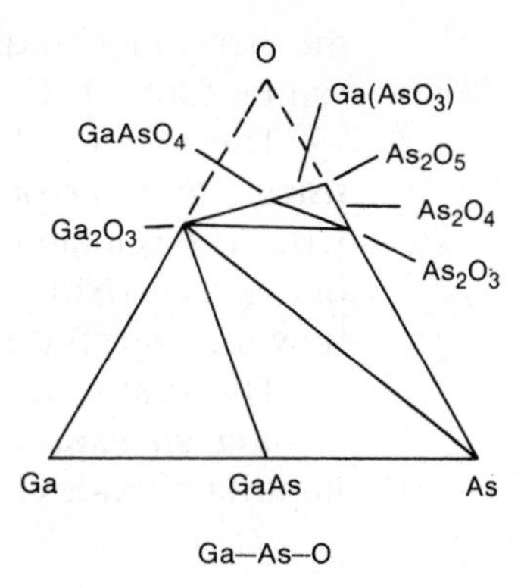

Fig. 8–1 Equilibrium phase diagrams. ***(After Schwartz*** **et al.**[17] ***and Thurmond*** **et al.**[22]***)***

are composed primarily of $InPO_4$.[18] For the very thin layers, the restraints of growth kinetics are minimal and thus the equilibrium reaction product can be formed. However, as the oxide layer thickens, a mixed oxide is formed and elemental P appears at the interface since the kinetics begin to dominate the growth.[13] In particular, it has been found that the diffusion rates of the three elements are different and fall into the following order: In, P, O_2. It has been shown that In and P diffuse to the surface,[19] where they form $InPO_4$ and In_2O_3. In_2O_3 forms because more In reaches the surface than P; some of the P remains at the interface.

Since the $InPO_4$/InP couple is thermodynamically stable and the thermal growth of oxide layers on InP is controlled by kinetics, the growth conditions can be altered to change the chemical composition of the oxide layer. There have been two successful approaches to varying the growth kinetics. One is anodic oxidation, in which a high field is applied to greatly increase the diffusion rate of both the In and P.[20] This generally leads to the formation of $In(PO_3)_3$. Recent work by Gann *et al.*[21] reports the use of very high pressure thermal oxidation in steam, which causes the diffusion rate of the oxidant to be greater than that of In and P. Thus, the oxidation takes place at the oxide-InP interface, where the equilibrium product $InPO_4$ is readily formed. The $InPO_4$-InP interface is stable and has similarities to that of the SiO_2-Si system.

For GaAs, there are again a number of stable oxides:[22] Ga_2O_3, As_2O_3, As_2O_5, and $GaAsO_4$. Unfortunately, the equilibrium phase diagram, Fig. 8–1B, indicates that the reaction between GaAs and O_2 yields Ga_2O_3 and elemental As. Thus, even if the kinetics are adjusted to form As_2O_3, As_2O_5, and/or $GaAsO_4$, the interface is unstable[3,23] since these compounds react with the GaAs substrate to form Ga_2O_3 and elemental As. This is an important conclusion, which forces the reader to realize that, independent of the growth technique or conditions, only two stable oxide-GaAs interfaces can be formed. These are (1) Ga_2O_3 − GaAs and (2) [Ga_2O_3 + As]-GaAs.

These latter conclusions must not be ignored when investigating or using a grown oxide on GaAs, no matter how much we might wish otherwise. The remaining sections of this chapter outline how the composition of the GaAs-oxide layers result from the various oxidation techniques, and how they reach the equilibrium compositions discussed previously.

The first oxides to be discussed are those resulting from chemical etching, air exposure, or oxidation in water. These are probably the most important oxides but, unfortunately, have the most complex kinetics.

8.3 Thin Native Oxides

Most device processing starts with the polishing, etching and rinsing of the GaAs surface. This is often followed by air exposure and/or water washing and subsequent heating in either air or vacuum. These procedures produce a native oxide on the GaAs surface which can strongly affect the fabrication steps that follow. The different etching/exposure procedures produce subtle changes in the composition of the native oxide that, if not understood and corrected for, can cause large differences in the final process step and the device characteristics. This section discusses the kinetics of the cleaning procedures and the subsequent air exposure and water rinsing.

8.3.1 The Etched GaAs Surface

There are several commonly used etches/polishes for GaAs. They are the following:

Bromine methanol
NaOCl
NaOH
HCl
$H_2SO_4/H_2O_2/H_2O$
H_2SO_4
HNO_3

Each of these solutions can yield a different surface composition. In order to understand why this is so requires a knowledge of the compounds formed during the etching process and the solubility of these compounds in both water and the etching solution itself.

An etch solution changes the chemical state of the GaAs surface atoms and forms soluble compounds or compounds that are weakly attached to the substrate.[9,24] There are two basic types of etches: those that form oxides and those that do not. Bromine methanol, for example, forms $GaBr_3$ and $AsBr_3$, both of which are soluble in methanol.[25] Since BrMeOH contains no oxidizer, the surface is left oxide-free unless subsequently exposed to air or another oxidant. $H_2SO_4/H_2O_2/H_2O$ is an example of the first type. In this etch, the H_2O_2 oxidizes the surface and the H_2SO_4 dissolves the oxidation products. Since an oxidizer is present, this etch could leave an oxide coating.

Details of the various factors which influence or modify the resultant oxides are discussed below.

8.3.1.1 Solubility of the Oxidation Products

First, it is realized that a number of compounds can form as a result of chemical etching. These include Ga_2O_3, $Ga(OH)_3$, $GaOOH{:}XH_2O$, As_2O_3, As_2O_5, $GaAsO_4$, H_3AsO_3, and elemental As. In general, the arsenic oxides are more soluble than the Ga oxides; however, the solubility of both are dependent upon the pH of the solution as outlined below:

As_2O_5 and H_3AsO_4 are highly soluble in aqueous solutions at all values of pH.[26a]

As_2O_3 is less soluble than As_2O_5 but will dissolve in aqueous solutions of all pH.[26a] The solubility increases rapidly in basic and strongly acidic solutions, i.e. $1 < pH < 8$.

Ga_2O_3 is insoluble in neutral water and for aqueous solutions in the range $3 < pH < 11$. The solubility increases rapidly outside of this range. Thus, Ga_2O_3 will only dissolve in concentrated acids and bases.[26b]

$Ga(OH)_3$ is many times more soluble in neutral water than Ga_2O_3 and becomes even more soluble in acidic and basic solutions.[26b]

Arsenic is insoluble in water and aqueous solutions of all pH that do not contain oxidizing agents.[26a] The addition of O_2 to the solution or the application of light causes the elemental As to oxidize to As_2O_3, which readily dissolves. Arsenic dissolves in strong acids such as HNO_3, HCl, H_2SO_4 and aqua regia.

$GaAsO_4$—solubility unknown.

8.3.1.2 Effects of Water Rinse

Two factors strongly affect the results of water rinsing: the concentration of dissolved O_2 and the intensity of the incident light. These factors control the conversion of elemental As to As_2O_3 and hence control the removal of elemental As from the GaAs surface.[5,26a] The concentration of dissolved O_2 is variable. It is higher in stagnant water than in running water. Providing a N_2 atmosphere over the water also reduces the O_2 content.

Water which contains no dissolved O_2 will not oxidize the GaAs surface. This has been shown by Aspnes and Studna,[27] who water-rinsed BrMeOH etched GaAs under flowing N_2. These results have been corroborated for $H_2SO_4/H_2O_2/H_2O$ etched surfaces by Massies and Contour.[28,29] Thus, oxide-free GaAs surfaces can be obtained by etching and rinsing in a dry box using O_2-free water.

Water containing no dissolved O_2 cannot remove elemental As from GaAs.[26a] Thus, if a surface preparation has left a residue of elemental As, it cannot be rinsed away with water until there is sufficient light present as discussed in the next section.

If the rinse water contains dissolved O_2, the GaAs surface forms an oxide layer composed primarily of Ga_2O_3 but containing some[28,29] As_2O_3. Although Ga_2O_3 and As_2O_3 are formed in equal amounts, the As_2O_3 is far more soluble than Ga_2O_3 and is rinsed away. Since these are water-formed oxides, $Ga(OH)_3$ and/or H_3AsO_3 may also be formed. This would modify the water rinsing results because of the differences in solubility. Any elemental As would also be removed by an O_2-containing rinse if it is not buried too deep below the oxide surface.

Since Ga_2O_3 is insoluble, it is often left as a residue on the GaAs surface. However, prolonged water rinsing in O_2-free water appears to remove the Ga_2O_3, perhaps by the removal of Ga^{+++} ions.[28] If the water contains dissolved O_2, then the oxide layer will continue to grow. Schwartz[30] reports 85 nm of Ga_2O_3 formed in six days when the GaAs was placed in water. However, the writer has observed that similar thicknesses can be obtained in less than an hour if O_2 is bubbled through the water.

8.3.1.3 Effects of Light

Incident light, with energy greater than the bandgap, creates electron-hole pairs within the GaAs. This provides photogenerated charge which collects at the surface and weakens the bond strength of surface atoms.[31] It also changes the position of the GaAs band edges relative to the decomposition potential of the electrolyte. The electrolyte could be neutral water or an aqueous acid or base. Thus, incident light enchances oxidation or even causes oxidation to proceed on a surface that cannot oxidize in the dark.

An important example of photo-oxidation is that of elemental arsenic. This is an important example, because elemental As often forms on the GaAs during both etching[32,33] and air oxidation at moderate temperatures. In oxygenated water, the As readily oxidizes to As_2O_3 or $HAsO_2$ and can be rinsed away.[26a] Unfortunately the GaAs surface can then be reoxidized by the oxygenated water, with the subsequent possibility of recontaminating the surface with elemental As. Using deoxygenated water with incident light provides a means of avoiding this problem and appears to be an important method of forming clean GaAs surfaces.[5]

8.3.1.4 Effects of Temperature

Elevated temperature drives an oxide film towards thermodynamic equilibrium. For example, if the oxide layer on GaAs contains As_2O_3 or As_2O_5, then these compounds will begin reacting with the GaAs substrate to form Ga_2O_3 and elemental As. Thus, before heating, the As_2O_3 can be rinsed away with water, but after heating, the elemental As is more difficult to remove. The elevated temperature can also evaporate the As_2O_3 and elemental As once the GaAs is removed from the etch.

Within the etching or rinsing solution, increasing temperature always increases the rate of chemical reaction and diffusion of the reactants, and the solubility of the reaction products. Thus, the processes discussed above are accelerated.

8.3.1.5 Surface Composition After Etching

HCl

HCl dissolves the oxides and hydroxides of Ga and As and thus can be used to remove anodic-, thermal-, water-, and air-grown oxides. However, it does not readily dissolve elemental As and thus can leave a residue of elemental As. The HCl etching of the GaAs substrate itself proceeds by one of the following reactions:[33]

$$GaAs + 3H^+ + 4Cl^- + H_2O \longrightarrow GaCl_4^- + 3H_2 + HAsO_2 \quad (8\text{–}1)$$

$$GaAs + 3H^+ + 4Cl^- \longrightarrow GaCl_4^- + 4H_2 + As \quad (8\text{–}2)$$

For concentrated HCl, reaction 8–2 appears to be thermodynamically more favorable, and again there is evidence that etching in concentrated HCl leaves elemental As on the surface. This has been demonstrated by Bertrand,[32] and Frese and Morrison.[33] Both of these investigations used concentrated HCl and rinsed with either ethanol or flowing water. Diluting the HCl with water provides a means of oxidizing the elemental As which can remove all or part of the elemental As layer. Diluting with ethanol will not remove the elemental As, but it appears to prevent the formation of an elemental As layer due to the HCl etch. This has been demonstrated by Vasquez and coworkers.[34] Any As_2O_3 or $HAsO_2$ formed in the etching process will dissolve in the HCl.

$H_2SO_4:H_2O_2:H_2O$

The results of this etch are highly dependent upon the pH and dilution. The etching proceeds by the growth of oxides formed by the H_2O_2 and the dissolution of these oxides by the H_2SO_4 and H_2O. Massies and Contour[28,29] have shown that no oxide is left on the GaAs by the etch. Oxides have been observed[35] on GaAs etched in concentrated $H_2SO_4:H_2O_2:H_2O$, but this has been shown to be result of air exposure or rinsing in static water.[28,29]

The following analysis was performed by Aspnes and Stocker[36] and, while it was performed on $In_{0.53}$ $Ga_{0.47}As$, it is very insightful and probably also applies to GaAs. In 30 percent H_2O_2 or highly diluted H_2O_2, without the H_2SO_4, a self-limiting gallium oxide of thickness 0.7 to 1.0 nm is left on the surface. Arsenic oxide is also formed by the H_2O_2, but is dissolved away by the solution. Diluting the H_2O_2 causes a small amount of elemental As to appear on the GaAs surface. If the dilute H_2O_2 is made basic with NH_4OH, a gallium oxide is again left on the surface but the thickness does not self-limit and will continue to grow with time. If the dilute H_2O_2 is made acidic with H_2SO_4, then the gallium oxide is also dissolved away, but a residue of elemental As remains on the surface.

The composition of the oxides grown in this dilute etch is not known, but it is most probably Ga_2O_3, although Ga(OH) is also possible. The oxide layer does not contain arsenic oxides. Vasquez *et al.*[35] identified the oxides formed on GaAs, after $H_2SO_4:H_2O_2:H_2O$ etching, but these apparently were grown by air exposure.

HNO_3

HNO_3 is a strong oxidizer that readily attacks GaAs to form As_2O_3 and Ga_2O_3. However, since this etch solution is highly acidic, the Ga_2O_3 is easily dissolved. The As_3O_3 also dissolves, but precipitates out onto the GaAs surface[30] in concentrated HNO_3. Diluting the HNO_3 with H_2O holds the As_2O_3 in solution and prevents precipitation.[24]

Schnell[37] first reported in 1957 results of $HNO_3:H_2O$ etching of GaAs. Later Schwartz[30] found that soaking a GaAs wafer in concentrated HNO_3 for five days converted the entire wafer to As_2O_3. Oda and Sugano[38] used Auger analysis of HNO_3 etched GaAs to observe an arsenic-rich oxide layer which they attribute to the dissolution of the Ga_2O_3 in the strong acid. Water rinsing would presumably wash away the As_2O_3.

Thus, no stable oxide film is grown on GaAs in concentrated HNO_3. High dilution with high illumination could possibly result in a stable Ga_2O_3 layer, but this has not been investigated.

Bromine Methanol

Bromine methanol does not form an oxide on GaAs, nor does it readily attack the native oxide of GaAs. Aspnes and Studna[27] report that the sharpest GaAs interface is obtained with a final rinse in 0.05-percent BrMeOH and thus does not leave oxides or metallic residues.

8.3.2 Air-grown Oxides

Generalizations about the composition and growth rate of air-grown oxides on GaAs must be carefully considered. As can be seen from the above discussion, the initial chemical composition of the GaAs surface is dependent upon the etching/rinsing history of the sample. In addition, the amount of water vapor in the air affects the initial growth rate. If the GaAs surface is covered with elemental As from an acid etch, then the GaAs may oxidize very slowly since arsenic is a noble metal. The addition of water vapor to the air increases the oxidation rate of the elemental As.

Normally, the initial oxidation rate is rapid and most air-exposed GaAs surfaces have 0.5 to 1.0 nm of oxide after a brief air exposure. The growth is logarithmic thereafter. Lukes[39] has shown that the growth after the initial layers is the same on etched and cleaved surfaces. The type of etch that he used was not given but his result is probably applicable to all etches. However, each etch would have a different "initial" growth time: e.g. acid treatments, such as concentrated HCl, leave a thin inner layer of elemental As. Oxidation of these surfaces may be slow initially, but would have the same growth rate once the elemental As is consumed. Lukes found the same logrithmic growth rate over a two-year period as shown in Fig. 8–2. Kuphal and Dinges[40] report similar results over an eight-year period.

The chemical composition of the air-grown oxides is somewhat controversial and there appears to be disagreement in the various publications. However, this may be due to the differing amount of exposure, H_2O in the air, and the temperature used to mount the sample for analysis. The work of Iwasaki *et al.*[41,42] shows that water vapor, at room temperature, by itself does not oxidize cleaned GaAs. This agrees with the previously discussed water rinsing of GaAs. If O_2 is added to the water vapor, then the oxidation rate is faster than with O_2 alone. Again this is in agreement with oxide growth when O_2 is bubbled through a water bath. Bertrand[32] reports that brief air exposure of etched GaAs results in chemisorbed O_2 and not a true oxide. The desorption experiments of Arthur[43] and Iwasaki *et al.*[41,42] support this conclusion because the energy for desorption, at least for As-O, is much higher than that required to evaporate As_2O_3.

Prolonged exposure to air or heating at low temperatures in air undoubtedly forms a mixed oxide. Bertrand[32] states that the oxide is a mixture of Ga_2O_3 and As_2O_3. This agrees with the analysis of Kowakzyk *et al.*,[44] who heated their samples in air in order to mount them with In. Massies and Contour[29] also mounted their samples with In and observed XPS peaks near that of Ga_2O_3 and As_2O_3. However, due to the thinness of the oxide layer (0.6 nm) and the Ga-As-O ratio, they concluded that the layer was probably composed of $Ga(OH)_3$ and H_3AsO_3. Webb and Lichtensteiger,[45]

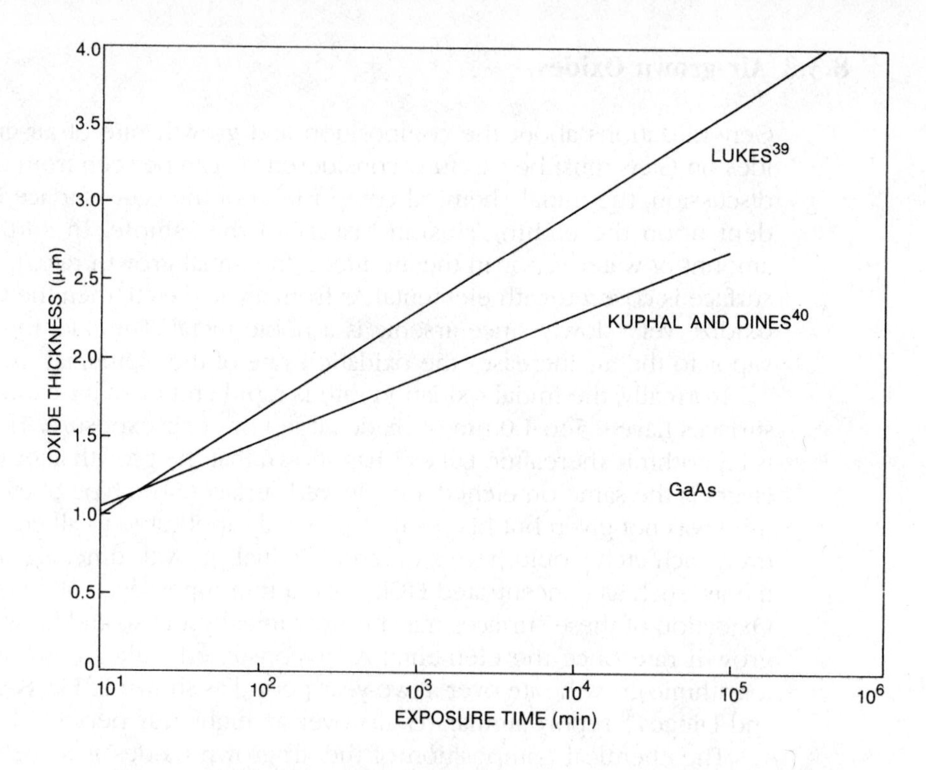

Fig. 8–2 Thickness of air-grown oxide on GaAs as a function of time. *(After Lukes*[39] *and Kuphal and Dinges*[40]*)*

and Demanet *et al.*[46] found direct evidence for $Ga(OH)_3$ in air-grown oxides. Thus, $Ga(OH)_3$ is present in the air-grown oxides.

When the air-grown oxide reaches a thickness of 2.5 to 3.0 nm, the composition becomes more complex. Demanet and coworkers[46,47] examined "as received" GaAs surfaces and report a layered composition. The outer layers contain $GaAsO_4$, As_2O_5, and $GaAsO_3$. The inner layer contains Ga_2O_3, As_2O_3, and elemental As. These results are similar to those of Vasquez *et al.*,[35] whose etched surfaces had oxides 2 to 3 nm thick. These samples were only briefly exposed to air, so there is some question as to how the oxide grew so thick. Whatever the cause, these results substantiate the observations of Demanet.[46,47]

The composition of air-grown oxides for various thicknesses is summarized in Fig. 8–3.

8.4 Thermal Oxidation

The thermal oxidation of GaAs has been investigated by many researchers.[10–12] The following sections distill this wealth of information into an

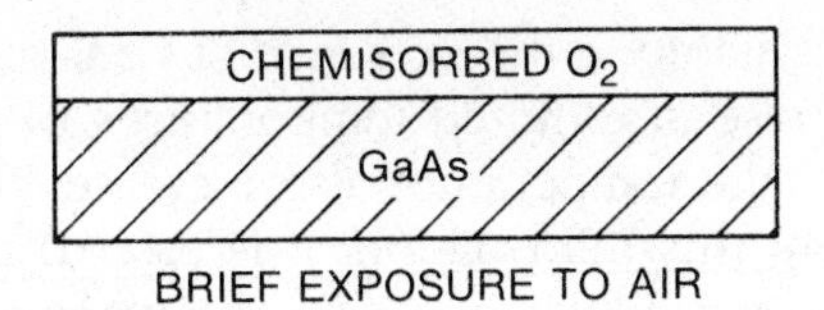

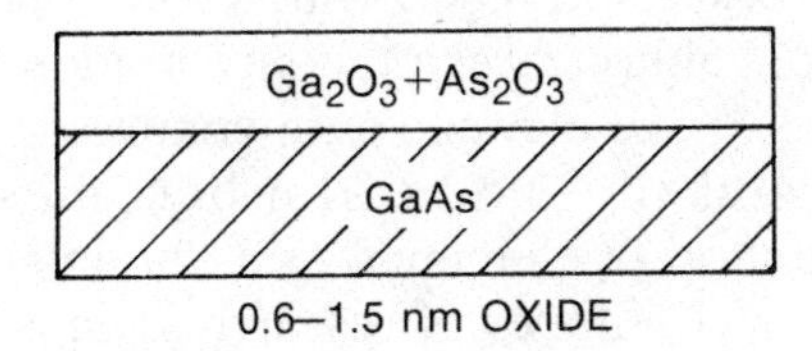

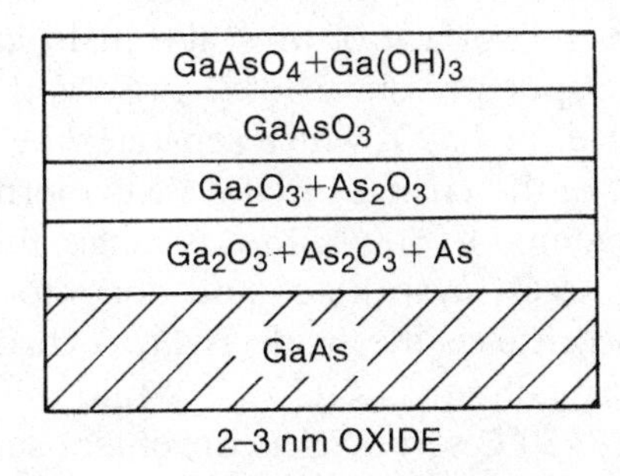

Fig. 8–3 Composition of air-grown oxides on GaAs for three thicknesses. (*After Demanet* et al.[46,47])

overview that provides a basic understanding of the composition and growth processes of GaAs thermal oxidation.

8.4.1 General Characteristics

Exposure of GaAs to an oxidizing ambient causes the growth of an initial oxide coating. In order to continue the oxidation process, either the oxidant molecules must diffuse through the oxide layer to the GaAs substrate or the Ga and As atoms must diffuse out to the oxide surface. The diffusion is enhanced by increasing the substrate temperature. For InP, it has been established that O_2 and H_2O diffuse slowly through the grown oxide and that the thermal oxidation proceeds by the outward diffusion of the substrate atoms.[19,21] For GaAs, it is not known which atoms/molecules diffuse, but this may not be important because of the thermodynamic instability of the oxidation products.

The Ga-As-O equilibrium ternary phase diagram[22] was discussed in a previous section. This diagram indicates that Ga_2O_3 is the only oxide which is stable in the presence of GaAs and that As_2O_3, As_2O_5, and $GaAsO_4$ decompose at elevated temperatures to form Ga_2O_3 + elemental As. Increasing the temperature also increases the evaporation rate of As_2O_3.

From this information, it is easy to predict the composition of thermally grown oxides on GaAs. At low temperatures, the decomposition and evaporation of As_2O_3 is slow. Thus, the oxide layer is composed of a mixture of Ga_2O_3 and As_2O_3 with a low concentration of elemental As at the interface,[48] although As_2O_3 concentration is small even at low growth temperatures.[49,50] At elevated temperatures, As_2O_3 both decomposes and evaporates and the oxide film becomes mostly Ga_2O_3 with a large concentration of elemental As at the interface. This composition has been verified by a number of research groups.[48,50–53] There is some recent evidence,[62] however, that indicates the elemental As is not localized at the interface, but this has not been corroborated. Some $GaAsO_4$ has also been found at the surface of most thermal oxides.[48–53] This occurs because of the low vapor pressure[54] of $GaAsO_4$ and the decomposition is prevented by the layer of Ga_2O_3 which separates the $GaAsO_4$ from the GaAs substrate.

Most of the changes in the GaAs thermal oxides occur in the temperature range 400 to 500°C. In this range, the growth rate increases rapidly[55] and the As_2O_3 evaporates and decomposes rather quickly. The loss of As_2O_3 also causes the oxide layer to change from an amorphous mixture to polycrystalline Ga_2O_3.[52,56,57] While there is no clear demarcation temperature, 450°C serves as a convenient separation between high- and low-temperature type thermal oxides of GaAs.

8.4.2 Kinetic Factors

The thermodynamics of the GaAs-oxide system has been shown to drive the oxide layer towards a composition of Ga_2O_3 + elemental As independently of the kinetics of growth. Even so, attempts have been made to alter the growth kinetics in order to grow "better" thermal oxides. These are discussed below.

8.4.1.1. Growth in As_2O_3

As_2O_3 has been found to thermally evaporate from oxide films grown in O_2. In an attempt to prevent the resulting As_2O_3 deficiency, several research groups[58–60] have thermally grown oxides in an atmosphere containing As_2O_3. Chemical analysis of films grown in this manner indicate that there is a significant increase in the concentration of As_2O_3 and $GaAsO_4$, particularly near the oxide surface but the distribution of As_2O_3 is nonuniform. However, there is still a high concentration of elemental As at the interface. The As is no longer isolated at the interface but penetrates somewhat into the oxide bulk. The interfacial elemental As is the result of the decomposition of As_2O_3 via the reaction $2GaAs + As_2O_3 \rightarrow Ga_2O_3 + 4As$. Since the reaction easily occurs, the use of As_2O_3 cannot prevent the interfacial elemental As.

8.4.2.2. High-Pressure Oxidation

Thermal oxidation of GaAs at 475 and 375°C in 500 atm of combinations of O_2, H_2O, As_2O_3, and As_2O_5 have been investigated.[61] It was found that high pressure, in and of itself, does not significantly change the chemical composition or compositional profile of the thermal oxides. For example, when As_2O_3 or As_2O_5 was added to the 500-atm oxidation ambient, the compositional profile was very similar to the films grown in 1 atm of As_2O_3 or O_2. In all cases the oxide layer was composed of Ga_2O_3, As_2O_3, and elemental As with $GaAsO_4$ at the surface.

8.5 Anodic Oxidation

Compared with thermal oxidation, anodization is very complex and has many more variables. Even so, the anodic oxides grown by various researchers on GaAs have many similar characteristics and thus some generalizations are possible. Previous reviews by Wilmsen[11–12] and Croydon and Parker[10] provide many of the details of the anodization process and compare the many reports of GaAs. The goal of this section is to summarize these results more concisely, update the material, and provide the reader with a general knowledge of the anodization process and the composition of the resulting oxide films.

8.5.1 General Characteristics

Anodization is an oxidation process which is driven by field-aided diffusion of the oxidant and/or substrate molecules. The electric field in the oxide during anodization is very high, being in the range of 5×10^6 V/cm. This high field significantly lowers the energy barrier between sites in the oxides and thus a rapid oxidation can take place at room temperature. The equilibrium thermodynamics governing the anodic oxides is the same as that for thermal oxides. However, the thermal energy required to drive the oxide to equilibrium is lacking. Therefore, the chemical composition of anodic oxides is somewhat different from that of thermal oxides.

Two ingredients are required for anodic oxidation: an ambient containing ionized oxidant molecules and a high field across the oxide layer. The high field can be obtained by applying a positive voltage to the GaAs substrate with respect to the oxidizing ambient. The ionized ambient can be obtained by immersing the GaAs in either a liquid solution containing OH^- or a gaseous oxygen plasma, which contains ionized O_2 or N_2O species. Oxidation in a gaseous ambient is called *plasma oxidation* or *plasma anodization*. The chemical compositions of anodic oxides grown in gaseous and liquid ambients are surprisingly similar to each other.

In both types of anodization ambients, an electric current must flow. In the oxide, this current is carried by both the ionized molecules (mass transport) and electrons/holes. It is the mass transport, of course, that results in the growth of an oxide film. Until recently, it was not known if the Ga and As outdiffuse or if the O_2^-/OH^- ions indiffuse. The early experiments of Coleman *et al.*[63] and others[64–65] seem to indicate that the Ga and As atoms diffused outward, which resulted in the oxidation occurring at the oxide-electrolyte interface. However, more recent marker experiments by Fischer and Canaday[66] show that 82 percent of the oxidation results from the inward diffusion of OH^- and 18 percent by the outward diffusion of Ga and As. In either case, the rate of oxide growth is proportional to the current and can be easily calculated if the ratio between mass-transport contribution and the electron/hole contribution is known. This is called the *anodization efficiency*. Efficiencies as high as 95 percent and as low as 34 percent have been reported.[67]

8.5.2 Initial Growth

When an oxide begins to grow on a surface, it either forms uniform monolayers, as does SiO_2 on Si, or it forms via isolated islands which eventually coalesce. Anodic oxides on GaAs form by the island stage mechanism. This was first shown by Szpak[68] and later, in more detail, by Makky *et al.*[69] These studies have shown that the nucleation of the islands continues until the islands completely coalesce into a continuous layer. During this stage, the islands grow outward as well as laterally, reaching a saturation thickness of about 20 nm. Thus, the minimum thickness of the continuous layer is $\sim$20 nm. This minimum thickness may change with electrolyte and current density, but it has not been investigated.

8.5.3 Chemical Composition

Over a wide range of growth conditions, the anodic oxides of GaAs are uniform layers of Ga_2O_3 and As_2O_3, except at the interface and surface.[11,48,70] Unlike thermal oxides, there appears to be little or no elemental As at the interface and generally no $GaAsO_4$ on the surface. There is also a high concentration of As_2O_3, which often appears in equal concentration to Ga_2O_3. The presence of the As_2O_3, particularly in films grown in aqueous electrolytes, is somewhat surprising because of the high solubility of As_2O_3. However, as stated previously, the oxide appears to grow by the inward diffusion of OH^- and thus the As_2O_3 is formed under the oxide film and away from the electrolyte.

Dissolution of the As_2O_3 and Ga_2O_3 can be a problem. In aqueous solutions, Cl^- and NO_3^- contaminates cause etching of the oxide, particularly in basic solutions.[67] Using a nonaqueous electrolyte, such as methanol,[67] removes the sensitivity to the ionic contaminates and good anodic films can be grown over a wider pH range. The addition of glycol to aqueous electrolytes greatly inproves,[1] but does not eliminate, the above etching problems. Glycol-based electrolytes do yield good oxide films and are most often used.

The composition of the anodic oxide-GaAs interface has been somewhat controversial. First we consider the presence of elemental As. The investigations which used ion milling, all report a low concentration of elemental As at the interface.[70–72] Those investigations not using ion milling, i.e. Raman scattering, ellipsometry or channeled RBS, found no elemental As.[73–75] Since these latter techniques are very sensitive to elemental As, it appears that the observed As was caused by the ion milling and not the anodization.

There is also controversy concerning a layer of Ga_2O_3 at the interface. XPS profiles of oxides in the thickness range 2 to 100 nm[71,72,76] thick indicate the presence of such a Ga_2O_3 layer and also that this layer appears to form during the island stage. High-resolution TEM also shows a narrow band at the interface that appears to be Ga-rich.[77] These results are in contrast to ellipsometry data that does not detect a change in composition at the interface.[74] However, since the difference in the index of refraction between As_2O_3 and Ga_2O_3 is small, ellipsometry is not sensitive to a thin layer of Ga_2O_3, and thus it is more probable that such a layer does exist.

8.5.4 Composition Changes due to Annealing

Anodic oxides on GaAs are grown far from equilibrium. Annealing provides the thermal energy necessary to drive the oxide layers toward equilibrium. Heating in a nonoxidizing atmosphere up to 300°C causes water to evaporate but does not measurably change the chemical composition of the layers.[78,79] Above 300°C, arsenic begins to evaporate slowly. At 400°C, this evaporation becomes rapid and, as with the thermally grown oxide, the remaining Ga_2O_3 begins to crystallize as the As_2O_3 is removed.[80] Elemental As also begins to collect at the interface in measurable quantities. Increasing the temperature accelerates these processes.

References

1. H. Hasegawa, K. Forward, and H. Hartnagel, "New anodic native oxide of GaAs with improved dielectric and interface properties," *Appl. Phys. Lett.,* 26:567–569 (1975).
2. C. R. Zeisse, L. J. Messick, and D. L. Lile, "Electrical properties of anodic and pyrolytic dielectrics on gallium arsenide," *J. Vac. Sci. Technol.,* 14:957–960 (1977).
3. G. P. Schwartz, G. J. Gaultieri, J. E. Griffiths, C. D. Thrumond, and B. Schwatrz, "Oxide-substrate and oxide-oxide reactions in thermally annealed films on GaSb, GaAs and GaP," *J. Electrochem. Soc.,* 127:2488–2599 (1980).
4. J. M. Woodall, P. Oelhafen, T. N. Jackson, J. L. Freeouf, and G. D. Pettit, "Photoelectrochemical passivation of GaAs surfaces," *J. Vac. Sci. Technol. B,* 1:795–798 (1983).
5. S. D. Offsey, J. M. Woodall, A. C. Warren, P. D. Kirchner, T. I. Chappell, and G. D. Pettit, "Unpinned (100) GaAs surfaces in air using photochemistry," *Appl. Phys. Lett.,* 48:475–477 (1986); N. A. Ives, G. W. Stupian, and M. S. Leung, "Unpinning of the Fermi level on GaAs by flowing water," *Appl. Phys. Lett.,* 50:256–258 (1987).
6. J. F. Dewald, "The kinetics and mechanisms of the formation of anodic films on single crystal InSb," *J. Electrochem. Soc.,* 104:244–251 (1957).
7. A. J. Rosenberg and M. C. Levine, "The oxidation of intermetallic compounds: I. high temperature oxidation of InSb," *J. Phys. Chem.,* 64:1135–1142 (1960).
8. H. T. Minden, "Thermal oxidation of GaAs," *J. Electrochem Soc.,* 109:733 (1962).
9. B. Schwartz, "GaAs surface chemistry: a review," *CRC Crit. Rev. in Solid State Sci.,* 5:609–624 (1975).
10. W. F. Croydon and E. H.C. Parker, *Dielectric Films on Gallium Arsenide,* New York: Gordon and Breach Science Publishers (1981).
11. C. W. Wilmsen, *Physics and Chemistry of III-V Compound Semiconductor Interfaces,* New York: Plenum (1985).
12. C. W. Wilmsen, "Chemical composition and formation of thermal and anodic/oxide III-V compound semiconductor interfaces," *J. Vac. Sci. Technol.,* 19:279–289 (1981).
13. C. W. Wilmsen, "Oxide layers on III-V compound semiconductors," *Thin Solid Films,* 39:105–117 (1976).

14. S. T. Pantelides, *The Physics of SiO_2 and Its Interfaces,* New York: Pergamon Press (1978).
15. A. S. Grove, *Physics and Technology of Semiconductor Devices,* New York: John Wiley & Sons (1967).
16. F. J. Grunthaner, P. J. Grunthaner, R. P. Vasquez, B. F. Lewis, and J. Maserjian, "Local atomic and electronic structure of oxide/GaAs and SiO_2/Si interfaces using high resolution XPS," *J. Vac. Sci. Technol.,* 16:1443–1453 (1979).
17. G. P. Schwartz, W. A. Sunder, and J. E. Griffiths, "The In-P-O phase diagram: construction and applications," *J. Electrochem. Soc.,* 129:1361–1367 (1982).
18. S. M. Goodnick, T. Hwang, and C. W. Wilmsen, "New model for slow current drift in InP metal-insulator-semiconductor field effect transistors," *Appl. Phys. Lett.,* 44:453–455 (1984).
19. C. W. Wilmsen, K. M. Geib, R. Gann, J. Costello, G. Hryckowian, and R. J. Zeto, "High pressure thermal oxide/InP interface," *J. Vac. Sci. Technol.,* B3:1103–1106 (1985).
20. A. Yamamoto and C. Vemura, "Anodic oxide film as gate insulator for InP MOSFET," *Electron. Lett.,* 18:63–64 (1982).
21. R. G. Gann, K. M. Geib, C. W. Wilmsen, J. Costello, G. Hrychowain, and R. J. Zeto, "High pressure thermal oxidation of InP in steam," *J. Appl. Phys.,* 63:506–509 (1988).
22. C. D. Thurmond, G. P. Schwartz, G. W. Kammlott, and B. Schwartz, "GaAs oxidation and the Ga-As-O equilibrium phase diagram," *J. Electrochem. Soc.,* 127:1366–1371 (1980).
23. G. P. Schwartz, "Analysis of native oxide films and oxide-substrate reactions on III-V semiconductors using thermochemical phase diagrams," *Thin Solid Films,* 103:3–16 (1983).
24. D. J. Stirland and B. W. Straughan, "A review of etching and defect characterization of gallium arsenide substrate material," *Thin Solid Films,* 31:139–170 (1976).
25. C. S. Fuller and H. W. Allison, "A polishing etchant for III-V semiconductors," *J. Electrochem. Soc.,* 109:880–8 (1962).
26. M. Pourbaix, *Atlas of Elecrochemical Equilibria in Aqueous Solutions,* New York: Pergamon Press (1966): (*a*) pp. 516–523 (*b*) pp. 428–435.
27. D. E. Aspnes and A. A. Studna, "Chemical etching and cleaning procedures for Si, Ge, and some III-V compound semiconductors,"*Appl. Phys. Lett.,* 39:316–318 (1981).
28. J. Massies and J. P. Contour, "X-ray photoelectron spectroscopy study of the effects of ultrapure water on GaAs," *Appl. Phys. Lett.,* 46:1150–1152 (1985).

29. J. Massies and J. P. Contour, "Substrate chemical etching prior to molecular-beam epitaxy: an X-ray photoelectron spectroscopy study of GaAs (001) surface etches by the $H_2SO_4 - H_2O_2 - H_2O$ solution," *J. Appl. Phys.*, 58:806–810 (1985).
30. B. Schwartz, "Preliminary results on the oxidation of GaAs and GaP during chemical etching," *J. Electrochem Soc.*, 118:657–658 (1971).
31. Heinz Gerischer, "Electrolytic decomposition and photodecomposition of compound semiconductors in contact with electrolytes," *J. Vac. Sci. Technol.*, 15:1422–1428 (1978).
32. P.A. Bertrand, "XPS study of chemically etched GaAs and InP," *J. Vac. Sci. Technol.*, 18:28–33 (1981).
33. K.W. Frese and S.R. Morrison, "Passivation and interface state study on n-GaAs," *Appl. Surf. Sci.*, 8:266–277 (1981).
34. R.P. Vasquez, B.F. Lewis, and F.J. Grunthaner, "Cleaning chemistry of GaAs (100) and InSb (100) substrates for molecular beam epitaxy," *J. Vac. Sci. Technol. B*, 1:791–794 (1983).
35. R.P. Vasquez, B.F. Lewis, and F.J. Grunthaner, "X-ray photoelectron spectroscopic study of the oxide removal mechanism of GaAs (100) MBE substrates in *in situ* heating," *Appl. Phys. Lett.*, 42:293–295 (1983).
36. D.E. Aspnes and H.J. Stocker, "Peroxide etch chemistry on (100) $In_{.53}Ga_{.47}As$," *J. Vac. Sci. Technol.*, 21:413–416 (1982).
37. H.A. Schnell, *Z. Metallkd*, 48:158 (1957).
38. T. Oda and T. Sugano, "Studies on chemically etched silicon, gallium, arsenide, and gallium phosphide surfaces by Auger spectroscopy," *Jpn. J. Appl. Phys.*, 15:1317–1327 (1976).
39. Lukes, F., "Oxidation of Si and GaAs in air at room temperature," *Surf. Sci.*, 30:91–100 (1972).
40. E. Kuphal and H.W. Dinges, "Composition and refractive index of $Ga_{1-x}Al_xAs$ determined by ellipsometry," *J. Appl. Phys.*, 50:4196–4200 (1979).
41. H. Iwasaki, Y. Mizokawa, R. Nishitani, and S. Nakamura, "X-ray photoemission study of the interaction of oxygen and air with cleaved GaAs (110) surfaces," *Jpn. J. Appl. Phys.*, 17:315–320 (1978).
42. H. Iwasaki, Y. Mizokawa, R. Mishitani, and S. Nakamura, "Effects of water vapor and oxygen excitation on oxidation of GaAs, GaP and InSb surfaces studied by X-ray photoemission spectroscopy," *Jpn. J. Appl. Phys.*, 18:1525–1529 (1979).
43. J.R. Arthur, "Adsorption and desorption of O_2 on GaAs (111) surfaces," *J. Appl. Phys.*, 38:4023–4028 (1967).

44. S. P. Kowalczyk, J. R. Waldrop, and R. W. Grant, "Interfacial chemical reactivity of metal contacts with thin native oxides of GaAs," *J. Vac. Sci. Technol.,* 19:611–616 (1981).
45. C. Webb and M. Lichtensteiger, "Formation of alternative surface oxide phases on GaAs by adsorption of O_2 or H_2O: a UPS, XPS and SIMS study," *J. Vac. Sci. Technol.,* 21:659–662 (1983).
46. C. M. Demanet, E. D. Rawsthorne, and C. M. Stander, "GaAs oxides formed at room temperature in air: comparison of X-ray photoelectron spectroscopic and ion microprobe mass analysis," *Surf. Interface Anal.,* 7:159–162 (1985).
47. C. M. Demanet and M. A. Marais, "A multilayer model for GaAs oxides formed at room temperature in air as deduced from an XPS analysis," *Surf. Interface Anal.* 7:13–16 (1985).
48. G. P. Schwartz, G. J. Gualtieri, G. W. Kammolott, and B. Schwartz, "An X-ray photoelectron spectroscopy study of native oxides on GaAs," *J. Electrochem. Soc.,* 126:1737–1749 (1979).
49. K. Löschke, G. Kühn, H. J. Bilz, and G. Leonhardt, "Oxidfilme auf $A^{III}B^{IV}$ Halbleitern," *Thin Solid Films,* 48:229–236 (1978).
50. C. W. Wilmsen, R. W. Kee, and K. M. Geib, "Initial oxidation and oxide semiconductor interface formation on GaAs," *J. Vac. Sci. Technol,.* 16:1434–1438 (1979).
51. K. Watanabe, M. Hashiba, Y. Hirohota, M. Nishino, and T. Yamashima, "Oxide layers on GaAs prepared by thermal, anodic and plasma oxidation: in depth profiles and annealing effects," *Thin Solid Films,* 56:63–73 (1979).
52. D. N. Butcher, and B. J. Sealey, "The thermal oxidation of GaAs," *J. Phys. D.: Appl. Phys.,* 11:1451–1456 (1978).
53. Y. Mizokawa, H. Iwasaki, R. Nishitani, and S. Nakamura, "In depth profiles of oxide films on GaAs studied by XPS," *Jpn. J. Appl. Phys.,* 17:327–333 (1978).
54. E. C. Shafer and R. Roy, "Studies of silica structure phases: I. $GaPO_4$, $GaAsO_4$ and $GaSbO_4$," *J. Am. Cerm. Soc.,* 39:330–336 (1956).
55. S. P. Muraka, "Thermal oxidation of GaAs," *Appl. Phys. Lett.,* 26:180–181 (1975).
56. F. Koshiga and T. Sugano, "Thermal oxidation of GaAs," *Jpn. J. Appl. Phys.,* 16(1977 suppl.): 465–469 (1977).
57. C. J. Bull and B. J. Sealy, "Studies of thermally oxidized GaAs by TEM," *Phil. Mag. A.,* 37:489–500 (1978).
58. H. Takagi, G. Kano, and I. Teramoto, "Thermal oxidation of GaAs in arsenic trioxide vapor," *J. Electrochem. Soc.,* 125:579–581 (1978).

59. H. Takagi, G. Kano, and I. Teramoto, "Thermal oxide gate GaAs MOSFET's," *IEEE Trans. Electron Dev.*, ED-25:551–552 (1978).

60. G. P. Schwartz, J. E. Griffiths, D. DiStefano, G. J. Gaultieri, and B. Schwartz, "Arsenic incorporation in native oxides of GaAs grown thermally under arsenic trioxide vapor," *Appl. Phys. Lett.*, 34:742–744 (1979).

61. C. W. Wilmsen and R. Zeto, unpublished.

62. N. T. McDevitt and J. S. Solomon, "Thermal oxide layers on GaAs studied by Raman and Auger spectroscopy," *J. Electrochem Soc.*, 133:1913–1917 (1986).

63. D. J. Coleman, D. W. Shaw, and R. D. Dobrott, "On the mechanism of GaAs anodization," *J. Electrochem. Soc.*, 124:239–244 (1977).

64. F. Koshiga and T. Sugano, "The anodic oxidation of GaAs in an oxygen plasma generated by DC electrical discharge," *Thin Solid Films*, 56:39–49 (1979).

65. K. M. Geib and C. W. Wilmsen, "Anodic oxide/GaAs and InP interface formation," *J. Vac. Sci. Technol.*, 17:952–957 (1980).

66. C. W. Fischer and J. D. Canaday, "Cation and anion transport numbers in anodic GaAs oxides," *J. Electrochem. Soc.*, 130:1740–1744 (1983).

67. C. W. Fischer and J. D. Canaday, "Galvanostatic anodization of GaAs in methanolic KOH," *J. Electrochem. Soc.*, 129:1016–1021 (1982).

68. S. Szpak, "Electro-oxidation of gallium arsenide: I. Initial phase of film formation in tartaric acid-water-propylene glycol electrolyte," *J. Electrochem. Soc.*, 124:107–112 (1977).

69. W. H. Makky, F. Cabrera, K. M. Geib, and C. W. Wilmsen, "Initial stages of anodic oxidation of GaAs," *J. Vac. Sci. Technol.*, 21:417–421 (1982).

70. Y. Mizokawa, H. Iwasaki, R. Nishitani, and S. Nakamura, "Quantitative chemical depth profiles of anodic oxides on GaAs obtained by X-ray photoelectron spectroscopy." *J. Electrochem. Soc.*, 127:454–461 (1980).

71. P. A. Breeze, H. L. Hartnagel, and P. M. A. Sherwood, "An investigation of anodically grown films on GaAs using X-ray photoelectron spectroscopy," *J. Electrochem. Soc.*, 127:454–461 (1980).

72. K. M. Geib and C. W. Wilmsen, "Anodic oxide/GaAs and InP interface formation," *J. Vac. Sci. Technol.*, 17:952–957 (1980).

73. G. P. Schwartz, G. J. Gualtieri, J. E. Griffiths, C. D. Thurmond, and B. Schwartz, "Oxide-substrate and oxide-oxide chemical reactions in thermally annealed films on GaSb, GaAs and GaP," *J. Electrochem. Soc.*, 127:2488–2499 (1980).

74. D. E. Aspnes, G. P. Schwartz, G. J. Gualtieri, A. A. Studna, and B. Schwartz, "Optical properties of GaAs and its electrochemically grown

anodic oxide from 1.5 to 6.0 eV," *J. Electrochem. Soc.,* 128:590–597 (1981).

75. C.J. Maggiore and R.S. Wagner, "Ion beam characterization of the GaAs-GaAs oxide interface for plasma and anodic oxides," *J. Vac. Sci. Technol.,* 19:463–466 (1981).
76. C.W. Wilmsen and R.W. Kee, "Analysis of the oxide/semiconductor interface using Auger and ESCA as applied to InP and GaAs," *J. Vac. Sci. Technol.,* 15:1513–1517 (1978).
77. O.L. Krivanek and S.L. Fortner, "HREM imaging and microanalysis of a III-V semiconductor/oxide interface," *Ultramicroscopy,* 14:121–126 (1984).
78. S.M. Spitzer, B. Schwartz, and G.D. Weigle, "Preparation and stabilization of anodic oxides on GaAs," *J. Electrochem. Soc.,* 122:397–402 (1975).
79. T. Ishii and B. Jeppsson, "Influence of temperature on anodically grown native oxides on GaAs," *J. Electrochem. Soc.,* 124:1784–1794 (1977).
80. B.L. Weiss and H.L. Hartnagel, "Crystallization dynamics of native anodic oxides on GaAs for device applications," *Thin Solid Films,* 56:143–152 (1979).

Chapter 9

Ohmic Contacts to GaAs

G. N. Maracas

9.1 Introduction

Semiconductor devices require metallic contacts with which to conduct current to external circuitry. The two types of metal-semiconductor contacts are ohmic and rectifying, the latter being characterized by a barrier to electron or hole transport which restricts current in one direction. An ohmic contact, in contrast, is a metal-semiconductor junction which ideally has no barrier to current and can thus, in principle, provide an essentially infinite number of carriers to a device structure. At present, ohmic contacts to compound semiconductor materials do not have zero resistance, but have approximately linear current-voltage characteristics and low contact resistance. For most GaAs devices the details of the current-voltage characteristics are not too important as long as the voltage drop across the contact region is much smaller than the drop across the active regions of the operating device.

A number of very good review articles on ohmic contacts are available. Rideout[1] reviews the theory and technology for ohmic contacts to III–V semiconductors and Braslau[2] the Au/Ge/Ni contact. Schroder and Meier[3] review the Schottky model of contacts in light of solar cell applications.

Requirements for an "ideal" ohmic contact are that it should have negligible contact resistance, be thermally and structurally stable under device processing and operation conditions, have highly reproducible electrical properties, and have the capability of being fabricated with arbitrarily small dimensions. While the industry standard Au/Ge/Ni alloyed ohmic contacts do not exhibit all of these desirable properties, successful large-scale device integration has been achievable with their use. As GaAs integration increases toward and beyond VLSI levels, the present alloyed contacts may no longer be appropriate because of limitations in one or more of the requirements mentioned above.

This chapter reviews the theory, properties, technology, and applications of today's ohmic contacts as well as methods of measuring contact resistivity. Novel contact schemes and requirements for contacts on future devices are also mentioned.

9.2 Metal-Semiconductor Contact Theory

When a metal and a semiconductor are brought into contact, a potential barrier is formed because of the dissimilar work functions of the two materials. The Schottky model is often used to describe the band structure near the heterointerface.[1,4–6] Figure 9–1A depicts the band structure at the junction of a metal to a moderately doped n-type semiconductor. A barrier is formed to electron flow in one direction causing the contact to be rectifying. Electrons traverse the barrier mainly by thermionic emission, although components due to quantum-mechanical tunneling and recombination through traps are also present. A space-charge region depleted of carriers is produced with spatial extent

$$w = \sqrt{2\epsilon_s(\Phi_B - V - kT/q)/qN_D}\,, \qquad (9\text{–}1)$$

where N_D is the donor concentration in the semiconductor, and V is the applied voltage. The term Φ_B denotes the barrier height or (in the Schottky model) the difference between the metal work function and the semiconductor electron affinity, or

$$\Phi_{Bn} = \Phi_M - \chi \qquad (9\text{–}2A)$$

for n-type material, and

$$\Phi_{Bp} = E_g/q + \chi - \Phi_m \qquad (9\text{–}2B)$$

for p-type material. The simple Schottky model predicts that the barrier height is independent of doping in the semiconductor and only depends on the metal work function. The electron affinity, χ, is the potential that an electron must achieve in order to free itself from the semiconductor. This is a bulk property of the semiconductor and is independent of the doping concentration. A problem with the Schottky model is that it does not describe the properties of the interface. It is not surprising, since it only considers bulk material properties. Interface states play an important role in the measured barrier heights and account for their presence.

The ideal barrier height is lowered by an amount $\Delta\Phi_B$, which is depicted by the curved, dotted line in Fig. 9–1. This barrier height lowering arises from an image force lowering of the interface potential. This force is present under an applied electric field which induces a positive charge

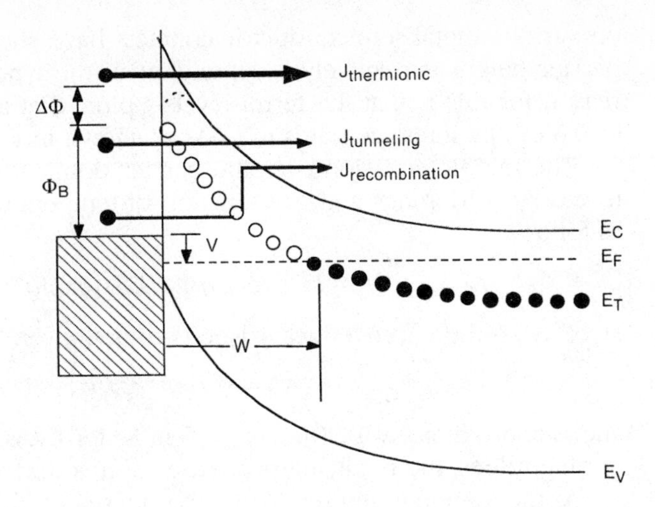

(A) For a moderately doped n-type semiconductor there is a barrier to electron flow in one direction.

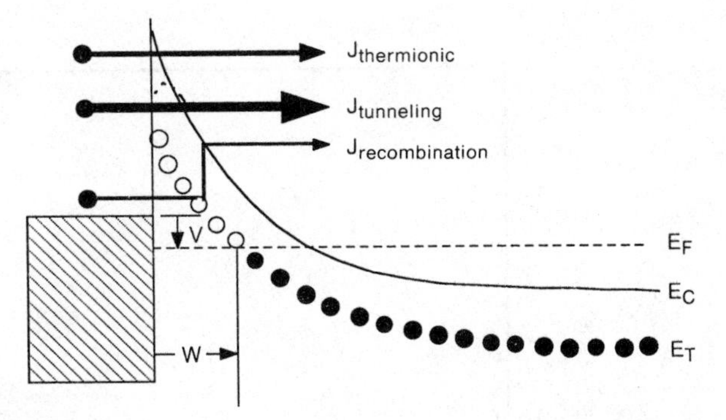

(B) As the doping in the semiconductor is increased, the barrier becomes thinner causing a larger component of tunneling current. The tunneling current is independent of barrier height and produces an "ohmic" contact.

Fig. 9–1. The band diagram of an "ideal" metal-n-type semiconductor junction according to the Schottky model.

on the metal to which electrons are attracted, thus lowering the effective barrier to electrons. The lowering of the Schottky barrier due to the image force is given by

$$\Delta\Phi = \sqrt{qE/4\pi\epsilon_s}\,, \tag{9–3}$$

where E is the applied electric field. Reductions in the barrier height are on the order of 0.1 eV for electric fields of only 10^5 V/cm. At higher fields, the deviation from the simple Schottky barrier height is much larger.

The simple Schottky model does not predict the behavior of the barrier height as a function of metal coverage. Experimental measurements on various metal-semiconductor contacts have shown that the Schottky barrier height is relatively independent of the type of metal used.[7] The reason for this is that the fermi level is pinned at approximately midgap (~0.6 eV) by interface states in GaAs as shown in Fig. 9–2.

Thermionic emission over the barrier dominates the current for low to moderate dopings and is a function of temperature and barrier height as follows:

$$J_{\text{thermionic}} = A^*T^2 \exp(-q\Phi_B/kT)[\exp(qV/kT) - 1]. \tag{9–4}$$

Here, A^* is the effective Richardson constant, given by

$$A^* = 4\pi q m^* k^2/h^2, \tag{9–5}$$

and is approximately $120(m^*/m_0)$ A/cm^{2}K^2 for GaAs. For low to moderate doping values, the rectifying properties of this device are quite good.

As the doping in the semiconductor increases, the space charge region width decreases according to the above expression. The potential barrier thus decreases in width, increasing the electron tunneling probability

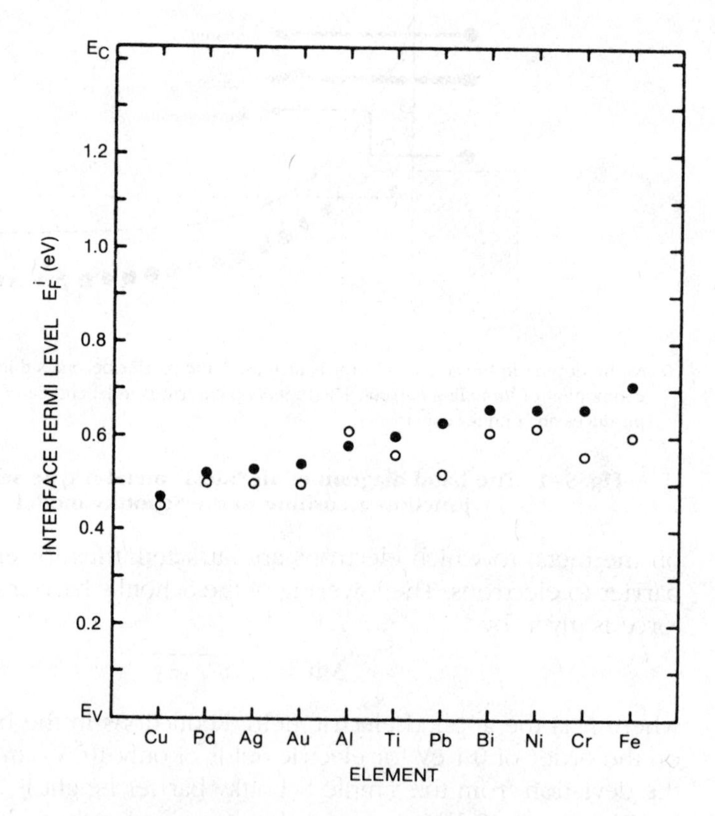

Fig. 9–2. In GaAs the interface states pin the Fermi level at approximately the middle of the bandgap. ***(After Waldrop[7])***

from the metal to the semiconductor, and thus increasing the current. This tunneling current has the form

$$J_{\text{tunnel}} \cong \exp(-q\Phi_B/E_\infty)\,, \tag{9–6}$$

where

$$E_\infty = \frac{q\hbar}{2}\sqrt{\frac{N_D}{\epsilon m^*}}. \tag{9–7}$$

Note that the tunneling current is not dependent on temperature. These two processes account for most of the current across the junction, but there can be an additional component due to interface states, which provide intermediate levels for tunneling across the barrier. This recombination current is directly proportional to the density of traps near the interface. For a contact to be "ohmic," it should ideally have no barrier to current. This situation is not practically attainable, but can be approached by using a tunneling contact. Formation of an ohmic contact is accomplished by degenerately doping the semiconductor so that its Fermi level lies in the conduction band. Figure 9–1B shows the case for a degenerately doped metal-semiconductor contact under a small forward bias, *V*. It is this tunneling current that dominates in "good" ohmic contacts. There are two basic methods for fabricating ohmic contacts to compound semiconductors. The easiest (and therefore the most widely used) method is to alloy a metal to the semiconductor by high-temperature processing. Usually, the metal is alloyed into a highly doped implanted or epitaxial region. The other method is to grow a layer of degenerately doped, small-bandgap material between the metal and the GaAs. A small-bandgap material has a lower effective barrier for a given metal thus forming a contact with lower contact resistance.

9.2.1 Metal Ohmic Contacts to GaAs

Alloyed ohmic contacts are formed by depositing layers of various metals, by sputtering or electron-beam or thermal evaporation, and then annealing in an inert atmosphere of a few hundred degrees Celsius for a few minutes. The most widely used metallization scheme for n-type GaAs is the Au/Ge/Ni system, so the discussion will concentrate on this system.

The earliest contacts to GaAs were fabricated with alloyed Sn. Wetting of the surface by the metal was difficult, which resulted in a nonuniform contact surface. It also was not stable with time and tended to migrate under applied electric fields. The AuGe/Ni ohmic metallization scheme was introduced in 1967 and has been used for many years.[8] In its initial configuration, a AuGe eutectic was first deposited and followed by a a Ni overlayer. Alloying was performed in an inert atmosphere at temperatures of approximately 450°C. This contact scheme produced a more uniform surface morphology, adhered better to the GaAs, and also produced better electrical properties. Other metal systems were attempted such as AgGe/In, AuGe/AgAu, and In/AuGe. These are investigated in depth by Christou in Ref. 9. Microstructure of the AuGeNi alloyed contact is discussed in Sec. 9.3 below.

9.2.2 Heterojunction Contacts

A second method is to epitaxially grow a semiconductor with a small bandgap on top of the GaAs layer requiring a contact. This method has been shown to produce low contact resistivities, but is more difficult to fabricate. An example is to insert a thin layer of degenerately doped Ge between the metal and the GaAs as shown in Fig. 9–3. The electrons see a reduced barrier because of the small bandgap of the Ge. Germanium is used in this case because its lattice constant is approximately equal to that of GaAs. Other small-bandgap semiconductors, such as InAs, have been used but because of the large lattice mismatch, the Fermi level becomes pinned at approximately the same location as in the GaAs case.[10] Figures 9–4A through 9–4C show the band structure at a metal/InAs/GaAs heterojunction. InAs has a small bandgap, and surface states pin the Fermi level near the conduction band, producing a small electron barrier. It does not make a low barrier to the GaAs, however. If compositionally graded

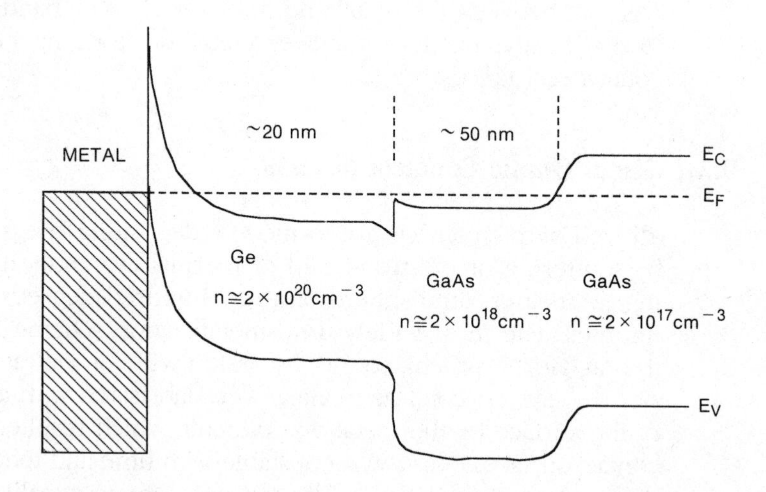

Fig. 9–3. A low-resistance contact can be fabricated by inserting a degenerately doped semiconductor, such as Ge, with a small bandgap between the metal and the GaAs. *(After Stall* et al.[30]*)*

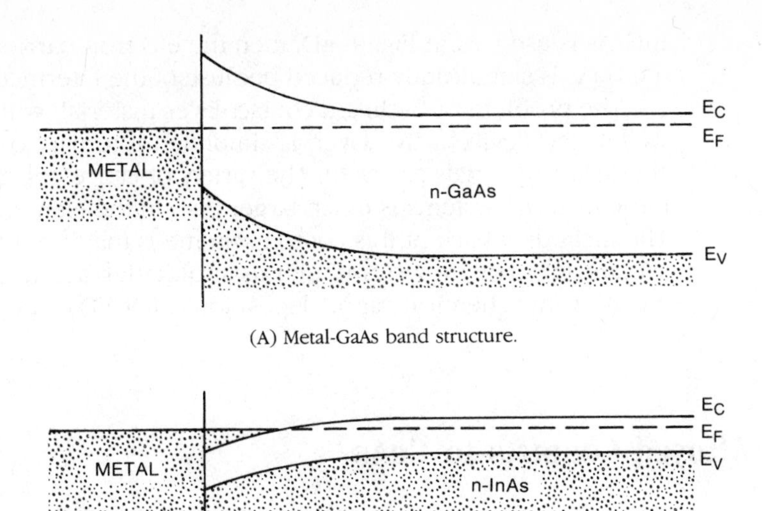

(A) Metal-GaAs band structure.

(B) The Fermi level is pinned near the conduction band at the interface between a metal and InAs.

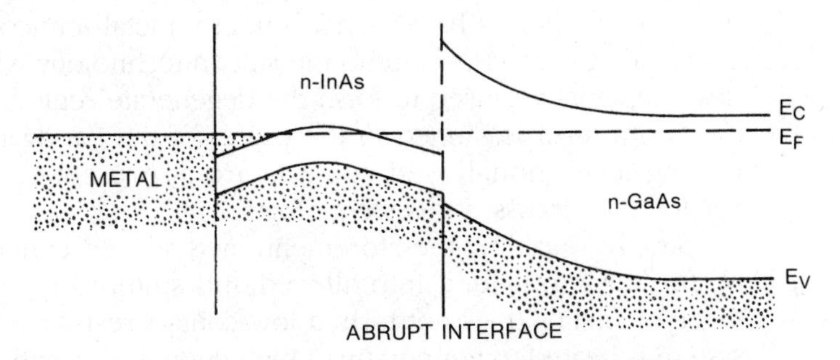

(C) Band structure of metal-InAs-GaAs contact system.

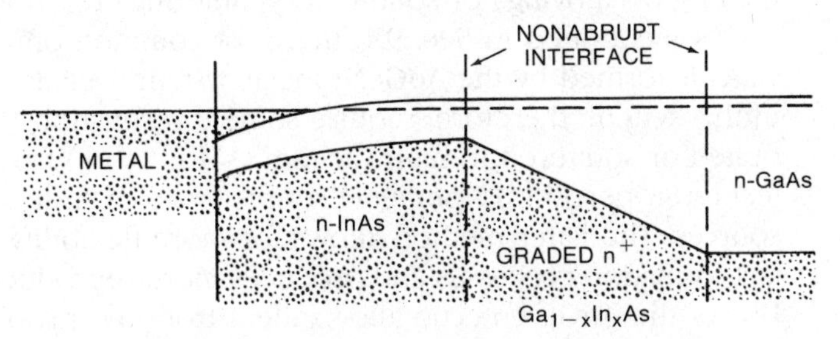

(D) An intermediate graded InGaAs layer further reduces the metal-to-GaAs barrier height.

Fig. 9–4. Band structure at heterojunction.
(After Woodall and Freeouf[10])

InGaAs is used, as in Fig. 9–4D, then the electron barrier from the InAs to the GaAs is significantly reduced because of the intermediate InGaAs layer.

The problem of finding a contact layer material, which lattice matches well to the device active layer, is simplified if ternary or quaternary semiconductor materials are used. The variety of lattice constants achievable in these material systems is much larger than in the elemental or binary case. The main drawback of this contact scheme is that it must be performed in a material growth system such as molecular-beam epitaxy (MBE) or metalorganic chemical vapor deposition (MOCVD).

9.3 Alloyed Contacts to GaAs

Metal-semiconductor ohmic contacts fall into two categories: nonalloyed and annealed (or alloyed). Nonalloyed ohmic contacts are formed by depositing metal directly onto very highly or degenerately doped semiconductor surfaces and have planar, smooth, metal-semiconductor interfaces. This type of contact is common in silicon technology where shallow, high-dose implants are used to form the degenerate regions. In GaAs, because the Fermi level is pinned in the center of the bandgap regardless of the doping level, nonalloyed contacts are at present difficult to fabricate. Section 9.4 discusses this technology.

Strictly speaking, the aforementioned alloyed contact category should be further subdivided into alloyed and sintered types. Deposition of a metal or metals does not form a low contact resistance structure until the system is heated to temperatures high enough that either liquid-phase (i.e. melting or alloying) or solid-phase (sintering) reactions occur.

As mentioned in Sec. 9.1, the most common ohmic contact to n-type GaAs is formed by the AuGeNi metal system. Details of contact formation in this system, therefore, require some investigation. The metals are evaporated or sputtered in high vacuum ($<10^{-5}$ torr) from a single AuGe eutectic (88 percent/12 percent) source or from individual Au, Ge, and Ni sources. The latter method allows for more flexibility in researching contact formation processes and produces more reproducible ohmic contacts. Use of the AuGe eutectic alloy suffers from the problem that Au and Ge evaporate (either in electron-beam or thermal evaporators) at different rates so that the eutectic composition is not maintained after extensive uses of the source material. This problem can be alleviated somewhat if the entire evaporator charge is used for each run. The practice of adding small quantities of Ge into the Ge-depleted charge is sometimes encountered but not recommended.

Contacts are then annealed in either N_2, H_2, forming gas (90 percent N_2 and 10 percent H_2) or in a vacuum at temperatures between 450 and

500°C. The alloying time varies from a few seconds to several minutes depending on the particular thicknesses and the order of which the metals are deposited. Different groups use different contact formation schemes. What is desired is a process that produces low contact resistance ohmics that are insensitive to variations in process parameters such as alloy time and temperature. Since metal thicknesses are fairly reproducible, this is not a limitation.

9.3.1 AuGeNi Contact Microstructure

Several studies on the microstructure and formation of AuGeNi alloyed contacts to GaAs have been performed.[11–15] These studies used X-ray diffraction, transmission electron microscopy (TEM), scanning transmission electron microscopy (STEM), Auger electron spectroscopy (AES), and X-ray photoelectron spectroscopy (XPS) to study the structural and chemical nature of the metal/semiconductor interface.

Figure 9–5 shows a TEM cross section of a cleaved, alloyed AuGeNi contact after Kuan.[13] In this work, TEM cross sections were obtained for contacts in various stages of annealing. The behavior observed is that a variety of grains or phases are formed from metal diffusing into the GaAs and Ga diffusing into the metal. The intermetallic phase, as identified by energy dispersive X-ray (EXD) analysis, are Au(GaAs), NiGe(Ga, As), NiGe(Ga, As) and NiAs(Ga, Ge) and are approximately 100 nm in size. The interface is far from uniform, as in typical nonalloyed contacts. Therefore the sheet resistance of the material immediately under the contact cannot be known to any degree of accuracy. This has implications in determining the specific contact resistance discussed in Sec. 9.5, because the semiconductor bulk resistance, R_s, must sometimes be known for its determination.

During high-temperature processing, Ga outdiffuses and forms AuGa alloys. Ge and Ni (which is used as a wetting agent) diffuse into the GaAs and form $NiAs_2$ and NiGeAs compounds. The latter is the preferred phase because Ni_2GeAs is a degenerately doped semiconductor and therefore forms a tunneling contact with the highly doped GaAs. Defects are also formed as the Au diffuses into the GaAs.[14] Ge, being amphoteric, can either occupy Ga or As sites. Ge on a Ga site is preferred because it dopes the GaAs n-type. Thus the number of outdiffused Ga atoms, leaving behind Ga vacancies, should be approximately equal to the number of indiffused Ge atoms to obtain the optimum contact resistance. This behavior has been suggested by Kulkarny,[16] who observed that the optimum contact resistance was obtained when the ratio of Ge to Ga was approximately unity. For typical alloy temperatures (~460°C), the NiGeAs compounds are located at the interface covered with AuGa as shown in Fig. 9–6A. At high alloy temperatures (~600°C), the low resistance (and low melting point)

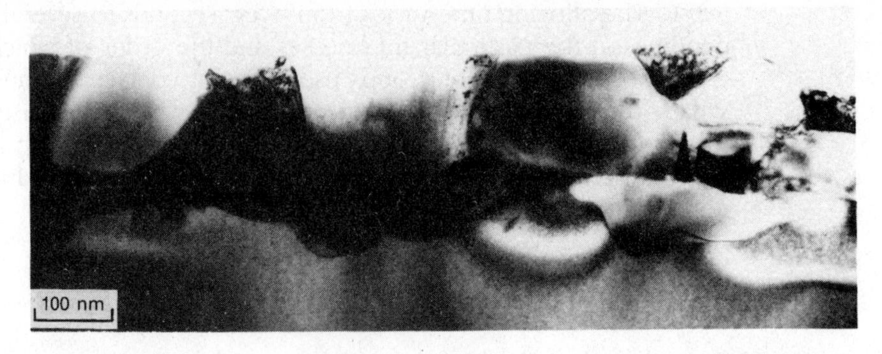

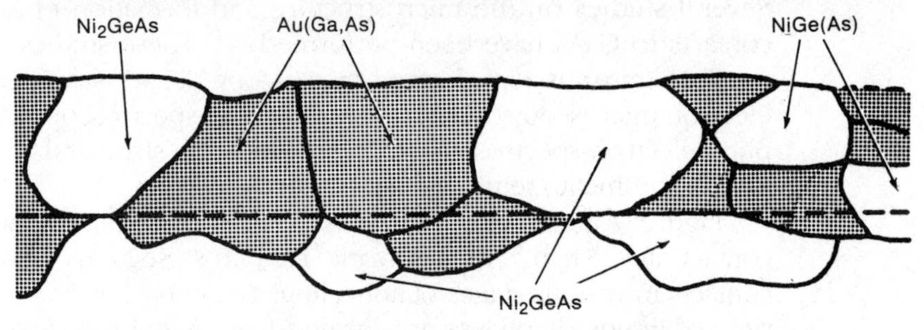

Fig. 9–5. A TEM cross section of an alloyed AuGeNi contact showing the various intermetallic phases. The chemical identification of the phases is shown in the schematic below the image. *(After Kuan* et al.[13]*)*

AuGa phase penetrates into the GaAs, reducing the area covered by the low-resistance NiGeAs phase and thus reducing the overall contact resistance.[15] Thus it is desirable to prevent the AuGa phase from either forming or penetrating the interface. Figure 9–6B shows the redistribution of the various elements at various annealing times measured by AES.

Control of the phase formation in AuGeNi contacts is thus essential in developing a technology capable of producing reproducible, low-resistance ohmic contacts. The grain sizes are fairly large (~1000 nm) and it may be difficult to adapt this technology to very small devices required for VLSI.

9.3.2 Hypoeutectic Alloys

Intermetallic grain formation control is necessary to make reproducible ohmic contacts that are insensitive to process parameters. In particular, because a variation in the annealing time required for minimum contact resistance is only on the order of a few seconds, any variation could cause a large difference in the resulting contact resistance. Johnson[17] used a AuGe

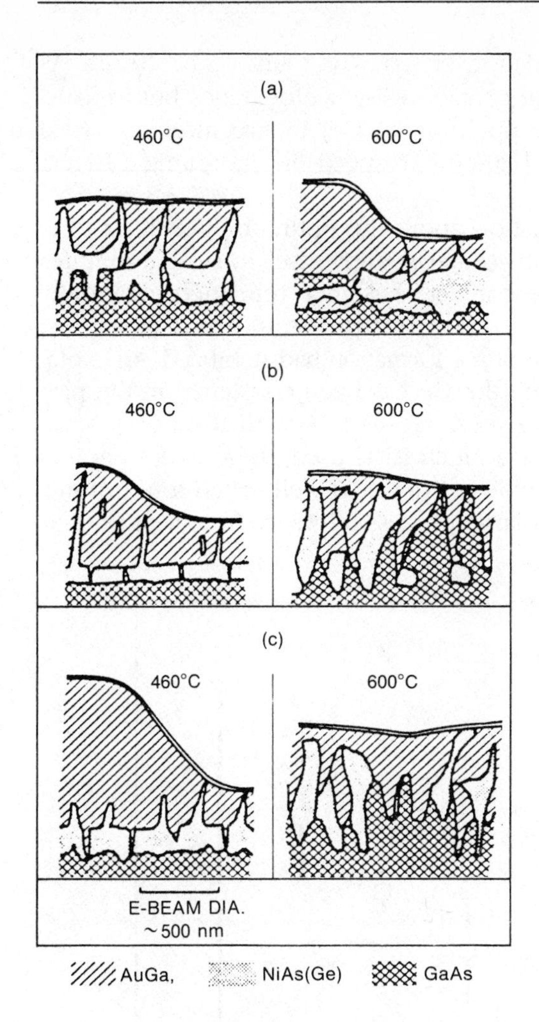

(A) Microstructure of three different alloyed AuGeNi/GaAs ohmic contacts inferred from X-ray diffraction and Auger electron spectroscopy. The top contact has AuGe eutectic as the first layer. The two bottom contacts have the Ni layer deposited first. Using Ni as the bottom layer produces a smoother interface than if either Ge or AuGe is deposited first.

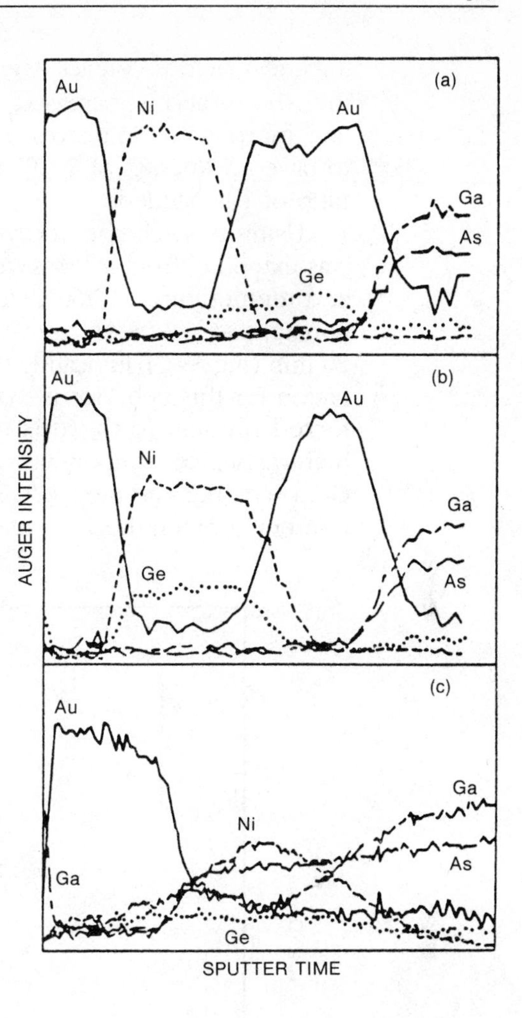

(B) Auger profiles for the top contact in (A) which shows the interdiffusion of the various elements. Top: before annealing; Middle: after 440°C for 2 min; Bottom: after 460°C and 2 min.

Fig. 9–6. Contact microstructures and Auger profiles.
(After Murakami et al.[15])

composition of 97.3 percent Au and 2.7 percent Ge, which is considerably different than the conventional eutectic. Ni is deposited first to wet the surface covered with Ge of an equal thickness. Placing the Ni at the semiconductor/metal interface reduces As outdiffusion, thus causing the Ge to occupy Ga (donor) sites instead of As (acceptor) sites. The minimum Ge thickness before contact degradation occurred is found to be 10 nm. As the Ni:Ge ratio is increased, the contact resistance degrades but the surface morphology improves. The optimum Ni/Ge/Au structure was found to have thicknesses of 2, 20, and 200 nm, respectively, indicating a Ni:Ge ratio of 1 is needed.

Using this scheme, the variation about the optimum annealing time was extended from a few seconds (eutectic alloy case) to several minutes at a temperature of 460°C. The contact resistance remained essentially constant at ~0.08 Ω · mm (R_s = 1000 Ω/sq, Z = 100 μm) from 5 to 20 min (Fig. 9–7), indicating that phase formation had stabilized. An explanation for this behavior is that all the Ge has been consumed in the preferred phases, so there is no excess Ge to mix with the Au to form high-resistance components. If too much Ni is used, there is not enough Ge to compensate the Ni. This type of contact is well suited to a production environment because of its large process tolerance in anneal time.

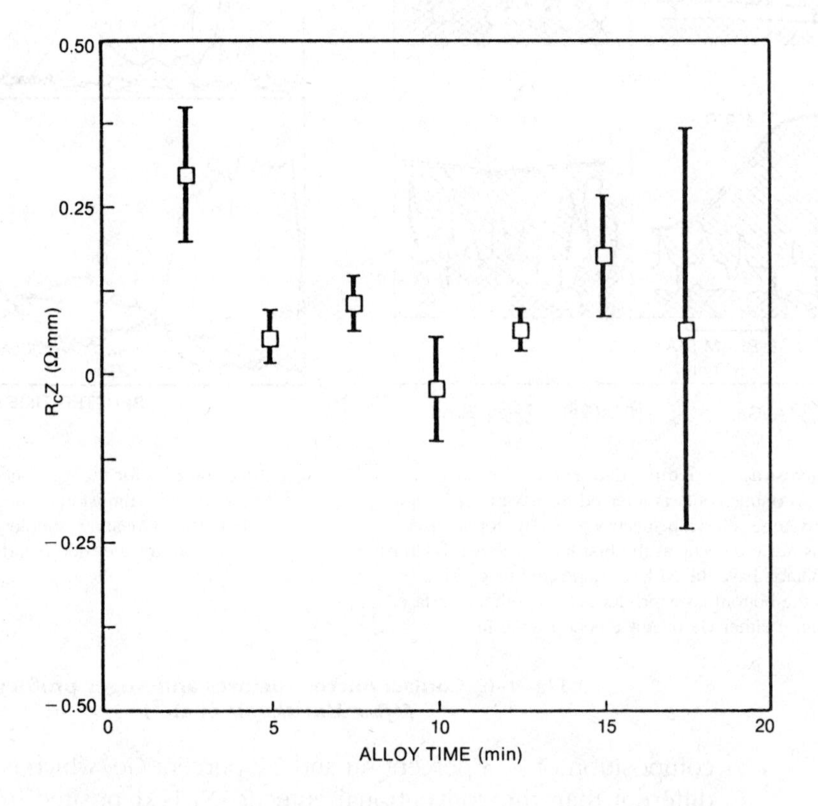

Fig. 9–7. Alloy behavior of a AuGeNi hypoeutectic ohmic contact. The contact resistance is essentially constant versus alloy temperature from 5 to 20 min. In the test structure used Z = 100 μm. *(After Johnson[17])*

9.3.3 Other Alloyed Contacts

The search for reproducible, reliable, low contact resistance ohmic contacts to GaAs has spurred considerable research. Various metals have been used and different annealing schemes attempted. Table 9–1 reviews the work on different alloyed contacts to GaAs and their annealing techniques. This is by no means an exhaustive listing of the contacts attempted to GaAs.

Thermal, flash-lamp, laser, and electron-beam annealing have been used successfully to obtain ohmic contacts. Heating of only the metal/semiconductor interface can be achieved by subbandgap laser illumination from the backside of the wafer. Since the GaAs is transparent to $\lambda = 1.06$-μm radiation, all of the energy is deposited at the contact.[18] Laser energies between 0.3 and 0.5 J/cm^2 produce reasonable contact resistances. Surface morphology is degraded at ~1.4 J/cm^2 and metal evaporation occurs at higher energy densities. Electron-beam annealing of AuGe/Pt, however, was found to produce excellent surface morphology, but poor reliability.[19]

In order to circumvent the problems encountered with low melting point metals such as AuGe, other material systems have been investigated. One such system is AlNiGe, described by Zuleeg.[20] Al has a higher melting point than Au requiring an anneal at 500°C. These contacts exhibited high reliability because of the high activation energy required for phase migration. Ni is used to prevent AlGe from forming and balling up to degrade the surface morphology. It is also speculated that the Al-based contact should decrease the sensitivity of the contact to radiation effects because of its low atomic mass.

Table 9–1. Review of Alloyed Ohmic Contacts to GaAs

Contact	Annealing Method	Annealing Conditions	Surface Morphology	ρ_c ($\Omega \cdot$ cm^2) at N_D (cm^{-3})	Comments	Reference
AuGe/Ni	Thermal	460°C 5–20 min	Good	8×10^{-6} (2×10^{17})	Hypoeutectic alloy insensitive to anneal time	17
AuGe	Nd/YAG $\lambda = 1.06$ μm	$0.3 < E < 0.5$ J/cm^2 in air	Rough	2×10^{-5} (min) (3×10^{17})	$E_{max} = 1.4$ J/cm^2	18
AlGe	Thermal	500°C	Good	1.4×10^{-6} (mid 10^{17})	High reliability Radiation hard	20
AuGe/Pt	e-beam in vacuum	20 keV, 100 ns $E = 0.4$ J/cm^2	Excellent	$<10^{-6}$ (5×10^{17})	Degrades at low temperatures (250°C)	19
Mo/Ge	Thermal	745°C	Good	$\sim 10^{-6}$ (mid-10^{17})	Requires high anneal temperature. Stable.	21
In/Pt	Thermal	300–500°C	Slightly rough	2×10^{-6} (1.5×10^{17})	In is soft so contact deforms during probing	23
AuGe/Nb	Thermal	390–420°C	Good (due to Nb)	2×10^{-6} (2×10^{18})	Linear I-V to 4.2 K Superconducting at $T < 8$ K	24

Refractory metals are desirable for use in ohmic contacts for several reasons. Contact formation occurs at higher temperatures and by sintering, as opposed to alloying. This type of contact does not penetrate into the GaAs as deeply, so is better suited to thin epitaxial structures. These also tend to be more thermally stable and can survive subsequent high-temperature processing steps after ohmic contact formation. The refractory metals can also be easily etched by dry processing techniques used in GaAs fabrication technology.

Mo/Ge contacts were successfully fabricated by Tiwari.[21] Very little (40 nm) intermixing of the MoGe-GaAs was observed at the relatively high anneal temperatures used (~745°C). The Ge diffused into the GaAs to form the usual n^+ region and $\rho_c \cong 10^{-6}$ was obtained without the thermal stability problem in the AuGeNi case.

A graded $In_xGa_{1-x}As$ heterojunction is formed when indium is alloyed on GaAs.[22,31] This small-bandgap material reduces the barrier to electron flow across the junction. Combining a refractory metal such as Pt with In should therefore produce a stable, low-resistance contact. In/Pt ohmic contacts have low contact resistance, but tend to be soft and distort upon wire bonding.[23] The surface morphology is usually rough because In has a tendency to form islands of varying composition, so that a uniform contact is not obtained even at temperatures as low as 350°C.

When devices are operated at low temperatures, it is desirable to not have the *I-V* characteristics degraded by the ohmic contacts becoming nonlinear. One attempt at forming a contact that will be linear down to cryogenic temperatures was done by Gurvitch *et al.*[24] A niobium overlayer on AuGe was used to form a superconducting, shallow contact. Contact resistivities were good and the *I-V*s were linear to ~4.2 K. Annealing was performed at 390 to 420°C for 1 to 5 s in a forming gas atmosphere.

9.4 Nonalloyed Contacts

Ultimately, it is desirable to use contacts that are not alloyed. Such contacts are shallow, interfacially uniform, and have excellent surface morphology. There are three methods for forming nonalloyed contacts: doping to extremely high levels, sintering, and using low barrier heterojunctions. Table 9–2 reviews some nonalloyed contact schemes.

The first method is used most extensively in silicon technology, where shallow implants in excess of 10^{20} cm^{-3} are obtainable and thus form good tunneling junctions. Silicon is usually used to dope GaAs and its solubility limit is only 2×10^{20} cm^{-3}. This level is difficult to obtain in practice but has been reported by Kirchner[25] with MBE using a Si filament source. Im-

Table 9–2. Review of Nonalloyed Ohmic Contacts to GaAs

Contact	Formed by	N_D (cm^{-3})	ρ_c ($\Omega \cdot cm^2$)	Comments	Reference
AuGe/Ni	Ion implantation through $SiSiO_xN_y$ cap	9×10^{18}	9×10^{-5} 6×10^{-6}	Dose 7×10^{13} cm^{-2} RTA ≅ 1100°C, 5 s after 300°C anneal	26
Ge-Pd(Sb)	Solid-phase epitaxy	$\sim 1 \times 10^{18}$	2×10^{-6}	Formation at 450°C Surface smooth and featureless	28
Pd/Ge/Au Ge/Pd/Au	VPE	2.2×10^{18} 2.2×10^{18}	2×10^{-5} 5×10^{-7}	Sintered contacts. Possibly formed by solid-phase epitaxy. $T \cong 450$°C, 30 s Metal deposition order is important.	29
Ag	MBE	2×10^{20}	1.310^{-6}	Used Si filament source to obtain doping at solubility limit of Si in GaAs	25
Ge/GaAs	MBE	2×10^{18}	$\sim 10^{-6}$	Heterojunction ohmic contact $\phi_B < 0.3$ eV	30
δ doped	MBE	2×10^{18}	2.5×10^{-6}	Controllable tunneling barrier	27
InAs/GaAs	MBE	1×10^{18}	$\sim 10^{-6}$ $\sim 10^{-7}$	Abrupt heterojunction Graded heterojunction	31

plantation through a Si/SiO_xN_y cap[26] has achieved a doping level of 9×10^{18} cm^{-3}. Both of these methods have achieved contact resistance in the low to mid 10^{-6} $\Omega \cdot cm^{-2}$ range. An effective doping higher than the solubility limit has also been reported[27] by delta doping the GaAs with Si. Thin Si layers with doping $N_D^{2D} = 5 \times 10^{13}$ cm^{-2} produce an effective doping of $N_D^{3D} = 3.5 \times 10^{20}$ cm^{-3}, which causes efficient tunneling. This scheme appears to produce low contact resistances.

Sintered contacts are formed by solid-phase migration and formation of metal constituents and thus tend to be more shallow than alloyed contacts. Ge/Pd(Sb) contacts were successfullly formed at 450°C.[28] The Sb doped Pd layer is deposited first. PdGe is formed and the remaining Ge is transported through the PdGe to grow epitaxially onto the GaAs. The Sb effectively dopes the Ge forming a highly doped Ge/GaAs heterojunction.

Pd/Ge/Au is also believed to be formed in the solid phase.[29] In this scheme, depositing Ge first produces a much lower contact resistance than if Pd is in contact with the GaAs (5×10^{-7} versus 2×10^{-5} $\Omega \cdot cm^2$). No dissociation of the GaAs is observed because the Pd and Ge are entirely consumed into PdGe compounds which form at ~250 °C.

Finally, the use of small-bandgap materials on GaAs produces a low effective barrier to electron flow. The Ge/GaAs heterojunction is the most popular because of the good lattice match between the two semiconductors and the high doping level achievable in the Ge.[30] InAs/GaAs heterojunctions have also been used.[31] The InAs (or InGaAs) layer must be very thin (on the order of a critical thickness) because the large lattice mismatch between the two materials will cause dislocations to form. These structures must be grown by either MBE or MOCVD.

9.5 Measurement of Contact Resistance

The development of a GaAs device technology relies heavily on the capability to fabricate low-resistance metal contacts. The end product of the metal deposition and anneal needs to be monitored to determine whether the process has resulted in ohmic contacts that fall within the tolerance levels for the particular fabrication process. Quick, simple methods of measuring the contact resistance or contact resistivity are therefore necessary. A variety of ohmic contact characterization schemes and test structures have been developed to this end. The measurement of contact resistance is performed by any one of three popular measurement techniques: the transmission line method, the Kelvin resistor, and the resistance of a field-effect transistor (FET). These measurement techniques and their associated problems and limitations are described in this section.

9.5.1 Definitions of Contact Resistance and Contact Resistivity

The distinction between contact resistance and contact resistivity was made by Berger[32] in 1972. Since then, a certain amount of confusion has appeared concerning the difference in the various measured quantities of contact resistance, specific contact resistance, and contact resistivity. Cohen[33] has attempted to clear up some of the confusion, thus his terminology will be used in this section.

Figure 9–8 shows a typical ohmic contact structure having area A. The current flows from one contact to the other in straight lines so the equipotential surfaces are all parallel to the metal electrodes. The metal layer has a resistance, R_m, which depends on the particular metal or combination of metals deposited. The interfacial layer between the metal and the semiconductor gives a contribution, R_i, to the total measured resistance. This resistance is affected by the condition of the surface during the metal de-

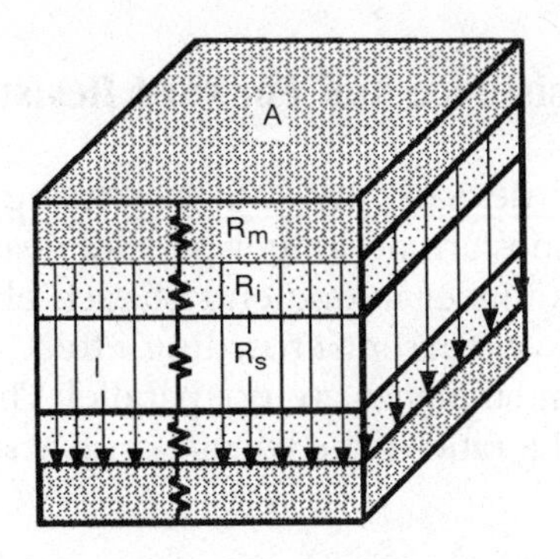

Fig. 9–8. An idealized ohmic contact resistor structure showing the various components of the contact resistance. Contributions are from the metal, the interfacial layer, and the semiconductor bulk. The current lines are parallel so the equipotential planes are parallel to the contacts. This structure has no current crowding.

position process and thus includes the effects of contamination and oxidation which cause the interfacial properties to deviate from the ideal behavior described in Sec. 9.2. The semiconductor bulk also contributes a resistance, R_s, which depends on the doping and defect concentrations in the bulk material.

When measurements of current and voltage are performed on this structure, the *contact resistance*, R_c, is measured. Therefore, strictly speaking, the contact resistance includes the contributions from the metal, interfacial region, and semiconductor bulk. In practice, the metal resistance is assumed to be negligible compared with the interfacial and bulk resistance, so the definition of contact resistance is therefore

$$R_c = R_m + R_i + R_s \cong R_i + R_s \quad (\Omega/\text{sq}). \tag{9–8}$$

The specific contact resistance, ρ_c, is obtained from the contact resistance by multiplying by the contact area:

$$\rho_c = R_c A \quad (\Omega \cdot \text{cm}^2). \tag{9–9}$$

This expression is only valid in the case when the *I-V* characteristic is linear and the contact has a uniform interfacial layer. The term ρ_c includes the interfacial layer resistance as well as any contribution from the spreading resistance or current crowding. The spreading resistance is a function of device geometry where the interfacial layer resistance is constant as the area of the contact varies. Therefore ρ_c has a nonlinear dependence on contact area while the specific interface resistance does not, being an inherent property of the material. A measurement of the latter property can be made by the use of Kelvin resistors, discussed later in this section.

9.5.2 End Resistance and Forward Resistance

In planar devices, the contact scheme is represented by Fig. 9–9A. The current lines are no longer parallel near the metal electrodes. The current density is higher at one end of each electrode than the other causing the well-known current crowding effect. Near the contacts, therefore, the equipotential planes are not parallel. The *front contact resistance*, R_f, is defined as the ratio of the voltage drop across the conducting layer to the total

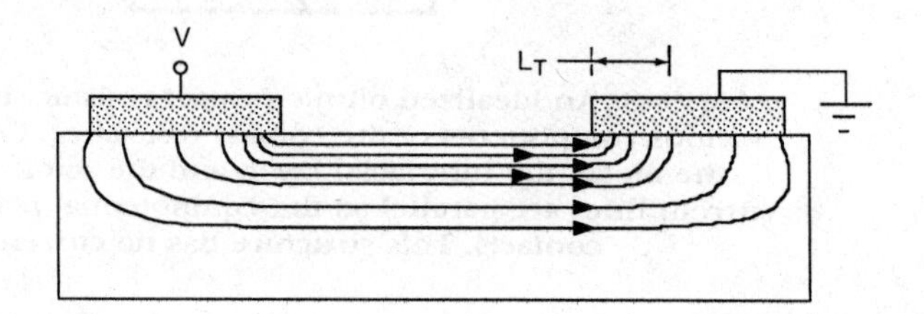

(A) Planar resistor structure showing the current crowding effect.

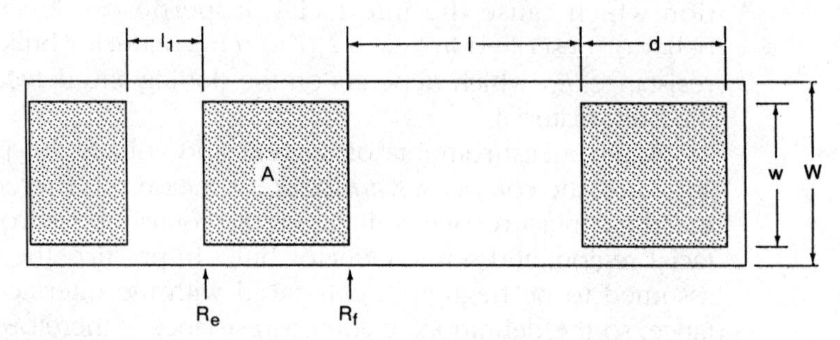

(B) The forward resistance is measured at the region between the two contacts while the end resistance is measured at the far end of the contacts.

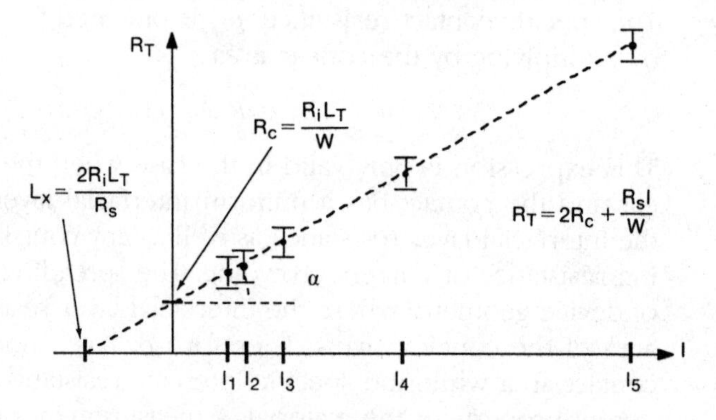

(C) The TLM method for measuring the contact resistance.

Fig. 9–9. Measuring forward, end, and contact resistances.

current I at the contact edge with the highest current density.[34] Alternately, the *end contact resistance,* R_e, is then the ratio at the edge having the lowest current density. These two quantities can be measured by using the transmission line, or transfer length, method (TLM) of contact resistance.

9.5.3 Transmission Line or Transfer Length Methods

The basic device structure shown in Fig. 9–9B is used in the two-terminal contact resistance measurement.[32] Two planar ohmic contacts are spaced a distance l apart on a bar of semiconductor. For this analysis, assume that the width of the bar is exactly equal to the contact width ($W = w$). The ohmic contact can be thought of as consisting of three layers: a top metal layer, a conducting region which is alloyed or diffused immediately under the metal (the interfacial layer), and a bottom bulk conducting layer. In all of the transmission line analyses, the contact resistance is assumed to be solely due to the interfacial layer.

If a bias, V, is placed between the contacts, then a current, I, will flow. Phenomenologically, we can say that since the contacts are planar the current will, on the average, flow vertically through the interface region for some distance, L_T, before flowing horizontally to the other contact. L_T is called the *transfer length* and is measured by the TLM method. This is sketched in Fig. 9–9C. Following this model, the total resistance measured between the two contacts spaced a distance, l, apart can be written as

$$R_T = \frac{R_s l}{W} + \frac{2R_i L_T}{W}, \qquad (9\text{–}10)$$

where R_s is the sheet resistance (Ω/square) of the conducting layer and R_i is the sheet resistance of the interfacial material immediately under the contact. For $l = 0$ the total resistance would be

$$R_T = \frac{2R_i L_T}{W}. \qquad (9\text{–}11)$$

The intercept on the x axis is obtained when $R_T = 0$. This defines the length

$$L_x = -\frac{2R_i L_T}{R_s}. \qquad (9\text{–}12)$$

If $R_i = R_s$ then the length is

$$L_x = -2L_T. \qquad (9\text{–}13)$$

This assumption is not strictly valid because the interfacial layer is usually degenerately doped by alloying or implantation so $R_i << R_s$. The final assumption is that the contact length, d, is much greater than the transfer length, L_T. For this case, a distribution line analysis of the current crowding effect has shown that nearly all of the current enters the contact within L_T of its front edge (see Fig. 9–9A).[35,36] This means that the effective contact area is approximately WL_T instead of Wd. Using this relation, Eq. 9–11 can be written in terms of the effective contact resistance,

$$R_c = R_i L_T / W \quad (\Omega/\text{sq}), \tag{9–14}$$

and from this the specific contact resistance can be found (using $R_i = R_s$) as

$$\rho_c = R_c A = R_c W L_T = R_s L_T^2 \quad (\Omega \cdot \text{cm}^2). \tag{9–15}$$

Therefore, the specific contact resistance can be approximated by measuring the total resistance of various sets of identical contacts. A plot of R_T versus l should yield a straight line whose y intercept equals $2R_c$ and whose x intercept equals $-2L_T$ (see Fig. 9–9C).[37] Notice that if only two different contact spacings are used, then the y intercept, and hence R_c, can be calculated from

$$2R_c = \frac{R_2 l_1 - R_1 l_2}{l_1 - l_2}, \tag{9–16}$$

which is the standard formula for extracting R_c from the three-terminal transmission line method.[35] Experimentally, Eq. 9–16 is very inaccurate since $R_c << R_2 l_1$ or $R_1 l_2$.

The fact that $R_i << R_s$ requires some consideration in this analysis. In practice, R_i is usually more than an order of magnitude smaller than R_s. Typical values[37] are $R_i \cong 20\ \Omega/\text{sq}$ and $R_s \cong 400\ \Omega/\text{sq}$ for AuGeNi alloyed contacts on an epitaxial layer doped at $5 \times 10^{16}\ \text{cm}^{-3}$. Corrections to the TLM analysis require a measurement of the end contact resistance, R_e. Proctor[34] gives expressions for the end and front contact resistances:

$$R_f = \frac{\sqrt{R_i \rho_c}}{w} \coth[(R_i/\rho_c)^{1/2} d], \tag{9–17}$$

$$R_e = \frac{\sqrt{R_i \rho_c}}{w} \operatorname{csch}[(R_i/\rho_c)^{1/2} d], \tag{9–18}$$

$$R_f = R_e \cosh[(R_i/\rho_c)^{1/2} d]. \tag{9–19}$$

The transfer length is implicit in these expressions as

$$L_T = \sqrt{\rho_c / R_i}\,. \tag{9–20}$$

A useful relation between the contact resistance and the end resistance is

$$R_c / R_e = \cosh(d/L_T), \tag{9–21}$$

from which L_T can be calculated and ρ_c obtained from Eq. 9–20.

The three-terminal TLM method for determining the various contact resistances is prone to certain experimental errors which limit its usefulness in certain applications.

Because most of the current under a contact is crowded into a region of approximately a transfer length, the voltage drop under the metal contact differs from one side to the other. Its exact variation cannot be determined in practical experiments. The TLM analysis assumes that the voltage drop under the contact is laterally uniform; thus, without correction for end resistance, the calculation of the various resistances is in error.

An additional assumption in deriving Eq. 9–16 is that all of the metal-semiconductor contacts are identical. In practice, variations among the various contacts can be significant because of surface contamination and material impurities. This consideration makes the determination of R_c with the TLM analysis suspect.

A major problem is that the method relies on measurements of large resistance values differing by only small amounts. For instance, the resistance between adjacent planar resistors having different lengths may be 100, 105, and 110 Ω. The 5-percent variation between successive resistors also has the inherent uncertainty in voltage and current measurements which produce the experimental error bars in Fig. 9–9C. These error bars produce large errors in the determination of the two intercepts and hence the contact resistivity and transfer length.

The exact spacing and dimensions of the photolithographically defined contacts are not known. Processing variations introduce uncertainties in resistor length (Δl) and contact length (Δd) (metal and implanted or diffused regions) which affect the contact resistivity measurements as

$$\frac{\Delta\rho_c}{\rho_c} = \frac{\rho_s}{\rho_c}\Delta l \quad \text{and} \quad \frac{\Delta\rho_c}{\rho_c} \cong \alpha\Delta d\,, \tag{9–22}$$

where α is shown in Fig. 9–9C.

Planar resistor structures are inherently susceptible to current spreading and crowding effects. This effect is nonlinear in distance and is greater for contacts where $W \neq a$. This prevents the use of the TLM for contacts with small geometries.

In summary, the three-terminal TLM method must be used carefully to minimize the inherent errors in the measurement. Correction factors must be applied on large-dimension, specially designed samples. It is not applicable to very small geometries encountered in VLSI devices. A more appropriate technique is the four-terminal contact resistivity measurement.

9.5.4 The Four-Terminal Kelvin Resistor

The interfacial contact resistance can be directly measured by using a four-terminal Kelvin resistor structure.[38] A typical test pattern is shown in

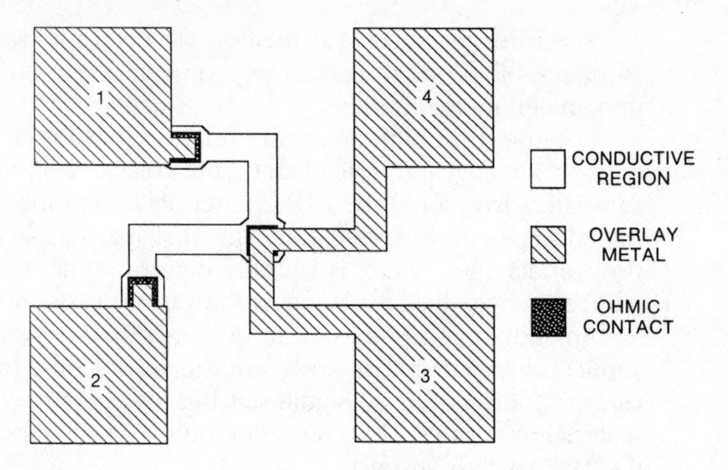

(A) A typical Kelvin resistor test pattern for measuring contact resistance. The conductive region is formed by mesa isolation, ion implantation or diffusion. A cross section of the center ohmic contact region is shown in (B).

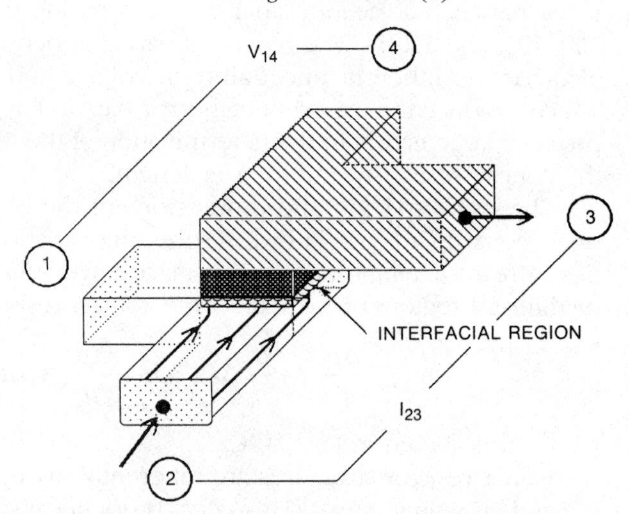

(B) Current in a four-point Kelvin resistor test structure. This nonplanar geometry has significantly reduced current crowding. The measured voltage is only the voltage drop across the interfacial ohmic contact layer.

Fig. 9–10. Using the four-terminal Kelvin resistor.

Fig. 9–10A. Unlike the TLM structure, current flows vertically and uniformly (as in Fig. 9–8) throughout the center ohmic contact between the underlying conducting region and the overlay metallization. The conducting region is formed by mesa isolation, implantation, or diffusion. Because the lateral and vertical current crowding effects are not present in this geometry, the measured resistance includes only the interfacial contact resistance. An additional benefit is that interfacial layer uniformity and end contact resistance can be directly obtained. The latter measurement requires a different test pattern (six-terminal) than the one shown here and is described by Proctor *et al.* in Ref. 34. The front contact resistance can then be obtained from R_e.

The measurement consists of applying a current between diagonally opposed contacts (e.g. 1 and 3) which flows perpendicularly and uniformly (Fig. 9–10B) through the center ohmic contact. A high-impedance voltage measurement is performed on the remaining two (2 and 4) pads, so no current flows out of these pads. The current polarity is then reversed and another voltage is measured. This is repeated for the other pads (e.g. I across 2 and 4 and V across 1 and 3) and the average of the four measurements is taken. If V_{jD} and V_{jM} are the voltages measured at the jth contact to the conductive region (bottom) contact and center ohmic (top) overlay metal respectively, then the averaged potential drop across the alloyed metal is[37]

$$V_2 - V_1 = \frac{1}{N}\sum_{j=1}^{N}(V_{jD} - V_{jM}), \tag{9–23}$$

where N is the number of contacts.

As discussed previously, all the current passes uniformly through the interfacial contact region, therefore

$$I = \sum_{j=1}^{N}\frac{V_{iD} - V_{jM}}{R_i} = \frac{N}{R_i}(V_2 - V_1). \tag{9–24}$$

Then, the contact resistance is simply

$$R_c = \frac{V_2 - V_1}{I} = \frac{R_i}{N} = \frac{\rho_c}{A}, \tag{9–25}$$

where the specific contact resistance is obtained from measurement of the area and Eq. 9–15.

For an ideal contact, the interfacial contact resistance should increase linearly with respect to decreasing contact area. Figure 9–11 shows the differences between a very uniform (pure) aluminum and Al/Si (98.5 percent/1.5 percent) contact to silicon.[34] The nonlinearity suggests either that the contact interface is inhomogeneous,[34] or for specific contact resistances of 1 $\mu\Omega \cdot cm^2$ or less, current crowding effects may be important. Inhomogeneity could indicate a fabrication process or materials system problem. The current crowding occurs since the diffusion sidewall tends to be slightly larger than the contact in the Kelvin resistor. Several models have been presented which can be used to correct for current crowding effects.[39–42]

In summary, the Kelvin measurement circumvents many of the problems inherent in the transmission line method, where the measured

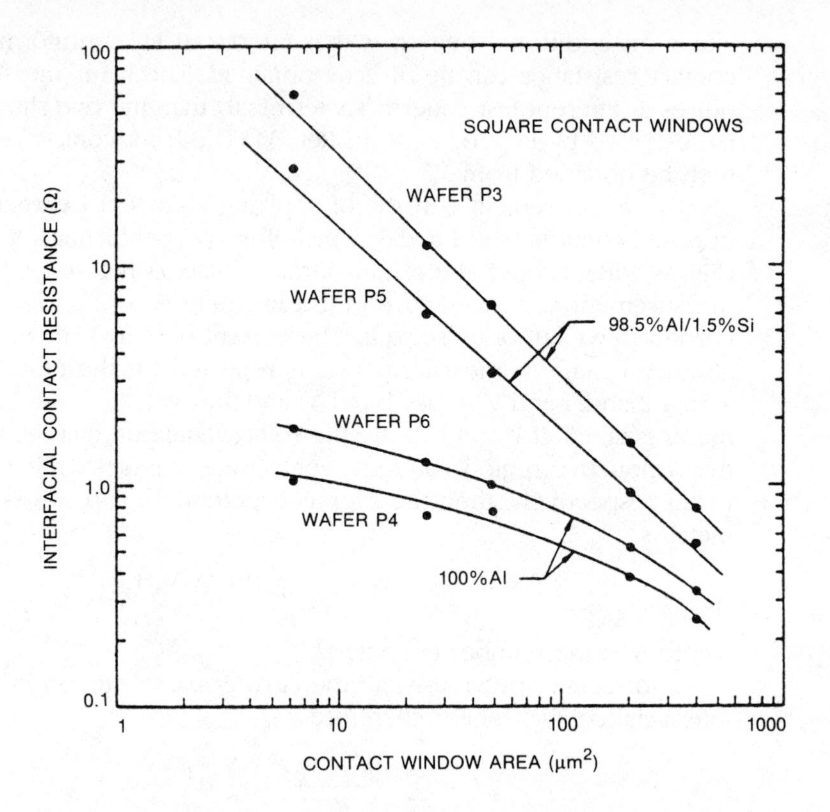

Fig. 9–11. Measured interfacial contact resistance R_c versus area for four-terminal Kelvin structures. Wafers P3 and P5 were sintered for 20 min at 425 and 500°C, respectively. Wafers P4 and P6 were sintered for 20 min at 425 and 500°C, respectively. *(After Proctor, Linholm, and Mazer[34])*

resistance has components associated with measurement error, current crowding, surface effects, and conductive layer resistivity. Another advantage is that an arbitrary contact area can be tested, which enables dimensional contact scaling effects to be directly measured—a very important consideration for developing contacts for GaAs VLSI.

9.5.5 The Cox and Strack Method

There is one other method of measuring the specific contact resistance, introduced by Cox and Strack[43] in 1967. The relatively simple geometry is a series of circular contacts having different radii on a conducting substrate. Adjacent devices on a back-metallized, conducting substrate must be isolated from each other to avoid fringing effects. The measured resistance for this geometry is

$$R_T - R_0 = \frac{\rho_s t + \rho_c}{\pi r^2} \quad \text{for } t << r\,, \tag{9–26}$$

where R_T is the total resistance, R_0 is the backside resistance, ρ_s ($\Omega \cdot$ cm) and t (cm) are the epitaxial layer resistivity and thickness, and r is the disk radius. For high doping and thin epitaxial layers, $\rho_s t$ is small and can be neglected. To extract the contact resistance, the measured resistance versus inverse contact radius is plotted. The specific contact resistance, ρ_c, is then obtained by fitting the data with Eq. 9–26.

This method has many drawbacks. The resistivity of the epitaxial material must be known in order to extract ρ_c with confidence. If this parameter is not known (as is the case for alloyed contacts), then the resistance versus $1/r$ curve must be fitted with ρ as another fitting parameter. To even a greater extent than in the TLM method, the errors arising from subtracting two large numbers to obtain a small one (ρ_c) are extremely large. A second source of error is that the dimensions of the ohmic dots must be well known because any error in the radius is squared in the denominator. Edge effects around the dots also change nonlinearly as the contact size changes. In addition, end and forward contact resistances cannot be extracted with this technique, so it is not a commonly used technique for measuring contact resistances.

References

1. V.L. Rideout, "A review of the theory and technology for ohmic contacts to group III-V compound semiconductors," *Solid State Electron.*, 18:541–550 (1975).
2. N. Braslau, "Alloyed ohmic contacts to GaAs," *J. Vac. Sci. Technol.*, 19:803–807 (1981).
3. D.K. Schroder and D.L. Meier, "Solar cell contact resistance—a review," *IEEE Trans. Electron Dev.*, ED-31(5): 637–647 (1984).
4. W. Schottky, *Naturwissenschaften*, 26:843–856 (1938).
5. W. Schottky, *Z. Phys.*, 113:367–415 (1939).
6. W. Schottky, *Z. Phys.*, 118:539–592 (1942).
7. J.R. Waldrop, "Schottky barrier heights of ideal metal contacts to GaAs," *Appl. Phys. Lett.*, 44(10): 1002–1004 (1984).
8. N. Braslau, J.B. Gunn, and J.L. Staples, *Solid State Electron.*, 10:381–384 (1967).

9. A. Christou, "Solid phase formation in Au:Ge/Ni, Ag/In/Ge, In/Au:Ge GaAs ohmic contact systems," *Solid State Electron.,* 22:141–149 (1979).
10. J.M. Woodall and J.L. Freeouf, "GaAs metallization: some problems and trends," *J. Vac. Sci. Technol.,* 19(3): 794–798 (1981).
11. W.T. Anderson Jr., A. Christou, and J. Davey, "Development of ohmic contacts for GaAs devices using epitaxial Ge films," *IEEE J. Solid State Cir.,* SC-13(4): 430–435 (1978).
12. D. Fathy, O.L. Krivanek, J.C. Spence, and W.M. Paulson, "X-ray microanalysis and high resolution imaging of Ge-Au-Ni metal layers on Gallium Arsenide," *Proc. Mat. Res. Soc.,* (1983).
13. T.S. Kuan, P.E. Batson, T.N. Jackson, H. Rupprecht, and E.L. Wilkie, "Electron microscopy studies of an alloyed Au/Ni/Au-Ge ohmic contact to GaAs," *J. Appl. Phys.,* 54(12): 6952–6957 (1983).
14. Z. Liliental and R.W. Carpenter, "Electron microscopy study of the AuGe/Ni/Au contacts on GaAs and GaAlAs," *Ultramicroscopy,* 14:135–144 (1984).
15. M. Murakami, K.D. Childs, J.M. Baker, and A. Callegari, "Microstructure studies of AuGeNi ohmic contacts to n-type GaAs," *J. Vac. Sci. Technol. B,* 4(4): 903–911 (1986).
16. A.K. Kulkarny and J.T. Lukowski, "Effect of annealing process parameters on the properties of AuGe contacts to GaAs," *J. Appl. Phys.,* 59(8): 2901–2904 (1986).
17. K. Johnson, "Hypoeutectic (Ge,Au) Ni ohmic contacts to n-GaAs," private communication (1986).
18. H. Oraby, K. Murakami, Y. Yuba, K. Gamo, S. Namba, and Y. Masuda, "Laser annealing of ohmic contacts to GaAs," *Appl. Phys. Lett.,* 38(7): 562–564 (1981).
19. C.P. Lee, B.M. Welch, and J.L. Tandon, "Reliability of pulsed electron beam alloyed AuGe/Pt ohmic contacts to GaAs," *Appl. Phys. Lett.,* 39(7): 556–558 (1981).
20. R. Zuleeg, P.E. Freibertshauser, J.M. Stephens, and S.H. Watanabe, "Al-Ge ohmic contacts to n-type GaAs," *IEEE Electron Dev. Lett.,* EDL-7(11): 603–604 (1986).
21. S. Tiwari, T.S. Kuan, and E. Tierney, "Ohmic contacts to n-GaAs with Germanide overlayers," *Proc. Intl. Electron Dev. Mtg.,* 115–118 (1983).
22. A.A. Lakhani, "The role of compound formation and heteroepitaxy in Indium-based ohmic contacts to GaAs," *J. Appl. Phys.,* 56(6): 1888–1891 (1984).
23. D.C. Marvin, N.A. Ives, and M.S. Leung, "In/Pt ohmic contacts to GaAs," *J. Appl. Phys.,* 58(7): 2659–2661 (1985).

24. M. Gurvitch, A. Kastalsky, S. Schwanz, and D. M. Hwang, "Ohmic superconducting, shallow AuGe/Nb contacts to GaAs," *J. Appl. Phys.*, 60(9): 3204–3210 (1986).
25. P. D. Kirchner, T. N. Jackson, G. D. Pettit, and J. M. Woodall, "Low resistance non-alloyed ohmic contacts to Si-doped MBE GaAs," *Appl. Phys. Lett.*, 47(1): 26–28 (1985).
26. M. Kuzuhara, T. Nozaki, and K. Kohzn, "Nonalloyed ohmic contacts to Si implanted GaAs activated using SiO_xN_4 capped infrared rapid thermal annealing," *J. Appl. Phys.*, 58(3): 1204–1209 (1985).
27. E. F. Schubert, J. E. Cunningham, W. T. Tsang, and T. H. Chin, "Delta-doped ohmic contacts to n-GaAs," *Appl. Phys. Lett.*, 49(5): 292–294 (1986).
28. E. D. Marshall, W. X. Chen, C. S. Wu, S. S. Lan, and T. F. Kuech, "Nonalloyed ohmic contact to n-GaAs by solid phase epitaxy," *Appl. Phys. Lett.*, 47(3): 298–300 (1985).
29. C. L. Chen, J. L. Mahoney, M. C. Finn, R. C. Brooks, A. Chu, and J. G. Mavoidas, "Low resistance Pd/Ge/Au and Ge/Pd/Au ohmic contacts to n-type GaAs," *Appl. Phys. Lett.*, 48(8): 535–537 (1986).
30. R. Stall, C. E. C. Wood, K. Board, and L. F. Eastman, "Ultra low resistance ohmic contacts to n-GaAs," *Electron. Lett.*, 15:800–801 (1979).
31. J. M. Woodall, J. L. Freehouf, G. D. Pettit, T. Jackson, and P. Kirchner, "Ohmic contacts to n-GaAs using graded bandgap layers of $Ga_{1-x}In_xAs$ grown molecular beam epitaxy," *J. Vac. Sci. Technol.*, 19:626–627 (1981).
32. H. H. Berger, "Contact resistance and contact resistivity," *J. Electrochem. Soc.*, 119(4): 507–514 (1972).
33. S. Cohen, "Contact resistance and methods for its determination," *Thin Solid Films*, 104:361–379 (1983).
34. S. J. Proctor, L. W. Linholm, and J. A. Mazer, "Direct measurements of interfacial contact resistance, end contact resistance, and interfacial contact layer uniformity," *IEEE Electron Dev.*, ED-30(11): 1535–1542 (1983).
35. H. H. Berger, "Models for contacts to planar devices," *Solid State Electron.*, 15:145–158 (1972).
36. H. Murrmann and D. Widmann, "Current crowding on metal contacts to planar devices," *IEEE Trans. Electron Dev.*, ED-16(12): 1022–1024 (1969).
37. G. K. Reeves and H. B. Harrison, "Obtaining the specific contact resistance from transmission line model measurements," *IEEE Electron Dev. Lett.*, EDL-3(5): 111–113 (1982).
38. S. J. Proctor and L. W. Linholm, "A direct measurement of interfacial contact resistance," *IEEE Electron Dev. Lett.*, EDL-3(10): 294–296 (1982).

39. T. F. Lei, L. Y. Leu, and C. L. Lee, "Specific contact resistivity measurement by a vertical Kelvin test structure," *IEEE Trans. Electron Dev.,* ED-34(6): 1390–1395 (1987).

40. M. Finetti, A. Scorzoni, and G. Soncini, "Lateral current crowding effects on contact resistance measurements in four terminal resistor test patterns," *IEEE Electron Dev. Lett.,* EDL-5(2): 524–526 (1984).

41. W. M. Loh, S. E. Swirhun, E. Crabbe, K. Saraswat, and R. M. Swanson, "An accurate method to extract specific contact resistivity using cross-bridge Kelvin resistors," *IEEE Electron Dev. Lett.,* EDL-6(9): 441–443 (1985).

42. T. A. Schreyer and K. C. Saraswat, "A two-dimensional analytic model of the cross-bridge Kelvin resistor," *IEEE Electron Dev. Lett.,* EDL-7(12): 661–663 (1986).

43. R. H. Cox and H. Strack, "Ohmic contacts for GaAs devices," *Solid State Electron.,* 10:1213–1218 (1967).

Chapter 10

Dry Etching

E. Hu

In recent years, reactive plasma etching techniques have rapidly moved from research to manufacturing of silicon based integrated circuits. The widespread acceptance of such dry etching processes results primarily from the enhanced resolution and degree of control over etch rates and etched profiles obtainable. Similar advantages hold true for fabrication of III-V based devices.[1–3] Reactive plasma etching has been applied to gate recess etching of GaAs MESFETs and heterostructure (HFET) devices, deeply etched via structures, and mirror facets for heterostructure lasers. The resolution of reactive plasma etching has been utilized to fabricate "quantum dots," zero- and one-dimensional structures etched into superlattice material. The most critical requirements for each of the above processes will dictate optimal choice of gas mixture, pressure, bias voltage, and other parameters; similar chemistries can produce etches that will be either selective or nonselective of the materials being etched.

10.1 Reactor Configurations

The most widely used dry etching techniques incorporate some form of reactive plasma chemistry: chemical reactions between the reactive gas or gases and the substrate are sustantially enhanced by physical, ion initiated sputtering processes. The sputtering component also helps produce profile anisotropies that are independent of the crystallographic orientation of the substrate. Dry etching techniques may often be referred to under the generic "plasma etch processes," since the activated etchants are formed as part of a plasma of ions, electrons, and neutral species. "Plasma etching" may also be used to refer to dry etch processing that takes place at pressures greater than a few hundred millitorr, while "reactive ion etching"

(RIE) describes processing that takes place at lower pressures. In the widely used RIE, a plasma is generated from the RF excitation of a reactive gas. The distribution of ion energies and densities is determined by the gas composition, pressure, and input power. Independent control of the ion energy and current density is possible in reactive ion-beam etching (RIBE),[4] where a collimated beam of reactive ions and neutrals is extracted from a plasma and directed at a sample. Finally, the reactive and sputtering components can be separately formed and controlled in chemically assisted ion-beam etching (CAIBE, also referred to as ion beam assisted etching, IBAE[5–7]). In this case, a low-energy ion beam and separate source of reactive flux are both coincident on the substrate. (For more general background on the formation and applications of plasmas to etching, see Refs. 8 through 10).

A "typical" planar parallel plate (or diode) reactor used to perform RIE is shown in Fig. 10–1. The plasma is initiated by coupling an RF signal at 13.56 MHz; the substrate sits on the RF-driven electrode. Although most RIE configurations have utilized a 13.56-MHz RF signal, other excitation sources at varying frequencies have been used.[11] In particular, there is a growing body of work incorporating microwave sources, such as is used in the RIBE system[4] shown in Fig. 10–2. In this case, the reactive plasma is extracted from an ion gun, forming the reactive ion beam. The bias voltage applied to the grid at the exit of the gun gives more sensitive control over the energies of the emerging ions. The Cl_2 plasma is excited by a 2.45-GHz microwave cavity. The particular plasma shown is also referred to as an ECR or electron cyclotron resonance plasma. In this case, a mag-

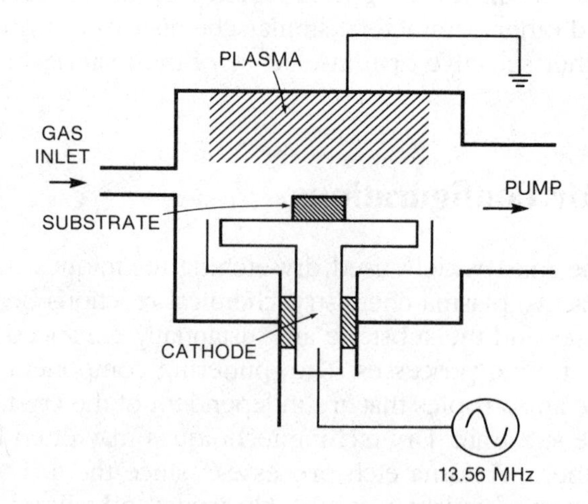

Fig. 10–1. Schematic of a reactive ion etching (RIE) chamber.

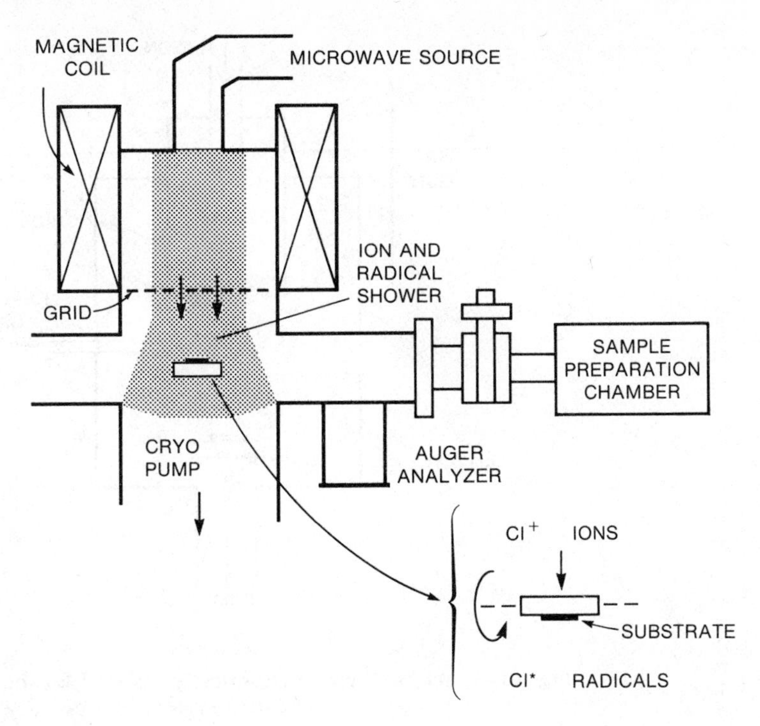

Fig. 10–2. Schematic of a reactive ion beam etching (RIBE) chamber *(After Asakawa[4])*.

netic field is applied that is resonant with the cyclotron (orbital) frequency of the electrons, increasing their mean free path in the plasma, and augmenting their ability to sustain a plasma excitation. Thus, the pressure in the gun may be $\sim 10^{-3}$ to 10^{-4} torr while the background pressure in the etch chamber may be much lower than that ($\sim 10^{-8}$ torr). For fully independent control over the reactive flux and the incident ion energy, etching may be carried out in a CAIBE system, indicated in Fig. 10–3. A Kaufman ion source provides the energetic ions that can promote the reaction between the reactive gas flux and the substrate. In fact, that reactive gas may be separately heated, forming a "hot jet" that will further enhance the etching process.[12,13] Although the above techniques are those that have been most commonly used for GaAs dry etching, the reactive species may be formed by means other than RF excitation and surface chemical processes stimulated by means other than ion bombardment activation. For example, there is a growing body of research on laser initiated and enhanced processes that have been applied to the etching of GaAs and other III-V materials. These techniques and processes will be further discussed in Sec. 10.6.

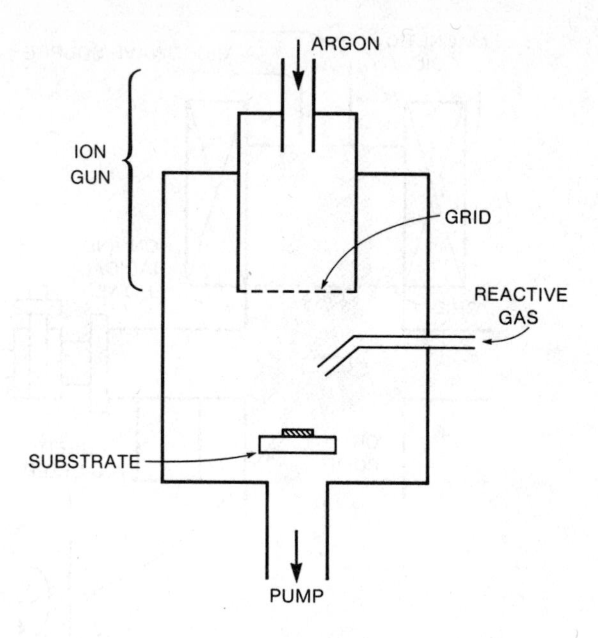

Fig. 10–3. Schematic of a chemically assisted ion-beam etching (CAIBE) system.

10.2 Etch Chemistry

A simple, and fairly general description of the etching of a solid by reaction with a gas is comprised of several stages, any one of which may limit the rate of the overall reaction: (1) transport of reactants to the surface, (2) adsorption onto the substrate surface, (3) reaction with the substrate and formation of the product molecules, (4) desorption of the product molecules, and (5) transport of products away from the surface. The sequence is schematically represented in Fig. 10–4. The physical sputtering component of reactive plasma etching can enhance the overall reaction by accelerating any one of the surface processes above: promoting adsorption of the reactant molecules, providing energy to drive forward the reaction of gas and substrate molecules, or aiding in the desorption of nonvolatile product molecules. Generally, the etching of III-V semiconductors is limited by the volatility of the group III halogen (i.e. the fluorides or chlorides of gallium, aluminum, indium). Since the chlorides are more volatile than the fluorides, chlorine-based etch gases have been utilized in these etch processes. Cl_2,[14] CCl_4,[15–18] HCl,[18] $SiCl_4$,[19,20] CCl_2F_2,[21–23] and

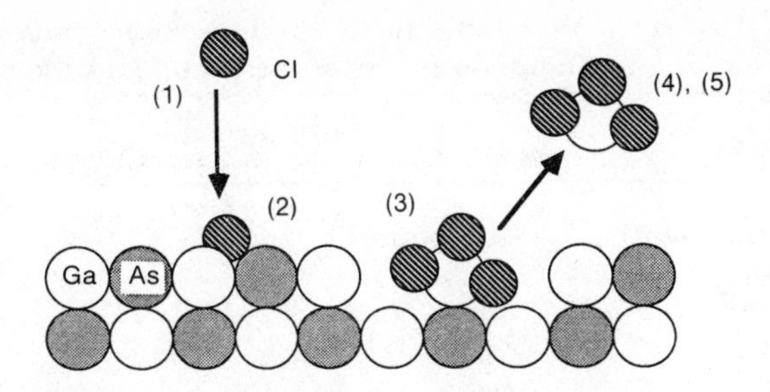

Fig. 10–4. Sequence of reactant-substrate interactions in dry etching. The steps are (1) transport to the substrate, (2) adsorption onto the substrate, (3) reaction, (4) desorption, and (5) transport away from the substrate.

BCl_3,[24,25] are among the gases that have served as the basis of the III-V etch processes. Recently, etches based on CH_4/H_2 have gained popularity, particularly for InP etching.[84] Other halogens, such as bromine and iodine, have also been used.[26] Table 10–1 gives a partial listing of such etch gases, with suitable references. Table 10–2 lists the boiling points and/or sublimation temperatures of a variety of group III chlorides and fluorides. These values can give an indication of the suitability of a particular halogen-containing gas in etching a given substrate. Such gases can be used alone, or in conjunction with other gases (such as argon or oxygen), which have been added to improve etch rate, profile, surface morphology, or to stabilize the plasma.

Table 10–1. Partial Listing of Etch Gases, with References

Etch Gas	References*
Cl_2	5, 7, 11–15, 25–33, 39, 47, 49, 50, 63, 66, 73, 74
CCl_2F_2	21–23, 34, 42–46, 56–61
$SiCl_4$	19, 20, 62, 75, 76
CCl_4	6, 15–18, 34
BCl_3	24, 25, 64, 65
HCl	6, 18, 82
HBr	80, 81
Br	26
CCl_3F	35
CF_4	34, 55

*While extensive, this list of references is not claimed as being inclusive. That is, other research in dry etching of GaAs has been published which is not represented here.

Table 10–2. Boiling Points and/or Sublimation Temperatures of Some Group III Chlorides and Fluorides*

Element	Fluorides		Chlorides	
	Compound	T_b (°C)	Compound	T_b (°C)
Al	AlF_3	1291 (T_s)	$AlCl_3$	177 (T_s)†
As	AsF_5	−53		
	AsF_3	−63	$AsCl_3$	130
Ga	GaF_3	~1000	$GaCl_2$	535
			$GaCl_3$	201
In	InF_3	>1200	$InCl_2$	560
			InCl	608
			$InCl_3$	300 (T_s)
Ni	NiF_2	1000 (T_s)	$NiCl_2$	973 (T_s)
P	PF_5	−75	PCl_5	−162 (T_s)
	PF_3	−101	PCl_3	75
Si	SiF_4	−86	$SiCl_4$	58

*From *Handbook of Chemistry and Physics,* CRC Press, Inc., 1980.
†T_s is the sublimation temperature.

10.3 General Etch Factors

10.3.1 Etch Species

Various processes will impose different critical requirements, such as rapid etch rate for deep via structures, highly vertical, smooth etched sidewalls for laser facets, or high etch rate ratios (selectivity) between different materials for the fabrication of HFET devices. In order to understand how the accessible parameters may be changed to produce the desired result, it is necessary to understand what the predominant etch species is, and what products are formed. Because of the complexity involved in trying to describe multigas mixtures or complex molecular reactants, detailed analyses have been done primarily on the "simpler" Cl_2/GaAs system.[27–30,†] Optical emission spectroscopy and laser induced flourescence have been used as *in situ* probes of the etching process.

From such measurements, the picture that emerges is one where etching is brought about by neutral, atomic chlorine radicals, with the forma-

† A great deal of analysis has been done on other gas systems as well. In particular, there has been recent excellent work on CCl_2F_2, motivated by interest in selective etching.

tion of chlorides of gallium and arsenic as the etch products. Ion bombardment will increase the etch rate, probably by aiding in the desorption of the $GaCl_3$ by-product. At high temperatures and pressures, etching by the chlorine radical Cl* will dominate, producing profiles which are determined by the crystallographic orientation of mask with respect to substrate. At low temperatures and pressures (i.e. for long mean free paths), the ion component may play a larger role in determining etch rates and profiles. Therefore, one can obtain both the highly vertical profile of Fig. 10–5, which was formed by RIE in 5 mtorr of Cl_2 at -350 V bias and room temperature, or the crystal-plane dependent profile more characteristic of wet chemical etching.

This picture of the GaAs etch process is borne out by recent experiments performed by Asakawa and Sugata.[4,30–32] The substrate holder of their etching apparatus (Fig. 10–2) may be turned to either directly face the ion gun, or be turned to face away from it. In this way, the effects of ions and radicals on the etch process can be separated out, for substrates facing away from the gun should be subject to radical etching only. *In situ* Auger analysis revealed that Cl neutrals (plasma excited, substrate facing away from gun) were far more easily chemisorbed onto the GaAs surface than Cl_2 molecules (plasma off). When the substrate faced away from the ion gun, no significant etching took place at room temperature. Increased substrate temperatures did give rise to etching by chlorine radicals; plots of etch rate versus temperature gave an apparent activation energy of ~500 kcal/mole. By turning the substrate to face the ion gun, the substrate was subjected to an additional, low-energy ion bombardment. The addition of the Cl^+ ions, extracted at 30 V, led to a total Cl* plus Cl^+ activation energy of 8.5 kcal/mole, far lower than the activation energy for Cl* alone. These data are displayed in Fig. 10–6.

10.3.2 Temperature

As the previous example has shown, exploration of the temperature dependence of the dry etch process is quite useful in elucidating the particular chemical reactions responsible. The substrate temperature may influence the gas-surface reactions, the surface mobility of the reactants, or the volatility of the etch products. Each of these processes is expected to have a characteristic activation energy. A plot of etch rate as a function of temperature should provide the activation energy of the rate determining step. For example, if the desorption of the $GaCl_3$ product is the rate limiting step in the process, we would expect that the characteristic activation energy of the Cl_2 etching of GaAs would have a value close to the latent heat of vaporization of the $GaCl_3$. In fact, Donnelly, *et al.*[33] found this to be true: their data gave an apparent activation energy of 10.5 kcal/mole for

(A) With straight-walled profile, etched in Cl_2 at 5 mtorr.

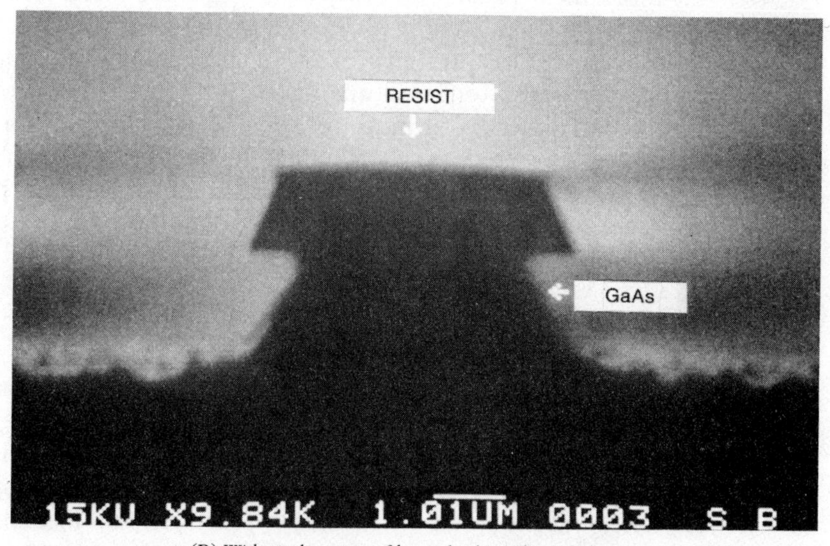

(B) With undercut profile, etched in Cl_2 at 40 mtorr.

Fig. 10–5. SEM photograph of GaAs.

GaAs etching in Cl_2, close to the latent heat of vaporization of $GaCl_3$ (12 kcal/mole). However, using the latent heat of vaporization to calculate an evaporation rate of the $GaCl_3$ (hence, etch rate of GaAs) resulted in etch rates that were orders of magnitude larger than those actually ob-

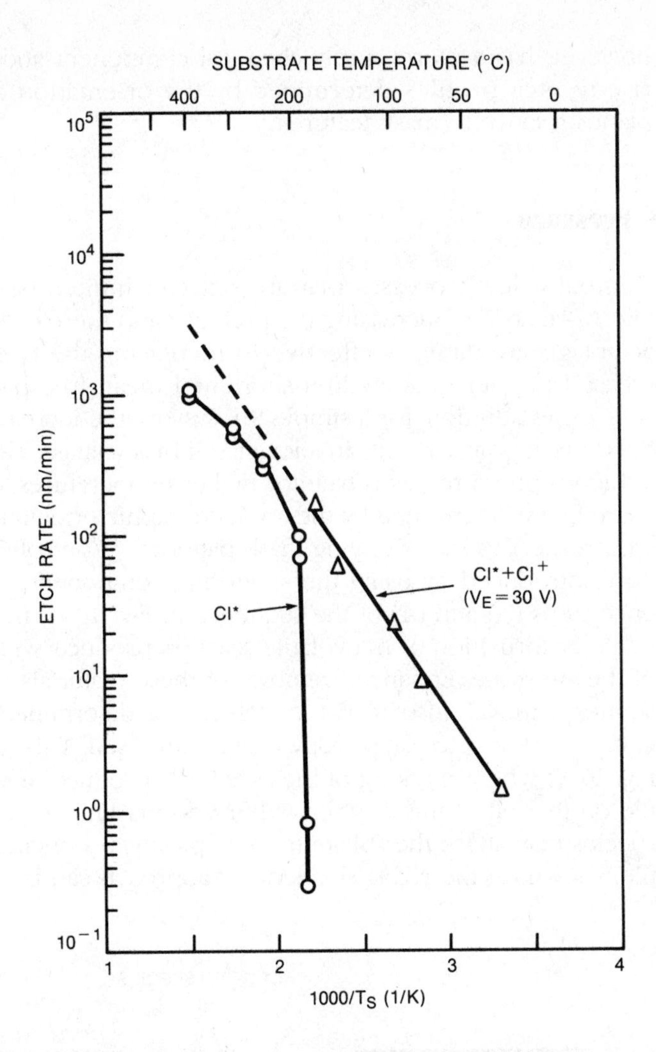

Fig. 10–6. Activation energies for etching of GaAs by Cl^+ (ions) and Cl* (radicals). *(After Sugata and Asakawa*[32]*).*

served. Recent thermodynamic analyses made by McNevin and Becker[28] also give much larger etch rates than those actually observed. The thermodynamic calculations set an upper limit to the etch rates; the actual etching mechanisms require additional consideration of the surface chemistry of product formation and desorption. For example, Donnelley *et al.* postulated that rather than the desorption of the $GaCl_3$, the rate limiting step was a slow chemical reaction at the surface. That reaction possibly involved the formation and evaporation of Ga_2Cl_6. Increasing the substrate temperature will generally result in an increase in etch rate, through the variety of mechanisms mentioned above. Etching at elevated temperatures, however, has a more purely chemical component and, as such, will give rise to etch profiles determined by the orientation of crystallographic planes relative to mask features.

10.3.3 Pressure

Generally, for processes that are reactant limited [steps (1) and (2) of Fig. 10–4, above], increasing the pressure and therefore the quantity of reactant gas available is effective in increasing the overall etch rate. Increased pressure results in a shortened mean free path for the reactive species. In addition, for a simple RIE system, the increase in pressure for a fixed input power results in a lower self-bias voltage. The end result is that at higher pressures, as is true for higher temperatures, etched profiles are more likely determined by the crystallographic orientation of the substrate with respect to the overlying mask pattern.[26] "Controlled" anisotropy may be reintroduced by using the sputtering component of the etch to influence steps (3) and (4) of the sequence in Fig. 10–4. In these cases, either polymer formation or nonvolatile reaction products will limit the etch rate of the substrate. Enhanced removal of these materials can be achieved by sputter removal; insofar as the etch rate is determined by the sputtering component, an etch anisotropy may be imposed. This idea is illustrated in Fig. 10–7, where masking of the side walls produces a straight-walled profile on the left of the diagram. Where no masking is present, neutral etch species may attack the substrate isotropically, producing an undercut profile, shown on the right. This kind of approach can be used for the deep,

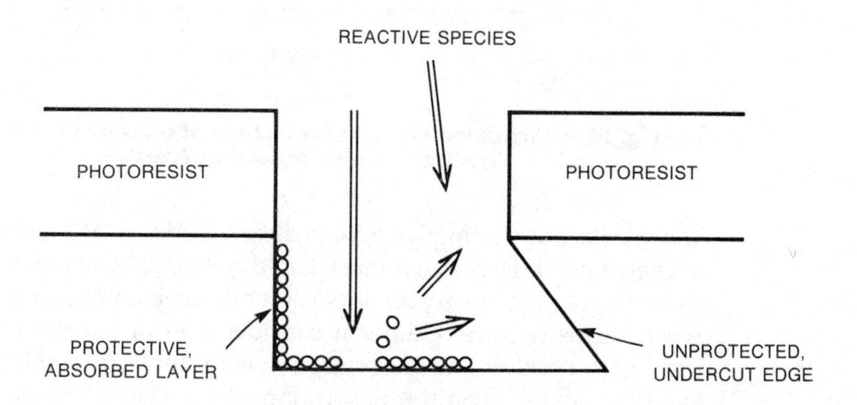

Fig. 10–7. Masked sidewall (left-hand side) or unmasked (right-hand side) etching, leading to either vertical or undercut profiles.

rapid etching of via structures having a modest degree of anisotropy. This will be further discussed in Sec. 10.4.1.

As mentioned previously, the etching of GaAs by Cl_2 comprises a *relatively* simple chemical system. Reactants are neutral or ionized molecular or atomic chlorine; products are the chlorides of Ga and As. Use of "more complex," Cl_2-containing gases, such as the Freons, will lead to a formation of a wider variety of product species that will influence etch rates and profiles.

10.3.4 Etch Products

10.3.4.1 Carbon Deposition From Halocarbon Gases

Other than Cl_2, Freon 12 (CCl_2F_2) has perhaps been the most widely used gas for the etching of GaAs based materials. As has been observed in the CF_4 etching of silicon, halocarbon gases leave carbon residues on the surface. This may be detected in Auger spectra of the surface after completion of the etch. Klinger and Greene[34] have made *in situ* optical emission studies of the GaAs surface under steady-state etching by CCl_4, CCl_2F_2, and CF_4. The total amount of carbon observed increased with increasing Cl to F ratio in the etch gas. That is, the steady-state carbon coverage was found to be greatest for CCl_4 and least for CF_4. In contrast, the lowest etch rates were obtained using CF_4, indicating that the rate limiting step was not removal of the C overlayer, but rather the removal of the Ga (or fluoride of Ga). The nonvolatile carbon and gallium fluoride product influenced the etched profile, resulting in a sloped profile whose angle increased with increasing F/Cl ratio in the etch gas. Finally, unsaturated carbon-halogen fragments may combine or polymerize in the plasma and be deposited as polymers onto the substrate surface.

These considerations and the variation of the F/Cl ratio in the etch gas chosen can be utilized to affect the angle of the etched profile, or the degree of etch rate selectivity.

10.3.5 Gas Additions

In order to further optimize a given etch process, additions of gases may be made to increase etch rate, or selectivity or to improve surface morphology. For example, argon and oxygen have been added to CCl_2F_2 in order to provide etching of GaAs that proceeded at a reasonable rate and produced a vertical profile with a smooth, clean surface.[21] Oxygen, hydrogen, and argon have been the most commonly studied gas additions to the basic, halogen-containing etchant. Boron trichloride has been incorporated in some cases in order to reduce the sensitivity of the process to residual oxygen or water vapor in the etch chamber.

10.3.5.1 Oxygen

The addition of small amounts of O_2 to Cl_2, CCl_4, or other Cl-containing gases results in an increased GaAs etch rate.[15,17,21,35] The mechanism is similar to that characterizing addition of O_2 to CF_4 for the etching of silicon.[36] Reaction of oxygen with fluorochloro-carbon fragments may release additional Cl to carry out the etching. Optical emission spectra show an increase in the intensity of the atomic chlorine line (837.6 nm) with the addition of O_2 to Cl_2, CCl_4, or CCl_3F, and an accompanying increase in the GaAs etch rate. The increased etch rate may also result from the ability of O_2 to react with unsaturated halocarbon fragments, preventing their polymerization and deposition/masking of the substrate.

10.3.5.2 Hydrogen

As O_2 has been described as a scavenger of C or C-containing fragments, H_2 scavenges chlorine from CCl_4, as it scavenges fluorine from CF_4.[36] Therefore, addition of H_2 to the etch gas would be expected to decrease the etch rate of GaAs. This result was in fact obtained for GaAs substrates etched in CCl_4/H_2 mixtures, which showed smooth surfaces and increasing anisotropy (high vertical to lateral etch rate ratio) as the relative ratio of H_2 in the gas mixture increased to 0.5. Interestingly, a large percentage of H_2 in the gas mixtures resulted in substrate surfaces that were identical with those which were treated by wet etch, according to the results of Raman spectroscopy.[17] This factor may be important in the design of an etch process which will minimize damage to the substrate, hence deleterious effects on device performance (see Sec. 10.5).

10.3.5.3 Argon

Argon can be used to enhance the sputtering component of the etch process. Chaplart *et al.*[23] utilized argon to aid in the removal of etch products (fluorides of gallium) that were impeding the etching of GaAs in CCl_2F_2. In using the Freon 12 alone, they found that, for a given pressure, the etch rate increased with increasing power. For a given power, the etch rate *decreased* with increasing gas pressure in the region from 5 to 10 mtorr. These results indicate that availability of reactant was not the limiting step in the etching process, rather, the limiting step might be removal of the reaction products. At pressures higher than 20 mtorr, a brownish deposit was found on the sample surface. The addition of Ar to the gas mixture produced an increase in etch rate, which peaked at a composition of 60-percent argon in CCl_2F_2. Presumably, further addition of argon to the gas mixture diluted the amount of reactant available for etching.

10.3.6 Variable Etch Rates: "Induction Effects"

10.3.6.1 Boron Trichloride

Gases may be added to make the plasma conditions more stable (e.g. addition of He to Cl_2) or to make the etch itself more reproducible. BCl_3 has been shown to be a scavenger of oxygen and water vapor.[37] It has long been used together with Cl_2 in reactive ion etching of aluminum, as the BCl_3 will etch oxides of aluminum far more effectively than will the Cl_2. Without the presence of a load-lock, or other means of keeping the background water vapor and oxygen low, strong *induction* effects will be observed, where the etched depth is not a linear function of etch time. Generally, the presence of a native oxide on the sample surface leads to a lower initial etch rate (or period of induction) than that found for etching at longer times. This is especially critical in shallow etching of the substrate material (etched depth of a few tens of nanometers) where high bias voltages are to be avoided to prevent damage to the material. Induction effects are also very likely observed when etching materials that incorporate aluminum, which is likely to react with oxygen to form an oxide. Figure 10–8 is a graphical illustration of induction in the etching of GaAs. Such considerations motivated Tamura and Kurihara[25] to use a mixture of Cl_2 and BCl_3 in the etching of GaAs and $Al_{0.45}Ga_{0.55}As$. The etched depth of $Al_{0.45}Ga_{0.55}As$ as a function of time extrapolated to zero at zero etch time for gas composition ratios of P_{Cl_2}/P_{BCl_3} of 0 to 0.2, at pressures between 60 and 150 mtorr, and incident power density of 0.42 W/cm^2. Gas compositions where the ratio of Cl_2 to BCl_3 was greater than 0.5 led to nonreproducible etching of the $Al_{0.45}Ga_{0.55}As$. Although there seemed to be no induction effect for $P_{Cl_2}/P_{BCl_3} = 0.2$, it was only in pure BCl_3 that true equirate etching of GaAs and $Al_{0.45}Ga_{0.55}As$ was observed. The slightly different etch rates of GaAs (0.6 $\mu m/min$) and $Al_{0.45}Ga_{0.55}As$ (0.4 $\mu m/min$) at gas compositions with greater amounts of Cl_2 indicate a residual effect of background oxygen or water vapor. The anisotropy and equirate etching possible with pure BCl_3 make it an attractive candidate for etching of high-resolution quantum-well structures (see Sec. 10.4.4).

10.3.6.2 Hydrogen Etching of GaAs Oxides

Hydrogen plasmas have also been used to etch the oxides of GaAs.[14,38] In the initial demonstration of the use of hydrogen, GaAs oxide etch rates of between 0.5 nm/s and 2.0 nm/s were obtained for pressures between 10 and 150 mtorr and input powers of between 100 and 300 W.[38] The plasma was initiated by RF power at 30 MHz coupled into the gas by a coil wound about the glass tube that contained the gas. The etch rate ratio between the oxides of GaAs and the GaAs itself was approximately 2:1. Hydrogen

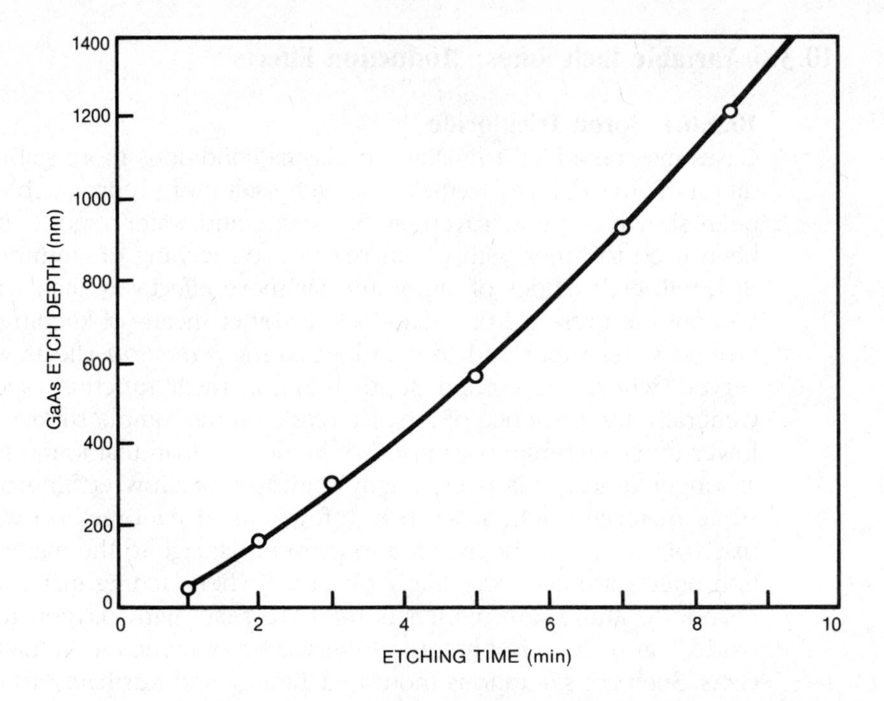

Fig. 10–8. Etch rate versus time of GaAs etched in CCl_2F_2. Extrapolation to zero etched depth at $t > 0$ illustrates "induction effect."

"pre-etches" have been used in conjunction with Cl_2 etching of GaAs and $Al_xGa_{1-x}As$, producing more reproducible etches which have more rapid etch rates at shorter etch times.[3] Use of a Cl_2 and Ar gas mixture (10 sccm and 40 sccm, respectively) at 5 mtorr pressure and −210 V bias yielded an average GaAs etch rate of 57.5 nm/min over 2 min. The same conditions preceded by a H_2 pre-etch of 35 s gave an average GaAs etch rate of 117 nm/min. Even with the use of an H_2 pre-etch, some induction effects are nevertheless observed. Although run to run etch rates were more uniform, a variation of etch rate with time is still observed. Residual H_2O in this non–load-locked system may result in persistent induction effects, after initial etching of the native oxide.

The best guarantor of reproducible etching is the reliable and reproducible preparation of the substrate surface prior to etching. Integration of a load-lock within the system helps control the background environment of the etch chamber. In addition, some means of *in situ* surface cleaning should be incorporated. Equirate etching of GaAs and $Al_xGa_{1-x}As$, a sensitive test of control over background oxygen and water vapor, has been achieved in the OJRL RIBE system described previously.[4] Equirate etching has also been observed for a standard RIE system that incorporated a load-lock,[38] as well as for a CAIBE system, also with load-lock.[50] In principle, sputter cleaning of a substrate should be able to remove any surface oxide or contaminated layer. The drawback of such an approach is the damage introduced to the sample. A preferable approach would be to use a highly selective chemical reaction whose products could be easily desorbed from the surface. Realizing the high chemical reactivity of beams of radicals (Cl*, H*), researchers have used ECR plasma-generated beams of H* and Cl* to remove surface oxygen and carbon.[40,41] Elevated substrate temperatures (400°C) ensured desorption of the beam-substrate products, leaving a "fresh" GaAs surface.

10.4 Dry Etching Applications

10.4.1 High Etch Rate Applications: Deep Vias, Chip Separation

There are a number of applications for dry etch-processing where the primary considerations are rapidity of etching, with only modest requirements of feature resolution. Such is the case for the etching of deep via structures to allow for low-impedance contacts in GaAs monolithic microwave integrated circuits (MMICs), and for the etching of features that will allow for easily made chip separation. In such cases, etch rates must be rapid enough that total etch times will be practical (on the order of micrometers or tens of micrometers per minute to etch through approximately 0.2 mm of material). In addition, etching of the substrate material must be sufficiently rapid with respect to the masking material so that deep etches may be made in the material without using impractically thick masking features. Finally, although via and chip separation features are generally fairly large (perhaps on the order of 100 μm or larger), nevertheless, a modest degree of control over the etched profile is desirable. As discussed in Sec. 10.3, increasing the concentration of reactant species and the temperature should generally result in the increase of etch rate.

Approaches used to achieve deep via etches have accordingly used relatively high gas pressures, often with elevated substrate temperatures to further promote chemical reactions.[42–44] Since these etches may extend several hundreds of micrometers deep, durable masks such as Ni films, are required, which will maintain their integrity throughout the total etching process. Geissberger and Claytor[42] explored a number of different gas mixtures for via etching. They chose a Freon 12-argon mixture (0.4/0.6) as being the most appropriate for their needs; however, they noted that gas mixtures containing $SiCl_4$ or BCl_3 provided smooth, anisotropic profiles, although with total etch rates diminished relative to the Freon 12-containing gases. All their experiments were done for substrates heated to between 60°C and 80°C, and at pressures between approximately 100 to

300 mtorr. Freon 12 alone has been used by a number of researchers within the same pressure range of 100 to 300 mtorr. Generally, etch rate was found to increase with increasing etch gas pressure, indicating a reactant limited process. However, at low incident power ($<$450 W), etch rates were *lower* at the higher value of pressure. This is attributed to the formation of polymers that mask the etching, but which can be sputtered off at sufficiently high bias voltage (i.e. power). Polymers masking the sides of the etched structures produced fairly steep structures. Such polymers were identified by Auger analysis to contain carbon, fluorine, chlorine, and gallium. As is the case for deep trench etching in silicon technology, there is a dependence of the etch rate on the dimensions of the via opening, with more rapid etching taking place in the wider dimensioned vias. This may be due to the slower influx of reactant species, or the less efficient removal of polymers from the via structures having higher aspect ratios (depth to width).

Contolini and D'Asaro[45] have explored the use of magnetron enhanced etching to allow deep, rapid transfer of patterned features without excessive sputter removal of masking material. The advantage offered by magnetron enhanced etching, as in the ECR plasma generation, is that the magnetic field confines the electrons, making them more efficient in the generation of ions. For a given power and bias voltage, more reactant species may be formed, augumenting the etching of the substrate GaAs while minimizing the sputter removal of the masking photoresist. Using Freon 12 at 40 mtorr pressure (-40 V bias), they were able to use a photo-resist mask (2.7 μm of AZ 1350J) to etch through GaAs wafers, at an etch rate of ~3 μm/min.

10.4.2 Etching of Laser Facets: Equirate Etching of GaAs and $Al_xGa_{1-x}As$

In the construction and interconnection of a number of optoelectronic components, micrometer or submicrometer features with predictable wall slopes are necessary. Some of the applications have been for the formation of waveguides, or for gratings of distributed feedback (DFB) lasers.[46,47] Mirror-quality, vertically etched facets are required to efficiently reflect and/or transmit an optical signal being channeled between on-chip components. Wet chemical etching has been used as an alternative to cleaving of mirror facets for diode lasers, producing satisfactory results. However, only dry etching techniques hold the promise of providing narrow, high-aspect–ratio grooves, desirable for efficient, partially transmissive mirrors.[4,38,48,49] The critical requirement of the etch process is to produce as smooth a facet surface as possible, with a high degree of control of the angle of the facet. In this case, it is important that the GaAs and

$Al_xGa_{1-x}As$ etch at *equal rates,* to prevent nonuniform sidewall formation or any delineation of the heterostructure interfaces.

In practice, for most "standard" reactive ion etching systems, the $Al_xGa_{1-x}As$ has had a slower etch rate than GaAs. This is generally attributed to the formation of an aluminum oxide through reaction of the aluminum in $Al_xGa_{1-x}As$ with oxygen from residual oxygen or water vapor in the etch chamber. The oxide is not readily chemically etched by the chlorine-containing gas, although it may be physically sputtered off. As mentioned in Sec. 10.3.6.1, gas mixtures of BCl_3 and Cl_2[25] have been used to provide a getter for the oxygen and to increase the etch rate of the $Al_xGa_{1-x}As$. Use of a load-lock system should provide the control over residual oxygen or water vapor that is necessary for equirate etching. Load-locks, incorporated into "standard" RIE[3,39] or RIBE[4] systems have been used to successfully obtain equirate etching of $Al_xGa_{1-x}As$ and GaAs, forming transverse junction stripe (TJS) lasers and multi–quantum-well lasers (MQW), respectively. The reflectivity of the etched facets was found to be within 50 to 75 percent of the cleaved facets. The threshold currents were ~10 percent (on the average) greater for etched than for cleaved-facet lasers. Figure 10–9 shows the smooth facet obtainable using Cl_2 etching at 0.5 mtorr in a load-locked RIE system.[39] CAIBE (with a load-lock) has also been used to dry-etch laser facets.[50]

It is useful to note here that earlier work, utilizing ion-beam etching of laser facets,[51] which consisted of ion milling or sputtering alone, rather than a chemically enhanced process, produced buried heterostructure

Fig. 10–9. SEM photograph of smooth sidewall in AlGaAs/GaAs structure produced by etching in Cl_2 at 0.5 mtorr, in a load-locked system. (*After Vawter* et al.[39]).

lasers whose threshold currents were 20 percent higher than cleaved facet lasers. This was true even though a wet chemical post-etch was done to remove the superficial layer of damage, and highlights the importance of the chemical component in obtaining the desired facet quality.

10.4.3 Application to HFET Fabrication

Since the postulation and demonstration of modulation doped materials that can exhibit quantum effects and extremely high mobilities at low temperatures,[52–54] a new branch of physics has arisen to encompass the study of these materials and a new area of technology to address the device potential of these structures. Figure 10–10A illustrates a typical HFET (heterostructure FET, also called high electron mobility transistor, HEMT) structure. The high mobility of the two-dimensional electron gas at the $GaAs/Al_xGa_{1-x}As$ interface results from the smoothness of the interface and the reduction of ionized impurity scattering, both made possible by high-resolution epitaxial techniques, such as MBE. In order to best take advantage of such devices in high-speed circuit applications, one must (recess) etch the material to the proper depth to define the threshold voltages of enhancement- and depletion-mode devices, and one must be able to do so accurately, uniformly and reliably. Errors in etched depth of only nanometers can result in errors of several tenths of a volt in the

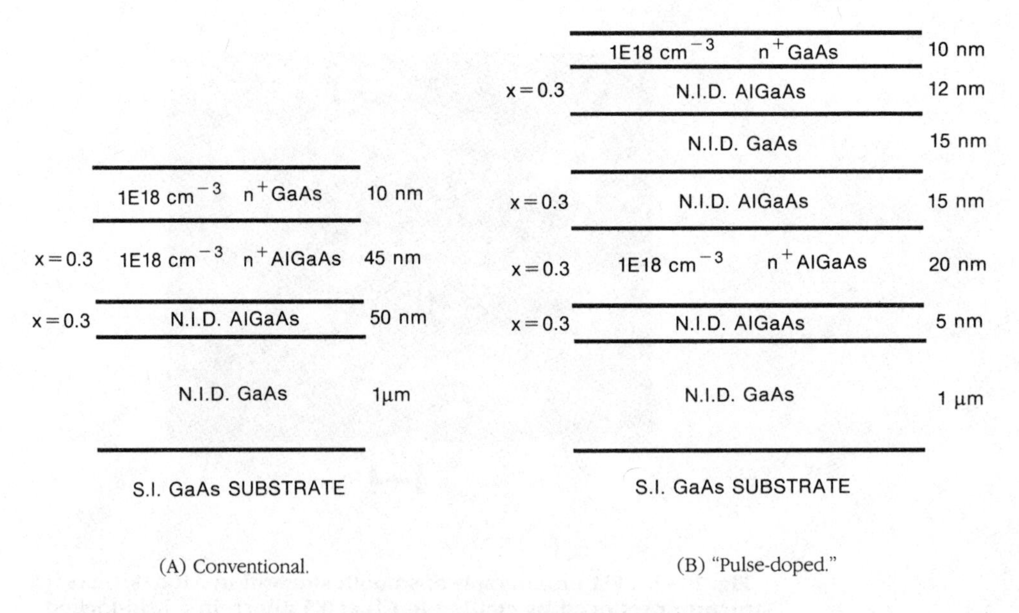

(A) Conventional.

(B) "Pulse-doped."

Fig. 10–10. Schematic of HFET structures.

threshold. The sensitivity of threshold voltage to the depth of the etched recess pertains to both the longer established GaAs MESFET[55] technology as well as the HFET[56–58] technology. However, since the HFET technology already incorporates precisely grown epitaxial structures, it is only a small process perturbation to grow layers of $Al_xGa_{1-x}As$ to serve as etch stop layers, which will be sufficiently thin so they will not affect device characteristics. Figure 10–10B depicts such a "pulse-doped" structure. The success of the technique, then, depends critically on the degree of selectivity or etch rate ratio between GaAs and $Al_x Ga_{1-x}As$ that is achievable. Note the thicknesses of the layers shown in Fig. 10–10; there is very little latitude for overetching. The use of an etch rate stop layer has been demonstrated by Cooper *et al.*[83]

We therefore need to establish the conditions where the GaAs and $Al_xGa_{1-x}As$, subject to the same plasma chemistry, will etch differently. From the considerations of equirate etching, above, it would seem that the addition of O_2 to Cl_2 etch gas would provide more rapid etching of GaAs than $Al_xGa_{1-x}As$. This has been shown to be true, but selectivities of only approximately 35:1 were obtained.[3] A number of reports have been made of the use of etches utilizing either CCl_2F_2 alone or with He.[59–61] The basis of achieving etch rate selectivity depends on the nonvolatility of the group III fluorides, particularly of AlF_3. Increasing the amount of F available fron the CCl_2F_2 etch gas should then lead to increased AlF_3 formation, hence suppression of the $Al_xGa_{1-x}As$ etch rate, since the product molecules are not desorbed from the surface. Various studies made on etch rate selectivity using CCl_2F_2 seem to support the above hypothesis.

Experiments performed using Freon 12 alone, or in a mixture with gases such as He, have found increased selectivity at higher total gas pressures. For example, Vatus *et al.*[57] found the etch rate selectivity of GaAs/AlGaAs in Freon 12 to range from approximately 20:1 at pressures of 10 to 20 mtorr up to 1000:1 at 50 mtorr (−50 V bias). Knoedler and Kuech[59] examined the etch rate selectivities of GaAs to $Al_xGa_{1-x}As$ for varying values of x ($x \in [0.04, 0.31]$). Some of their results are illustrated in Fig. 10–11. At a given value of x, higher selectivities were obtained at higher gas pressures, consistent with the above mentioned results. At a given pressure, higher selectivities were obtained for higher values of x. All results confirm a model in which increased formation of AlF_3 would more effectively retard etching of the AlGaAs. Recent work has more directly confirmed the role of AlF_3.[61] XPS, together with sensitive optical emission and mass spectroscopy, was used to analyze the GaAs and AlGaAs surfaces after reactive ion etching in Freon 12/He. The analysis indicates that the etch-stop layer is less than 30 nm of (nonvolatile) AlF_3 and $GaCl_xF_y$.

The use of highly selective dry etching in the fabrication of HFETs has produced device characteristics that are more uniform than was the case for wet etching. The Freon 12/He process used by Hikosaka *et al.*[60] gave a

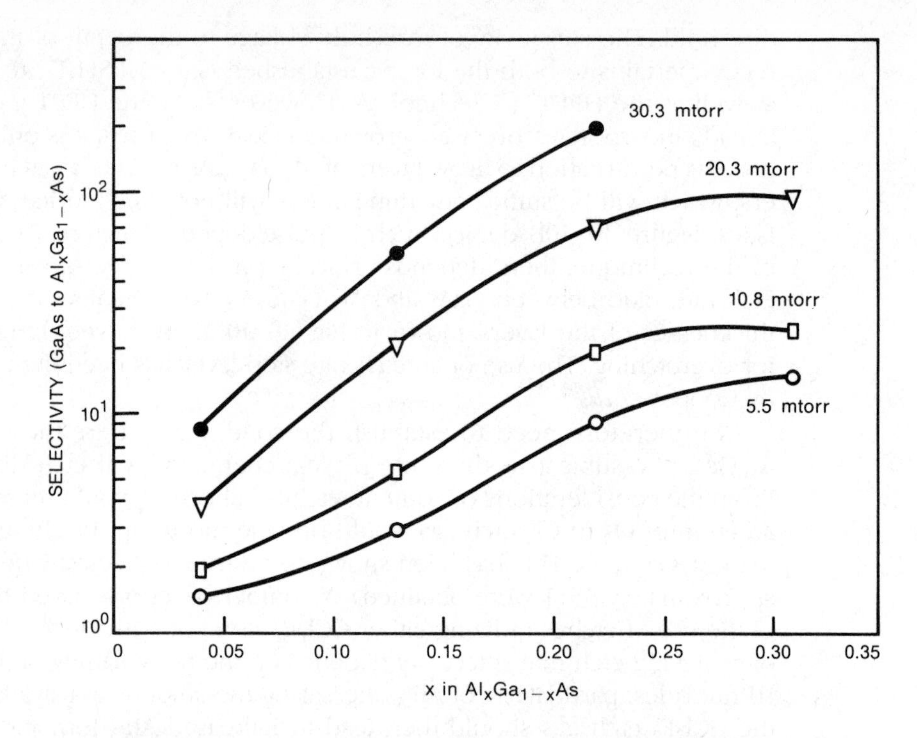

Fig. 10–11. Selectivity as a function of x for $Al_xGa_{1-x}As$, at various pressures of CF_2Cl_2. *(After Knoedler and Kuech[59]).*

28-mV variation in threshold voltages, as compared to the 132-mV variation observed for devices fabricated with wet chemical etching. Later application to the fabrication of enhancement-mode and depletion-mode HEMTs gave $V_T = 1.11 \pm 0.02$ V (depletion) and 0.02 ± 0.012 V (enhancement)† Vatus *et al.*[57] employing Freon 12 alone, used device saturation currents to monitor process uniformity, and found an $I_{DSS} = 70 \pm 5.2$ mA for 100 transistors.

The CCl_2F_2-based etch therefore shows great promise in the fabrication of high-speed HFETs that require large etch rate selectivities. A number of problems remain, however. Increasing the etch gas pressure to promote the formation of AlF_3 can also lead to the crossover point between etching and deposition onto the $Al_xGa_{1-x}As$. It has been observed that too high a ratio of CCl_2F_2 to He resulted in the formation of a "contaminated," brown surface layer, presumably rich in carbon (see Sec. 10.3.4.1). Some experiments have indicated a total GaAs etched depth which is less than the thickness of the starting material. GaAs control sam-

† Although values were more uniform, the mean values of the threshold voltages were different, with dry etching producing V_{th} approximately 0.2 mV higher than wet chemical etch.

ples etched at the same time may have etched to nearly a micrometer depth, while the GaAs cap layer of an HFET structure may have shown etching of only 15 nm of a 20-nm cap. This may be due to polymer formation and deposition onto the substrate surface. Moreover, induction effects may be especially critical in these applications, where only shallow etches (few tens to few hundred nanometers) are required. Without a load-lock system, techniques such as incorporation of BCl_3, or initiation with a H_2 etch, or presputtering at high bias will be required.

10.4.4 Formation of Quantum Structures

A promise inherent in the controlled processing of dry etching techniques is the ability to fabricate "nanostructures." Accurate pattern transfer at dimensions on the order of tens of nanometers is important not only for the fabrication of high-speed devices, but also for the exploration of the physics of structures at dimensions that are small compared with inelastic scattering lengths, or at dimensions where quantum confinement can take place. The primary considerations in this case are accuracy and control of structure dimensions. For these reasons, such etches are usually carried out at low pressures. A combination of e-beam patterning and RIE in $SiCl_4$ has produced 20-nm–diameter GaAs columns on 50-nm centers, etched ~240 nm deep.[62] Ion divergence and scattering effects were used to form 20 nm wide, deep trenches in GaAs, using CF_3Br at −500 V and 2-mtorr pressure.[63] The original masked pattern was a grating with 320-nm period.

The ability to fabricate atomically smooth interfaces to confine two-dimensional electron gases has naturally provoked researchers to further explore one-dimensional (quantum wires) and even zero-dimensional (quantum dots) structures. Both Reed *et al.*[64] and Scherer and Craighead[65] have used etch gases based on BCl_3 (either pure BCl_3, in the former case, or a 9:1 mixture of Ar:BCl_3 in the latter case). Fairly slow (30 nm/min), controlled etching of GaAs and $Al_xGa_{1-x}As$, with vertical profiles was achieved. Reed *et al.* observed a luminescent peak for the quantum dots (0.25 × 0.25 μm wide, greater than 0.25 μm deep) that was not present for the higher-dimensioned structures. Direct spatial quantization from discrete electronic levels in the quantum dots was not observed; it was postulated that the variation in dot size across the probed array was sufficient to smear out any fine structure in the quantum-well luminescence. Scherer and Craighead fabricated quantum "ribbons" 40 ± 50 nm wide, and quantum disks with diameters 45 ± 10 nm. The etched depths were ~0.2 μm. These structures should be extremely sensitive to damage that may be introduced by the etching process itself; interestingly, the initial conclusion for both experiments was that there was no substantial damage incurred. Reed *et al.* base their conclusion on inspection of the photoluminescent intensities of the various quantum-well, quantum-wire, and quantum-dot spectra. The loss of intensity with lower dimensionality was accounted for simply by loss of sample volume, indicating that nonradiative loss mechanisms due to damage to the sidewalls of the structure were minimal. Kash *et al.* in fact observe enhanced luminescence from their quantum disks, which are 50 to 100 times as bright, per unit volume of quantum-well material, as the as-grown material for an excitation energy of 1.73 eV.[85] They concluded from this evidence that the free surfaces created by the RIE therefore did not function as efficient nonradiative recombination sites for photoexcited carriers.

Total etched confinement is not always necessary to achieve spatial confinement of carriers; often, surface depletion may be effective in confining an electrical conduction path. This approach is illustrated in Fig. 10–12, where a partial etch into the AlGaAs is made. Depletion of the AlGaAs surface in the etched region is sufficient to confine the two-dimensional electron gas at the AlGaAs/GaAs interface. The possibility of degradation in electrical transport, introduced by degradation of the etched sidewall surface, is therefore minimized. This has been utilized to fabricate narrow conductive structures[66] and a narrow annulus for the study of the Aharonov-Bohm effect.[67,68]

10.5 Process-induced Damage

Notwithstanding the conclusions of the quantum-dot experiments, the effects of the dry-etch process itself on the performance of the fabricated device or structure remain an important, not totally understood issue. The damage can be brought about by chemical modification of the etched surfaces or by the physical bombardment of the material by energetic ions.

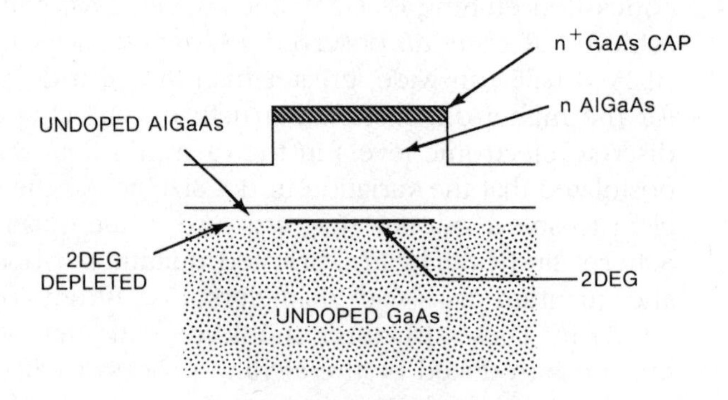

Fig. 10–12. Partial etch-confinement of a two-dimensional electron gas. *(After van Houten* et al.[66]*).*

Process induced damage may be manifested in many ways: surface modification from impurity deposition (e.g. carbon) has been observed to prevent epitaxial regrowth of materials by liquid-phase or molecular-beam epitaxy. Striations in etched laser facets will reduce the efficiency of those surfaces as mirrors, and raise the threshold current of the laser. Trap states introduced into the material may degrade the mobility of electronic devices. Fonash has used a simple model to describe dry-etch damage in Si, which may serve as a useful guide for discussion of damage in GaAs.[69] The model comprises three layers formed in the etched substrate, which are not sharply defined (Fig. 10–13). The uppermost layer has been called the "R-layer" or "residue" layer, such as would be formed from the deposition of carbon as a by-product of etching. These are expected to be quite thin layers (less than ~10 nm) located at the substrate surface. The next layer is termed the "P layer" or "permeated" layer, where the etch species themselves or other impurities have been driven into the substrate. This layer is expected to be from a few to several tens of nanometers in depth. The third layer is the "D layer" for "intrinsic bonding damage" layer. This may be coincident with the P layer, or may extend farther into the substrate. The D layer is characterized by dislocations, vacancies and interstitials.

Because the extent of the damage may only extend a few tens or a few hundred nanometers into the substrate, electrical measurements, such as deep-level transient spectroscopy (DLTS), or *I-V* and *C-V* characteristics of Schottky diodes, may have limited utility. If the damaged layer falls within the surface depletion width, it will not be easily probed. Surface analytical techniques, such as Auger electron spectroscopy (AES), reflection high energy electron diffraction (RHEED), and Raman spectroscopy can be sensitive probes of the surface crystal quality, composition, and doping. Rutherford backscatter spectroscopy has been used to determine lower lying crystalline disorder. In photoluminescence measurements, one must ensure that the optical absorption length is sufficiently small that the dominant luminescence is emanating from the damage layer alone, without excessive contribution from the larger volume of material under it. Early work on Ar ion sputtering of GaAs (at energies ranging from 100 eV

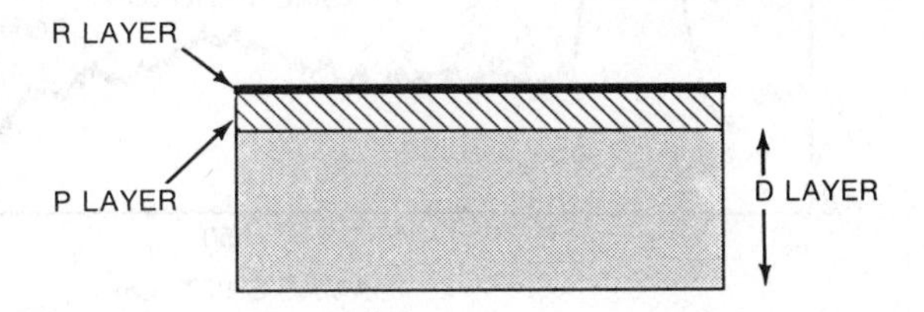

Fig. 10–13. Schematic of the dry-etch damage model. *(After Fonash[69]).*

to 2 keV) utilized a combination of techniques to deduce the presence of a thin (<10 nm) damaged, amorphous layer, and a deeper distribution of defects (extending ~200 nm into the material).[70] RBS and electron diffraction allowed estimation of the depth of the amorphous region, while the intensity of the photoluminescent peak allowed mapping of the deeper defect level. Using the photoluminescent peak intensity as a monitor, it was shown that low ion energy damage (100 eV) could be annealed out (450°C), but recovery of the material quality by annealing is more difficult at greater incident ion energies.

Studies of damage on GaAs made by Pang indicate that the nature and degree of the damage is quite different whether using reactive or inert etch species, even at the same bias voltage (−500 V).[71–73] The use of reactive gases such as Cl_2, $SiCl_4$, CCl_2F_2, or CF_4 produced Schottky *I-V* characteristics showing much less change in barrier height and breakdown voltage, compared with the control sample, than were obtainable using Ar ion-beam milling.

Raman spectra can give information on crystalline quality, from examination of the phonon peaks, and on electrical properties of doped substrates such as carrier concentration, mobility and depletion depth, from the study of the plasmon peaks. Figure 10–14 is the Raman spectrum of a

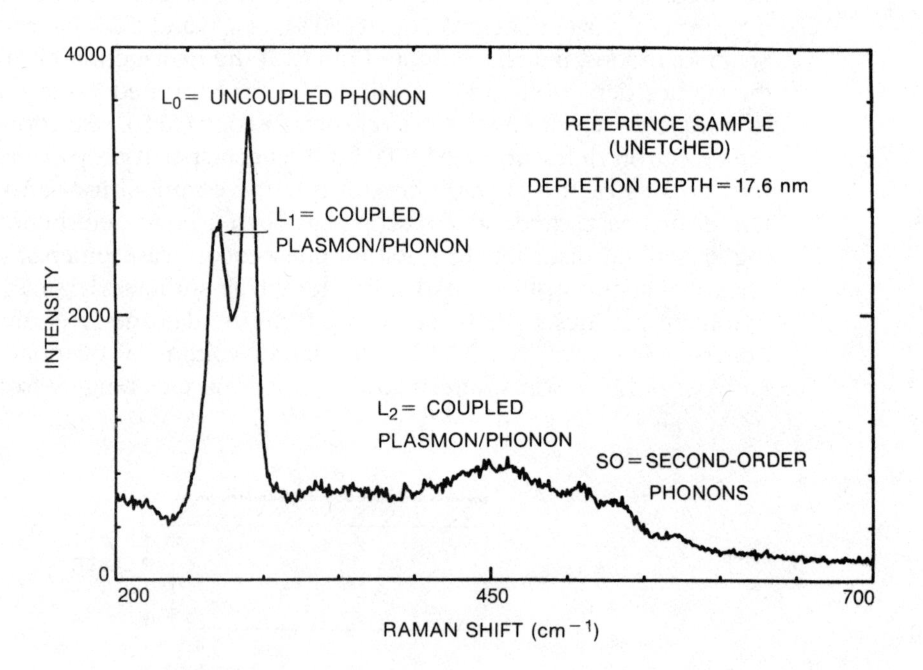

Fig. 10–14. Raman spectrum of unetched (100) n-type GaAs, Si-doped 1.4×10^{18} cm^3.

reference sample, 1.4 × 10^{18} cm^{-3} n-type (100) GaAs, showing the peaks due to longitudinal optical phonons (LO) and coupled plasmon-phonon modes (L_1, L_2). For the particular scattering geometry used to obtain Fig. 10–14, Raman group-theoretical selections rules prohibit observation of the peak due to transverse optical (TO) phonons. Appearance of the TO peak after etching of the substrate is interpreted as degradation of the initial crystal symmetry, hence introduction of damage. Such an identification of the appearance of the TO peak with material damage formed the basis of several studies of dry-etched damage to GaAs and $Al_xGa_{1-x}As$ ($x = 0.22$) etched in Ar and $SiCl_4$ plasmas.[74–76,86] Generally, an expected increase in the TO/LO peak intensity was observed for higher incident power (50 W, 100 W) or bias voltage.

As the phonon scattering gives information on the structural properties of the material, the plasmon peaks yield information on its electrical properties. For example, the depletion layer thickness can be determined from the intensity ratio I_{LO}/I_{L1}, and this in turn may serve as a monitor of damage. Systematic studies of both downstream (see afterglow etches in Sec. 10.6) and RIE etching used the depletion width as an indicator of damage.[87] Downstream "etching" in argon produced no observable change in depletion width; RIE in argon at bias voltages of −350 V resulted in damage that was located within the first 30 nm of the surface. The depletion depths increased with increasing bias voltage, with the rate of change of the depletion depth increasing above approximately −300 V (see Fig. 10–15). Either downstream or reactive ion etching in a chemically active gas, such as Cl_2 or CCl_2F_2, produced depletion depths less than the control sample, indicating significant differences in the surface electronic properties, as compared with the cases using inert ambient gases.

Comparisons of dry-etched and cleaved laser facets have deen discussed in Sec. 10.4.2. Although the best results achieved show that dry-etched facet quality can be nearly comparable to cleaved facets, there is still area for improvement. The slight degradation in reflectivity of the dry-etched facets may be due either to the composition or the morphology of the etched surface, and the remedies to be taken will vary accordingly. A gentle, *in situ* post-etch clean may be desirable (perhaps using the radical beams of Asakawa and Sugano) or greater attention given to the quality of the edge smoothness of the etch mask.

Data on electronic devices have shown a degradation in the transconductance of the as-etched transistors. That degradation, in fact, tracks the increase in self-bias voltage, and can in some cases be annealed out (400°C, 30 min).[56] A 4-min anneal at 470°C, after the etch process was completed, produced devices with a room temperature transconductance, g_m, of 230 mS/mm for a 0.7 × 300 μm gate.[57] The higher the self-bias voltage during the dry-etch, the more difficult it becomes to anneal out the damage.

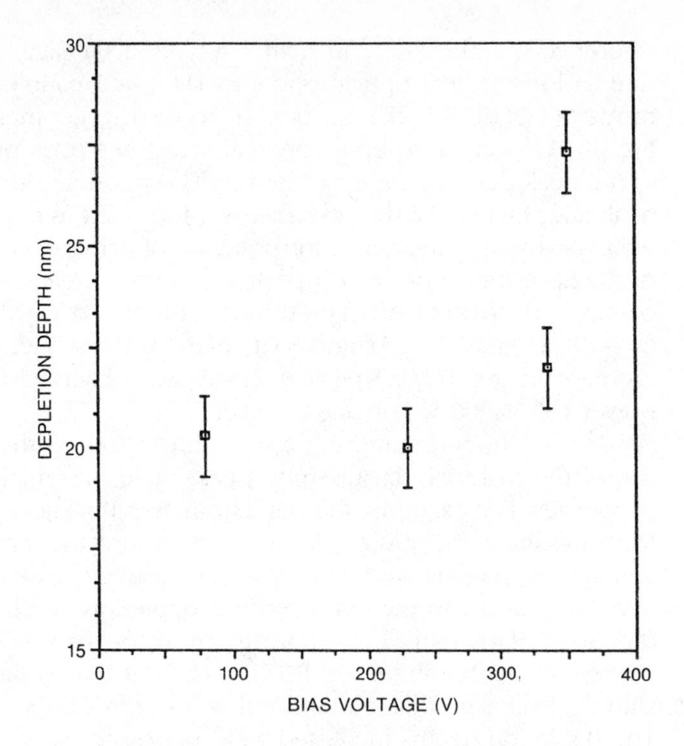

Fig. 10–15. Depletion depth versus RIE bias voltage for GaAs samples "etched" in argon.

Finally, interest in quantum structures (see Sec. 10.4.4) has prompted the use of appropriate structures as probes of dry-etch damage. Measurements have been made on the conductivity as a function of width of semiconductor "wires" formed by etch delineation.[88,89] As the wires become more narrow, the conductivity will ultimately be cut off, ostensibly at a width w_0 equal to twice the depletion width from the two sidewalls. The results obtained using wet-etch delineation, compared to dry-etch delineation, show a larger cutoff width for the dry-etched wires, suggesting additional damage to the sidewalls. Moreover, the choice of etch gas can affect the cutoff width dramatically: CH_4/H_2 gives a w_0 = 400 nm, compared with 100 nm for $SiCl_4$ RIE and 40 nm for wet-etched wires.

Such results of the deep penetration of a light-mass etching species are also borne out by recent cathodoluminescence studies, using selectively placed GaAs quantum wells (QWs) as probes of etch damage.[90] The probe structure is shown in Fig. 10–16. Since each quantum well is of a different width, placed at varying distances from the substrate surface, the cathodoluminescence peak of a particular QW provides information on the material quality at a particular depth, and the spectrum of peaks will provide a depth distribution of damage. Figure 10–17 shows a typical un-

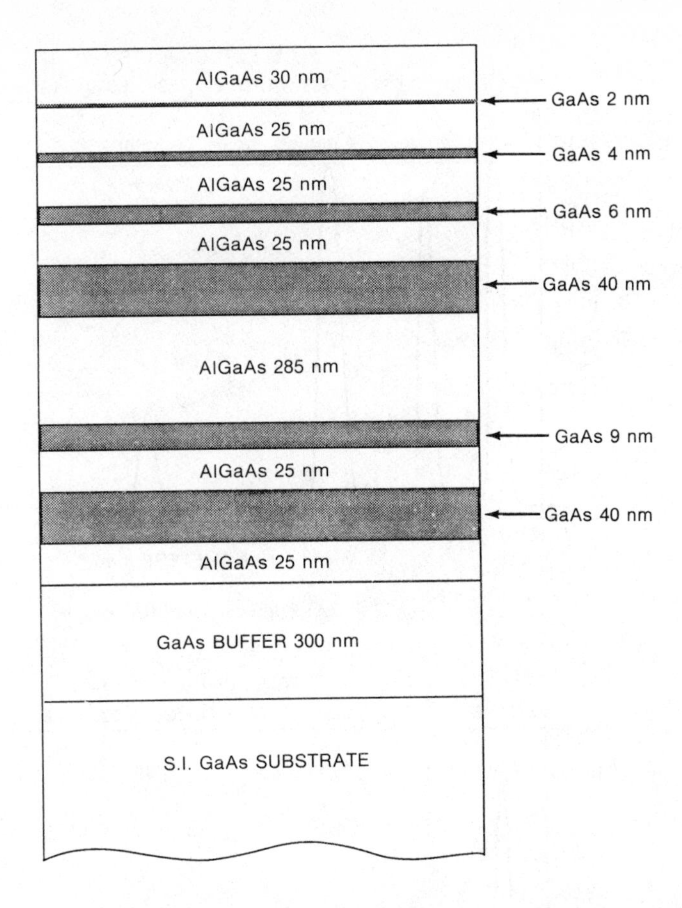

Fig. 10–16. Cross section of GaAs/AlGaAs multiple-quantum-well structure.

etched spectrum compared to one taken for material that was subjected to argon RIE at −500 V for 1 min. Figure 10–18 displays those data in another way, comparing the depth distribution of damage to argon RIE with that due to helium RIE carried out at the same bias voltage. Earlier studies have indicated that treatment in hydrogen plasma[73] may passivate shallow level impurities and deep defect levels in GaAs; this might provide an important post-etch treatment. Further work will have to be done to ascertain the benefits or liabilities of plasma treatments using light-mass gases.

10.6 Other Dry-Etch Processes

The desire to eliminate the materials damage brought about by the physical sputtering component of the dry etching processes described above may make other dry etching schemes attractive. Afterglow etchers, in

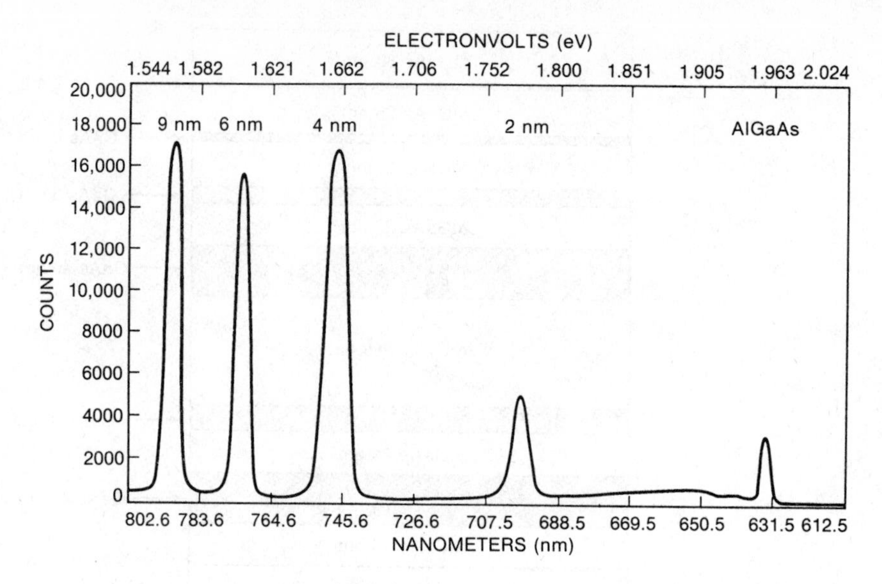

(A) Reference, unetched sample.

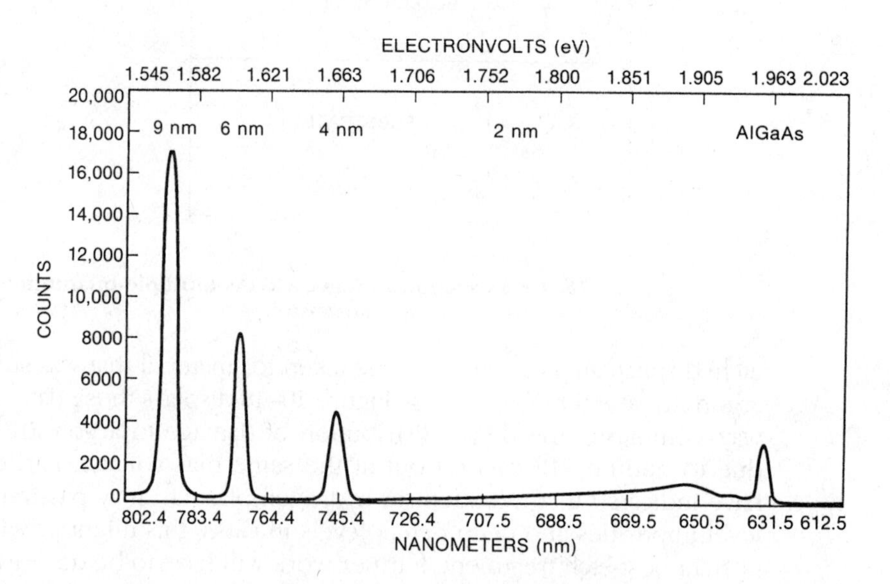

(B) Sample subject to argon RIE at −500 V for 1 min.

Fig. 10–17. Cathodoluminescence spectra.

which the substrate is placed downstream, out of physical contact with the plasma, are gaining popularity for oxygen plasma stripping of photoresist. The technique may be extended to other gases and other materials, including GaAs. Geis[12,13] has injected Cl_2 gas into a tungsten tube heated to 1500–2000°C. The high temperature cracks the molecular Cl_2 to form the

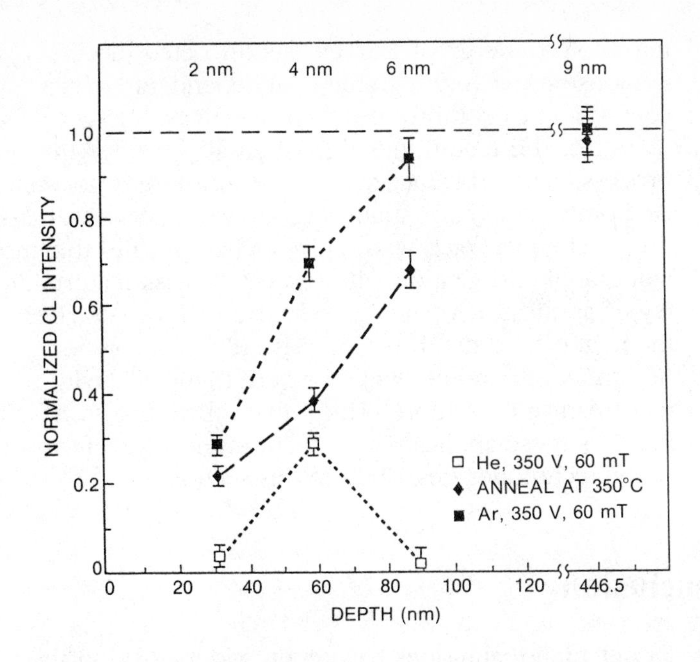

Fig. 10–18. Normalized CL intensity versus quantum-well depth for samples subjected to RIE in argon and helium respectively, for 1 min at −350 V bias. Also shown are data for the helium-treated samples after an anneal at 350°C for 20 min.

reactive etch species, without initiation of a plasma. A similar method has been reported recently, where the "cracking" of the Cl_2 is now brought about by a microwave cavity.[91] This radical-beam, ion-beam etching (RBIBE) has shown controllably high etch rates (550 nm/min) at low bias voltages (−200 eV), producing smooth, anisotropic etched profiles. Lasers may also be used to dissociate the molecular etch gas into the activated species needed to carry out etching. Takai *et al.*[78] have used a focused argon ion laser in a CCl_4 atmosphere at 80 torr to pyrolytically etch GaAs. The laser power density was 233 kW/cm^2, and etch rates between 0.3 and 40 μm/scan were obtained for scan speeds between 23 and 12 μm/s. Laser pyrolysis of methyl halides and HBr have also been used to etch GaAs.[79–81]

Very promising work on the laser-enhanced, photochemical selective etching of GaAs and other III-V compound semiconductors has been demonstrated.[82] The assumption here is that electron-hole pairs, created by the incident laser radiation, are confined at the surface for some finite time and are influential in the surface chemistry that determines the etch process. Ashby[82] has demonstrated etching of the GaAs (E_g at 400 K = 1.38 eV), $GaAs_{0.2}P_{0.8}$ (E_g = 1.58 eV) and $GaAs_{0.63}P_{0.37}$ (E_g = 1.80 eV) in HCl strongly diluted (0.2 percent) in He. Use of an argon ion laser with photon energy of 2.41 eV produced etching in all samples, while a dye laser having photon energy of 1.62 eV failed to etch the $GaAs_{0.63}P_{0.37}$. The process demonstrates a very high material-dependent etch rate, with no sputtering component to compromise etch selectivity.

All of the techniques described above offer the possibility of lower process-incurred damage, since the substrate is not subject to direct physical bombardment, or sputtering. However, as was discussed in Sec. 10.3, a purely chemical etch would give etched profiles that may depend on crystallographic orientation with respect to mask features. Yet there are a variety of applications (such as selective etching for HFET structures) where the total etched depth required is only on the order of a few tens to a few hundreds of nanometers and where high selectivity and reduction of material damage are more critical than obtaining an absolutely vertical profile. It is these applications that could most benefit from the development of the techniques described in this section.

10.7 Conclusion

Dry etching techniques have been widely used in the fabrication of GaAs devices. Dry-etch processes can be tailored to address the needs of rapidly performed, deep etches, high resolution pattern transfer, high etch rate selectivity and controllably etched profiles. Fuller understanding of the surface chemistry, etch-induced damage and methods of *in situ* monitoring and evaluation is required. Techniques need to be developed to allow adequate monitoring of etching of layers only tens of nanometers deep. Full understanding is needed of all parameters that contribute to and modify the etch environment over time, so that uniformity and reproducibility of etching can be achieved. Methods of minimizing ion-induced damage are needed. The capabilities thus far demonstrated by dry etching techniques are widespread and impressive, but the potential of dry etching has not yet been fully realized.

References

1. G. Smolinsky, R. P. Chang, and T. M. Mayer, "Plasma etching of III-V compound semiconductor materials and their oxides," *J. Vac. Sci. Technol.,* 18:12 (1981).
2. V. M. Donnelly, D. L. Flamm, and D. E. Ibbotson, "Plasma etching of III-V compound semiconductors," 29th National Symposium of the AVS (November 16–19, 1982).

3. E.L. Hu and L.A. Coldren, "Recent developments in reactive plasma etching of III-V compound semiconductors," *Proc. SPIE Sym. Adv. Process. Semiconductor Dev.,* Baypoint, Florida (1987).
4. K. Asakawa and S. Sugata, "GaAs and GaAlAs equi-rate etching using a new reactive ion beam etching system," *Jpn. J. Appl. Phys.,* 22:L653 (1983).
5. M.W. Geis, G.A. Lincoln, N. Efremow, and W.J. Piacenti, "A novel anisotropic dry etching technique," *J. Vac. Sci. Technol.,* 19:1390 (1981).
6. J.D. Chinn, A. Fernandez, I. Adesida, and E.D. Wolf, "Chemically assisted ion beam etching of GaAs, Ti and Mo," *J. Vac. Sci. Technol. A,* 1:701 (1983).
7. Y. Ochiai, K. Gamo, and S. Namba, "Characteristics of ion beam assisted etching of GaAs using focused ion beam: dependence on gas pressure," *Jpn. J. Appl. Phys.,* 23:L400 (1984).
8. B. Chapman, *Glow Discharge Processes.,* New York: John Wiley & Sons (1980).
9. J.L. Vossen and W. Kern (ed.), *Thin Film Processes,* New York: Academic Press (1978).
10. T. Sugano (ed.), *Applications of Plasma Processes to VLSI Technology,* New York: John Wiley & Sons (1983).
11. V.M. Donnelley, D.L. Flamm, and G. Collins, "Laser diagnostics of plasma etching: measurement of Cl_2^+ in a chlorine discharge," *J. Vac. Sci. Technol.,* 21:817 (1982).
12. M.W. Geis, N.N. Efremow, G.A. Lincoln, "Hot jet etching of GaAs and Si," *J. Vac. Sci. Technol. B,* 4:315 (1986).
13. M.W. Geis, N.N. Efremow, S.W. Pang, and A.C. Anderson, "Hot jet etching of Pb, GaAs and Si," *J. Vac. Sci. Technol. B,* 5:363 (1987).
14. E.L. Hu and R.E. Howard, "Reactive ion etching of GaAs in a chlorine plasma," *J. Vac. Sci. Technol. B,* 2:85 (1984). So much of the literature of GaAs dry-etching involves Cl_2 that all references will not be separately listed here.
15. R.H. Burton and G. Smolinsky, "CCl_4 and Cl_2 plasma etching of III-V semiconductors and the role of added O_2," *J. Electrochem. Soc.,* 129:1599 (1982).
16. R.A. Gottscho, G. Smolinsky, and R.H. Burton, "Carbon tetrachloride plasma etching of GaAs and InP: a kinetic study utilizing nonperturbative optical techniques," *J. Appl. Phys.,* 53:5908 (1982).
17. S. Semura, H. Saitoh, and K. Asakawa, "Reactive ion etching of GaAs in CCl_4/H_2 and CCl_4/O_2," *J. Appl. Phys.,* 55:3131 (1984).

18. G. Smolinsky, R.A. Gottscho, and S.M. Abys, "Time-dependent etching of GaAs and InP with CCl_4 or HCl plasmas," *J. Appl. Phys.,* 54:3518 (1983).
19. M.B. Stern and P.F. Liao, "Reactive ion etching of GaAs and InP using $SiCl_4$," *J. Vac. Sci. Technol. B,* 1:1053 (1983).
20. J.Z. Li, I. Adesida, and E.D. Wolf, "Evidence of crystallographic etching in (100) GaAs using $SiCl_4$ reactive ion etching," *J. Vac. Sci. Technol. B,* 3:406 (1985).
21. E.L. Hu and R.E. Howard, "Reactive ion etching of GaAs and InP using $CCl_2F_2/Ar/O_2$," *Appl. Phys. Lett.,* 37:1022 (1980).
22. R.E. Klinger and J.E. Greene, "Reactive ion etching of GaAs in CCl_2F_2," *Appl. Phys. Lett.,* 38:620 (1981).
23. J. Chaplart, B. Fay, and Nguyen Linh, "Reactive ion etching of GaAs using CCl_2 F_2 and the effect of Ar addition," *J. Vac. Sci. Technol. B,* 1:1050 (1983).
24. G.J. Sonek and J.M. Ballantyne, "Reactive ion etching of GaAs using BCl_3," *J. Vac. Sci. Technol. B,* 2:653 (1984).
25. H. Tamura and H. Kurihara, "GaAs and GaAlAs reactive ion etching in BCl_3-Cl_2 mixture," *Jpn. J. Appl. Phys.,* 23:L731 (1984).
26. D.E. Ibbotson, D.L. Flamm, and V.M. Donnelly, "Crystallographic etching of GaAs with bromine and chlorine plasmas," *J. Appl. Phys.,* 54:5974 (1983).
27. M. Balooch and D.R. Olander, "The thermal and ion-assisted reactions of GaAs (100) with molecular chlorine," *J. Vac. Sci. Technol. B,* 4:794 (1986).
28. S.C. McNevin, "Chemical etching of GaAs and InP by chlorine: the thermodynamically predicted dependence of Cl_2 pressure and temperature," *J. Vac. Sci. Technol. B,* 4:1216 (1986).
29. R.A. Barker, T.M. Mayer, and R.H. Burton, "Surface composition and etching of III-V semiconductors in Cl_2 ion beams," *Appl. Phys. Lett.,* 40:583 (1980).
30. K. Asakawa and S. Sugata, "Optical emission spectrum of Cl_2 ECR plasma in the GaAs reactive ion beam etching (RIBE) system," *Jpn. J. Appl. Phys.,* 23:L156 (1984).
31. S. Sugata and K. Asakawa, "GaAs radical etching with a Cl_2 plasma in a reactive ion beam etching system," *Jpn. J. Appl. Phys.,* 23:L564 (1984).
32. S. Sugata and K. Asakawa, "GaAs and AlGaAs crystallographic etching with low-pressure chlorine radicals in ultrahigh-vacuum system," *J. Vac. Sci. Technol. B,* 5:894 (1987).

33. V. M. Donnelly, D. L. Flamm, C. W. Tu, and D. E. Ibbotson, "Temperature dependence of InP and GaAs etching in a chlorine plasma," *J. Electrochem. Soc.,* 129:2534 (1982).

34. R. E. Klinger and J. E. Greene, "Reactive ion etching of GaAs in $CCl_{4-x}F_x$ ($x = 0, 2, 4$) and mixed $CCl_{4-x}F_x$/Ar discharges," *J. Appl. Phys.,* 54:1595 (1983).

35. R. H. Burton, C. L. Hollien, L. Marchut, S. M. Abys, G. Smolinsky, and R. A. Gottscho, "Etching of gallium arsenide and indium phosphide in rf discharges through mixtures of tricholorfluoromethane and oxygen," *J. Appl. Phys.,* 54:6663 (1983).

36. J. W. Coburn and E. Kay, "Some aspects of the fluorocarbon plasma etching of silicon and its compounds," *IBM J. Res. Develop.,* 23:33 (1979).

37. D. Hess, *Plasma Chem. Process.,* 2:141 (1982).

38. R. P. H. Chang, C. C. Chang, and S. Darack, "Hydrogen plasma etching of semiconductors and their oxides," *J. Vac. Sci. Technol.,* 20:45 (1982).

39. G. A. Vawter, L. A. Coldren, J. M. Merz, and E. L. Hu, "Non-selective etching of GaAs-AlGaAs double heterostructure laser facets by conventional Cl_2 reactive ion etching," *Appl. Phys. Lett.,* 51:719 (1987).

40. K. Asakawa and S. Sugata, "Damage and contamination-free GaAs and AlGaAs etching using a novel ultrahigh-vacuum reactive ion beam etching system with etched surface monitoring and cleaning method," *J. Vac. Sci. Technol. A,* 4:677 (1986).

41. A. Takamori, S. Sugata, K. Asakawa, E. Miyauchi, and H. Hashimoto, "Cleaning of MBE GaAs substrates by hydrogen radical beam irradiation," *Jpn. Appl. Phys.,* 26:L142 (1987).

42. A. E. Geissberger and P. R. Claytor, "Application of plasma etching to via hole fabrication in thick GaAs substrates," *J. Vac. Sci. Technol.,* 3:863 (1985).

43. L. G. Hipwood and P. N. Wood, "Dry etching of through substrate via holes for GaAs MMIC's," *J. Vac. Sci. Technol. B,* 3:395 (1985).

44. K. P. Hilton and J. Woodward, "Via holes for GaAs MMIC's fabricated using reactive ion etching," *Electron. Lett.,* 21:962 (1985).

45. R. J. Contolini and L. A. D'Asaro, "High rate magnetron enhanced reactive ion etching of GaAs," *J. Vac. Sci. Technol. B,* 4:706 (1986).

46. P. Buchmann, H. Kaufmann, H. Melchior, and G. Guekos, "Reactive ion etched GaAs optical waveguide modulators with low loss and high speed," *Electron. Lett.,* 20:295 (1984).

47. H. Yamada, H. Ito, and H. Inaba, "Microprocessing of GaAs cylindrical columns for integrated optical device fabrication by Cl_2-Ar reactive ion etching," *Electron. Lett.,* 20:591 (1984).

48. L. A. Coldren, K. Iga, B. I. Miller, and J. Rentschler, "GaInAs/InP stripe geometry lasers with reactive ion etched facet," *Appl. Phys. Lett.,* 37:681 (1980).
49. T. Yuasa, M. Mannoh, K. Asakawa, K. Shinozaki, and M. Ishii, "Dry-etched-cavity pair-groove-substrate GaAs/AlGaAs multiquantum well lasers," *Appl. Phys. Lett.,* 48:748 (1986).
50. P. Tihanyi, D. K. Wagner, A. J. Roza, H. J. Volhmer, C. M. Harding, R. J. Davis, and E. D. Wolf, "High-power AlGaAs/GaAs single quantum well lasers with chemically assisted ion beam etched mirrors," *Appl. Phys. Lett.,* 50:1640 (1987).
51. N. Bouadama, J. Riou, and A. Kampfer, "Low threshold GaAs/GaAlAs BH lasers with ion-beam etched mirrors," *Electron. Lett.,* 21:566 (1985).
52. R. Dingle, H. L. Stormer, A. C. Gossard, and W. Weigmann, "Electron mobilities in modulation-doped semiconductor heterojunction superlattices," *Appl. Phys. Lett.,* 33:665 (1980).
53. T. Mimura, S. Hiyamizu, T. Fujii, and K. Nanbu, "A new field-effect transistor with selectively doped GaAs/n-AlGaAs heterojunctions," *Jpn. J. Appl. Phys.,* 19:L225 (1980).
54. D. Delagebeaudeuf, P. Delescluse, P. Etienne, M. Laviron, J. Chaplart, and N. T. Linh, "Two-dimensional electron gas MESFET structure," *Electron. Lett.,* 16:667 (1980).
55. F. J. Ryan, M. F. Chang, R. P. Vahrenkamp, D. A. Williams, W. P. Fleming, and C. G. Kirkpatrick, "New dry recess etching technology for GaAs digital ICs," *Tech. Dig. 1985 GaAs IC Symp.,* 45 (1985).
56. S. Kuroda, T. Mimura, M. Suzuki, N. Kobayashi, N. Nishiuchi, A. Shibatomi, and M. Abe, "New device structure for 4Kb HEMT SRAM," *Tech. Dig. 1984 GaAs IC Symp.,* 125 (1984).
57. J. Vatus, J. Chevrier, P. Deslescluse, and J.-F. Rochette, "Highly selective reactive ion etching applied to the fabrication of low-noise AlGaAs GaAs FETs," *IEEE Trans. Electron Dev.,* ED-33:934 (1986).
58. B. J. F. Lin, S. Kofol, H. Luechinger, J. N. Miller, D. E. Mars, B. White, and E. Littau, "Threshold voltage control of MODFET IC," *GaAs IC Symp.,* 51 (1986).
59. C. M. Knoedler and T. F. Kuech, "Selective GaAs/$Al_xGa_{1-x}As$ reactive ion etching using CCl_2F_2," *J. Vac. Sci. Technol. B,* 4:1233 (1986).
60. K. Hikosaka, T. Mimura, and K. Joshin, "Selective dry etching of AlGaAs-GaAs heterojunction," *Jpn. J. Appl. Phys.,* 20:L847 (1981).
61. K. L. Seaward, N. J. Moll, D. J. Coulman, and W. F. Stickle, "An analytical study of etch and etch-stop reactions for GaAs on AlGaAs in CCl_2F_2 plasma," *J. Appl. Phys.,* 61:2358 (1987).

62. M.B. Stern, H.G. Craighead, P.F. Liao, and P.M. Mankiewich, "Fabrication of 20-nm structures in GaAs," *Appl. Phys. Lett.,* 45:410 (1984).

63. S.W. Pang, J.N. Randall, and M.W. Geis, "Sub-100-nm-wide, deep trenches defined by reactive ion etching," *J. Vac. Sci. Technol. B,* 4:341 (1986).

64. M.A. Reed, R.T. Bate, K. Bradshaw, W.M. Duncan, W.R. Frensley, J.W. Lee, and H.D. Shih, "Spatial quantization in GaAs-AlGaAs multiple quantum dots," *J. Vac. Sci. Technol. B,* 4:358 (1986).

65. A. Scherer and H.G. Craighead, "Fabrication of small laterally patterned multiple quantum wells," *Appl. Phys. Lett.,* 49:1284 (1986).

66. H. van Houten, B.J. van Wees, M.G.J. Hejiman, and J.P Andre, "Submicron conducting channels defined by shallow mesa etch in GaAs-AlGaAs heterojunctions," *Appl. Phys. Lett.,* 49:1781 (1986).

67. R. Behringer, P.M. Mankiewich, and R.E. Howard, "Fabrication of ultrahigh resolution structures in compound semiconductor heterostructures," *J. Vac. Sci. Technol. B,* 5:326 (1987).

68. Timp, A.M. Chang, J.E. Cunningham, T.Y. Chang, P. Mankiewich, R. Behringer, and R.E. Howard, "Observation of the Aharonov-Bohm effect for $\omega_c \tau > 1$," *Phys. Rev. Lett.,* 58:2814 (1987).

69. S.J. Fonash, "Damage effects in dry etching," *Solid State Technol.,* 28:201 (1985).

70. M. Kawabe, N. Kanzaki, K. Masuda, and S. Namba, "Effects of ion etching on the properties of GaAs," *Appl. Opt.,* 17:2556 (1978).

71. S.W. Pang, "Dry etched induced damage in Si and GaAs," *Solid State Technol.,* 249 (1984).

72. S.W. Pang, "Surface damage on GaAs induced by reactive ion etching and sputter etching," *J. Electrochem. Soc.,* 133:784 (1986).

73. S.W. Pang, M.W. Geis, N.N. Efremow, and G.A. Lincoln, "Effects of ion species and adsorbed gas on dry etching induced damage in GaAs," *J. Vac. Sci. Technol. B,* 3:398 (1985).

74. S. Sugata and K. Asakawa, "Investigation of GaAs surface morphology induced by Cl_2 gas reactive ion beam etching," *Jpn. J. Appl. Phys.,* 22:L813 (1983).

75. D. Kirillov, C.B. Cooper III, and R.A. Powell, "Raman scattering study of plasma etching damage in GaAs," *J. Vac. Sci. Technol. B,* 4:1316 (1986).

76. D. Kirillov, C.B. Cooper III, and R.A. Powell, "Dry etching damage in GaAs and $Al_{.22}Ga_{.78}As$, *Mat. Res. Soc. Proc. Vol. 76: Symposium on Science and Technology of Microfabrication,* ed. R.E. Howard, E.L. Hu, S. Namba, and S. Pang, 163 (December 1986).

77. W. C. Dautremont-Smith, J. C. Nabity, V. Swamiathan, M. Stavola, J. Chevallier, C. W. Tu, and S. J. Pearton, "Passivation of deep level defects in molecular beam epitaxial GaAs by hydrogen plasma exposure," *Appl. Phys. Lett.,* 49:1098 (1986).
78. M. Takai, J. Tokuda, H. Nakai, K. Gamo, and S. Namba, "Laser induced local etching of gallium arsenide in gas atmosphere," *Jpn. J. Appl. Phys.,* 22:L757 (1983).
79. D. J. Ehrlich, R. M. Osgood, Jr., and T. F. Deutsch, "Laser-induced microscopic etching of GaAs and InP," *Appl. Phys. Lett.,* 36:698 (1980).
80. P. D. Brewer, D. McClure, and R. M Osgood, Jr., "Dry, laser-assisted rapid HBr etching of GaAs," *Appl. Phys. Lett.,* 47:310 (1985).
81. P. D. Brewer, D. McClure, and R. M. Osgood, Jr., "Excimer laser projection etching of GaAs," *Appl. Phys. Lett.,* 49:803 (1986).
82. C. I. Ashby and R. M. Biefeld, "Composition-selective photochemical etching of compound semiconductors," *Appl. Phys. Lett.,* 47:62 (1985).
83. C. B. Cooper, III, S. Salimian, and H. F. MacMillan, "Use of thin AlGaAs and InGaAs stop-etch layers for reactive ion etch processing of III-V compound semiconductor devices," *Appl. Phys. Lett.,* 51:2225 (1987).
84. R. Cheung, S. Thoms, S. P. Beaumont, G. Doughty, V. Law, and C. D. W. Wilkinson, *Electron. Lett.,* 23:857 (1987).
85. K. Kash, A. Scherer, J. M. Worlock, H. G. Craighead, and M. C. Tamargo, "Optical spectroscopy of ultrasmall structures etched from quantum wells," *Appl. Phys. Lett.,* 49:1043 (1986).
86. M. Watt, C. M. Sotomayor-Torres, R. Cheung, C. D. W. Wilkinson, H. E. G. Arnot, and S. P. Beaumont, "Raman scattering of reactive-ion etched GaAs," *J. Mod. Opt.,* 35:365 (1988).
87. D. G. Lishan, D. Green, H. F. Wong, E. L. Hu, J. L. Merz, and D. Kirillov, "Dry etch induced damage in GaAs investigated using Raman scattering spectroscopy," *J. Vac. Sci. Technol. B,* 7:556 (1989).
88. R. Cheung, S. Thoms, I. McIntyre, C. D. W. Wilkinson, and S. P. Beaumont, "Passivation of donors in electron-beam lithographically defined nanostructures after methane/hydrogen reactive ion etching," *J. Vac. Sci. Technol. B,* 6:1911 (1988).
89. S. W. Pang, W. D. Goodhue, T. M. Lyszczarz, D. J. Ehrlich, R. B. Goodman, and G. D. Johnson, "Dry etching induced damage on vertical sidewalls of GaAs channels," *J. Vac. Sci. Technol. B,* 6:1916 (1988).
90. H. F. Wong, D. L. Green, T. Y. Liu, D. G. Lishan, M. Bellis, E. L. Hu, P. M. Petroff, P. O. Holtz, and J. L. Merz, "Investigation of RIE-induced damage in GaAs-AlGaAs quantum-well structures," *J. Vac. Sci. Technol. B,* 6:1906 (1988).
91. J. A. Skidmore, L. A. Coldren, E. L. Hu, and J. L. Merz, "Radical beam/ion beam etching (RBIBE) of GaAs," *J. Vac. Sci. Technol. B,* 6:1885 (1988).

Chapter 11

Characterization of GaAs by Photoluminescence and Fourier Transform Infrared Spectroscopy

S. G. Bishop

11.1 Introduction

As the quality of semiconductor materials such as GaAs has improved in recent years, increasing demands have been placed on traditional characterization techniques for the determination of crystal quality and purity. For example, with background impurity levels at the $\sim 10^{13}$ cm^{-3} level in state of the art GaAs (e.g. material grown by molecular-beam epitaxy, MBE) the limits of detectability for some techniques have been challenged. However, in the case of optical techniques such as photoluminescence (PL) and Fourier transform infrared spectroscopy (FTIR), improvements in crystal quality and purity have enhanced the applicability of these fundamental research tools as characterization techniques.

With the reduction of defect and impurity concentrations, both the efficiency and spectral detail exhibited by GaAs PL spectra[1] have improved dramatically making it possible to identify impurities on the basis of the sharp line near-band-edge PL spectra attributable to the recombination of excitons bound to impurities,[2] free carriers recombining with carriers bound at impurities,[3] and so-called donor-acceptor pair recombination.[4] Correspondingly, it is now possible to identify donors and acceptors in GaAs on the basis of FTIR spectra of intraimpurity electronic transitions[5–7] and local vibrational mode (LVM) absorption bands[8] characteristic of light atom impurities. Such experiments rely on the very low background impurity concentrations achievable in state of the art epitaxial growth techniques, and upon the ability to back-dope epitaxial films under highly controlled conditions.

The PL technique is far more widely applied as a characterization tool than is FTIR. It is convenient and less costly to implement than FTIR, there are fewer (and less stringent) constraints on the materials to which it is applicable, and a somewhat broader scope of phenomena (excitations, defects, recombination mechanisms, etc.) can be investigated. For these reasons the discussion of PL in this review is far more detailed than that of FTIR spectroscopy. The major objective in the discussion of both techniques is to familiarize the reader with the important experimental results which serve as the basis for the application of these spectroscopies to the characterization of GaAs. The presentation is purposely unencumbered by theoretical detail, with the emphasis on recognition of the wide variety of spectral features which are observed, especially in the case of PL, and identification of the impurities or defects from which they arise.

It is hoped that this paper will provide the nonspecialist with an easily readable and conveniently useful description of the characterization of GaAs by PL and FTIR spectroscopy, including a relatively comprehensive bibliography to facilitate more detailed study.

11.2 Characterization of GaAs by Photoluminescence Spectroscopy

Characterization of semiconductors by PL techniques is based upon the measurement and interpretation of the spectral distribution of recombination radiation emitted by the sample. Electrons and holes which are optically excited across the forbidden energy gap can recombine radiatively by a variety of mechanisms[9] (see Fig. 11–1) whose relative efficiencies are determined largely by the defect and impurity content of the sample. Electrons and holes usually become localized and bound at an impurity or de-

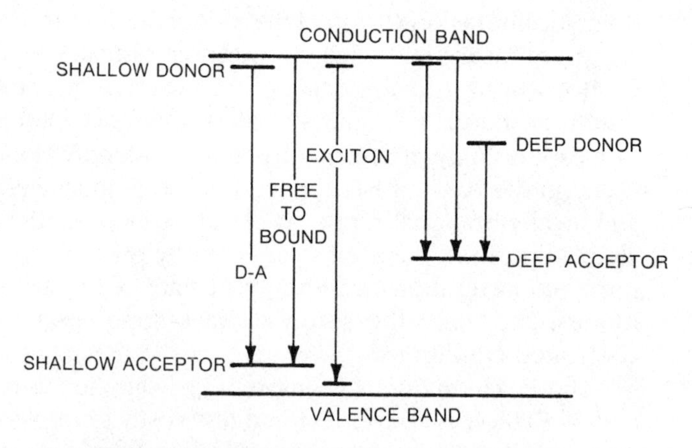

Fig. 11–1. Energy level diagram for a semiconductor indicating possible radiative recombination processes.

fect before recombining, and the identity of the localized center to which they are bound can often be determined from the spectral distribution of the emitted light (PL spectrum). Relatively sharp line, near-band-edge PL spectra arise from the recombination of electron-hole pairs which form bound excitons (BE)[2] at impurity sites, or the free-to-bound (FB) transitions which involve recombination of free electrons (holes) with holes (electrons) bound at neutral shallow acceptors[3] (donors), while lower-energy and much broader PL bands arise from the recombination of carriers localized in deep traps.[1] The efficiency, line widths, and resolvable spectral details of the narrow, near-band-edge PL features provide a qualitative assessment of crystal quality, and the identities of impurities can often be determined from the binding energies inferred from the spectral positions (energies) of bound exciton or free-to-bound transitions.

A description of the various experimental configurations and procedures employed in PL spectroscopy is presented below. This is followed by a brief, qualitative treatment of the application of effective mass theory to the description of shallow (hydrogenic) bound states in semiconductors. The ensuing discussion of the specific applications of the PL technique to the characterization of GaAs is organized on the basis of the spectral energy of the relevant recombination mechanism, beginning with the near-band-edge phenomena and proceeding to discussions of recombination involving deeper (near midgap) levels.

11.2.1 Experimental Techniques

In order for photoluminescence spectroscopy to provide the detailed spectral information required to characterize fully a relatively narrow gap, small effective mass semiconductor such as GaAs, the measurements must be carried out at or near liquid helium temperature. Only the most approximate, qualitative information can be derived from PL spectra obtained at liquid nitrogen temperature. The most useful liquid helium cryostats are those which provide variable temperature capability with temperatures ranging from ~1.6 K to ~200 K, and have optical access through two or three windows. Samples are mounted in a nearly strain-free configuration if possible, and on a cold finger which can be immersed in liquid helium or cooled by flowing helium gas.

Optical excitation is usually provided by a visible or near-infrared laser such as an argon or krypton ion laser (a variety of UV and visible wavelengths are available) or a Nd:YAG laser (1.06 μm). In the standard PL configuration an image of that portion of the sample to be excited is focused on the entrance slit of a grating monochromator by an objective lens or by a focusing (spherical or toroidal) front surface mirror. The luminescence excited in the semiconductor is then dispersed or analyzed by the monochromator and detected by a photomultiplier tube (PMT) or one of a variety of photoconductive detectors, with the choice being determined by the wavelengths of interest. The most commonly used detectors include the Ge PIN diode detector for wavelengths out to about 1.8 μm, the PbS photoconductor out to 3.5 or 4 μm, and the InSb photovoltaic detector out to about 5 μm. The PMT offers the highest detection efficiency of all the detectors but is limited to wavelengths shorter than about 0.9 μm (for a GaAs photosurface) or 1.1 μm (for an S-1 photosurface). However, this does allow high-resolution studies of sharp line, near-band-edge PL features in GaAs ($\sim$1.52 to 1.45 eV) such as bound exciton recombination, acceptor free-to-bound, and donor-acceptor pair transitions, to be carried out with the PMT. The PMT also offers the advantage of photon counting, which facilitates digital data acquisition and multiscan signal averaging. When photoconductors are employed the laser beam is chopped by a mechanical chopper and the analog signal is recorded with synchronous, phase-sensitive detection by a lock-in amplifier. This signal is often processed with an analog-to-digital converter which allows digital data processing, including signal averaging.

In the standard PL experimental configuration the sample is excited by a laser of fixed photon energy, usually with energy greater than the bandgap, and the spectral distribution of the emitted luminescence is recorded as a function of wavelength by scanning the analyzing grating monochromator. However, there is another important class of PL experiments referred to as photoluminescence excitation (PLE) spectroscopy. In the PLE experiment, the intensity of the PL is measured as a function of the wavelength of the exciting light. In its simplest form, one takes several PL spectra in the usual manner for each of several different exciting wavelengths. Far more detailed information, however, is acquired by utilizing an excitation source whose wavelength can be varied (scanned) continuously. Examples include a broadband lamp coupled with a monochromator which provides a low-intensity, relatively low resolution source over a broad spectral range (e.g. a tungsten lamp), or a tunable dye laser which provides a high-power, high-resolution source over a relatively restricted spectral range (for a given dye). In this PLE experiment, the wavelengths of the detected PL are chosen by a fixed selection of optical transmission filters or by a monochromator set at a fixed wavelength, and the intensity of the observed PL is recorded as a function of the scanned wavelength of the exciting light. One measures, in effect, how efficiently various absorption processes excite the PL. The PLE experiment can often detect weak extrinsic (below gap) absorption processes which contribute strongly to the excitation of a particular PL band but which are too weak to be detected in an optical absorption experiment.

11.2.2 Effective Mass Theory

One of the most important aspects of the characterization of semiconductors by photoluminescence and infrared techniques is the identification of chemical impurities and certain other defects through the detection of the bound electronic states which they produce in the crystal. Identification is usually achieved through the comparison of binding energies inferred from spectroscopic data with catalogs of empirically determined binding energies for various impurities, where available, or by comparison with theoretical calculations of binding energies. The most useful approach to the description of shallow bound states in semiconductors is effective mass theory (EMT),[10] in which it is assumed that the bound states of, for example, the electron at a shallow donor or the hole at a shallow acceptor can be treated by analogy to the states of a hydrogen atom. The lowest or ground state of this series of hydrogenic bound states is given by

$$E = -e^4 m^* / 2\epsilon^2 \hbar^2, \tag{11–1}$$

with effective Bohr radius

$$r = \epsilon \hbar^2 / e^2 m^*, \tag{11–2}$$

where the strength of the Coulomb interaction has been reduced by the macroscopic dielectric constant, ϵ, and the free electron mass is replaced by an effective mass, m^*, whose definition depends upon the electronic system being described, e.g. a carrier bound to an impurity or an exciton bound to a charged or neutral impurity, etc. In addition to this ground state, the hydrogenic description includes a series of bound excited states.

The larger the orbit of the bound carrier, the more accurately the EMT describes the bound states. For large orbits, the potential is very nearly Coulombic and the wavefunction of the state has a slow spatial variation. These two conditions contribute to the accuracy of the EMT. However, even for very shallow, spatially extensive bound states, the simple hydrogen model cannot provide a quantitative description of the series of bound states. More sophisticated calculations are required and the review article of Ramdas and Rodriguez[5] and the work of Baldereschi and Lipari[11,12] are recommended for those requiring more detailed knowledge of the quantitative theoretical description of shallow levels in semiconductors. For present purposes it is sufficient to point out some of the general features of bound shallow impurity states which are highly significant in the context of impurity identification or characterization. First, it is important to note that the p-like (bound) excited states are identical for all of the shallow acceptors (donors) in a given semiconductor. The orbits of these excited states are larger than those of the ground states and the p-like excited state wavefunction of the hole (electron) vanishes at the impurity core. The *s*-like ground states have a larger interaction with the impurity core leading to significant differences in the ground state binding energies for the various impurities. Thus it is the measurement of the ground state binding energy (or, in some cases, the separation of the *s*-like ground state from the *p*-like excited states) which is required for the identification of an impurity.

Photoluminescence and infrared spectroscopies provide complementary means for the measurement of these ground state to excited state separations for donor and acceptor impurities in GaAs. These energy separations, which are of the order of tens of meV for shallow acceptors and a few meV for shallow donors in GaAs, can manifest themselves as two electron and two hole replicas of donor and acceptor bound exciton PL spectra,[3] and in the PLE spectra of donor-acceptor pair PL bands,[13–15] each of which has near-band-edge photon energies detectable by high-sensitivity photomultiplier tubes. Correspondingly, the free-to-bound (FB) PL transitions involving acceptor ground states can provide a measure of the ground state binding energy and a direct identification of the acceptor.[3] Shallow donor binding energies in GaAs are too small (~5.7 meV) to allow the discrimination of different chemical donors on the basis of shifts in donor free-to-bound PL transitions under routine experimental conditions. However, very high resolution PL spectroscopy[16] and PL in high magnetic fields has been used recently to identify shallow donors in GaAs[17,18] and InP.[19] In complementary fashion, the ground to excited state separations can be observed by excitation of an electron (hole) from the ground to an excited bound state (usually a 1*S* to 2*P* transition) of the donor (acceptor) by the absorption of a photon with energy in the far-infrared spectral range (~4 meV or 30 cm^{-1} for donors and ~20 meV or 150 cm^{-1} for acceptors). These transitions can be observed directly in far-infrared absorption spectroscopy[5] or by the more sensitive photothermal ionization spectroscopy (PTIS),[5–7] a form of extrinsic phtotconductivity.

11.2.3 Excitons

The highest energy features in the PL spectrum of high-quality GaAs samples are the sharp line spectra attributed to free and bound Wannier-Mott excitons.[20] In contrast to the Frenkel excitons[21] observed in ionic materials with low mobility and dielectric constant which have internal binding energies ~1 eV, the Wannier-Mott free exciton binding energies in GaAs are of the order of a few meV as a consequence of the covalent bonding, high mobility, and relatively large dielectric constant of this compound semiconductor. The wavefunction describing these weakly bound electron-hole pairs extends over many unit cells of the GaAs crystal and the small binding energy can be treated[2] by the EMT in an expression similar to Eq. 11–1, where m^* is now the reduced mass for the interacting electron-hole pair given by $1/m^* = 1/m_e^* + 1/m_h^*$. The electron and hole effective masses, m_e^* and m_h^*, are those which characterize the conduction and valence band edges at the appropriate bandgap (the zone-center direct bandgap for GaAs).

The scope of the present discussion is limited to the role of excitonic PL spectra in the characterization of GaAs. In this context, the most useful spectral features are the narrow, symmetrical PL bands attributable to bound excitons (BE) which are localized at impurities or defects in the semiconductor. [For a comprehensive account of the nature and properties of BE in semiconductors, the reader is referred to the excellent review of Dean and Herbert.[2] The determination of the lowest-energy configuration of these BE states, their optical transition oscillator strengths, the splittings produced by electron-hole *j–j* coupling, the effects of phonon coupling, and the use of a variety of experimental tools (polarization studies, magneto-optics, uniaxial stress, etc.) to identify transitions and measure binding energies are all discussed in this reference.] The characteristic sharp-line PL spectra of the spatially extensive excitonic states can easily be broadened beyond recognition by inhomogeneous internal strains and the occurrence of bound exciton states can be precluded by the screening which arises even at relatively moderate impurity concentrations. Thus the study of the strengths and line widths of excitonic PL spectra in semiconductors can provide a qualitative assessment of crystal perfection and purity. For example, the strength and lineshape of the *free* exciton band can provide a qualitative measure of crystal quality. The free exciton or exciton-polariton emission is the highest-energy feature in the excitonic PL spectrum, and is located at 1.5155 eV in the near-band-edge PL spectrum for GaAs shown in Fig. 11–2. (Although the free exciton emission in GaAs will not be considered in detail, it should be pointed out that a frequently observed doublet lineshape for this band is the subject of some controversy.[22,23]) In addition, more specific information such as the nature or identity of the crystal imperfection which binds the exciton is afforded by the BE PL spectra which are discussed below.

Bound excitons (BE) are observed in all semiconductors which have the required purity and crystal perfection. They are characterized by typical binding energies $E_{BX} < 10$ meV and radii greater than 3 nm. The additional binding energy of an exciton at a donor, acceptor, or isoelectronic center shifts the radiative recombination spectrum of the bound exciton to lower energy by an amount determined by the particular impurity which binds the exciton. In some circumstances it is possible to identify the impurities or other defects which are present in the crystal on the basis of the bound exciton PL spectra. Haynes[24] was first to observe (in Si) that in some semiconductors there is a linear variation of exciton binding energy

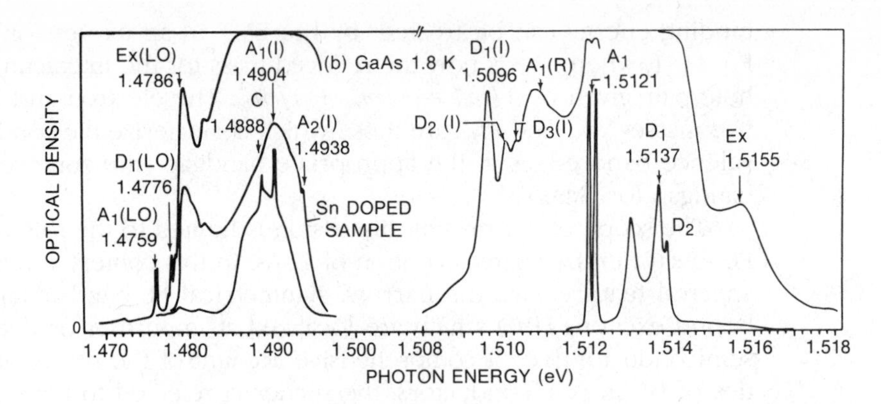

Fig. 11–2. Photoluminescence spectrum of GaAs at 1.8 K showing the following near-band-edge features: A, exciton bound to neutral acceptor (doublet emission); D, exciton bound to neutral donor; superscript (I), two-carrier transition (recombination of exciton leaving impurity center in excited state I); (R), resonant coupling to acoustic phonons (replica); (LO), phonon replica; Ex, free exciton. Different traces refer to different exposure times except that showing the Sn doped sample. Note change of gain scale at photon energy 1.508 eV. *(After White* et al.[27]*)*

(E_{BX}) with increasing impurity binding energy (E_i) of the form

$$E_{BX} = a + E_i b \,, \tag{11–3}$$

where a is a constant. This relationship, often referred to as Haynes's rule, is attributable to the central cell corrections to E_{BX} whose magnitude is otherwise dominated by the long-range Coulomb interaction. For Si Haynes found that $b \cong 0.1$ and $a \cong 0$, but these parameters do not apply to all semiconductors. In general, the dependence of E_{BX} on E_A or E_D cannot be expected to pass close to the origin as it does for Si, and an offset of one sign for acceptor BEs requires an offset of the opposite sign for donor BEs. For shallow acceptors in direct gap semiconductors such as GaAs there is a particularly low sensitivity[2,25] of E_{BX} to E_A. This circumstance severely limits the utility of acceptor BE PL spectra for the direct identification of acceptor impurities in GaAs.[3] However, it is possible to distinguish different acceptor impurities on the basis of acceptor free-to-bound transitions because the acceptor binding energies are ~25 to 30 meV (see Sec. 11.2.4.1). Although the sensitivity of the donor BE in GaAs to changes in E_D ($E_D \cong 5.7$ meV) is apparently not nearly as low as in the case of acceptors, the differences in the GaAs shallow donor binding energies are so small[26] that it is usually not possible to distinguish different donor impurities on the basis of either the donor bound exciton or donor free-to-bound PL transitions. (For exceptions see Refs. 16 to 18 and the discussion in Sec. 11.2.3.1, under Donor Bound Excitons.)

The detailed theoretical calculation of the binding energy of excitons localized at impurities in semiconductors is reviewed by Dean and Herbert.[2] Effective mass theory is the basic approximation employed, with the Hamiltonian containing a kinetic energy term for free particles with an effective mass, and a screened interparticle and impurity or defect potential. However, many corrections are required for a quantitative comparison of theory and experiment, as was the case for the hydrogenic impurity states described above. The central cell corrections arising from differences in impurity and host atom core potentials, strain fields associated with the impurity, and valence charge redistributions are the most difficult to treat quantitatively. Additional corrections are required for the electron-phonon interaction for binding energies of the order of or greater than the polar-phonon energy.

Binding energy is not the only characteristic of the BE which contributes to the identification of the impurity center. Experimentally observed BE PL bands exhibit fine structure which originates from exchange and correlation interactions[2] which are neglected in the EMT approximations mentioned above. In the following section, examples of PL spectra due to excitons bound to neutral impurities in GaAs are presented and their features are identified on the basis of binding energy and fine structure.

11.2.3.1 Excitons Bound to Neutral Impurities

Acceptor Bound Excitons

The PL spectrum for the near-band-edge region of GaAs obtained by White *et al.*[27] at 1.8 K is shown in Fig. 11–2. The strong doublet near 1.5121 eV (labeled A_1) has been attributed[27,28] to the recombination of excitons bound to neutral acceptors (denoted $A^\circ X$). This characteristic splitting of the neutral acceptor-bound exciton PL is universally explained on the basis of j–j coupling between the two holes and one electron localized at the acceptor site. For the zinc-blende semiconductor GaAs, the two holes each with $j = 3/2$ can couple to give the two values of the total angular momentum allowed by the Pauli exclusion principle,[2,29] $J = 0$ or 2, with $J = 2$ as the ground state. The additional, relatively small interaction with the electron yields $J = 1/2, 3/2$, and $5/2$. It is presumed that the latter two levels are closely spaced and unresolved in the PL spectra giving rise to the observed doublet. This doublet splitting due to the hole-hole exchange interaction observed in GaAs is ~0.19 meV. This assignment of these transitions has been confirmed in InP by comparison with absorption spectra[30] and from the Zeeman splitting[29] of the $J = 1/2$ state which, as pointed out above, lies uppermost for $A^\circ X$ bound to the shallowest acceptors in InP and GaAs. The splitting of the $J = 3/2$ and $J = 5/2$ BE states, due to electron-hole exchange, is simply too small for convenient analysis. Although the $A^{\circ}X$ splitting is expected to increase with increasing E_A, there are no examples of intermediate binding energy to provide a detailed test. The well known Sn BE PL in GaAs (1.5094 eV) is the only example to be studied[31] in detail and the Sn acceptor ($E_A \cong 148$ meV) is too deep to be effective mass-like.

The lowest energy features shown in the PL spectrum of Fig. 11–2 include two replicas of the A_1 doublet. These are the LO_Γ phonon replica, labeled A_1 (LO), which is displaced to lower energy relative to A_1 by one LO_Γ phonon energy (36.7 meV for GaAs), and the peak labeled $A_1^{(I)}$ at 1.4904 eV. The $A_1^{(I)}$ band is a relatively poorly resolved example of a "two-hole" transition or replica of the A_1 acceptor bound exciton line. Its displacement below A_1, 21.7 meV, is about three quarters of the hydrogenic acceptor binding energy and such two-hole (TH) replicas are attributed to transitions in which the recombination of the acceptor bound exciton leaves the hole of the neutral acceptor in an excited state. Referring to the energy level diagram for hydrogenic shallow acceptors shown as an inset in Fig. 11–3, it can be seen that the principal $A^{\circ}X$ transition leaves the neutral acceptor in its ground ($1S_{3/2}$) state and the TH transition leaves it in the first excited ($2S_{3/2}$) state. (It should be pointed out that because crystal field effects are small for such shallow states, parity selection rules allow these transitions.) The spectral shift of the TH transition from the principal or fundamental $A^{\circ}X$ transition represents the excitation energy of the $2S_{3/2}$ level of the shallow acceptor. These shifts are particularly important because the transitions from which they derive are parity forbidden in direct optical excitation of the hole bound to the neutral acceptor. Furthermore, because the central cell corrections or chemical shifts of the *s*-like ground states give rise to appreciable differences in the ground state binding energies of different acceptors, the energy positions of the TH transitions are a highly useful complement to acceptor FB transitions in the identification of acceptor impurities from PL spectra. Examples of TH PL spectra observed for C, Be, and Mg shallow acceptors in GaAs are shown in Fig. 11–3 and the energies of the TH transitions for a variety of acceptors in GaAs are listed in Table 11–1. Figure 11–3 and Table 11–1 are from the paper of Ashen *et al.*[3] This work, published in 1975, remains one of the most important steps taken in the process of establishing PL spectroscopy as a reliable, quantitative tool for the characterization of acceptor impurities in GaAs. The authors thoroughly documented the confusion which pervaded the literature on acceptor related PL prior to their systematic study, analyzed the reasons for this confusion, and presented the first reliable identification catalog of acceptor binding energies in GaAs as established through complementary studies of acceptor FB, DA pair PL, and TH replicas of acceptor bound exciton PL spectra on carefully prepared VPE and LPE samples of GaAs. The results of this valuable study are also discussed in Sec. 11.2.4 on FB and DA pair PL spectra.

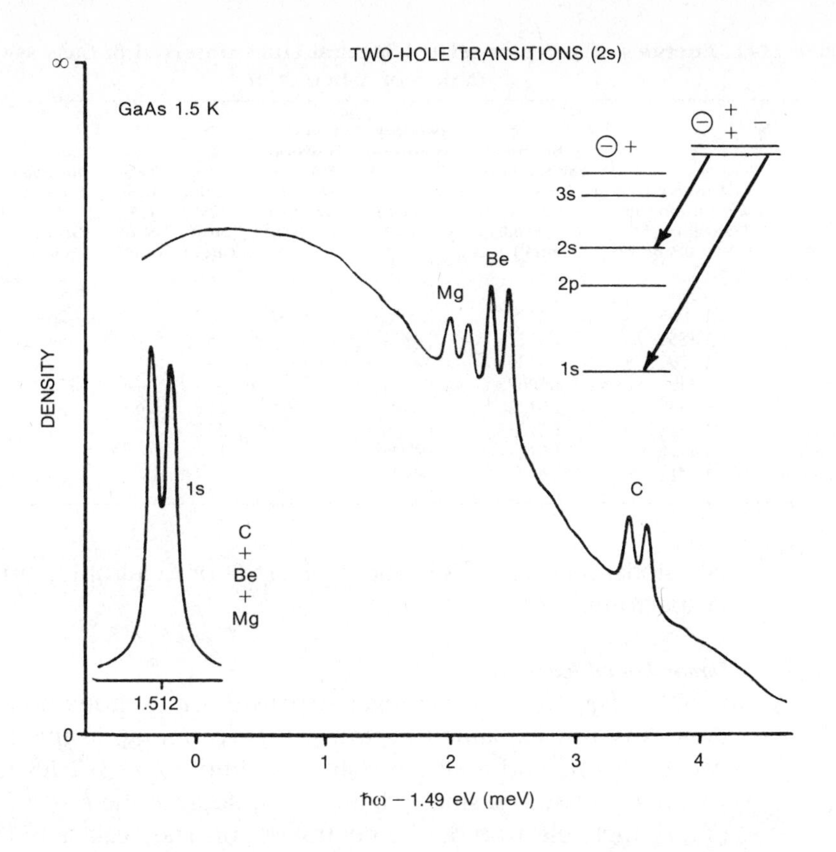

Fig. 11–3. Two-hole (TH) replicas for C, Be, and Mg in LPE spectra. The lower left inset shows that the fundamental 1S transition is common to all three elements and is therefore of no value for chemical identification. The upper right inset shows the relevant experimentally observed transitions. The commonly observed TH transition terminates in a 2*S* level. *(After Ashen* et al.[3]*)*

It should be emphasized that the PL spectrum of Fig. 11–2, with its rich detail, is typical only of the highest-quality epitaxial samples (in this case VPE). The strongest and sharpest features of this VPE spectrum are often not resolved in PL spectra of bulk or epitaxial GaAs. Factors contributing to the broadening of the BE spectra include the presence of more than one acceptor or donor species, high-impurity concentrations, and inhomogeneous internal strain. White *et al.*[27] demonstrated that the presence of internal strain in high-quality VPE samples of InP can cause shifts in the energy of the A_1 emission doublet (occurring at about 1.4144 eV in InP) and that the doublet can undergo additional strain splittings into a triplet or quadruplet form as the focused exciting laser beam is traversed across the sample. Such internal strain splittings probably explain the

Table 11–1. Energies of Acceptor-related Spectral Lines Observed in GaAs as Compiled in Table 3 of Ashen *et al.*[3]

Acceptor	Observed Free-to-Bound Transition (5K) (eV ± 0.3 meV)	Bound Exciton (1S) (1.5-K Doublet Centre (eV ± 0.05 meV)	Two-Hole Transition (2S) (1.5-K Doublet Centre) (eV ± 0.07 meV)	Two-Hole Transition (3S) (1.5-K) Doublet Centre) (eV ± 0.15 meV)	Two-Hole (2S) Shift (meV)	Two-Hole (3S) Shift (meV)	Binding Energy $1S_{3/2}$ (Ground State)	Binding Energy $2S_{3/2}$	Binding Energy $3S_{3/2}$
Carbon	1.4935	1.5124	1.4938	1.4900	18.5	22.4	26.0	7.4	3.6
Silicon	1.4850	1.5123	1.4871	—	25.2	—	34.5	9.3	—
Germanium	1.4790	1.5126	—	—	—	—	40.4	—	—
Tin	1.349	1.5067	—	—	—	—	171	—	—
Zinc	1.4888	1.5122	1.4904	1.4862	21.8	26.1	30.7	8.9	4.6
Cadmium	1.4848	1.5123	1.4869	—	25.3	—	34.7	9.4	—
Beryllium	1.4915	1.5124	1.4926	—	19.7	—	28.0	8.2	—
Magnesium	1.4911	1.5124	1.4922	—	20.1	—	28.4	8.3	—

occasional reported observations of triplet or quadruplet forms of the $A^{\circ}X$ Pl spectrum.[32,33]

Donor Bound Excitons

The expected energy level structure for excitons bound at neutral donors is also explained in terms of j–j coupling, in this case involving two electrons and a single hole.[29] A simple $J = 3/2$ level is expected which gives rise to a single PL line upon decay to the $J = 1/2$ ground state of the single electron on the neutral donor. The peak at 1.5137 eV labeled D_1 in Fig. 11–2 is attributed to the recombination of excitons bound at neutral donors ($D^{\circ}X$). This peak occurs in the expected energy range for $D^{\circ}X$. (On the basis of the electron and hole mass ratio, it is expected that $E_{BX}/E_D \cong 0.2$ compared with the observed ratio ~0.22.) The observed Zeeman splitting[29] of the principal peaks observed in this region is consistent with the $D^{\circ}X$ interpretation. In addition, the occurrence of a replica spectrum, $D_1^{(I)}$ in Fig. 11–2, at an energy which is 4.42 meV below the D_1 peak provides further evidence for the $D^{\circ}X$ assignment.[34] Such replicas are attributed to "two-electron" transitions (the counterpart of the two hole transitions discussed above) in which the recombination of the exciton bound at the neutral donor leaves the donor in its first orbital excited state.[27,29,34] More recently, Almassey *et al.*[16] have utilized a very high resolution spectrometer to resolve two-electron replica transitions from three different neutral donor bound excitons in high-quality VPE films of GaAs. Donor binding energies estimated from these PL spectra agreed quite well with the results of FTIR spectroscopy (see Sec. 11.3).

Note that the D_1 line in Fig. 11–2 is an apparent doublet with a secondary peak (D_2) at slightly higher energy. Such features were first interpreted in terms of the presence of more than one species of donor impurity.

However, if these features represented $D^\circ X$ for distinct donors, their separation would imply differences in donor binding energy of the order of 1 meV, whereas IR spectroscopy[6,35,36] has established that the binding energies of the shallow donors present in GaAs as inadvertent contaminants all fall within an energy interval of about 0.1 meV. Subsequently, it has been shown that the multiple sharp lines in the $D^\circ X$ spectral range represent excited states of the neutral donor bound exciton which are related to the different angular momentum states of the $j = 3/2$ hole.[2,29] The high-resolution spectroscopy of Almassey *et al.* [16] also provided a well resolved representation of these transitions which originate from donor bound exciton complexes that are initially in excited states.[30] Spectral shifts in the $D^\circ X$ lines due to differences in the ground states of different donor species are unresolvable by the normal, zero-field PL spectroscopy exemplified by the spectrum of Fig. 11–2. (See Sec. 11.2.3.3 for a discussion of the identification of donor species on the basis of magneto-optical studies of donor bound exciton PL bands.[17–19])

11.2.3.2 Exciton Binding at Ionized (Charged) Impurities

An electron and a hole bound to an ionized donor or acceptor is the simplest bound exciton and, accordingly, this system has received much theoretical attention. One of the general conclusions of these calculations is that for an exciton bound to an ionized donor the localization energy decreases monotonically with increasing mass ratio $\sigma = m_e^*/m_h^*$, and that there exists a critical mass ratio σ_c above which the bound complex is unstable.[37,38] For ionized donors, the exciton complex becomes unstable[39] for $\sigma > 0.43$. (If the mass ratio is inverted, an analogous conclusion for excitons bound at ionized acceptors obtains.) It was Hopfield who first pointed out that whereas neutral centers should always bind an exciton, ionized donors or acceptors will bind excitons only under the conditions described above. Thus if $m_e^* \cong m_h^*$ as in Si, Ge, and GaP, excitons bound at ionized donors and acceptors will not be observed.

In an early study of the exciton spectra of GaAs, Bogardus and Bebb[40] identified features at 1.5133 and 1.49 eV as the recombination of excitons bound to ionized donors and ionized acceptors, respectively. However, in subsequent work on refined GaAs these assignments were called into question.[9,41] Doubts regarding the occurrence of a PL transition due to excitons bound to ionized acceptors are consistent with the theoretical expectation that for such a complex to be stable σ^{-1} must be less than unity,[9] a condition which does not obtain for GaAs. In addition, the 1.5133-eV PL line originally assigned to D^+X was later claimed to be a donor free-to-bound (FB) transition, that is, a free hole recombining with the electron bound at a neutral donor.[41]

The confusion between the donor FB transitions and those associated with excitons bound to ionized donors is understandable in that their expected photon energies coincide very closely in GaAs. Probably the most definitive work to date is that of Almassy *et al.* [16] in which they reported the observation of $D^{+}X$ transitions in the vicinity of 1.513 eV associated with three different residual ionized donors in high-purity VPE GaAs. Their interpretation was supported by the simultaneous observation of two electron replicas of $D^{\circ}X$ corresponding to three distinct donors. These workers were able to measure both the localization energy of the exciton bound to the ionized donor and the ionization energy of the donor for each of the three residual donors observed. They found a "Haynes Rule"-type linear dependence of the exciton binding energy on the donor binding energy, E_D.

11.2.3.3 Magneto-Optical Photoluminescence Studies of Donor Bound Excitons

The work of Almassey *et al.* [16] demonstrates that in the case of the very best VPE films of high-purity GaAs, it is possible to distinguish different chemical donors through high-resolution PL spectroscopy of the donor bound excitons. However, the utility of this high-resolution PL spectroscopy for donor identification in GaAs can be enhanced considerably through the use of magneto-optical photoluminescence spectroscopy (MPS). This technique was first demonstrated by Dean and coworkers[19] in InP. The quality of the best VPE InP is still not adequate to allow the resolution of different donor species in zero-field PL. In a high-intensity applied magnetic field, which compresses the wavefunctions of the bound electronic particles, the PL lines sharpen and the spectral shifts for different chemical donors are accentuated and Dean *et al.* were able to distinguish different donor species in InP on the basis of resolved substructure in the two-electron satellites of resonantly excited donor bound excitons.

Subsequently, Reynolds and coworkers[17] used the MPS technique to identify three residual donors (Si, S, Ge) and three residual acceptors (Ge, C, Zn) in VPE GaAs. They pointed out that a unique feature of the MPS technique is its ability to identify simultaneously residual donors and residual acceptors in either n- or p-type material. In addition, the technique provides a qualitative evaluation of the compensation of the sample. The various forms of infrared spectroscopy which are used to identify donors and acceptors in GaAs can only identify donors in n-type material and acceptors in p-type material (with the exception of infrared absorption spectroscopy under simultaneous intrinsic photoexcitation). Furthermore, donor and acceptor infrared (IR) spectroscopy are carried out in quite different spectral ranges requiring different experimental conditions, while donors and acceptors can both be studied in the same PL spectrum by the MPS technique.

Recently, Bose *et al.* [18] have correlated the results of high magnetic field MPS with far-infrared PTIS of donors in high-purity epitaxially grown

GaAs and have identified the two-electron satellites of the donor bound exciton PL transitions corresponding to the shallow donors S, Si, Ge, Sn, and Te. Two-electron PL spectra for three GaAs samples recorded at a magnetic field of 9.0T by Bose *et al.*[18] are shown in Fig. 11–4. The spectra for samples *C* and *D* were obtained using dye laser excitation resonant with a principal bound exciton line, while the spectrum for sample *B* was recorded with above-bandgap excitation. The correlation of these MPS spectra with the infrared photothermal ionization spectroscopy (PTIS) measurements on the same samples reported in Ref. 18 places the identification of the two-electron MPS features on a firm basis which permits them to be used for the identification of residual donor impurity species in high-purity GaAs.

Thus the MPS technique has greatly increased the usefulness of PL spectroscopy for shallow donor identification in GaAs. This circumstance is clearly demonstrated by the successful use[42] of PL and PLE spectroscopy under a magnetic field to identify shallow donors in LEC grown crystals of GaAs having electrically active impurity concentrations of $\sim 10^{16}$ cm^{-3}.

11.2.3.4 Excitons Bound at Isoelectronic Centers

The simplest form of isoelectronic center which can bind an exciton in semiconductors is the isoelectronic (or isovalent) impurity. In this type of center the short-range impurity potential may bind a hole (electron) if the electronegativity of the isovalent substituent is smaller (greater) than that of the major constituent atom which it replaces. The Coulomb potential of this trapped carrier can then bind a carrier of the opposite charge in an effective mass-like state. Because the one particle is tightly bound while the other is in an effective mass-like state, splittings due to j–j coupling of the two particles are a small proportion of the total exciton binding energy in contrast to the situation discussed above for excitons bound to neutral acceptors. This piecemeal approach to the binding of excitons at isoelectronic traps was first treated theoretically by Faulkner.[43] (For a complete description of the theoretical treatment of bound states at isoelectronic centers, see the reviews of Dean and Baldereschi[2,44] and references therein.)

The most celebrated example of an exciton bound to such an isoelectronic center, and the example for which the properties of neutral isoelectronic centers were first established,[45] is the nitrogen-bound exciton in GaP:N. In addition, it is this trap which makes possible the high radiative efficiency in GaP and GaAsP light-emitting diodes. However, the number of these point defect isoelectronic traps which occur in semiconductors is strictly limited. For example, none have been observed for isoelectronic cation substituents in III-V or II-VI compounds.[44] Narrow direct-gap semiconductors such as GaAs and InP have small electron effective mass and this makes it difficult for the short-range potential of an electron-attractive

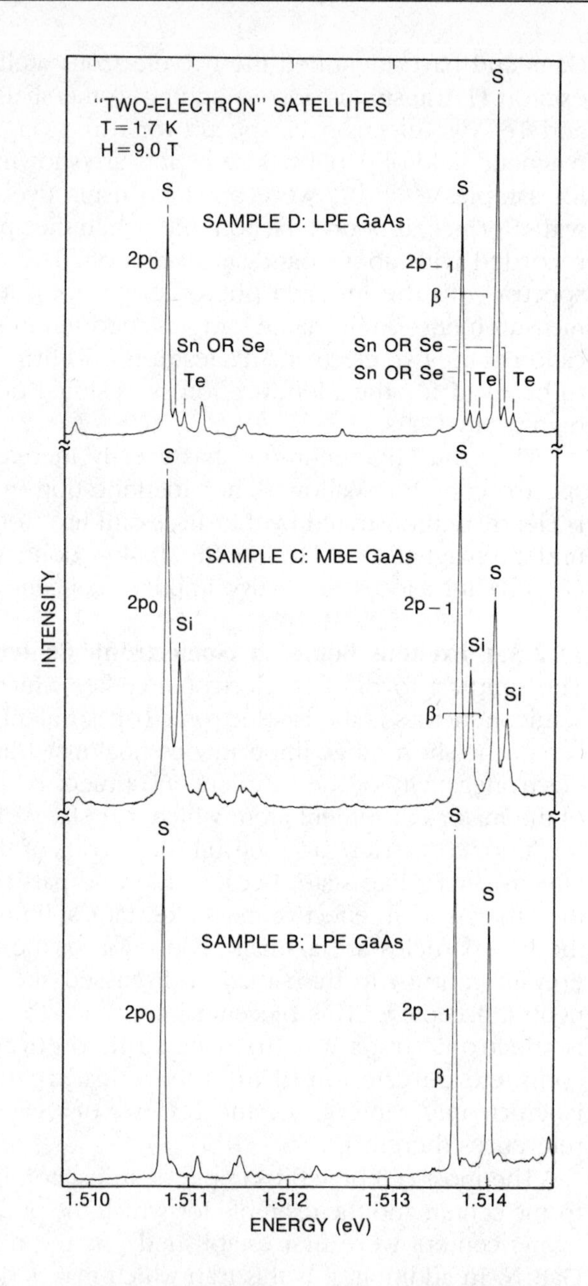

Fig. 11–4. Two-electron luminescence spectra for three GaAs samples recorded at a magnetic field of 9.0 T. The principal lines occur at nearly the same energies for these samples. Spectra for LPE and MBE GaAs samples show that the peaks due to S donors match in energy position and the peaks corresponding to Sn (or Se) and Te donors lie on both sides of the Si peak. *(After Bose* et al.[18]*)*

point defect isoelectronic impurity to produce a bound state. Many of these narrow gap semiconductors, and GaAs in particular, contain anions from the lower part of the periodic table which are placed low on the scale of electronegativity.[46] Thus hole-binding, point defect isoelectronic traps, which are favored from the standpoint of the larger hole effective mass, are nonetheless unlikely to occur because most anion isovalent impurities will have larger or, at best, only slightly smaller electronegativity than the host anion being replaced. The one well-established example of an anion impurity, isoelectronic, hole-binding trap in a narrow direct-gap semiconductor is Bi in InP, which is understandable because the host anion, P, is from the upper part of the periodic table.[44,47,48]

In view of the preceding discussion, it is not surprising that point defect isoelectronic impurities do not play a large role in the PL characterization of GaAs. However, Wolford *et al.*[49] have recently observed sharp line PL spectra in nitrogen-doped VPE GaAs (grown in an NH_3 atmosphere) under hydrostatic pressure which are attributable to excitons bound at N isoelectronic traps analogous to those in GaP:N. In addition to elucidating the nature of these highly unusual recombination centers, these experiments proved that GaAs can be doped intentionally with N impurities. Subsequently, Leroux *et al.*[50] reported the observation of the nitrogen-bound exciton PL spectra in GaAs:Si under hydrostatic pressure, along with additional excitonlike PL lines which they attributed to complexes involving isovalent impurities. This work apparently provides evidence for the nonintentional introduction of nitrogen as an impurity in MBE grown GaAs. Even more interesting is the observation by Schwabe *et al.*[51] in nitrogen-doped VPE GaAs, *without* hydrostatic pressure, of a series of unusual PL lines in the 1.509- to 1.440-eV spectral range which they attribute to nitrogen bound excitons. These workers presented Zeeman splitting and PL decay data which are consistent with an exciton bound to an isoelectronic center with C_{3v}-symmetry. Because this work[51] constitutes the first reported observation of an exciton bound to a classic point defect isoelectronic impurity in GaAs (although possibly coupled to a symmetry lowering neighboring defect), it is likely that it will stimulate further investigations of GaAs:N.

In addition to the simple point defect isoelectronic impurity, there exist somewhat more complicated isoelectronic substituents which involve axial centers formed from associated pairs of simple donors and acceptors. Inclusion of such "molecular" isoelectronic traps widens the scope of excitons bound to isoelectronic centers. The concept was first verified by the observation of exciton PL in GaP which was characteristic of single particle binding for the Zn_{Ga}-O_P molecular isoelectronic trap (electrons).[52] A variety of such complexes have since been observed in GaP, especially in association with Cu and Li doping of GaP.[53] However, in the case of GaAs the ability of such isoelectronic complexes to produce bound exciton emission is just as limited as for the point defect isoelectronic impurity.

Copper doping in GaAs induces some characteristic bound exciton emission lines, as well as the deep acceptor levels and deep PL bands discussed below in Sec. 11.2.7.1. The so-called C_0 (1.5026 eV) and F_0 (1.4832 eV) BE lines shown in Fig. 11–5 have been studied by a number of workers.[54–60] Trigonal and orthorhombic symmetries were inferred for the C and F centers, respectively, from piezospectroscopic studies, leading to suggestions of various Cu-vacancy associates.[56,57] Recently published work[58–60] has disproved some of the originally proposed and traditionally accepted associations of the C_0 and F_0 bound exciton PL lines with the well known 0.156- and 0.45-eV Cu-related acceptor levels and with certain deep PL bands observed in GaAs:Cu.[53,54] This recent work[58] supports the assertion that the ubiquitous Cu-related 1.36-eV PL band represents transi-

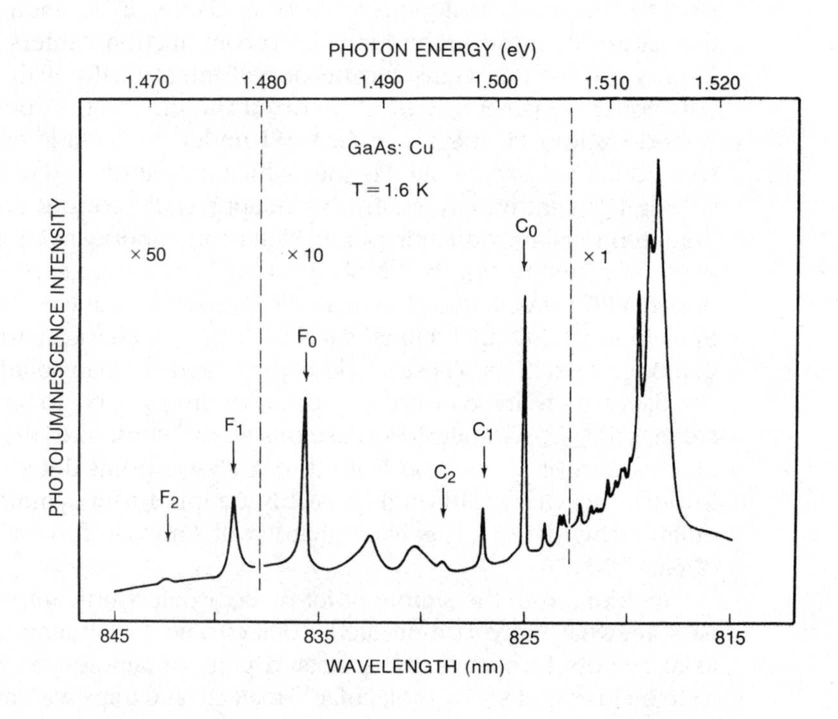

Fig. 11–5. Photoluminescence spectrum of a copper-doped GaAs crystal diffused for 7 min at 700°C. In the magnified parts of the spectrum the background has been subtracted. $C_{0,1,2}$ and $F_{0,1,2}$ are recombination lines of excitons bound at two different complex copper acceptors and their resonant phonon replicas. *(After Willmann* et al.[54]*)*

tions between the conduction band (or shallow donors) and the 0.156-eV Cu acceptor, but concludes that although the C_0 PL line is due to an acceptor-bound exciton there is no simple relationship between the 0.156-eV Cu acceptor and the C_0 bound exciton. In addition, it is concluded that the F_0 bound exciton is not related to any other Cu-related PL band. In another paper,[60] these authors conclude that although the neutral complex defects which occur with copper doping in GaP are also formed in GaAs: Cu, such defects usually will not bind excitons in GaAs as they do[61] in GaP. This absence of bound excitons is attributed to strong compressive strain fields in Cu-related neutral complex defects in GaAs which decrease the binding energy of the BE effective mass-like electron states. (Note that for these neutral complex defects in GaAs, the central cell potential which is responsible for the tightly bound particle is *hole*-attractive.) However, in this same paper[60] the first identification of an exciton deeply bound to an isoelectronic complex defect in GaAs is reported! A sharp line emission at 1.4285 eV is observed in Cu-diffused, Zn-doped GaAs; this PL band (see Fig. 11–6) is ascribed to the decay of an exciton bound to a neutral Cu-Zn or Cu_{Ga}-As_{Ga} complex.[60,62] Zeeman data for this BE PL line indicate that the defect gives rise to a *tensional* local strain field which increases the binding energy of the exciton.[62]

11.2.3.5 Defect-induced Bound Exciton Lines in MBE-Grown GaAs

Photoluminescence spectroscopy is frequently employed in the characterization of the high-quality thin-film GaAs grown by the MBE technique. Detailed study of the exciton region of the PL spectrum can provide a valuable qualitative indication of the quality of the material. In addition to the usual donor and acceptor bound exciton features, which can be observed in exquisite detail in MBE material, a remarkable series of defect or impurity related PL lines are observed in the 1.504- to 1.511-eV spectral region (see Fig. 11–7). These so-called defect-induced bound exciton lines are apparently a unique characteristic of the MBE grown GaAs. First observed by Kunzel and Ploog,[63] the highly resolved spectra have since been studied by a number of workers[64–68] with as many as 50 spectral features observed in the best samples.[64,65] Perhaps the most striking feature of the spectra is that the sharp lines are polarized along one of the ⟨110⟩ directions lying in the (100) growth plane.[65,66]

At this writing the understanding of the origin of these sharp line spectra is the subject of considerable controversy.[68] The spectra have been interpreted variously in terms of exciton recombination at neutral single acceptor pairs of varying separation,[66] distant donor-acceptor pair recombination,[64] and exciton recombination at double acceptor-isoelectronic center pairs.[65,67,68] Skolnick[68] has reviewed the current understanding of the recombination mechanisms and the recent experimental work with par-

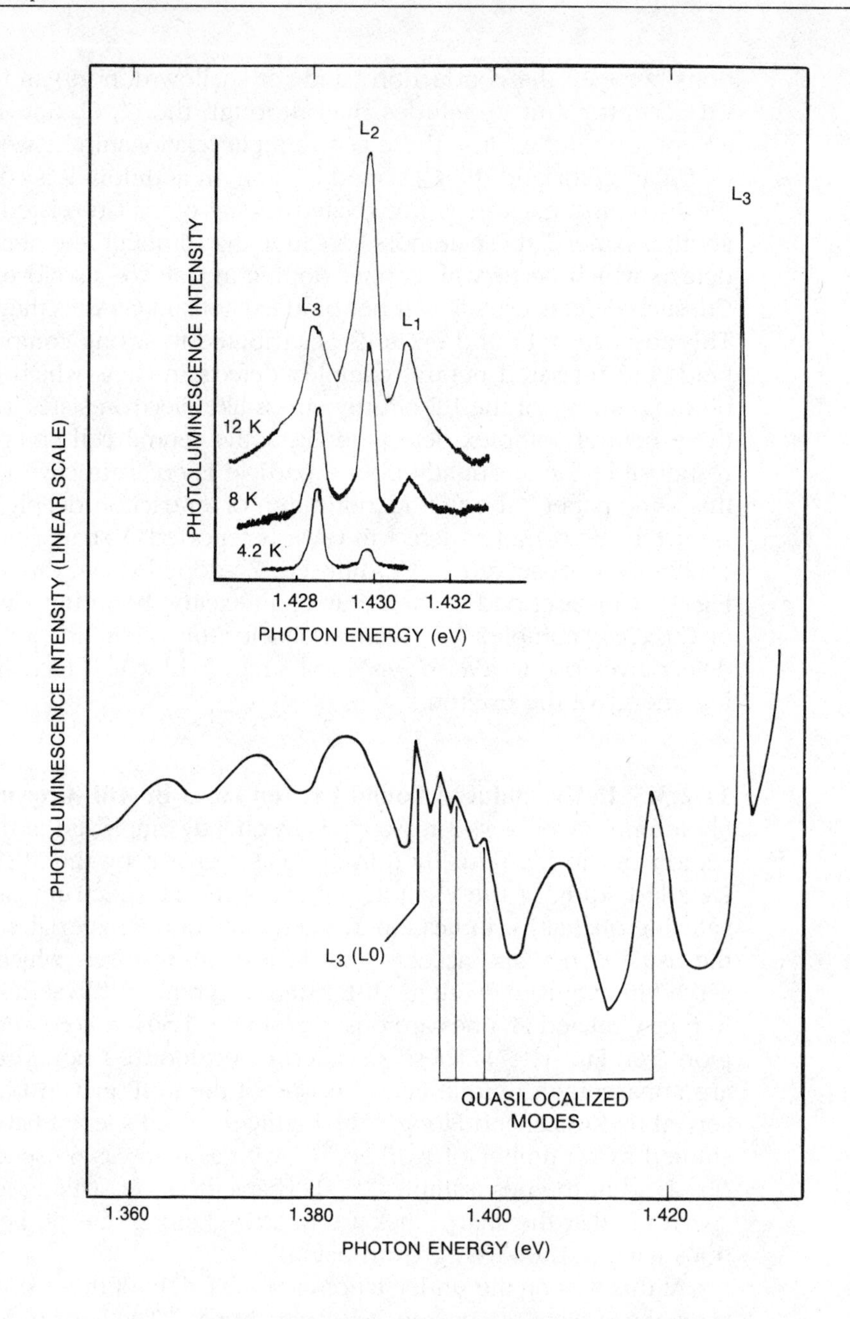

Fig. 11–6. Photoluminescence spectrum at 2 K for a Zn-doped horizontal Bridgman-grown bulk GaAs sample ([Zn] = 1.3×10^{16} cm^{-3}), Cu diffused at 500°C for 19 h, and rapidly quenched in water. Only the lowest-energy electronic line L3 at 1.4285 eV is seen at 2 K, together with a phonon wing towards lower energy. In the inset PL spectra are also shown for higher temperatures where three electronic lines can be resolved. *(After Monemar* et al.[62]*)*

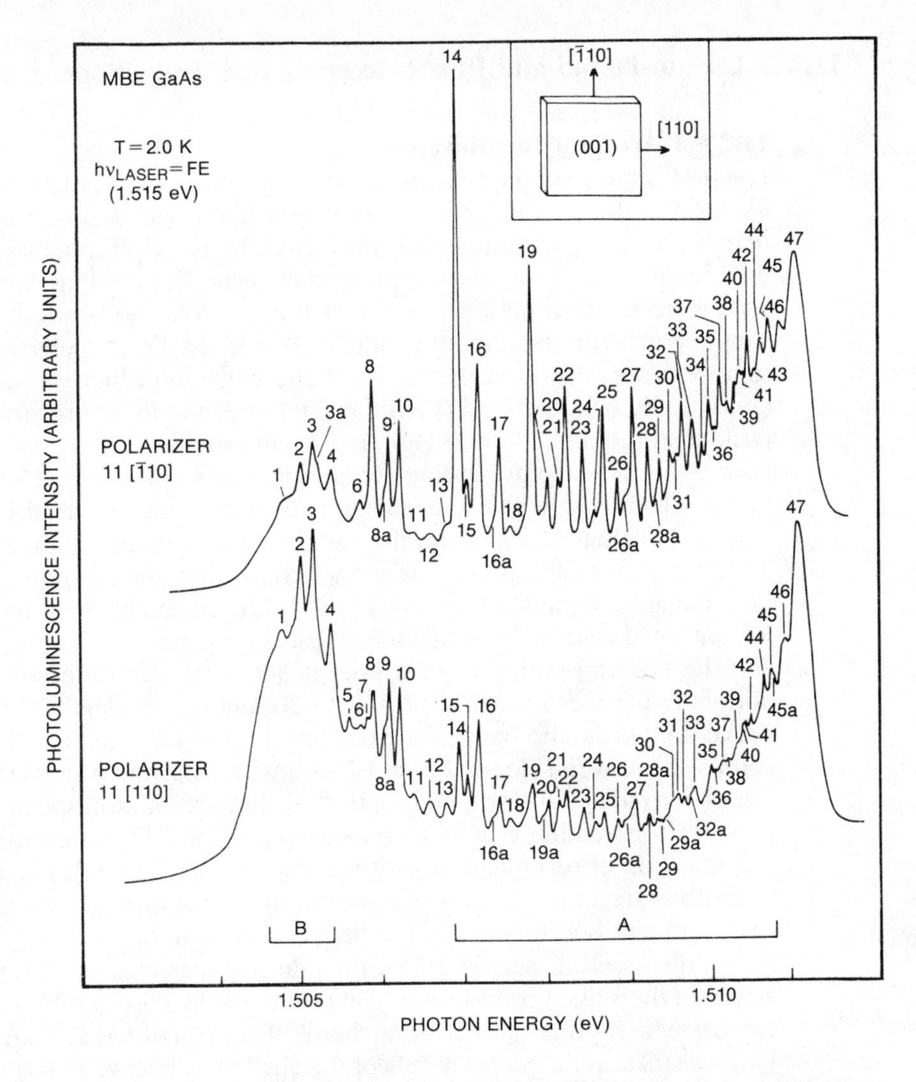

Fig. 11–7. Low-temperature (2-K) PL spectra for MBE GaAs excited with dye laser light resonant with the free-exciton energy at 1.515 eV. Spectra are shown for [110] and [1̄10] polarizations. The labeling of the peaks 1 to 47 is for reference purposes only. The inset shows a schematic diagram of the sample. *(After Skolnick* et al.[67]*)*

ticular attention to the symmetries of the defects as revealed by the polarization and Zeeman spectroscopy of the PL spectra. The present brief discussion is intended only to make the reader aware of the unique excitonic PL spectra shown in Fig. 11–7 which may be encountered when characterizing MBE-grown GaAs. A more complete description is provided by the review of Skolnick[68] and references therein.

11.2.4 Free-to-Bound and Donor-Acceptor Pair Transitions

11.2.4.1 Acceptor Identification

Free-to-bound (FB) PL transitions involving the recombination of a free electron at the bottom of the conduction band with a hole trapped at a neutral acceptor comprise the most directly useful PL phenomenon for the identification of acceptor inpurities in GaAs. The binding energy of the hole is deduced by subtracting the PL energy from the bandgap energy; a simple correction for the initial kinetic energy of the free electron and its capture cross section at the acceptor site is made (which amounts to increasing the bandgap by $kT/2$ in GaAs). For example, the relatively broad peak centered at 1.49 eV in Fig. 11–2 is attributable to acceptor FB transitions. Shallow acceptor binding energies for GaAs are in the ~25- to 40-meV range, which means that the acceptor FB transitions are displaced in energy and distinguishable from the excitonic transitions, and their variation from acceptor to acceptor is large enough to allow the FB transitions from different acceptor impurities in GaAs to be resolved clearly in most cases.[3] In contrast, the small binding energy of shallow donors in GaAs (~5.7 meV) and the correspondingly small energy separation of different chemical donors (<0.2 meV) make it difficult to distinguish or identify different residual donors on the basis of the donor FB transitions. In Sec. 11.2.3.2, the confusion between donor FB and D^+X transitions was discussed and the use of zero-field[16] and magneto-optical[17–19] high-resolution spectroscopy of two-electron satellites of $D^\circ X$ transitions and of D^+X transitions for the identification of residual donors was treated in Sec. 11.2.3.3. The donor FB transitions are in most cases of little use in the routine characterization of GaAs and will be discussed no further in this section.

As discussed in Sec. 11.2.2 on the effective mass theory, it is the interaction of the s-like ground states with the impurity cores of the various acceptors which gives rise to these chemical shifts. Baldereschi and Lipari[12] have calculated the energy levels of the shallow (effective mass) acceptors in narrow gap III-V semiconductors. Their classification scheme for the energy levels, in which the total angular momentum of the hole (3/2) is coupled to the orbital angular momentum associated with the hydrogenic envelope function, has been adopted almost universally and their results for the p-like excited states have been enormously useful in the interpretation of PLE and FIR spectroscopy of acceptor excited states. However, theoretical treatments of the chemical shifts of the s-like states have not been particularly successful. These shifts are best determined experimentally by PL and FIR spectroscopy of intentionally doped samples to create empirical catalogs of acceptor binding energies.

The identification of acceptors in GaAs on the basis of FB PL spectra is complicated by the presence of the closely related donor-acceptor pair

(DAP) transitions. At low temperatures most of the electrons in relatively pure samples of GaAs are localized at shallow donor sites. Under these conditions the intensity of the FB transitions decreases as recombination occurs via the DAP mechanism in which electrons bound at neutral donors recombine with holes bound at neutral acceptors. The energy of the DAP transitions is given by

$$E_{\mathrm{PL}} = E_G - (E_A + E_D) + e^2/\epsilon R + J(R)\,, \qquad (11\text{–}4)$$

where E_G denotes the bandgap energy, E_A and E_D are the acceptor and donor ionization (binding) energies, respectively, R is the spatial separation of the donor and acceptor pair involved in the recombination, and $J(R)$ is a correction term which describes the deviation of the luminescence energy E_{PL} from a simple Coulomb law because of the interaction of the donor and acceptor wavefunctions.[4] Basically, Eq. 11–4 states that the photon energy of a DAP transition is given by the bandgap energy reduced by the sum of the isolated impurity binding energies, modified by corrections for the Coulomb interaction between the charged final states of the impurities and the interaction between the neutral initial states of the impurities. Because both of the correction terms are functions of pair separation, R, the DAP transitions can occur over a range of photon energies determined by the distribution of pair separations occurring in the crystal. In some semiconductors, especially wide, indirect-bandgap materials such as GaP and cubic SiC, sharp line spectra due to discrete PL transitions on close pairs are observed;[4] with increasing pair separation, the energy interval between these lines decreases, eventually merging into a broad continuum PL band due to the distant pairs. In the relatively narrow, direct-gap III-V semiconductors, such as GaAs and InP, only the broad distant pair PL bands are observed (in spectral positions displaced a few meV below the FB transitions). (Donor-acceptor pair transitions and other interimpurity recombinations are discussed in great detail in the excellent, comprehensive review of Dean.[4] This review is especially useful to those interested in the interpretation of sharp lines of close DAP spectra occurring in materials other than GaAs.)

It is easy to see how the distant DAP PL bands can lead to mistaken acceptor identifications when interpreting the low-temperature PL spectra of GaAs. In some cases the FB spectrum of one acceptor species can coincide in energy with a distant DAP band which involves *another* acceptor species and the background donors. Therefore, when making an acceptor identification on the basis of acceptor-related PL spectra it is important to distinguish FB from DAP bands. The relative intensities of the FB and DAP bands in GaAs depend upon the background donor concentration and the temperature at which the PL spectra are obtained. As demonstrated in Fig. 11–8, the FB and DAP spectra in a given sample can be distinguished by studying the temperature dependence of the PL spectra in the 2- to 20-K

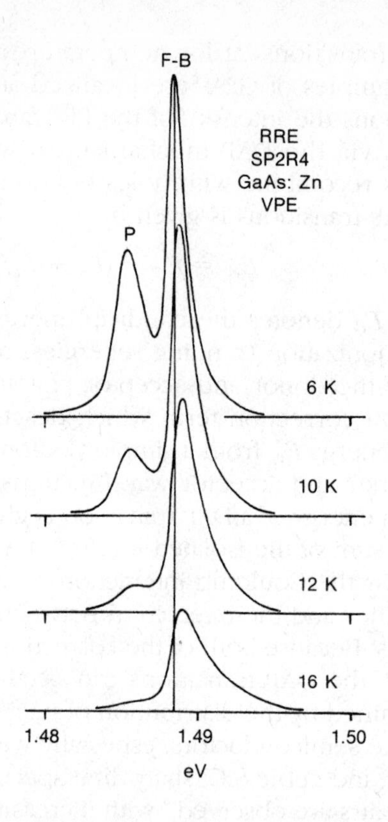

Fig. 11–8. The evolution of the free-to-bound band from the donor-acceptor pair band with increasing temperature in VPE GaAs which contains only one significant residual shallow acceptor, Zn. (*After Ashen* et al.[3])

range. With increasing temperature the donors are ionized and the FB transitions dominate the PL spectrum.

It is also possible to distinguish FB from DAP bands on the basis of their radiative lifetimes, which manifest themselves in time resolved PL spectroscopy and in a contrasting dependence on excitation intensity.[4] Whereas the FB transitions are characterized by a single radiative lifetime, the lifetime of DAP transitions varies with the pair separation, R. The greater the separation the less overlap between donor and acceptor wave-functions and the longer the recombination time. The photon energy of the DAP transitions also depends upon R, so that the photon energy decreases and the radiative lifetime increases with increasing pair separation. Because of the long lifetime of the low-energy, more distant pair portion of the DAP band, it is possible to saturate the PL from the more distant pairs at high exciting light intensity and induce a shift of the observed DAP peak to higher energy. In contrast, the peak position of the FB band is relatively insensitive to excitation intensity. Correspondingly, time resolved

PL spectroscopy, in which PL spectra are obtained as a function of delay time relative to pulsed excitation, reveals a shift in the peak of the DAP spectrum to lower energy with increasing delay time.

These lifetime effects in DAP recombination are exemplified in the time resolved PL spectra of Bishop *et al.*[69] shown in Fig. 11–9 for a p-type sample of bulk LEC-grown GaAs. The strong PL band at 1.443 eV is often observed in p-type conducting or semi-insulating GaAs grown from Ga-rich melts.[69–73,75,76] The 78-meV acceptor level which gives rise to the PL band has been attributed by various workers to the gallium antisite defect[70,71,73,76] (Ga_{As}) or the boron antisite defect[74] (B_{As}). Boron is a pervasive background impurity in LEC GaAs grown in boron oxide encapsulant. When incorporated on the Ga site it is isovalent, but if it transfers to an As site it, like the gallium antisite defect, forms a double acceptor. A variety of PL[69] and infrared[69,77] experiments have clearly demonstrated the double acceptor character of the 78-meV acceptor responsible for the 1.443-eV PL band, and that its first ionization (negative charge) state lies 203 meV above the top of the valence band (PL band at ~1.31 eV). The time resolved spectra[69] of Fig. 11–9 clearly indicate that the 1.443-eV double acceptor PL band shifts to lower energy with increasing delay time after pulsed excitation thereby revealing that the PL involves a DAP recombina-

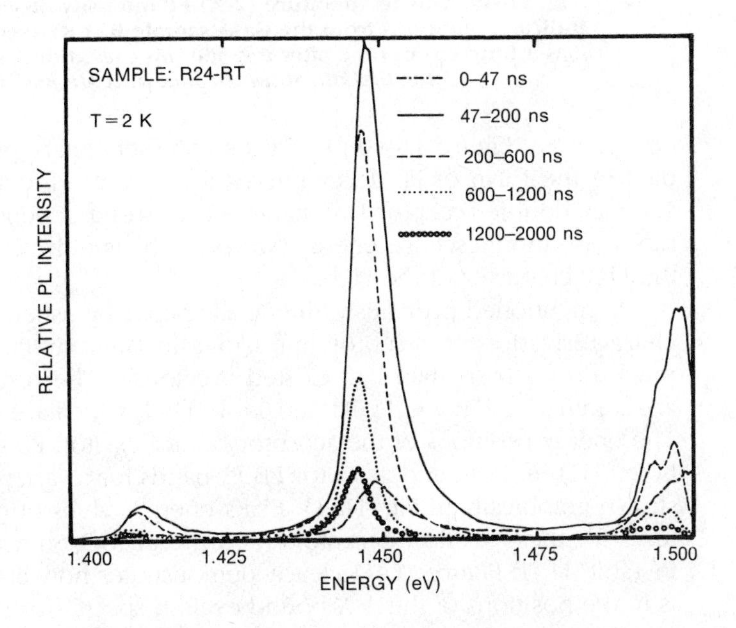

Fig. 11–9. Low-temperature (2-K) time resolved PL spectra obtained from a sample (R24-RT) of LEC GaAs grown from a Ga-rich melt. The relative PL intensities shown have been obtained by summing the PL intensities observed in the indicated time-delay windows. *(After Bishop, Shanabrook, and Moore[69])*

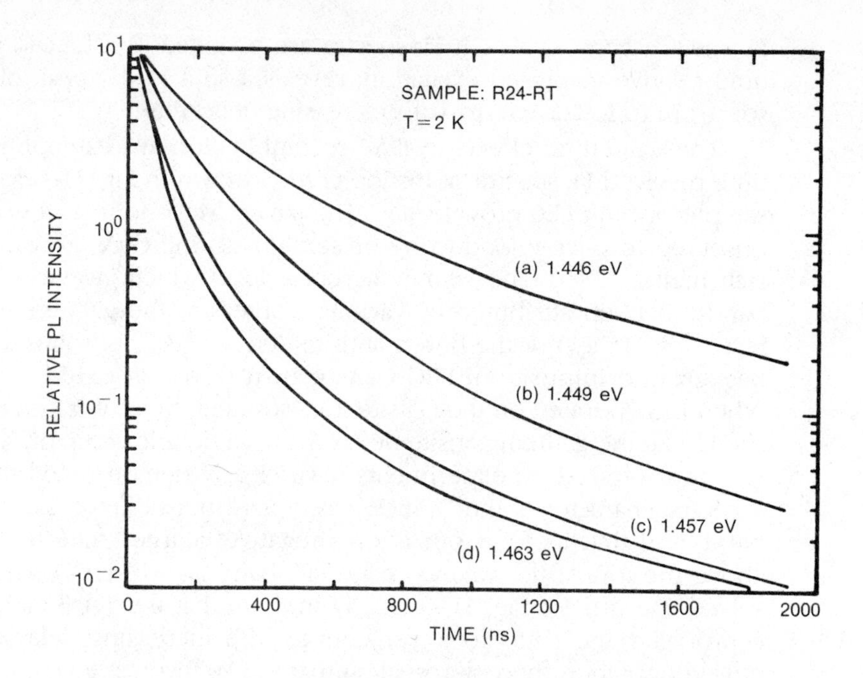

Fig. 11–10. Low-temperature (2-K) PL intensity observed at the indicated energies from the GaAs sample R24-RT (see Fig. 11–9) as a function of time after the 580-nm excitation laser pulse. *(After Bishop, Shanabrook, and Moore[69])*

tion process. Figure 11–10 provides a complementary presentation of the data in the form of PL decay curves for several PL energies within the 1.443-eV double acceptor DAP band.[69] The strong dependence of PL decay rate upon luminescence energy (wavelength) is quite evident, confirming the DAP character of the PL band.

As mentioned previously, the classic paper by Ashen *et al.*[3] placed the characterization of acceptor impurities in GaAs by PL techniques on a much firmer basis than had existed previously. The results of their work are summarized in Fig. 11–11 and Table 11–1, which are taken from Ref. 3. The energy positions of the acceptor bound exciton PL spectra, their two-hole (TH) replicas, and acceptor FB PL bands for a variety of acceptors are shown graphically in Fig. 11–11. Exact energy values of these spectral features for these shallow acceptors (and the deeper Sn acceptor) are listed in Table 11–1. Figure 11–11 clearly demonstrates how little variation there is in the positions of the $A^{\circ}X$ bound exciton spectra for the various acceptors. Note that there is a far greater variation in the positions of the acceptor FB transitions and that they can serve in most cases as a basis for identification of the particular acceptor. The identifications based on the acceptor FB PL can be strengthened or verified if the TH replicas of the BE spectra are also observed with shift corresponding to that expected for the

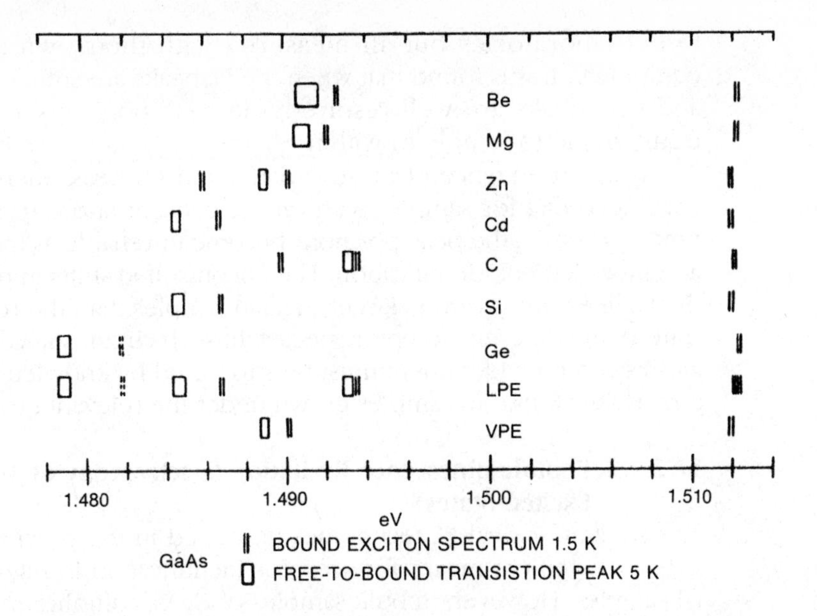

Fig. 11–11. The energies of the broad free-to-bound bands and the two-hole satellites (to the left) of the neutral acceptor bound exciton components (right) in the band edge PL of refined GaAs, lightly back-doped with the indicated shallow acceptors. The lower two rows illustrate spectra typical of undoped VPE and LPE GaAs. *(After Ashen* et al.[3]*)*

acceptor identified from the FB position. These replicas are much sharper and more clearly resolved than the FB spectra and their identification is usually not confused by the presence of DAP bands.

There are practical limitations and exceptions for this acceptor identification procedure which are discussed in detail by Ashen *et al.*[3] Several effects can lead to fluctuations in the measured FB peaks and therefore to false identifications of the acceptor impurities. For example, the kinetic energy distribution of the electrons in the conduction band can broaden the FB transition line width and shift the peak position to higher energy by $kT/2$, a correction which was calculated for GaAs by Eagles[78] and verified experimentally by Ashen *et al.*[3] Other problems which cannot easily be corrected for include Stark effect or dielectric screening shifts due to high concentrations of neutral or ionized impurities, and homogeneous and inhomogeneous strain effects resulting from growth techniques, substrate problems, or lattice parameter variation due to incorporation of high concentrations of impurities. Ashen *et al.*[3] compiled a listing of FB band energies measured in a number of laboratories for GaAs lightly doped with several of the most common acceptor impurities to document the level of variation in measured binding energies encountered in the literature. They also listed FB peak energies for samples grown in many different laboratories but all measured with their own (Ashen *et al.*[3]) equipment. It was found that when the PL peaks are sufficiently narrow, FB and DAP peaks are well resolved; the peak positions are reproducible from sample to sample to within ~1 meV.

As a consequence of these variations in FB peak energy, Ashen *et al.* concluded that for samples with carrier concentrations greater than a few times 10^{16} cm^{-3}, the peak positions become unreliable as the sole means of acceptor element identification. The uncontrolled shifts in peak energy and the PL line widths can be larger in such samples than the relative chemical shifts of the different acceptor species. In such circumstances, PL peak energies as acceptor identifiers must be supported by knowledge of the acceptors usually found in samples grown under the relevant growth conditions.

11.2.4.2 Photoluminescence Excitation Spectroscopy of Acceptor Excited States

The straightforward PL techniques described in the previous section provide a convenient means for acceptor identification in high-quality epitaxial samples. However, in bulk samples of GaAs, complicating factors which produce the line broadening and spectral shifts discussed above are very important and the FB PL bands have only limited usefulness for acceptor identification. Photoluminescence excitation (PLE) spectroscopy and its complementary technique, selectively excited luminescence (SEL), which utilize tunable dye lasers to excite directly the shallow impurities in the below-gap spectral region, can be used for the identification of acceptor impurities in bulk samples of semiconductors. First demonstrated in GaP,[79] the PLE and SEL techniques have been applied successfully to ZnTe,[80,81] ZnSe,[82] and to bulk samples of the closely related III-Vs, InP[13,14] and GaAs.[15]

The application of the technique to GaAs is described in Ref. 15. Below-gap absorption of photons from a tunable dye laser creates an electron in the ground state of a donor and a hole in an excited state of a nearby acceptor. The holes quickly relax to their ground state and donor-acceptor pair radiative recombination ensues. Now the photon energy at which the resonant exciting absorption process occurs depends upon the separation of the acceptor excited state from its ground state, and upon the separation, R, of the donor and acceptor, as described in Eq. 11–4. In the PLE experiment the luminescence energy is fixed by setting the observing monochromator to an energy within the DA pair PL band of the GaAs sample. The dye laser wavelength is then scanned and resonant absorptions are observed in the PLE spectrum when the PL energy (donor and acceptor ground state separation) *for the particular value of R selected by the chosen PL energy* differs from the dye laser energy by the separation of an acceptor excited state from its ground state (for sufficiently separated pairs). Such PLE spectra for shallow acceptors in GaP and InP are shown in Refs. 79 and 13–14, respectively.

In the SEL experiments reported for GaAs in Ref. 15, the exciting laser wavelength is fixed in the region of the resonant below-gap absorption and the energy of the observing monochromator is scanned through the broad DA pair PL band. When the resonant absorption conditions described above obtain for a given acceptor excited state at a value of R selected by the PL energy, relatively sharp peaks occur which are superimposed on the broad DA pair background PL. SEL spectra obtained by Hunter and McGill[15] for bulk GaAs samples are shown in Fig. 11–12. As is typical for such experiments, the PL intensity is plotted as a function of the difference between the photon energies of the emitted light and the laser. Thus the peak positions give directly the separations of the $1S_{3/2}$ ground state and various S and P excited states of the acceptor as labeled in Fig. 11–12. The shifts obtained from the SEL results were compared in Ref. 15 with the results of far-infrared photoconductivity,[7] measurements of two-hole replicas of acceptor bound exciton PL,[3] and the calculations of Baldereschi and Lipari[12] in order to identify the Zn and C acceptors in the undoped GaAs samples under study.

11.2.5 Photoluminescence from Deep Levels

In the previous sections we have discussed near-band-edge PL involving the recombination of weakly bound carriers. Such weakly bound carriers are relatively spatially delocalized, as in the case of carriers bound at shallow donor or acceptor impurities whose orbital radii can be as large as 1 to 10 nm. We now address lower-energy PL bands which originate from the recombination of more strongly bound, spatially localized carriers. In the most strongly bound examples, the orbital radius may be of the order of or smaller than the lattice constant of the host crystal.

The strong binding of carriers in deep levels has two major consequences for the spectral characteristics of the associated PL bands. It is apparent that large carrier binding energies reduce the energy of radiative transitions between such a deep level and the opposite continuum band edge. Thus photoluminescence from deep levels implies PL bands whose energies are significantly less than the bandgap energy. In addition, the broadening of PL bands attributable to the electron-phonon interaction is strongly dependent upon the spatial extent of the localized carrier.[83] For the case of the large electron orbits of a shallow donor in a covalent semiconductor the electron density at any point is quite small, the coupling to the lattice is weak, and the PL spectra are usually sharp. In the opposite extreme, the case of the atomiclike states of a rare earth impurity which are well shielded from the lattice, the impurity-phonon coupling is also small and the absorption and emission spectra due to electronic transitions between the bound states (f-electron states) are characterized by narrow

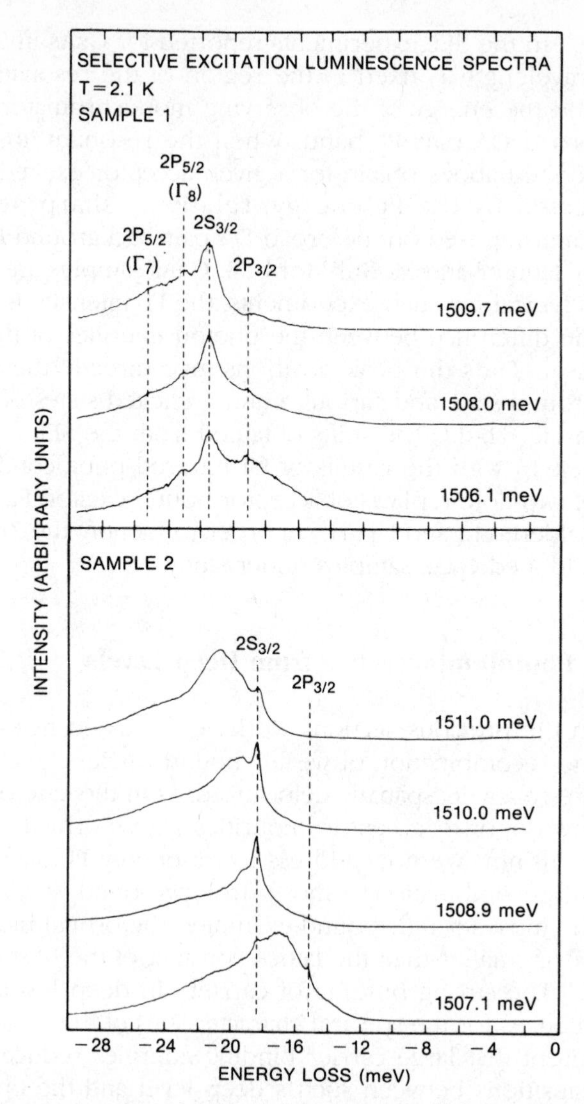

Fig. 11–12. Selective excitation luminescence spectra for two samples of bulk-grown GaAs. Intensity of emitted light is plotted against the difference in photon energy of emitted light and that of the laser. Photon energies of the laser for each spectrum are shown at the right. *(After Hunter and McGill[15])*

lines. Although the interaction is somewhat larger for transition metal impurities whose *d*-electrons are less shielded from the lattice, some of the 3*d* transition metal impurities exhibit sharply structured *d–d* transitions in their optical spectra and these are discussed in Sec. 11.2.6.1. However, the electron-phonon interaction is much larger for electrons directly involved in bonding to the ligands and in such cases, for example the vacancy-impurity complexes to be discussed in Sec. 11.2.5.2, the deep PL bands can be strongly broadened and featureless.

Section 11.2.5.1 comprises a brief qualitative discussion of the broadening of optical spectra by the electron-lattice interaction. In the subsequent sections examples of the PL and absorption spectra associated with a variety of deep levels are presented and their relative importance in the characterization of GaAs is discussed.

11.2.5.1 The Electron-Lattice Interaction

Obviously, a detailed description of the interactions among the large number of particles, electrons, and atoms involved in the coupled electron-lattice system is an extremely difficult problem. However, we can simplify the problem substantially by employing the adiabatic approximation[84] (referred to as the Born-Oppenheimer approximation in the case of molecules) in which it is assumed that the motion of the relatively heavy atoms can be treated separately from that of the light electrons. The consequences of this approximation are that the motion of the electrons can be determined on the assumption that the nuclei are fixed, and that the nuclei can be thought of as moving in a potential determined by the average motion of the electrons. For a detailed discussion of the electron-lattice coupling under the adiabatic approximation the reader is referred to Ref. 84, Chapter 2.

In order to illustrate how the electron-phonon coupling leads to spectral broadening, we introduce the configuration coordinate model,[85] which is often used to provide a qualitative description of optical processes at impurity centers. In the configuration coordinate diagram shown in Fig. 11–13 the total energy of the system, including both atomic and

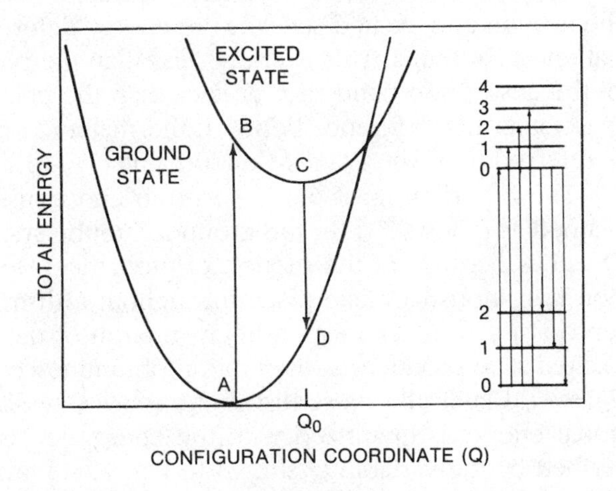

Fig. 11–13. Illustration of absorption and emission of light from an electronic center in a solid including the effects of lattice relaxation (electron-lattice interaction).

electronic terms, is plotted on the vertical axis for the ground and excited states. The configuration coordinate Q, a parameter specifying the positions of the atoms surrounding the center, is plotted on the horizontal axis. The broad applicability of the model stems from the fact that Q can specify the positions of a large number of atoms in the vicinity of a diffuse center such as a shallow donor or it might specify the positions of the nearest neighbor atoms only for a highly localized center. The absorption of a photon is indicated by the vertical transition from the equilibrium position, A, of the ground state to the excited state at B. The transition is vertical as a consequence of the Franck-Condon principle,[86] which states that the optical (electronic) transition occurs in a short time compared with the time required for significant motion of the atoms. After the transition to the excited state is completed, the lattice distorts to the configuration specified by Q_0 to minimize the electron energy. The energy difference between E_B and E_C is given off as lattice vibrations. Having reached its new equilibrium position, the system can return to the ground state by emitting a photon with energy $E_C - E_D$ (vertical transition), and the ground state again relaxes to the equilibrium position by phonon emission. Because of the energy transferred to lattice vibrations, the emitted photon energy is smaller than the absorbed photon energy by an amount referred to as the Stokes shift. The size of the Stokes shift obviously depends upon the strength of the interaction between the center and the neighboring atoms, that is, the electron-phonon coupling. If both the PL spectrum and PLE spectrum are obtained, the energy separation of the peaks in the phonon broadened PL and PLE spectra provides a measure of the Stokes shift. These two spectra will form a "mirror image" reflected about the ZPL (zero phonon line). The thermally activated energy separation between the ground and excited states is the energy difference between the potential energy minima. This is always less than the photon energy at the peak of the absorption band (but greater than the energy of the luminescence peak) and this difference between thermal and optical activation energies is referred to as the Franck-Condon shift.

The broadening of the absorption and emission spectra can be explained as follows.[83–85] In the ground state the system has a distribution of Q values because of the motion of the lattice. Hence the vertical absorption transitions can take place throughout a distribution of energies centered at $E_B - E_A$. Correspondingly, motion of the system about Q_0 in the excited state produces a distribution of luminescence energies centered at $E_C - E_D$. Classically, these distributions are Gaussians, but the actual vibrational energy is quantized with the energy levels for a single mode described by the variable Q shown in Fig. 11–13. Absorptive transitions can take place from the lowest ground state level to each of the excited state levels with intensities determined by the range of Q values which describe the motion in each level. The envelope of the absorption distribution in these transitions will still be given by a Gaussian for our assumption of a single vibrational mode. However, a real lattice will have an entire spectrum of vibrational modes and the line shape will be determined by the detailed coupling of each of these modes to the defect. The lowest energy absorption transition, which is called the zero phonon line (ZPL), will remain sharp but the rest of the distribution will consist of a convolution of broadened modes which reflects the phonon density of states.[86]

For recombination at centers with binding energies greater than a few tenths of an eV, most of the emission intensity is in the vibrational part of the spectrum. This is evident in the well known example of the 0.839-eV Cr-related emission[87–90] from GaAs:Cr shown in Fig. 11–14. (The origin of the PL spectrum in Fig. 11–14 is discussed in Sec. 11.2.7.2.) A qualitative representation of the shape of the broad vibrational portion of the emission spectrum can be obtained with the theory of Hopfield.[91] If it is assumed (as it was in the example of the 0.839-eV PL band in GaAs:Cr) that the phonon coupled band can be described in terms of contributions from the LO(Γ) and TA(X) modes and their combinations, then the relative intensities of the contributions from various phonon combinations are given by[88,91,92]

$$I(E) = \sum_{n_1,n_2} \frac{\overline{N}_{\mathrm{TA}}^{n_1}}{n_1!} \frac{\overline{N}_{\mathrm{LO}}^{n_2}}{n_2!} \frac{1}{1 + (E - E_0 - E_{n_1+n_2})^2/\gamma^2}, \tag{11–5}$$

where n_1 and n_2 are the numbers of TA and LA phonons emitted, respectively, the Ns are the average phonon coupling parameters for the respective phonons, γ is a phenomenological broadening parameter, E_0 is the ZPL energy, E is the emitted photon energy, and

$$E_{n_1+n_2} = n_1 E_{\mathrm{TA}} + n_2 E_{\mathrm{LO}} \tag{11–6}$$

is the energy of the multiphonon emission.

Another alternative is to use a theoretical model[93] in which two nondegenerate electronic states of the localized center are assumed to couple linearly to a *single* normal mode of the lattice (phonon) with energy $h\nu_{\mathrm{PH}}$. At low temperature the broad phonon coupled emission band is then composed of n phonon sidebands with position E_n given by

$$E_n = E_0 - nh\nu_{\mathrm{PH}}, \tag{11–7}$$

where E_0 is again the energy of the ZPL. The transition probability of the nth phonon sideband is given by

$$I_n = \exp(-S) \cdot S^n/n!, \tag{11–8}$$

where S is a coupling parameter representing the strength of the localized center-phonon interaction,[93,94] usually referred to as the Huang-Rhys factor.[95]

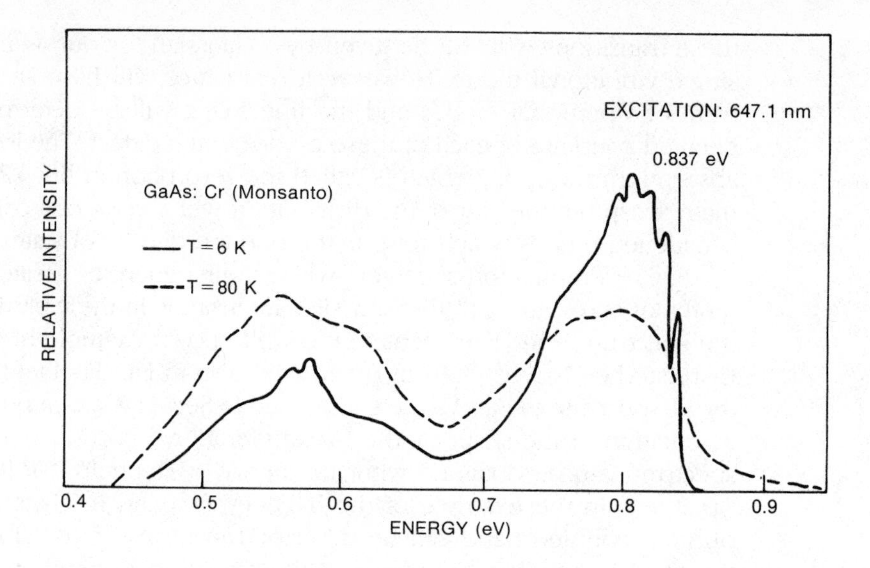

(A) PL spectra at 6 K and 80 K with 647.1-nm krypton laser excitation.

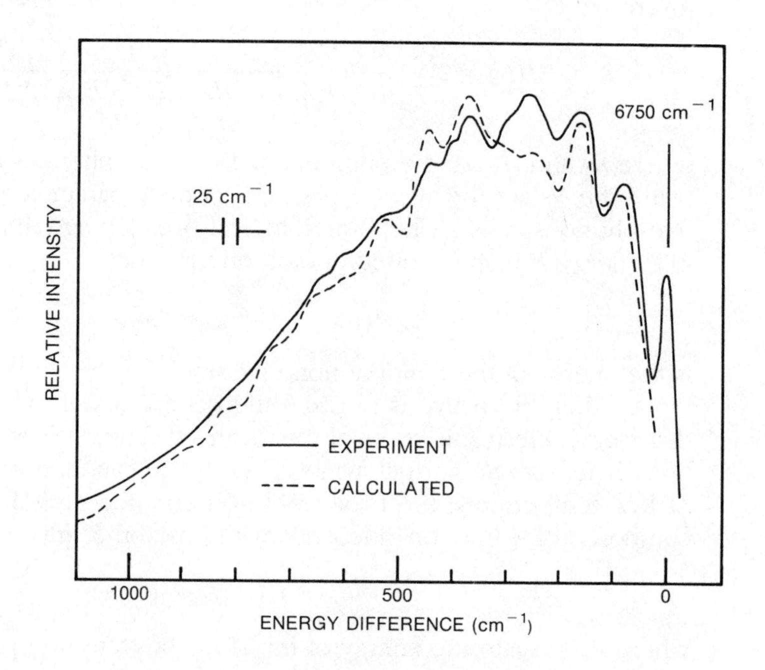

(B) Distributions for the 0.8-eV Cr-related band in GaAs. The ZPL at 6750 cm^{-1} is shown in higher resolution in Fig. 11–27.

Fig. 11–14. Photoluminescence spectra of GaAs:Cr and luminescence intensity distribution for Cr-related band in GaAs. *(After Koschel, Bishop, and McCombe[88])*

In many instances the electron-lattice coupling becomes so strong (Huang-Rhys factor $S > 5$) that all resolution of individual phonon combinations in the vibrational spectrum is lost and a broad, featureless band is observed at energies below the ZPL. The ZPL may be weakened to the point of being unobservable. This is true of the PL bands associated with vacancy-impurity complexes discussed[96–99] in Sec. 11.2.5.2. The broad, approximately Gaussian line shapes of such spectra and their temperature dependent line widths can be analyzed within the configuration coordinate model to obtain the Huang-Rhys factor and the phonon energy.[85,86–99] The temperature dependence of the linewidth predicted by the configuration model is given by[85]

$$W(T) = (8 \ln 2)^{1/2} S^{1/2} h\nu_{PH}[\coth(h\nu_{PH}/2kT)]^{1/2}. \qquad (11\text{–}9)$$

The constant multiplying the cotanh function is the low-temperature limit (W_0) of the linewidth. Thus a fit of Eq. 11–9 to the measured temperature dependence of the linewidth, including the low-temperature limit, will yield S and $h\nu_{PH}$. The Franck-Condon shift is given by $\Delta_{FC} = Sh\nu_{PH}$.

11.2.5.2 Vacancy-Impurity Complex Photoluminescence Bands

In the spectral range below the lowest-energy PL band due to isolated shallow acceptors (e.g. ~1.477 eV for Ge_{As}) a variety of broad, featureless PL bands have been observed in GaAs.[96–101] Early studies[96] of these bands referred to them as self-activated luminescence because they exhibit Gaussian lineshapes and temperature dependences of linewidths, peak energies, and intensities which are similar to the self-activated luminescence bands which have been studied so extensively[102] in the wide gap II-VI compounds ZnS and ZnSe. Most of these bands have now been attributed to various native defect-impurity complexes and they will be labeled here as complex PL bands rather than self-activated luminescence.

Because of the strong interaction between defect-impurity complexes and the crystal lattice, their associated radiative recombination bands are severely broadened and devoid of sharp spectral features which might aid in the identification of the recombination centers. Any identification of such a center solely on the basis of the measurement of the peak energy of a broad PL band must be considered highly unreliable. It is notoriously difficult to correct exactly the measured spectral distribution of intensity in a broad PL band for the spectral dependence of the sensitivity of the optical spectrometer, including such factors as grating efficiency and detector response. The uncertainties inherent in the measurement of the peak energy of a broad PL band are therefore much greater than those associated with the study of sharp line spectra. Consequently, the experimenter must rely on studies of PL lineshapes, temperature dependence, and excitation spectra to characterize the centers. In addition, valuable information can be obtained by measuring the effects of intentional doping, heat treatment or annealing, and departures from stoichiometry on the broad complex PL bands. However, even such studies are vulnerable to misinterpretation resulting from unintentional dopants or inadvertent departures from stoichiometry that are introduced during annealing.

One of the first broad, deep level PL bands to be associated with a complex in GaAs was the ~1.2-eV gallium vacancy-donor complex band which is commonly observed[96–98] in n-type GaAs with donor concentrations greater than $\sim 1 \times 10^{16}$ cm^{-3}. (Readers familiar with the high purity of currently available semi-insulating LEC GaAs will realize that the complexes discussed here are usually not observed in such material. Most studies of the vacancy-impurity complexes are carried out in the doped material with impurity concentrations as high as $\sim 1 \times 10^{18}$ cm^{-3}. However, annealing conditions characteristic of device processing sometimes induce complex PL bands in nominally high-purity material.) In the 1960s it was found that the 1.2-eV PL band was observed in Te-doped GaAs grown by Bridgman or floating-zone techniques but not in crystals grown from a gallium solution.[55,104] In addition, diffusion with copper or doping with zinc or cadmium, all of which would reduce the gallium vacancy concentration by substituting on the gallium sublattice, also eliminated the 1.2-eV PL. It was therefore postulated that the PL center was a gallium vacancy-donor impurity complex.[55,103,104]

Williams[96] studied the 1.2-eV PL band (Fig. 11–15) in GaAs doped with the group IV donors C, Si, Ge, and Sn (substituting on the gallium site) and the group VI donors S, Se, and Te (substituting on the As site). He re-

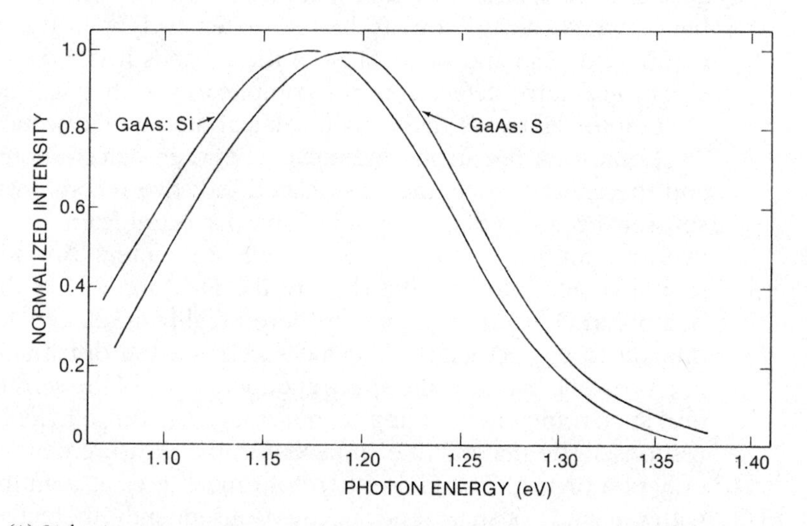

(A) SA luminescence in GaAs:Si and GaAs:S at 74 K. The separation between peaks is 0.018 eV.

Fig. 11–15. Self-activated (SA) luminescence in GaAs doped with various donors. *(After Williams [96])*

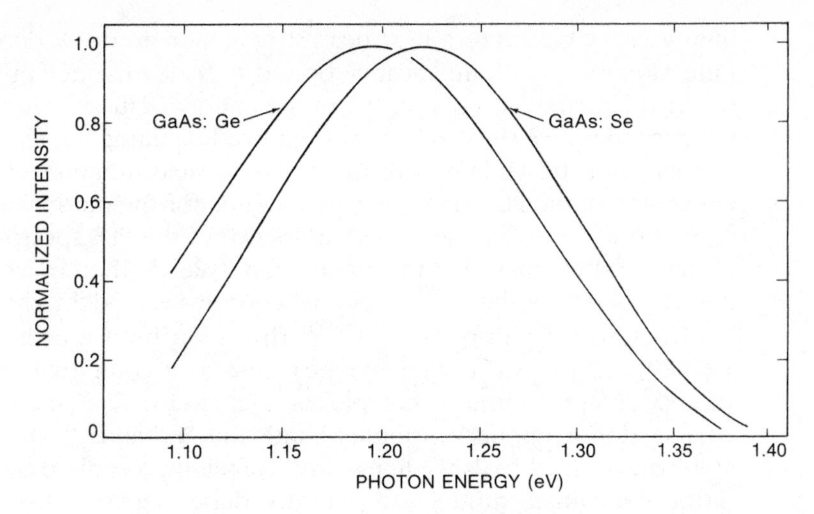

(B) SA luminescence in GaAs:Ge at 78 K and GaAs:Se at 74 K. The peak separation is 0.025 eV.

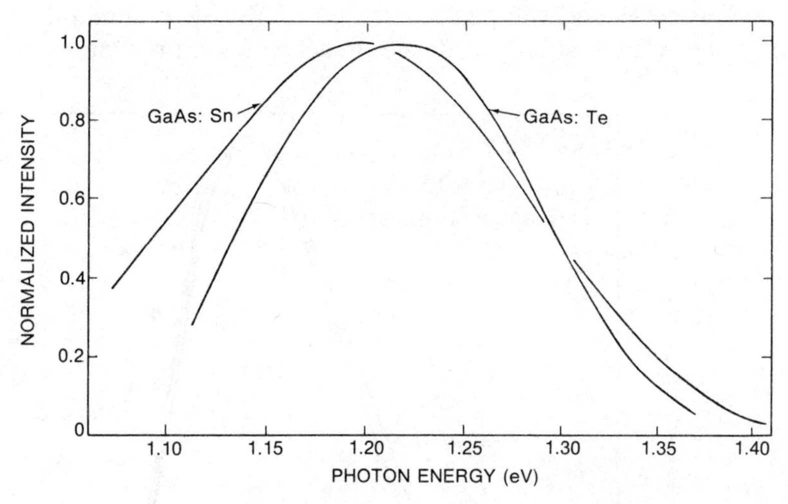

(C) SA luminescence in GaAs:Sn at 74 K and GaAs:Te at 76 K. Peak separation is 0.020 eV.

Fig. 11–15. *continued*

garded the similarity of the Gaussian lineshapes and temperature dependence of the linewidth (it follows the configuration coordinate equation) exhibited by this PL band to those of vacancy-impurity complex bands observed in ZnS as evidence that the 1.2-eV band in GaAs is attributable to a gallium vacancy-donor impurity complex. On the basis of the shifts to lower energy of each of the group IV-related PL bands relative to the bands for the group VI donors of the same period (see Fig. 11–15) it was suggested that the complexes responsible for the 1.2-eV PL bands in GaAs are analogous to those proposed by Prener and Williams[105] for ZnS: a gallium vacancy bound to a next nearest neighbor group IV donor on a gallium site and a gallium vacancy bound to a nearest neighbor group VI donor on an arsenic site. A value of the energy of the vibrational mode of the excited state $h\nu = 0.022$ eV was inferred by fitting the temperature dependence of the PL linewidth to Eq. 11–9. Measurements of the thermal quenching of the PL intensity yielded a value for the thermal activation energy of 0.18 eV. Subsequent measurement[98] of the PLE spectrum for some of the PL bands provided the Stokes shift, 0.28 eV, thereby enabling all of the parameters of the configurational coordinate model to be calculated.

In similar fashion, Hwang[99–101] first identified a broad PL band at 1.37 eV in Zn or Cd doped (p-type) GaAs as a complex band due to As vacancy-acceptor impurity complexes (Fig. 11–16). As in the case of the Ga vacancy-donor impurity complex bands the PL bands from these centers are characterized by large linewidth, half-width described by the configuration coordinate model, temperature dependence of the peak energy which does not follow the bandgap shift, and a large Stokes's shift between PL and PLE peaks. Hwang[99–101] also determined the parameters for the configuration coordinate model with the procedures described above.

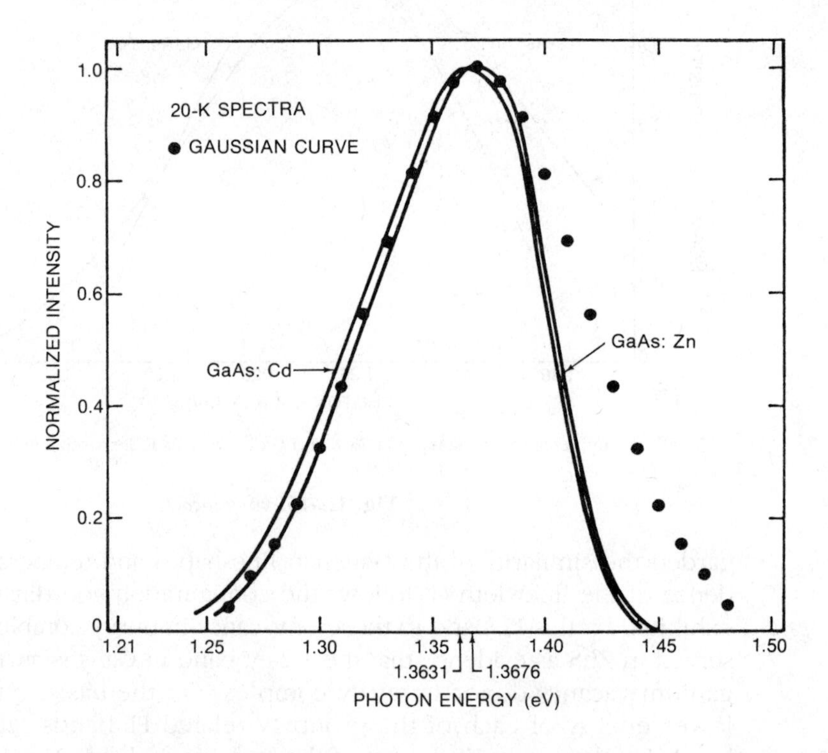

Fig. 11–16. The 1.37-eV band in GaAs:Zn and GaAs:Cd at 20 K. The solid circles are points of a Gaussian curve computed to get the best fit of the GaAs:Zn curve. *(After Hwang[99])*

Note that no detailed information concerning the identity of the recombination center is provided by the PL spectra of Figs. 11–15 and 11–16 and the association with particular vacancy-impurity complexes is inferred rather indirectly. This technique of systematically varying annealing conditions such as temperature, duration, and arsenic overpressure for doped and undoped samples has allowed many conclusions to be drawn concerning the chemistry of native defect formation and complexing with impurities which are responsible for the appearance of various broad complex PL bands in GaAs. However, such indirect methods have their limitations and one repeatedly encounters confusion and conflicting results in the extensive literature concerning the PL characterization of doped and annealed GaAs. One of the principal sources of confusion is the participation of residual or unintentional impurities in the formation of radiative recombination centers when GaAs, doped or undoped, is subjected to various annealing procedures. This point is illustrated in Sec. 11.2.6.2, in which the assignment of PL bands associated with transition metal impurities and their complexes is discussed.

11.2.5.3 Photoluminescence and Absorption Studies of EL2

The high resistivity of undoped semi-insulating GaAs is believed to be due to electrical compensation in which the Fermi level is pinned at the celebrated midgap deep donor level,[106] EL2. Semi-insulating behavior obtains when the EL2 concentration exceeds that of the carbon acceptors which are the dominant residual shallow impurity centers.[107,108] The literature describing the investigation of the optical and electrical properties of EL2 is dauntingly voluminous, and no attempt will be made to review it within the scope of this article. However, the reader has recourse to several recent review articles[108–111] if background information is required. The present discussion will focus on the extrinsic optical absorption which extends from a ~0.8-eV threshold to the band edge and on the broad photoluminescence bands in the 0.5- to 1.0-eV range which are observed in LEC grown GaAs. Other properties of EL2 will be discussed only as they relate to the efforts to associate these absorption and PL bands with the EL2 center.

Examples of some of the broad deep PL bands which have been observed in LEC GaAs are presented in Fig. 11–17A. Those in the range of 0.63- to 0.68-eV range have received perhaps the most attention, and the nearly 25 year long discussion of their origin has produced, correspondingly, the most confusion. Consistent with early work which associated the deep donor in GaAs with the presence of oxygen, a broad PL band peaking in the 0.62- to 0.64-eV range was attributed to oxygen by some workers.[112–114] Others have assigned this 0.62- to 0.64-eV band[115–119] or the 0.68-eV band[120–122] or both[111] to EL2. For example, Mircea-Roussel and Makram-Ebeid[115] found that the linewidth (130 meV) and temperature de-

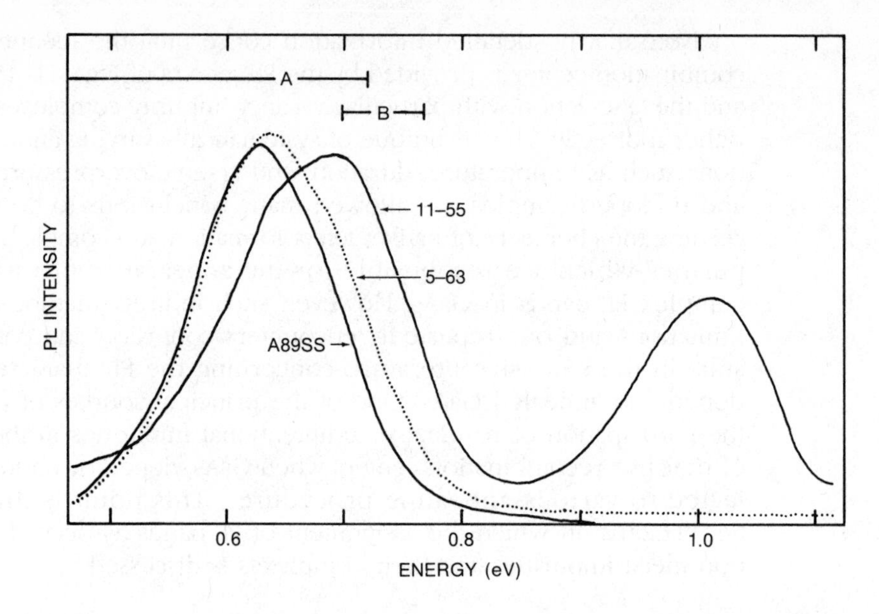

(A) PL spectra obtained from various semi-insulating GaAs samples. The line segments labeled *A* and *B* denote the different spectral ranges detected in the PLE measurements in Fig. 11–17B.

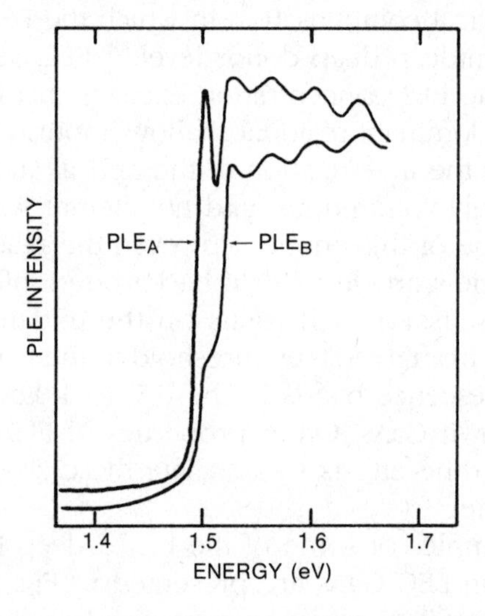

(B) PLE spectra obtained for PL occurring in the energy ranges *A* and *B*.

Fig. 11–17. Photoluminescence (PL) and photoluminescence excitation (PLE) spectra. *(After Shanabrook* et al.[117])

pendence of a 0.645-eV PL band which they commonly observed in bulk semi-insulating GaAs were consistent with the Franck-Condon shift (140 meV) and associated phonon energy ($h\nu$ = 20 meV) of the main electron trap EL2 as determined by phonon assisted tunnel emission.

Obviously, the identification of recombination centers solely on the basis of the peak energies of such broad, featureless PL spectra is fraught with the difficulties discussed in the previous sections. Consequently, spectroscopists have attempted to utilize two other phenomena to achieve additional specificity in their studies of these broad PL bands, the PLE spectra and the quenching or fatiguing property of the PL.

Shanabrook *et al.*[117,119] reported distinct differences in the PLE spectra of the 0.635-eV and 0.68-eV PL bands in the vicinity of the band edge (Fig. 11–17B). Because of the overlap of these two PL bands selective optical filtering and selection of samples in which the desired PL band is accentuated are required in order to obtain their separate PLE spectra. They also reported an interesting oscillatory phenomenon in the above gap PLE which was interpreted in terms of resonant cooling of hot electrons by LO phonon emission and indicated that PLE involved capture of electrons by an intermediate shallow donor state before localization of the carriers by some of the deep donor levels.[117,119] Similar effects in the PLE spectra of the deep PL bands have been reported subsequently by others.[123–126] Large variations in the strength of the below gap (extrinsic) portion of the PLE (compare Refs. 116–117 and 119) and the observed fatigue or quenching[116–119,122] of the PL were eventually explained in terms of the optical quenching (bleaching) of the extrinsic absorption, discussed below.

Bois and Vincent[127] discovered a quenching behavior in the transient photocapacitance of EL2 which they attributed to an optically induced metastable state of EL2 which corresponded to an atomic configuration different from that of the normal EL2. Vincent *et al.*[128] measured the spectral distribution of the cross section for this optical transformation. It peaks near 1.1 eV, which is close to the 1.17-eV photon energy of the Nd:YAG laser. This optically induced metastability has become an important factor in the identification of those portions of the extrinsic optical absorption spectrum of GaAs which are associated with EL2.

Martin *et al.*[129,130] identified an EL2-related portion of the extrinsic optical absorption spectrum in non-Cr-doped semi-insulating and n-type bulk GaAs by comparing the shape of the absorption spectrum to that of the electron photoionization spectrum of the (occupied) EL2 centers. He established a quantitative relationship between the EL2-related absorption coefficient and the concentration of EL2 in the material.[130] (In this work he noted the importance of avoiding confusion due to the possible presence of Cr and an associated near-infrared absorption, an important point where characterization of GaAs by infrared absorption is concerned.) Subsequently, Martin[131] firmly established the fingerprint of EL2 in the absorption spectrum when he observed that the extrinsic optical absorption features could be completely quenched (bleached) by a few minutes of exposure to light in the photocapacitance quenching band of Vincent *et al.* [128]

These findings had serious implications for both PLE spectroscopy and the use of PL quenching or fatigue to identify an EL-2-related broad PL band. Because the EL2 level is the main source of extrinsic photo-electron and -hole production in high-purity bulk GaAs, the quenching or bleaching of this extrinsic absorption will quench or fatigue the extrinsic excitation of *all* deep PL bands excited by below gap light.[117,124,132,133] Thus great care must be taken in using the fatiguing process to single out any of the deep PL bands as that which is attributable to EL2. This was clearly demonstrated by the work of Samuelson *et al.* [132,133] They obtained low-temperature PLE spectra of the 0.635-eV PL band in GaAs before and after quenching of the EL2 centers. In recording the "before" spectrum, care was taken to avoid quenching of the EL2 centers by the PLE light itself by adjusting the power density and scanning rate. These PLE spectra are shown in Fig. 11–18. The

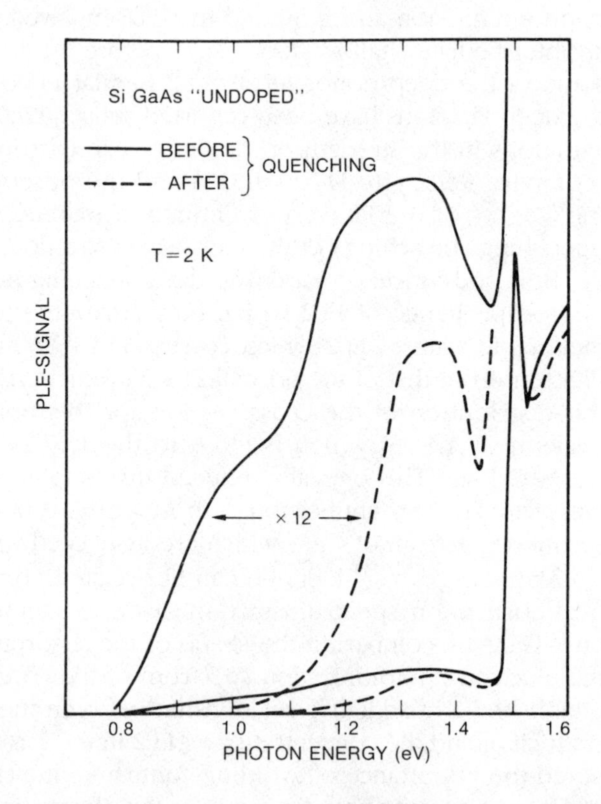

Fig. 11–18. PLE spectra of the 0.635-eV PL band in semi-insulating GaAs obtained before (solid line) and after (broken line) quenching of the EL2 centers. ***(After Samuelson, Omling, and Grimmeis[124])***

below gap portion of the unquenched spectrum is typical of the EL2 absorption spectrum. The intrinsic or bandgap excitation portion of the PLE spectra (above 1.47 eV) including the 1.5-eV peak and oscillatory features (not shown) is unchanged by the quenching process, and it is similar to that reported by Shanabrook *et al.*[117,119]. When excited by bandgap (intrinsic light) the efficiency and line shape of the 0.635-eV PL band are totally unaffected by the quenching of EL2. Thus the 0.635-eV band cannot be related to EL2. Furthermore, *any* deep level PL band in a semi-insulating GaAs:EL2 sample which is excited via the below gap EL2 absorption band will exhibit the optical quenching and thermal recovery effects typical of EL2, and this behavior cannot be regarded as evidence for its direct association with EL2.

A 0.67-eV PL band with 120-meV width has been observed by Tajima[134] under 1.32-μm excitation from a YAG:Nd laser. Although this wavelength is below gap and provides extrinsic excitation, it is not within the optical quenching band for EL2. This band was observed in a wide variety of GaAs crystals which under above gap excitation exhibited a variety of broad PL bands with peak energy between 0.65 and 0.8 eV (Fig. 11–19). On the basis of the intensity distribution across a wafer and the photoquenching properties of this PL band, he concluded that it involves EL2. It would appear, however, that the photoquenching experiments are subject to the criticisms outlined above. There is an EL2-related extrinsic absorption at the 1.32-μm exciting wavelength which would be quenched by the 1.06-μm light[124,132,133] so the observed photoquenching behavior does not require a direct connection between the PL center and EL2.

More recently, Tajima *et al.*[125] have shown that the 0.63-eV and 0.67-eV PL bands have quite different excitation spectra; the 0.63-eV band is excited strongly by photon energies above about 1.4 eV and the 0.67-eV band is excited best in the 0.85 to 1.3- to 1.4-eV range. They found that the low-energy portion of the PLE spectrum of the 0.67-eV PL band is a mirror image of the PL band itself. This was interpreted in terms of a Stokes shift due to strong coupling to the lattice. They estimated a Huang-Rhys factor of 5 and a ZPL of 0.78 eV. Tajima[135] has also observed a ZPL in the 0.67-eV band at 0.758 eV with accompanying TA phonon sidebands under 1.32-μm excitation (Fig. 11–20). Analysis of the vibrational structure in the PL spectrum indicates that the broad 0.67-eV band and the vibrational structure cannot be fitted by the same phonon coupling. It was postulated that two different phonon modes couple to the zero phonon transition. Tajima concluded that this PL band corresponds to transitions between the dominant midgap donor (EL2) and the valence band.

Kaminska *et al.*[136] have noted a systematic difference between the shape of the extrinsic optical absorption spectrum and the electron photoionization spectrum in n-type GaAs. This difference, shown in Fig. 11–21, corresponds to an intracenter absorption whose spectral position (1.15 eV

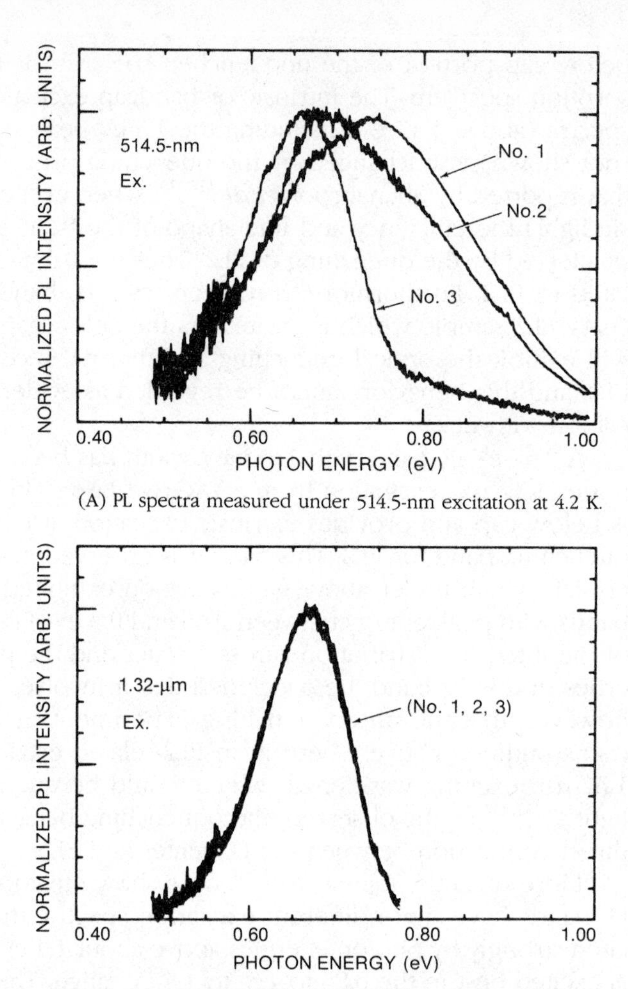

(A) PL spectra measured under 514.5-nm excitation at 4.2 K.

(B) PL spectra of the same samples measured under 1320-nm excitation at 4.2 K. All spectra are normalized at their respective peak heights.

Fig. 11–19. Photoluminescence spectra of undoped semi-insulating LEC GaAs crystals. *(After Tajima[134])*

peak) identifies it as the optical transition from the fundamental to the metastable state of EL2. A careful examination of this band at low temperature revealed spectral details corresponding to a 1.0395-eV ZPL and a cascade of 11-meV phonon replicas. Uniaxial stress experiments on the ZPL indicate that the crystal field symmetry of the excited state is T_2 and that of the fundamental initial state is A_1.

An important lesson to be learned from this discussion is that great care must be exercised in any attempt to infer the concentration of EL2 and its spatial distribution from the measurement of the absorption coefficient of a GaAs substrate at a single near-infrared wavelength (e.g.

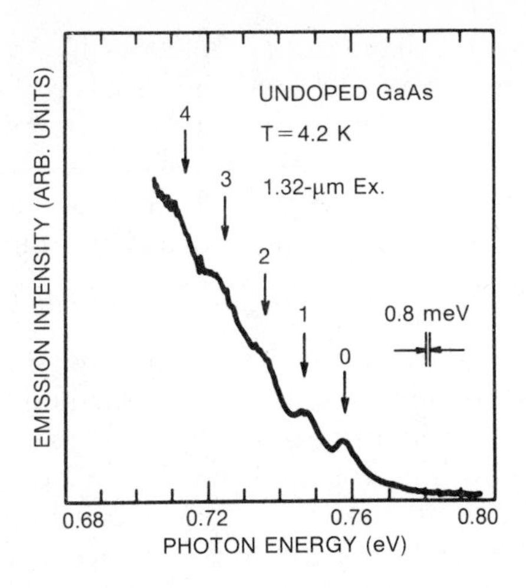

Fig. 11–20. Detail of 0.67-eV PL band in semi-insulating GaAs showing the 0.758-eV ZPL and accompanying TA phonon sidebands. *(After Tajima[135])*

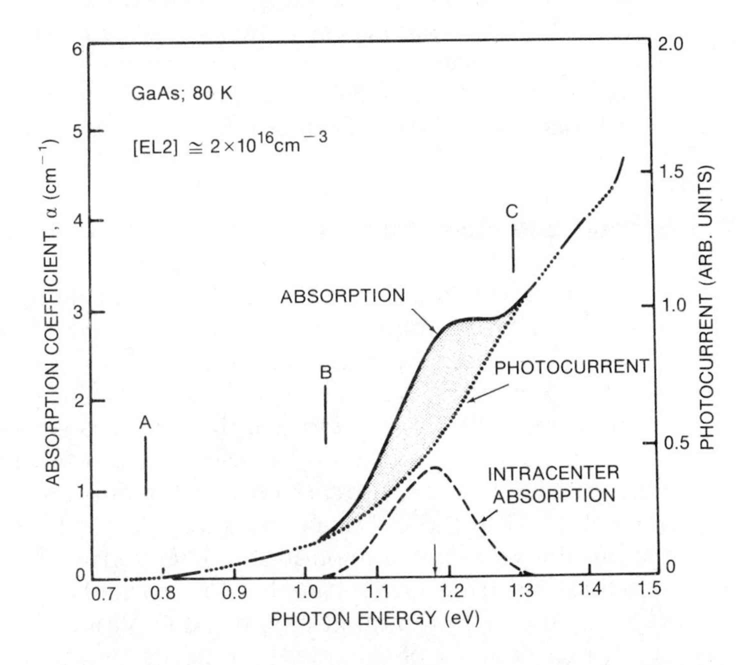

Fig. 11–21. Optical absorption and photocurrent spectra measured on a sample of melt grown GaAs with EL2 concentration of about 2 × 10^{16} cm^{-3}. The difference between the two spectra defines the intracenter absorption band which extends from 1.03 to 1.32 eV and is responsible for the characteristic shape of the EL2 absorption (shaded area). *(After Kaminska[136])*

1.06 μm). The literature is replete with such measurements including the determination of EL2 concentration profiles across wafers by spatially scanning a near-infrared beam.[137–141] The most reliable spectroscopic procedure is to measure the low-temperature optical transmission in the EL2 absorption band before and after optical quenching of the EL2 centers and evaluate the EL2 absorption coefficient from the relationship[132]

$$\alpha_{EL2}(h\nu) = (1/d)\ln(T_Q/T), \tag{11–10}$$

where T and T_Q are the optical transmission before and after quenching and d is the sample thickness. This method avoids interference from background absorptions (e.g. Cr) and it is independent of assumptions about reflection coefficients and internal reflections (see Samuelson *et al.*[132]). Similarly, efforts to determine the spatial distribution of EL2 centers from the measurement of the intensity of the deep broad PL bands across a GaAs wafer are unavoidably complicated by the confusion regarding the assignment of these overlapping bands. Again, one must proceed carefully.

This advice also holds true for the imaging of defect distributions in substrates which is being carried out by IR photography, scanning laser beams, IR vidicons, cathodoluminescence images produced in electron microscopes, etc. (see papers in Ref. 142). The most convincing evidence for the association of the absorption features observed in such images with the EL2 defects is the vidicon work of Skolnick *et al.*[143,144] which clearly demonstrates the effects of optical quenching and thermal recovery.

11.2.6 Transition Metal Impurity Luminescence: *d–d* Transitions

The $3d$ transition metals such as Ti, V, Cr, Mn, Fe, Co, Ni, and Cu are pervasive inadvertent impurities in GaAs at concentrations of the order of 10^{15} cm^{-3} or lower. They can be introduced directly as impurities in the source materials or as contaminants during the growth procedure. Some of these impurities give rise to sharply structured PL or absorption spectra due to intraconfigurational d–d transitions characteristic of the open-shell core configuration of $3d$ transition metal impurities on substitutional cation sites.[145–152] If these sharply structured PL bands are cataloged for the various transition metal impurities in intentionally doped samples, the PL spectra of uncharacterized samples can provide a sensitive means for the detection of inadvertent transition metal impurities.[150]

However, some of the transition metals, notably Cu,[55,153–155] Cr,[87–90,152] and Mn,[156–158] behave quite differently as impurities in III-V semiconductors. For a variety of reasons, the PL spectra of these impurities in GaAs do not exhibit sharply structured d–d transitions. Consequently, the association of particular PL bands in GaAs with the presence of these impurities and the determination from the PL spectra of the details of their incorporation in the lattice (isolated or complexed) are uniquely challenging tasks and are discussed in Sec. 11.2.7.

The details of the splitting of the transition metal ion *d*-bands are determined by the symmetry of the crystal field which the ion experiences[159, 160] (tetrahedral in the case of zinc blende semiconductors) and the charge state of the variable valency transition metal ion. Thus, the spectral distribution of the optical transitions between the crystal field split *d*-band levels can be used to determine the charge state of the transition metal impurity. Identification is based upon a detailed comparison of the *d*-band absorption spectra with the sequences and relative energy separations of the *d*-shell energy states predicted by crystal field theory.[145–152]

It is also possible to use PLE spectroscopy for this purpose. The transition metal PL transitions, which occur between the first excited state *d*-band manifold and the ground state manifold, can be excited by absorption of above gap light which excites electron-hole pairs or by extrinsic absorption involving transitions from the ground state to the various excited *d*-band states produced by the crystal field splitting. Consequently, PLE spectroscopy, which traces the spectral distribution of those absorption processes which lead to the excitation of the selected PL band, can provide the same information as an absorption spectrum and lead to the identification of the charge state of the emitting transition metal ion. This technique has been applied successfully to the study of a number of transition metal dopants in wide gap III-V and II-VI semiconductors.[146, 148, 150, 161] However, its usefulness in relatively narrow gap semiconductors such as GaAs and InP is sharply curtailed because many of the ground state to excited state separations for the impurity *d*-bands are greater than the bandgap energies.

11.2.6.1 Iron

One of the most important examples of the detection of the presence of a transition metal impurity by its characteristic *d–d* PL band is that of Fe.[148] On the basis of EPR spectra[162] it has been determined that Fe is incorporated into GaAs as Fe_{Ga}, with the neutral charge state being Fe^{3+}. However, optical excitation produces a highly distinctive, sharply structured, four-line PL spectrum[163] at about 0.37 eV in GaAs:Fe (Fig. 11–22). This spectrum is characteristic of the PL spectra observed in Fe-doped samples of most zinc blende, III-V,[148] and II-VI[164, 165] semiconductors and is attributed to transitions from the $^5T_2$ excited state to the ^5E ground state of the ^5D term of Fe^{2+} ($3d^6$) ions in the tetrahedral crystal field. The four sharp lines are attributed to spin-spin and second order spin-orbit splitting of the ^5E ground state.

This four-line PL spectrum is a virtual "fingerprint" for the presence of Fe in GaAs. The distinctive nature of the PL spectrum and the fact that Fe forms a dominant trap or shunting recombination mechanism for

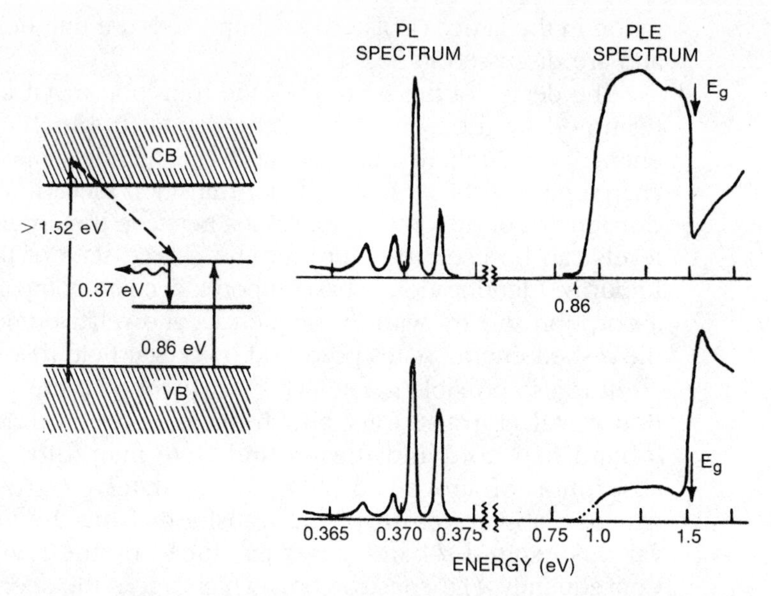

Fig. 11–22. Fe^{2+} PL and PLE spectra of bulk GaAs: Fe (top) and an annealed, undoped GaAs sample (bottom). The level diagram indicates possible optical excitation mechanisms for above- and below-gap excitation of the Fe PL. (*After Nordquist* et al.[163])

photoexcited electron-hole pairs[148,166] make it possible to observe the Fe^{2+} PL band in high-quality GaAs and InP at Fe concentrations as low as 1×10^{15} cm^{-3}. In fact, the excitation mechanism for the 0.37-eV PL band in InP is so efficient that laser oscillations have been achieved by Klein and coworkers;[166] these workers have also provided a detailed analysis and explanation of the excitation mechanism for the Fe^{2+} PL.

The PLE spectrum for the 0.35-eV Fe^{2+} PL band in GaAs: Fe is shown in Fig. 11–22. As described in detail elsewhere,[163,167,168] this PLE spectrum and similar spectra obtained for the Fe^{2+} PL bands in InP and GaP have been used to derive energy level diagrams which fix the positions of the ^5E ground state of Fe^{2+} relative to the top of the valence band. The energy level diagram for GaAs: Fe is also presented in Fig. 11–22. (The reader is cautioned that the "energy level" of the Fe^{2+} ground state in such a diagram actually represents the energy absorbed in the process of converting the Fe ion with its variable valency from Fe^{3+} to Fe^{2+}. For a more complete discussion of this point see Refs. 90 and 169.)

11.2.6.2 Titanium

Titanium doped III-V compounds received relatively little attention before 1986 and the available literature presented conflicting and confusing results (see the 1985 review of transition metals in III-Vs by Clerjaud, Ref. 151). As a result of the recent increased interest in semi-insulating

GaAs:Ti and InP:Ti, an improved understanding of the incorporation of Ti in these compounds is now emerging. DLTS measurements in Ti-doped GaAs have detected two electron traps which are 0.23 and 1.00 eV below the bottom of the conduction band, and represent different charge states of the same center.[170] In high-resistivity GaAs:Ti, Hennel *et al.*[170] observed a characteristic absorption spectrum at 4.2 K (Fig. 11–23) with two ZPLs at 0.566 eV and 0.569 eV and a broad phonon sideband peaking at 0.64 eV. (A similar spectrum is observed in n-type InP:Ti with all features shifted to slightly lower energy.[170]) This spectrum is interpreted as a ^2E ground state to $^5T_2$ excited state transition (with a ~3-meV spin-orbit splitting of the excited state) associated with the $Ti^{3+}(3d^1)$ neutral charge state of substitutional Ti_{Ga}. Observation of this spectrum requires that the Fermi level be positioned above the 1.00-eV midgap electron trap.

In n-type GaAs:(Ti, Si), Hennel *et al.*[10] obtained broad absorption bands due to the $^3A_2$ to $^3T_1$(F) and $^3A_2$ to $^3T_1$(P) intracenter transitions within the $Ti^{2+}(3d^2)$ ionized titanium "acceptor" state. The excited states for both these transitions are resonant with the conduction band. Interestingly, similar broad absorption spectra were reported by Ulrici *et al.*[171] with the same attribution. However, they also observed the two ZPLs

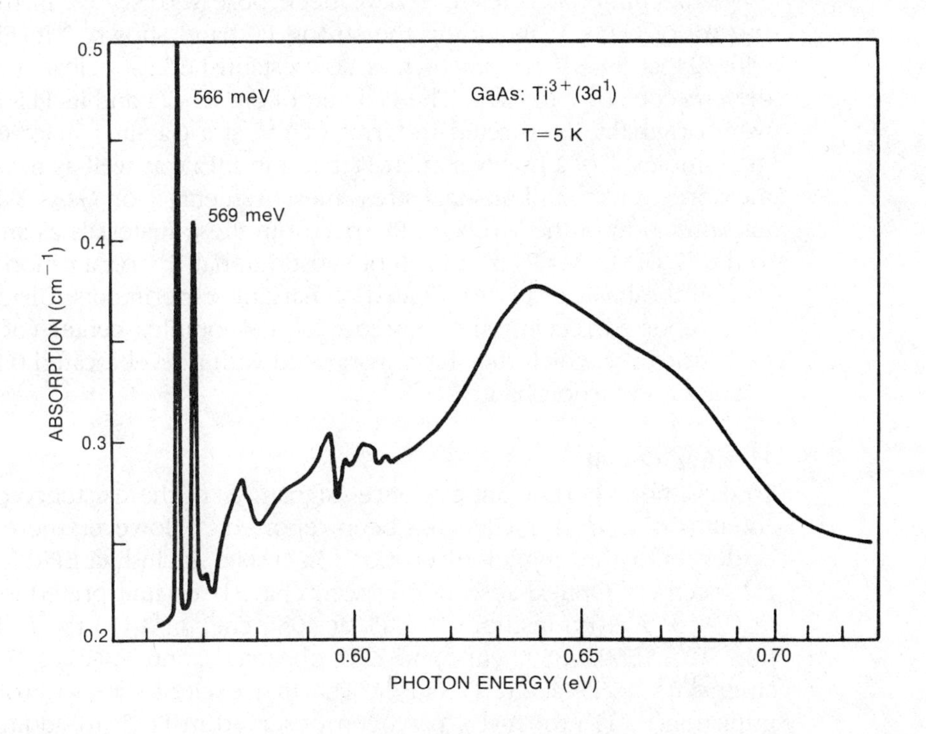

Fig. 11–23. Optical absorption spectrum of lightly n-type Ti-doped GaAs at 5 K. *(After Hennel* et al.[170])

which characterized the $Ti^{3+}(3d^1)$ spectrum of Hennel *et al.* In addition, the 0.566-eV ZPL was observed in PL by Ulrici *et al.* These authors[171] also found no Zeeman splitting of the 0.566-eV ZPL in the absorption spectrum, indicative that both ground and excited states of the transition are spin singlets. Thus they concluded that the 0.566-eV ZPL is not due to Ti^{2+}. It would appear that Ulrici *et al.* were able to see manifestations of both the Ti^{2+} and Ti^{3+} charge states in the absorption spectrum of some of their GaAs:Ti samples.

11.2.6.3 Vanadium

Semi-insulating V-doped GaAs is being marketed[151] and is purported[172] to be superior in terms of thermal stability to semi-insulating GaAs:Cr and undoped GaAs (EL2). The Clerjaud review[151] describes a number of confusing transport measurements on GaAs:V which provided no evidence for a midgap acceptor. More recent experiments[173–175] have established that the isolated V_{Ga} has its V^{2+}/V^{3+} level 0.14 eV below the conduction band and its V^{3+}/V^{4+} level near the top of the valence band. Thus it appears that V_{Ga} can provide no mechanism to explain the fact that some of the V-doped material has high resistivity.

Vanadium-related features have been observed[151, 176–179] in the optical spectra of GaAs:V, including the strong PL band shown[178] in Fig. 11–24 with ZPL at 5958.7 cm^{-1}, which is now established as a clear indicator of the presence of V in GaAs. The PL band of Fig. 11–24 and its PLE spectrum were originally interpreted in terms of V^{2+} at a Ga site.[178] However, Zeeman studies[180] of a highly similar PL band in InP:V as well as new Zeeman measurements[181] and uniaxial stress measurements[182] on GaAs:V led to reinterpretation of the V-related PL spectra in these materials as attributable to the $^3T_2(F)$ to $^3A_2(F)$ transition of substitutional V^{3+} on a cation site.

On the basis of photoinduced recharging experiments, Ulrici *et al.*[179] have reported recently the existence of a V-complex center (of unspecified structure) which they have associated with a level located 0.23 eV below the conduction band.

11.2.6.4 Cobalt

To date, no experimental evidence (signature) of the existence of neutral cobalt ($Co^{3+}(3d^6)$) in GaAs has been reported.[151] However, there is ample evidence for the singly ionized $Co^{3+}(3d^6)$ state, including EPR[183] and optical spectra.[184] Optical absorption spectra have been interpreted in terms of $^4A_2(F)$ to $^4T_2(F)$ transitions[184] (ZPL at 4035 cm^{-1}), $^4A_2(F)$ to $^4T_1(F)$ transitions[145, 185] (7000–8500 cm^{-1}, no ZPLs observed), and $^4A_2(F)$ to $^4T_1(P)$ transitions[145, 185] (ZPL at 11,317 cm^{-1}). The first excited state to ground state transition, $^4T_2(F)$ to $^4A_2(F)$, has been observed in PL.[186] In addition, a DA pair PL band supposedly involving the Co^{3+}/Co^{2+} level has been reported at 1.3 eV.[184, 187] Hall effect,[188] photoionization,[189] and photoconductivity[190]

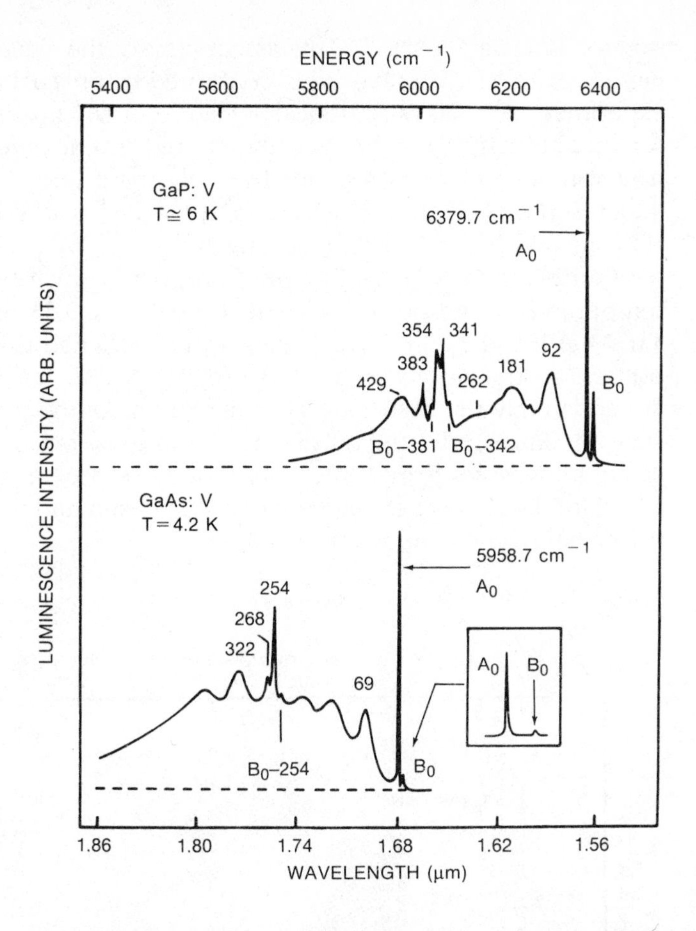

Fig. 11–24. Luminescence spectra (514-nm excitation) of LEC grown vanadium doped GaP and GaAs. Numbers associated with the phonon sidebands represent their energetic separation from the ZPL A_0. The $B_0 - X$ sidebands are replicas of the ZPL B_0. The luminescence is assigned to an internal crystal-field transition of isolated $V^{2+}3d^3$. *(After Kaufmann* et al.[178])

measurements place the Co^{3+}/Co^{2+} level 0.14 to 0.16 eV above the valence band. Finally, it is noted that the double acceptor Co^{+} state (Co^{3+}/Co^{2+} level) is apparently located more than 0.1 eV *above* the conduction band edge at atmospheric pressure, and it can be observed optically and electrically only under hydrostatic pressure.[191]

11.2.6.5 Nickel

The presence of neutral Ni^{3+} ($3d^7$) has been established from EPR spectra.[192] Until quite recently the only reported *d–d* transition associated unambiguously with the presence of Ni in GaAs was a 4613 cm^{-1} ZPL corresponding to the $^2T_2$ to ^2E transition of the $Ni^{+}(3d^9)$ double acceptor state of Ni on a Ga site.[184] This assignment of the sharp line spectrum of Fig. 11–25 to $Ni^+(3d^9)$ has been confirmed by uniaxial stress and Zeeman experiments.[193] This d–d optical signature of $Ni^{+2}(3d^9)$ has not been observed in PL apparently because the excited state for the transition is resonant with the conduction band. The only sharp line PL spectra related to the presence of Ni are the ZPLs reported at 4697 and 4424 cm^{-1} due to the Ni^+_{Ga}-Si_{Ga} and Ni^+_{Ga}-S_{As} pairs, respectively.[194]

Quite recently, Ulrici *et al.*[195] reported the observation of low-temperature absorption bands in GaAs:Ni at 0.55 eV, 1.10 eV (ZPL at 1.0703 eV), and 1.22 eV. These bands were assigned to the intraconfigurational d–d transitions from $^3T_1(F)$ to $^3T_2$, $^3T_1(P)$, and $^3A_2$ of Ni^{2+}_{Ga}. These are apparently the first optical manifestations of the presence of the Ni^{2+} state in GaAs. Hall effect and photoionization measurements by these same authors confirmed the conclusions of previous workers that the Ni^{2+}/Ni^{3+} level is 0.2 eV above the valence band and the Ni^+/Ni^{2+} level is 0.4 eV below the conduction band.

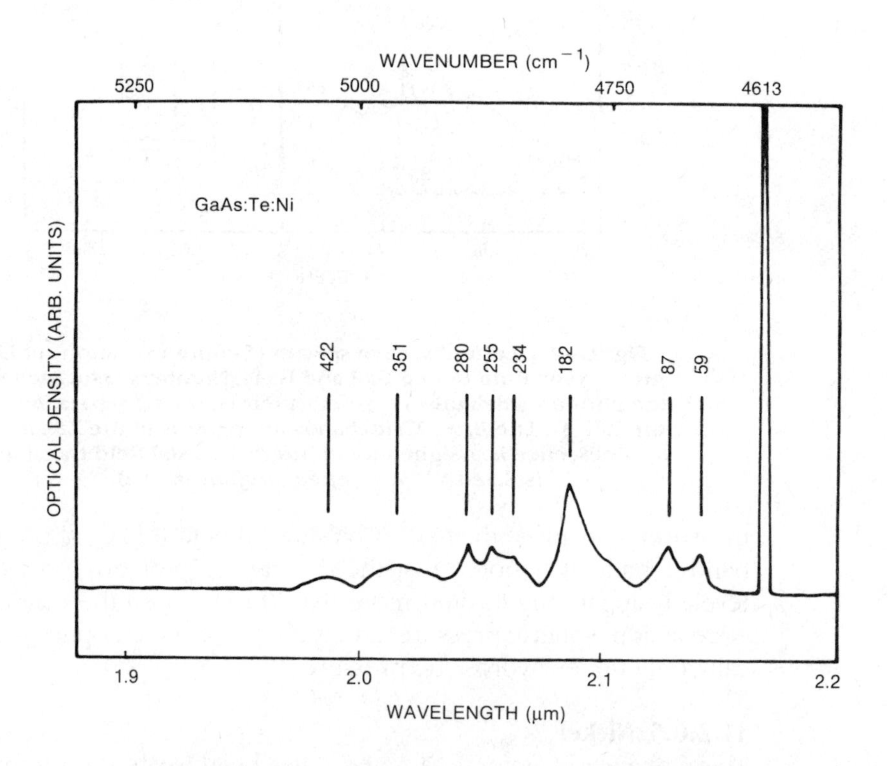

Fig. 11–25. Optical absorption of Ni-diffused n-type GaAs at 6 K. The band arises presumably from the $\Gamma_7(^2T_2) \rightarrow \Gamma_8(^2E)$ transition of $Ni^+(3d^9)$. *(After Ennen, Kaufmann, and Schneider[184])*

11.2.7 Transition Metal Complexes

11.2.7.1 Copper

The free copper atom has the $3d^{10}\ 4s$ outer electron configuration, and the stable configuration for Cu substituting on the Ga site in GaAs (or In in InP) is apparently $3d^{10}$ with one electron contributed to bonding orbitals.[154] Substitutional Cu would therefore behave as a double acceptor and because of the full *d*-shell the *d–d* crystal field transitions characteristic of open-shell $3d$ transition metal impurities are not observed for Cu in III-V semiconductors.

Although Cu exhibits no *d–d* transitions in GaAs, Cu-related complexes give rise to several characteristic PL bands whose classification and identification have produced a vast and confusing body of literature during the past two decades.[53,55,104,196,197] Because Cu is such a pervasive contaminant in the III-V compounds and exhibits rapid diffusion[154] at temperatures as low as 300 to 400°C, it is important to understand its incorporation in the crystal lattice, to catalog the PL bands which can both signal its presence and identify the details of its incorporation (isolated, complexed, etc.) and, where possible, to associate the PL bands with the electrically active Cu centers. However, this has proved to be a difficult task.

Copper-doped GaAs exhibits two well known acceptor levels at 0.156 and 0.45 eV.[55,197,198] Early workers[196,197] interpreted these as the two ionization levels of the expected Cu double acceptor, but it is now believed that there is no unambiguous evidence to support this conclusion.[58] Over 20 years ago, a 1.36- (1.356-) eV Cu-related PL band was reported[55] and associated with the 0.156-eV acceptor level of substitutional Cu. This band (Fig. 11–26) is strongly coupled to the lattice and usually has intense phonon replicas. The 1.356-eV band (often referred to as the 1.36-eV band) has been studied intensely over the past 20 years and its assignment to isolated Cu_{Ga} has often been questioned (see, for example, Guislain *et al.*[53]). However, recent piezospectroscopic studies[199] of the 1.36-eV Cu band led to its assignment (and the 0.156 acceptor level) to a ⟨100⟩ oriented Jahn-Teller distorted Cu_{Ga}. The assignment of the 0.45 eV Cu-related acceptor level remains uncertain, although a $Cu_{Ga}V_{As}$ associate has been suggested.[53,55]

Hwang[101] demonstrated that Cu diffusion of Zn-doped (p-type) GaAs can result either in the replacement of the As vacancy-Zn acceptor complex band at 1.37 eV (see Sec. 11.2.5.2) by the 1.36-eV Cu_{Ga} PL band or the appearance of a broad PL band near 1.37 eV with ZPL at 1.429 eV, which he attributed to As vacancy-Cu_{Ga} or D_{As}-Cu_{Ga} pairs. Correspondingly, Cu diffusion of Si-doped (n-type) GaAs causes a shift of the 1.22 eV V_{Ga}-Si_{Ga} donor-

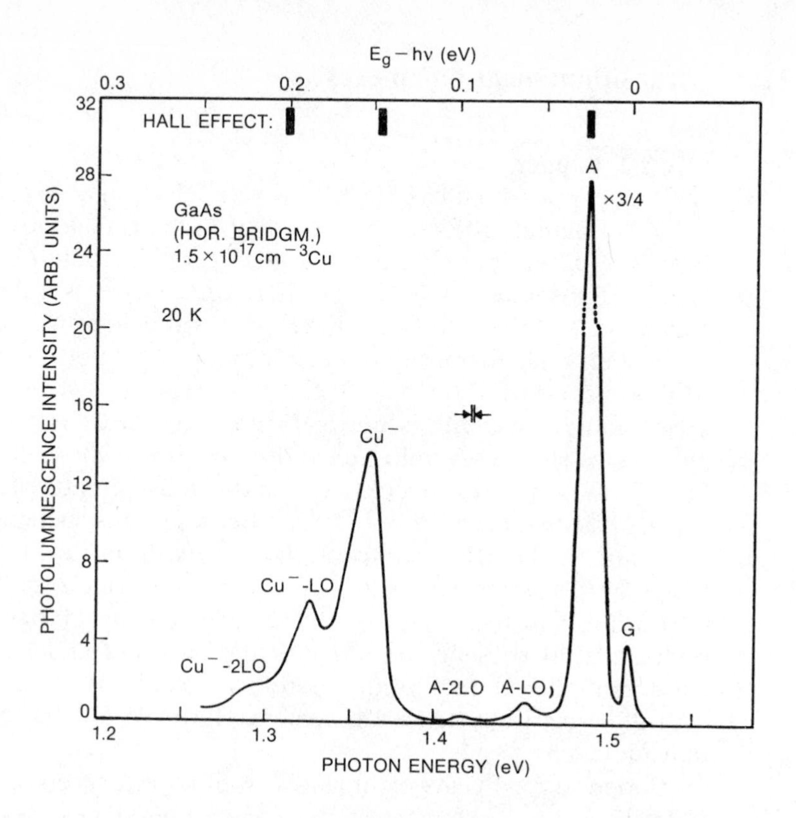

Fig. 11–26. Photoluminescence spectrum at 20 K of Cu-diffused GaAs crystal. Bars on upper abscissa show levels detected by Hall-effect measurements. The intensity of the phonon replica A-2LO is exaggerated. *(After Queisser and Fuller[55])*

acceptor complex PL band to 1.27 eV, which is attributed to the formation of Cu_{Ga}-Si_{Ga} complexes.[104,200] These broad PL bands and the bound exciton lines C_0 and F_0, at 1.5030 and 1.4839 eV, respectively, are the PL features most often associated with the presence of Cu in GaAs. The attributions of the 1.356-eV band to isolated substitutional Cu and the 1.27-eV band to the Cu_{Ga}-Si_{Ga} donor-acceptor pair are now on a fairly firm basis. But the assignments of the 1.37-eV band with its 1.429-eV ZPL and the C_0 and F_0 bound excitons are being reevaluated[58–60] as discussed in Sec. 11.2.5.2.

11.2.7.2 Chromium

Chromium has played a prominent role in the materials development of semi-insulating GaAs substrates. In addition to being a frequently encountered residual impurity in III-V semiconductors it has been the preferred deep acceptor dopant for the compensation of residual shallow donors in

GaAs for many years.[152] The ~0.8-eV electron trap introduced by Cr doping is attributed to Cr_{Ga}, which is loosely referred to as a double acceptor.[130,152,188] A provocative discussion of the conceptual difficulties associated with the description of the incorporation of Cr ($3d^5 4s^1$) as an "acceptor" in GaAs and the resulting confusion in the literature has been provided by A. M. White in Ref. 90. These difficulties will be avoided by confining this discussion to the optical (PL) manifestations of the presence of Cr in GaAs.

A broad 0.8-eV PL band associated with Cr in GaAs was first reported twenty years ago.[87] About ten years later, sharp line structure was observed[88] in this band (Fig. 11–14B) with a ZPL at 0.839 eV, prompting the tentative assignment of the 0.8-eV PL to intraconfigurational d–d transitions (^5E–$^5T_2$) of the Cr^{2+} on Ga sites. Subsequent to this observation, intensive study of the luminescence properties of GaAs:Cr has led to a more complete understanding of the 0.8-eV band and a very different attribution. High-resolution studies[201–205] of the 0.839-eV ZPL revealed considerable fine structure (Fig. 11–27) which first prompted alternative explanations of the origin of the Cr-related band. It was determined that the ground state of Cr^{2+}, as detected by EPR,[206] could not be reconciled with the ground state inferred from the fine structure of the 0.839-eV ZPL. In addition, an independent absorption line associated with the isolated Cr^{2+} was observed at 0.825 eV.[207] The Zeeman spectroscopy of Eaves *et al.*[208] (Fig. 11–27) and Killoran *et al.*[209] then established the axial symmetry (⟨111⟩) of the 0.839-eV center and it became clear that a Cr complex and not isolated Cr_{Ga} was involved. The trigonal symmetry of the defect can be inferred qualitatively by noting that the multiplicity of the Zeeman splitting shown in Fig. 11–27 assumes its simplist form (a doublet at high fields) when **B** is parallel to ⟨100⟩ and all four ⟨111⟩ axes are equivalent.

Subsequent ODMR studies,[210] Zeeman spectroscopy,[211–213] and stress perturbed optical spectroscopy[214,215] confirmed the trigonal symmetry of the defect and provided further spectroscopic details. These experimental results have led to general acceptance of the Cr^{2+}-X trigonal center proposed by Picoli *et al.*,[205] Cr^{2+} (on a Ga site) coupled to a donor on an As site, as the recombination center giving rise to the 0.839-eV PL band in GaAs:Cr. A succession of refinements[211–217] to this basic model have produced increasingly detailed and accurate representations of the available spectroscopic data. Note that the broad 0.8-eV Cr PL band, even with its 0.839-eV ZPL but without resolved fine structure (as observed in low-quality material or at 77 K), could not provide an identification of the Cr-related recombination center. The fine structure in the ZPL and the symmetry information conveyed by its splitting under applied perturbations such as magnetic field and uniaxial stress were required in order to effect a reliable identification.

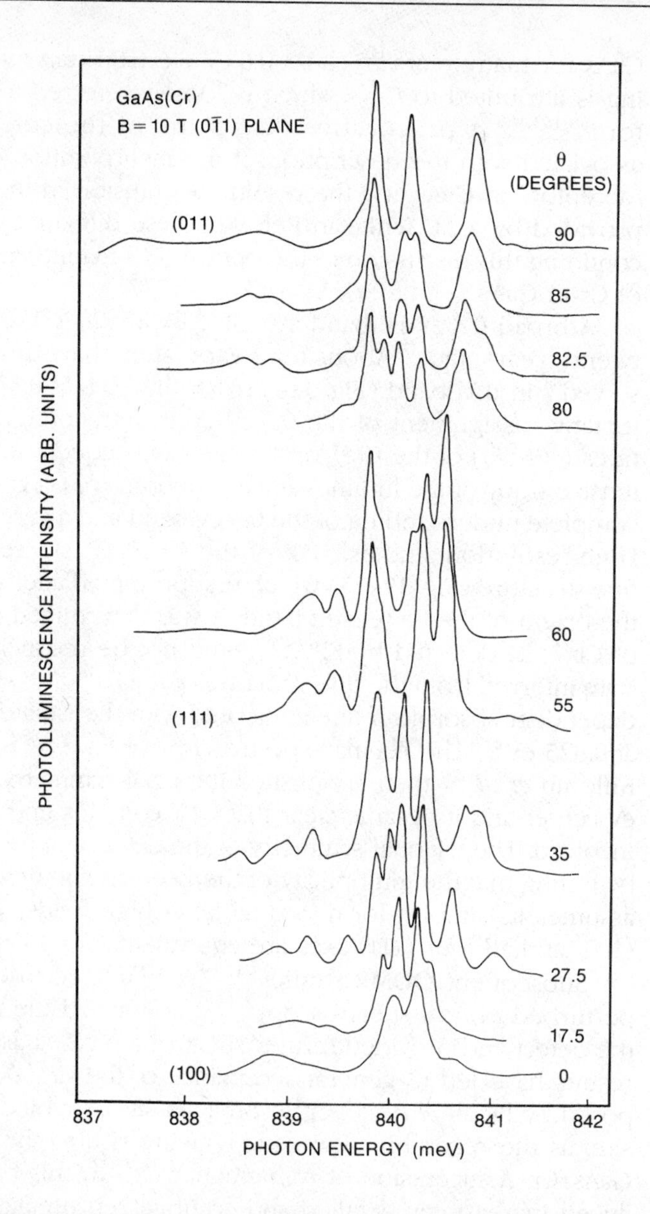

Fig. 11–27. The Zeeman spectra of the 0.84-eV PL zero phonon structure of GaAs: Cr at 2 K with B = 10 T rotated in the (011) plane. (*After Eaves* et al.[208])

From the standpoint of characterization, the ~0.8-eV PL bands shown in Fig. 11–14 provide a clear indication of the presence of Cr in GaAs. The broad band at 0.56 eV (Fig. 11–14A) has also been associated with Cr in GaAs,[88, 201, 205] and it is best observed when the 0.839-eV line is very strong and overlapping broad bands not associated with Cr are weak. Under these circumstances a ZPL at 0.574 eV and a replica of 0.535 eV have been observed.[201,218–220] Although Koschel *et al*[88] and Deveaud *et al.*[220] have suggested an association with the internal transitions of Cr^{3+}, Allen[152] has concluded that the nature of the center has not yet been established. The absence of a corresponding absorption band suggests that only a small fraction of the chromium atoms is involved. Because of its overlap with several other broad bands in this spectral range which are frequently observed in GaAs (see Sec. 11.2.5.3), the 0.574- to 0.535-eV PL system is somewhat less useful as an indication of the presence of Cr.

In recent years it has been shown that the PL due to intraconfigurational d–d transitions (^5E–$^5T_2$) of *isolated* Cr^{2+} on Ga sites in GaAs (to which the 0.839-eV band was erroneously attributed at first) can only be observed under hydrostatic pressure.[220,221] The ^5E excited state lies above the conduction band minimum, and the Cr^{2+} internal transition can only be observed in PL when this excited state is brought below the conduction band edge by the application of hydrostatic pressure.[221] For this reason PL associated with isolated Cr^{2+} is of little use for routine characterization of GaAs.

11.2.7.3 Manganese

Manganese is another $3d$ transition metal impurity whose behavior differs from that of most $3d$ substituents in GaAs. It is incorporated as Mn_{Ga}, acts as a deep acceptor[156–158,222] ($E_A = 0.113$ eV) and gives rise[156–158] to classic DA pair and free-to-bound PL transitions at ~1.41 eV (see Fig. 11–28). Perhaps because of the well known d–d transitions associated with Mn in the II-VI compounds, there was some initial reluctance to accept the assignment of the 1.41-eV DA pair PL band in GaAs to Mn. The fact that the Mn was introduced by diffusion[156] in the early work raised the possibility of the simultaneous introduction of native defects. Subsequent PL studies in which Mn dopants were introduced during MBE growth,[158] liquid-phase epitaxial growth,[157] or by ion implantation[223] left little doubt that there is a 1.41-eV band associated with the presence of Mn in GaAs. The electrical properties of heavily Mn-doped samples, as determined by Hall effect measurements,[158] were found to be dominated by a 110-meV acceptor; these samples also exhibited a Mn PL band with ZPL at 1.406 eV, corresponding to an acceptor 113 meV above the valence band edge. In addition, it was concluded that the Mn is incorporated on a simple cubic site and is not involved in a complex.[158]

Because Mn is a persistent residual impurity in GaAs whose redistribution (diffusion) during heat treatment can alter the electrical characteristics of the material,[224–227] the 1.41-eV PL signature of Mn has become an important factor in the characterization of GaAs.

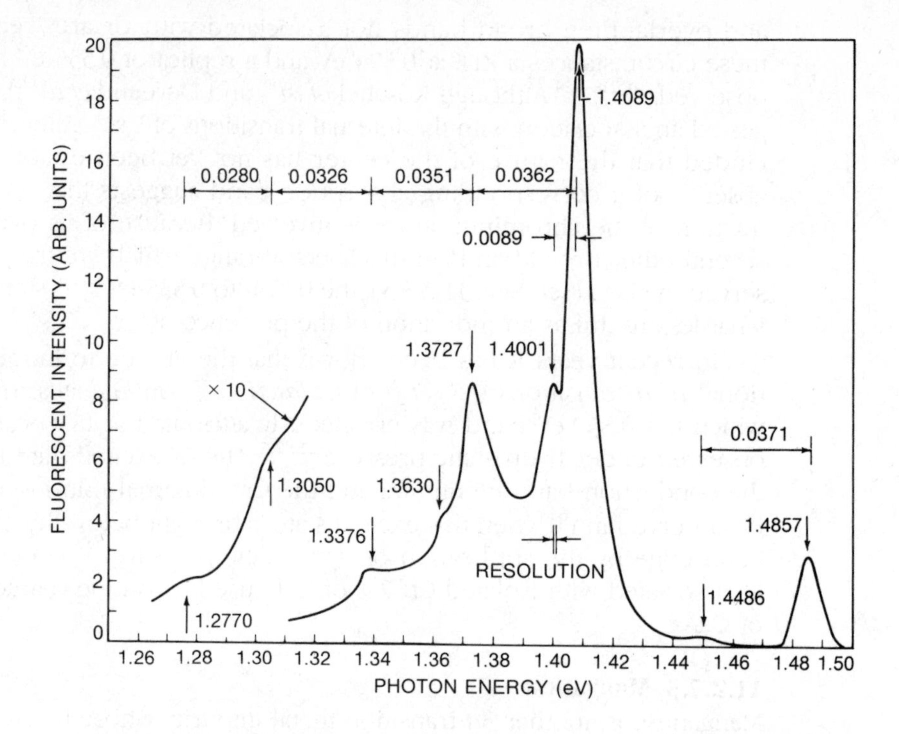

Fig. 11–28. Near-band-edge photoluminescence of manganese-diffused GaAs at 4.2 K. ***(After Lee and Anderson[156])***

11.2.7.4 The Effects of Annealing

Studies of the effects of heat treatment on GaAs have made a substantial contribution to the vast and confusing literature concerning impurity and defect related PL spectra in GaAs. Such studies are complicated by the presence of background impurities in the GaAs and by the inadvertent introduction of impurities from the experimental system. Impurities can diffuse into the crystal at a higher rate than native defects formed at the surface during the annealing procedure. Under some circumstances, background impurities whose concentrations are below PL detection limits can diffuse to the surface (redistribute) during heat treatment and form a concentrated layer which exhibits a strong PL spectrum.[163, 228] The relative importance of such processes is a strong function of the starting material (e.g. bulk or epitaxial, doped or intrinsic, majority carrier type, concentrations of background impurities) and the annealing conditions (e.g. temperature, capped or uncapped, As overpressure, hydrogen flow, sources of contamination in the experimental system). In view of these circumstances, the level of confusion in the literature is not surprising and demonstrates the need for careful work in such investigations. Well characterized starting material, carefully controlled annealing conditions, determination of diffusion profiles by secondary ion mass spectroscopy (SIMS) or repeated PL measurements after successive etches, parallel electrical measurements, confirmatory FIR, EPR, or ODMR measurements where possible, and sufficient repetitions of the experiments to ensure reproducibility are among the conditions which can increase the probability of obtaining meaningful results.

One of the best known and technologically significant examples of the effects of impurity redistribution during heat treatment of GaAs is the phenomenon of type conversion[229–232] (sometimes referred to as thermal conversion) of the near surface region of semi-insulating GaAs substrates subjected to ~700°C annealing. The term type conversion refers to alterations in the electrical properties of the surface, for example the creation of a highly conducting p-type layer on the surface of a semi-insulating crystal. In some cases this type conversion can be caused by the redistribution or diffusion of transition metal impurities which act as either deep (Cr, Fe) or moderately shallow (Mn, Cu) acceptors.[224–228, 232, 233] Both intentional dopants such as Cr and residual background impurities such as Mn, Fe, and Cu have been shown to diffuse in GaAs at elevated temperatures[232] (e.g. annealing stages prior to epitaxial growth or postimplantation anneals). Under other circumstances the indiffusion or outdiffusion of native defects such as vacancies or the EL2 complex[234] will be the dominant factor in the alteration of the electrical characteristics of the sample.

The characterization of type-converted surfaces of GaAs substrates has been undertaken by a large number of investigators, many of whom have implicated the shallow acceptor Mn in the thermal conversion process.[163, 224–227, 233] An excellent example is provided by the work of Klein *et al.*,[163, 226] which utilized PL, PLE spectroscopy, SIMS, EPR, and van der Pauw measurements. This work demonstrated that Mn and Fe which occur as uniformly distributed background impurities in the GaAs substrates at concentrations below 1×10^{15} cm^{-3} can diffuse to the surface during heat treatment and form a thin surface layer (1 to 3 μm thick) in which the Mn or Fe concentrations are $\sim 1 \times 10^{17}$ cm^{-3}. The relatively shallow Mn acceptors can cause the surface layer to convert to p-type. The results of this study are summarized in Fig. 11–29, which compares PL, SIMS, and transport measurements for Mn-doped GaAs, undoped semi-insulating GaAs before heating, after heating in H_2 at 740°C for 90 min, and after heating and polishing to remove 20 to 50 μm of the heated surface. The GaAs:Mn sample is characterized by a uniform Mn distribution as determined by the SIMS and the presence of a strong 1.41-eV Mn PL band. The unheated, undoped sample exhibits only a carbon acceptor free-to-bound PL band, but after heating, the presence of a Mn-rich surface layer is manifested in

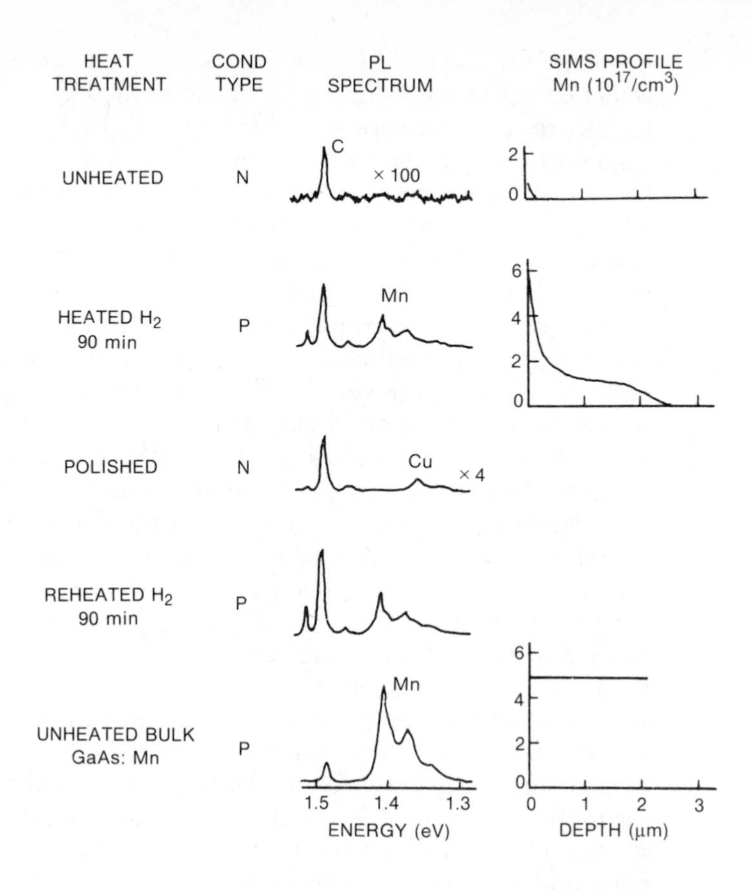

Fig. 11–29. Comparison of PL, SIMS, and transport measurements on GaAs at the indicated stages of the heating cycle. Results of similar measurements on bulk GaAs : Mn are included for comparison. *(After Norquist* et al.[163]*)*

both the PL spectrum and the Mn SIMS profile, and the surface has converted to p-type. Polishing away the surface layer removed the 1.41-eV PL band and restored the original semi-insulating properties of the GaAs surface. Temperature-dependent Hall measurements showed that the p-type conductivity of the surface layer on the heated GaAs sample and bulk Mn-doped GaAs crystal were dominated by an acceptor with a 0.1-eV binding energy, which has been associated with Mn_{Ga} in GaAs.[158,222] For these particular samples and this set of annealing conditions, the concerted PL, Hall effect and SIMS measurements clearly correlate the surface type conversion with the presence of Mn.

It should be emphasized that Mn is not always implicated in the surface type conversion of heat treated GaAs. For example, an earlier investigation of thermal conversion in GaAs concluded solely on the basis of PL spectra that the p-type converted layer was due to the formation of As vacancy-acceptor impurity complexes.[235–238] However, the present author, who was a collaborator in this early work, now believes that in the absence of any corroborating measurements by other techinques the assignment of the observed 1.409- to 1.413-eV PL bands was highly speculative, and it is likely that what was actually observed was the Mn PL band. More recently, Look *et al.*[239] reported a detailed study of thermal conversion in a large number of GaAs substrates which utilized temperature dependent Hall effect, differential Hall effect, PL, spark source mass spectroscopy, and SIMS. On the basis of their observations, they concluded that a 0.5-eV acceptor formed during 900°C anneals is attributable to Fe which has accumulated at the surface, and that the 0.1-eV acceptor, which is widely reported after 750°C anneals, is not *always* due to Mn. An alternative defect origin for this acceptor was suggested, possibly complexes involving V_{Ga}. Although their suggestion of a Ga vacancy complex as the 0.1-eV acceptor which is responsible for surface conversion in their samples was admittedly highly speculative, they presented conclusive evidence that the acceptor was *not* Mn.

In summary, it can be said that there is strong evidence for the involvement of Mn in the formation of p-type conducting surface layers in GaAs during heat treatment,[224–227] but there also is evidence[239] that it is not the only cause of thermal conversion. This should not be surprising because many impurity atoms and defects can be expected to redistribute during an annealing process. Changes in conductivity can be caused by the introduction of a shallow impurity (e.g. Mn or Si) or by the removal of a deep compensating impurity or defect (e.g. Cr or EL2), and the electrical characteristics of the surface are determined by the combination of these diffusions.

11.3 Characterization of GaAs by Fourier Transform Infrared Spectroscopy

In the following, a relatively brief description of the application of Fourier transform infrared (FTIR) spectroscopy to the characterization of GaAs is presented. As a characterization technique, FTIR spectroscopy is complementary to PL spectroscopy, offering some advantages and superior capabilities and having some disadvantages and limitations. Both the intracenter electronic transitions[5–7, 18, 240] and the local vibrational mode (LVM)[8] absorption processes which will be discussed can yield a quantitative assessment of the concentration of detected impurities which PL cannot provide. It is somewhat easier to resolve the closely spaced binding energies of the shallow donors in GaAs by FTIR than PL spectroscopy although, as described in Sec. 11.2.3.3, recent developments in magneto-PL techniques are beginning to diminish this advantage. In addition, LVM spectroscopy can in some cases detect the presence of isovalent impurities in GaAs which are usually not observed in PL.

However, FTIR spectroscopy places far more restrictions on required sample characteristics than does PL. For example, under routine experimental conditions for FTIR, intracenter electronic transitions can be observed for donors (acceptors) only in n-type (p-type) samples. Furthermore, different donor species can be resolved only in samples having donor concentrations in a very limited range just above the state-of-the-art $\sim 10^{14}$ cm^{-3} background level for the best epitaxial samples. Magneto-PL techniques for donor spectroscopy[17–19] have now been successful even in bulk samples.[42] In addition, the high detection efficiencies, powerful lasers, and other experimental conveniences available in the visible and near-infrared spectral ranges make PL a much simpler characterization tool to implement.

Nevertheless, FTIR spectroscopy enjoys a position of considerable importance in the hierarchy of characterization techniques. It can be expected to play an increasingly expanded role as the sensitivity and convenience of the available equipment and the quality of the GaAs material continue to improve.

11.3.1 Experimental Techniques

All of the measurements described in this section are carried out at cryogenic temperatures with samples mounted in liquid helium cryostats. The exact configuration of the sample mounting differs depending upon the choice of IR transmission spectroscopy or photoconductive spectroscopy. In the case of direct transmission measurements, the samples are mounted immediately in front of a detector, usually a gallium-doped germanium extrinsic photoconductor (for the 80 to 300 cm^{-1} range) or a germanium bolometer (for frequencies less than about 100 cm^{-1}), on the cold finger of a liquid helium dewar. For photoconductive measurements, often referred to as photothermal ionization spectroscopy (PTIS),[6,240] the sample itself serves as detector and it is usually mounted directly on the cold finger of the dewar. Often it is necessary to employ light pipes rather than focusing optics; this is especially true when measurements are to be made in a solenoidal superconducting magnet. In this case, the sample (and detector when required) are usually mounted below a metallic cone in the light pipe which concentrates the light onto the surface of the sample. The light pipe is sealed in a tubular container which is first evacuated and then filled with helium exchange gas to provide thermal contact with the liquid helium bath in which the tubular unit is immersed.

The so-called Fourier transform infrared (FTIR) spectrometer is based on the Michelson interferometer. Light from a broad-band IR source such as a globar or a mercury arc lamp passes through the evacuated or dry nitrogen purged interferometer, including either a Mylar or KBr beamsplitter, and then through the sample. An interferogram is generated as a function of the displacement of the movable mirror. This analog signal is digitized and the Fourier transform is performed by an on-line computer to provide the IR spectrum. Most modern instruments are rapid scan units and it is possible to scan repeatedly and signal average to enhance the signal-to-noise ratio. The best available instruments are capable of resolutions of the order of 0.002 cm^{-1} with mirror displacements of about 1.5 m.

In IR transmission spectroscopy the intensity of the light transmitted through the sample is measured by an infrared detector. Absorption processes in the sample such as intraimpurity electronic transitions and LVM absorption bands are detected as a diminution of the intensity of the transmitted light. This technique requires relatively strong absorption mechanisms and thick samples and is used primarily for the characterization of wafers of GaAs. For thin epitaxial films of high-purity GaAs with impurity concentrations at or below the 10^{15} cm^{-3} level photoconductive techniques must be employed. The PTIS method involves a two-step ionization process.[240,241] First the absorption of a photon causes a transition from the ground state of a hydrogenic impurity to a bound excited state. This is the same absorption process which is observed in transmission spectroscopy. A lattice phonon then excites the electron (hole) into the conduction (valence) band resulting in a change in the conductivity of the sample which is recorded as a function of photon wavelength. At sufficiently high photon energies, the bound carriers are excited directly from the ground state of the impurity to the continuum band and extrinsic photoconductivity is observed.

In the discussion of FTIR spectra of GaAs which follows, examples of intracenter electronic transitions observed by PTIS and transmission spectroscopy will be cited. Because local vibrational mode spectroscopy does not involve electronic transitions it is usually observed only in transmission spectroscopy.

11.3.2 Characterization of Donors in GaAs by FTIR Spectroscopy

As explained in our discussion of PL characterization, spectroscopic identification of hydrogenic impurities in GaAs relies upon slight differences in the chemical shifts or central cell corrections to the 1*s* ground state for various impurity species, with the excited states having nearly identical energies (see Secs. 11.2.2 and 11.2.4). Because of the small conduction band effective mass and large dielectric constant for GaAs the effective Rydberg constant for shallow donors is only 5.72 meV and the differences in ground state binding energy for the different chemical donors are as small as ~0.01 meV.[5, 26, 240, 242] Because of these extremely small differences and the broadening due to internal strains and charged centers, the central cell structure in the $1s$–$2p$ transitions (differences in chemical shifts) is barely resolvable by FTIR spectroscopy even in GaAs samples of the very highest quality. Successful resolution of donor central cell structure in GaAs usually relies on one of several techniques which can narrow the IR absorption lines. These include[6] the application of a magnetic field which shrinks the electronic wavefunction thereby narrowing the lines and increasing central cell splittings, optical excitation (bandgap light) of electrons and holes which can neutralize ionized donors and acceptors and reduce the associated electric fields which broaden the lines, reduction in temperature between 4 and 2 K, and the application of hydrostatic pressure which increases the bandgap, effective mass, donor binding energy, and chemical shifts.

The usual approach to donor impurity characterization involves the measurement by IR spectroscopy of chemical shifts for most of the common donors in purposely doped samples in order to establish a catalog of characteristic energies which can be used to identify the donor contaminants which occur in high-purity GaAs grown by various techniques (VPE, OMCVD, MBE). These investigations are difficult in GaAs because there is a limited range of donor concentrations between the background level obtained in the purest material grown by a given technique and that for which the effects discussed above begin to broaden and distort the IR spectra. These difficulties are evident in the fact that in spite of careful study of the donor central cell structure by several research groups over a period of ten years some of the common donor contaminants occurring in high-purity GaAs remain unidentified and the assignments of the three dominant residual impurities in VPE GaAs have been changed within the past few years.[6, 18, 243–245] The assignment of these three donors, now referred to as X_1, X_2, and X_3 by most workers (Cooke *et al.* had labeled them as b, c, and h), is discussed below.

Donor impurity spectra are usually obtained with FTIR spectroscopy, using the highly sensitive photothermal ionization technique[240, 241] for detection, and in an applied magnetic field. The sensitivity of this photoconductive technique is required because the infrared absorption due to a neutral donor concentration of $\sim 10^{14}$ cm^{-3} in a 10 μm thick film of GaAs is far too weak to be detected directly in IR transmission.

The principal spectral features observed for IR magneto-optical spectroscopy of donors in GaAs are the $1s$–$2p$ transitions. For a simple parabolic conduction band the $2p$ level of hydrogenic donors splits in a magnetic field into three magnetic sublevels characterized by orbital quantum numbers $m_1 = -1$, 0, and $+1$. These splittings are linear in H and are governed, for a hydrogenic donor, by the Zeeman relationship $\Delta E = e\hbar H/m_e^*c$. Stillman *et al.*[246] fitted their magnetic field splitting data and obtained a value of $0.0665 \pm 0.0005 m_0$ for the conduction band effective mass in GaAs.

The $1s$–$2p$ ($m = -1$) transition has a relatively narrow linewidth and large amplitude at high magnetic fields, which makes it the preferred transition for resolving the closely spaced multiplet of peaks due to different donor species occurring in a sample. Because the $1s$–$2p$ ($m = -1$) transition energies are a strong function of magnetic field, donor identification in a test sample is accomplished by comparing its FTIR spectrum with that of a well characterized reference sample taken at exactly the same magnetic field. Usually the superconducting magnet is operated in persistent mode. The FTIR donor spectra obtained by Low *et al.*[243] for an ultrapure $AsCl_3$-VPE-grown reference sample ($\mu_{77} = 201{,}000\ cm^2/V \cdot s$, $n_{77} = 4.5 \times 10^{13}\ cm^{-3}$) and three MBE grown samples are compared in Fig. 11–30. These MBE samples were the best that had been reported at that time with $\mu_{77} \cong 100{,}000\ cm^2/V \cdot s$ and $n_{77} \cong 5 \times 10^{14}\ cm^{-3}$. The X_1, X_2, and X_3 peaks characteristic of the three dominant background donors in VPE GaAs are evident in the reference sample spectrum. These lines are well separated in this spectrum and their relative intensities are expected to correspond to relative concentrations of the donor contaminants. The present assignments of the X_1, X_2, and X_3 peaks are Si, S, and Ge, respectively. It should be noted that conflicting assignments of these lines will be found in the literature prior to papers published by Ozeki *et al.*[247] in 1977 and Low *et al.*[243–245] in 1982. (Specifically, the peaks labeled X_1, X_2, and X_3 (or b, c, h) had been assigned previously to a Ga vacancy, Si, and C, respectively.)

Comparison of the VPE GaAs reference spectrum in Fig. 11–30A with the MBE-grown GaAs spectra of Figs. 11–30B through 11–30D indicates the presence, in various combinations, of X_1 (Si) and X_2 (S) peaks, as well as features attributed to Pb and Sn. In the less pure samples, especially that of Fig. 11–30C, the peaks broaden due to the electric fields from charged centers (compensated donors and acceptors) and it is difficult to resolve properly the peaks due to the individual neutral donor species. An envelope fitting procedure, which produced the dashed theoretical spectral components shown in Fig. 11–30C, was used to determine the relative intensities of the donor peaks for this sample. In the other cases, deconvolution of the intensities was more straightforward. The relative intensities of the peaks obtained from the FTIR spectra of Fig. 11–30 were used in conjunction with total donor (and acceptor) concentrations derived from Hall mobility analysis[248] to produce the concentrations of the various donor impurities in the VPE- and MBE-grown samples of GaAs listed in

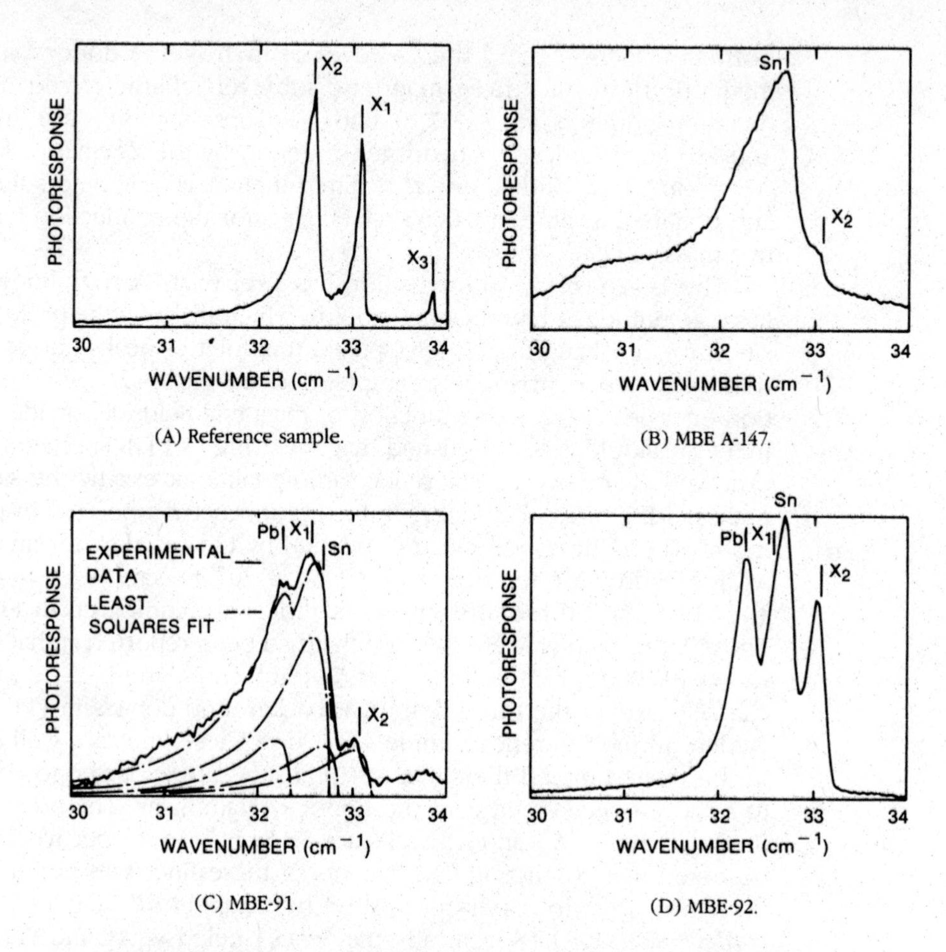

(A) Reference sample.

(B) MBE A-147.

(C) MBE-91.

(D) MBE-92.

Fig. 11–30. Photothermal ionization spectra of the 1*s*–2*p* (*m* = −1) transitions for MBE samples and a reference sample of GaAs at a magnetic field of 2.92 T. (*After Low* et al.[243])

Table 11–2, which has been reproduced from Ref. 243. Note that peaks associated with Pb, Si, Sn, and S donors are observed in each of the MBE samples studied although each was grown in a different laboratory by a different technique. The X_3 donor, which has been identified as Ge, has a concentration at least a factor of 20 smaller than the total donor concentration and there is no FTIR spectroscopic evidence for its presence in these MBE samples. Two interestingly contrasting results are that Ge, which is absent from the MBE samples, is evident as an impurity in GaAs grown by a variety of VPE and OMCVD techniques, while the Pb donor is not present in unintentionally doped GaAs grown by any of the above techniques except MBE.

Table 11–2. Impurity Concentrations Derived from the Photothermal Ionization Spectra and the Hall Mobility Analysis of Wolfe *et al.*[248], as listed in Table I of Low *et al.*[243]

	Impurity Concentrations ($\times 10^{14}$ cm^{-3})						
Sample	**N_D (total)**	**N_A (total)**	**N_D(Pb)**	**N_D(X_1 = Si)**	**N_D(Sn)**	**N_D(X_2)**	**N_D(X_3)**
Reference	0.89	0.45	—	0.47	—	0.36	0.06
A-147	5.8	1.8	<0.3	<0.3	3.9	1.3	<0.1
MBE-91	4.9	3.9	0.9	2.5	0.8	0.7	<0.2
MBE-92	3.4	2.8	0.8	<0.15	1.3	1.2	—

In the more recent work of Bose *et al.*,[18] which was discussed in Sec. 11.2.3.3 of this chapter, magneto-photoluminescence spectra (MPS) of residual donors in high-purity epitaxial GaAs were compared with photothermal ionization spectra (Fig. 11–31) of the same samples. These correlated measurements resulted in the identification of the MPS and PTIS signatures for the shallow donors S, Si, Ge, Sn, and Te, and the apparently represent the definitive spectroscopic study of shallow donors in GaAs at the time of this writing.

Although photothermal ionization (photoconductive) spectroscopy provides very high sensitivity for the detection of donor transitions, it can introduce distortions of the observed lines. For example, a shift in apparent line position occurs when the photoconductive signal is due to tunneling between adjacent impurity sites instead of the photothermal ionization mechanism. Several other potential problems have been discussed by Armistead *et al.*[6] Of these, perhaps the most important artifact which can occur is the "notch effect" reported by Low *et al.*[244] This effect can lead to the mistaken identification of a "notched" line as two central cell components (two different donor species). Proper technique for reliable assignment of multiple $1s$–$2p$ transitions to multiple donor species includes the investigation of the magnetic field dependence of the various components and the reproducibility of the individual lines from sample to sample.

It should be apparent from the foregoing discussion that the identification of different donor species by IR spectroscopy should not at present be categorized as routine characterization. The stringent requirements for sample purity and controlled doping conditions, the relative difficulty of the spectroscopy (compared, for example, with PL), and the careful data interpretation required place IR donor spectroscopy in the realm of fundamental research. However, the recent developments in magneto-optical spectroscopy, both in the IR and near-band-edge PL spectral ranges (Sec. 11.2.3.3), show promise of placing spectroscopic identification of donors on a firmer, readily accessible basis.[18]

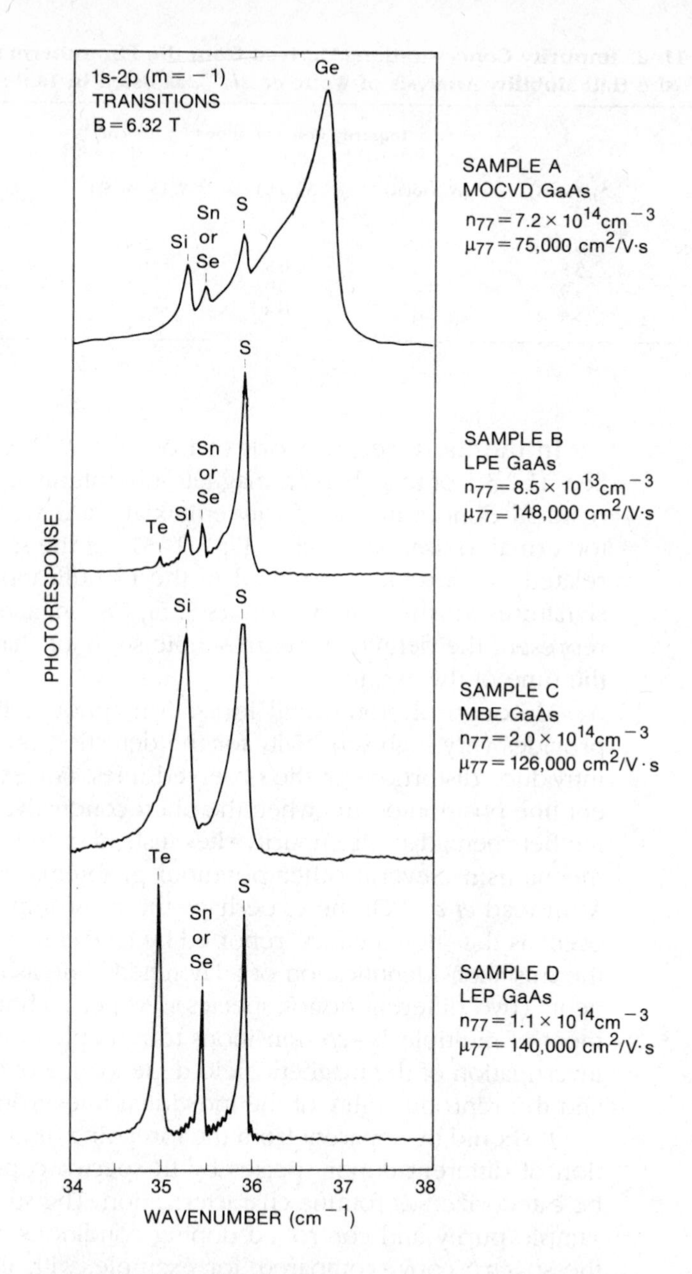

Fig. 11–31. Photothermal ionization spectra recorded at a magnetic field of 6.32 T for four GaAs samples grown by LPE, MBE, and MOCVD. (*After Bose* et al.[18])

11.3.3 Characterization of Acceptors in GaAs by FTIR Spectroscopy

In Sec. 11.2.4.1, on the identification of acceptors by PL techniques, it was explained that acceptor binding energies in GaAs are in the ~25- to 40-meV range and their variation from acceptor to acceptor is large enough to allow PL transitions from different acceptor impurities to be resolved clearly. Correspondingly, this relatively large variation in the chemical shifts of the ground state energies for various shallow acceptors in GaAs makes it possible to distinguish readily different acceptor species on the basis of variations in the ground state to excited state transitions observed in IR spectroscopy. As in the case of donor spectroscopy, a catalog of characteristic acceptor absorption spectra has been obtained from studies of purposely doped samples. The most comprehensive study of this nature is that carried out by Kirkman *et al.*,[7] which established ground state binding (ionization) energies for carbon (26.9 meV), magnesium (28.7 meV) zinc (30.6 meV), and silicon (34.8 meV) acceptors in epitaxially grown films of GaAs. If there is an IR counterpart of the important paper of Ashen *et al.*[3] concerning PL characterization of acceptors in GaAs it would probably be this 1978 paper of Kirkman *et al.*[7]

For the study of thin epitaxial films of high-quality GaAs, photothermal ionization spectroscopy (PTIS) is employed. Because of the larger binding energies of the acceptor excited states which form the final states of the IR absorption transitions, the PTIS measurements must be carried out at somewhat higher temperatures than those employed in donor spectroscopy in order to ensure thermal activation of the holes from the acceptor excited states to the valence band. In fact, the applicability of the PTIS mechanism is clearly indicated by studies of the photoconductive spectra as a function of temperature (2 to 20 K).[7] At lower temperature, the intensity of the lowest-energy transitions (corresponding to transitions to the most strongly bound or deepest acceptor excited states) is reduced relative to that of higher-energy transitions to bound states and direct to the valence band continuum. In thick samples of suitable quality IR transmission (absorption) spectroscopy can be employed with highly sensitive, cooled IR detectors such as Ga-doped Ge photoconductors or Ge bolometers. The IR transmission technique is especially important for the characterization of acceptors in bulk GaAs substrate material.[77]

Far-infrared PTIS spectra of acceptors in GaAs exhibit a characteristic series of sharp line spectra between 120 and 240 cm^{-1} which are attributable to transitions from the *s*-like acceptor ground state to several of the *p*-like excited states. When the photon energy exceeds the binding energy of the shallow acceptor levels, transitions from the bound states directly to the valence band continuum cause an increase in the photoconductive background signal. Another limitation on the spectroscopy is the high reflectivity of the GaAs sample in the restrahlung region between the TO and LO phonon frequencies (273 to 296 cm^{-1}) which causes the photoconductive response to drop to zero.

The characteristic FTIR absorption spectra reported by Kirkman *et al.*[7] for four different shallow acceptors (C, Zn, Mg, and Si) in GaAs are shown in Fig. 11–32. Although the peak spacings are similar for all four acceptors, the entire series of lines is shifted in each case by an energy corresponding to the central cell shift for the particular acceptor. The labeling of the *G*, *D*, and *C* lines of Fig. 11–32 is based upon the convention established by several groups which have calculated theoretical binding energies of the hydrogenic levels of acceptors in GaAs. The assignments are as follows:

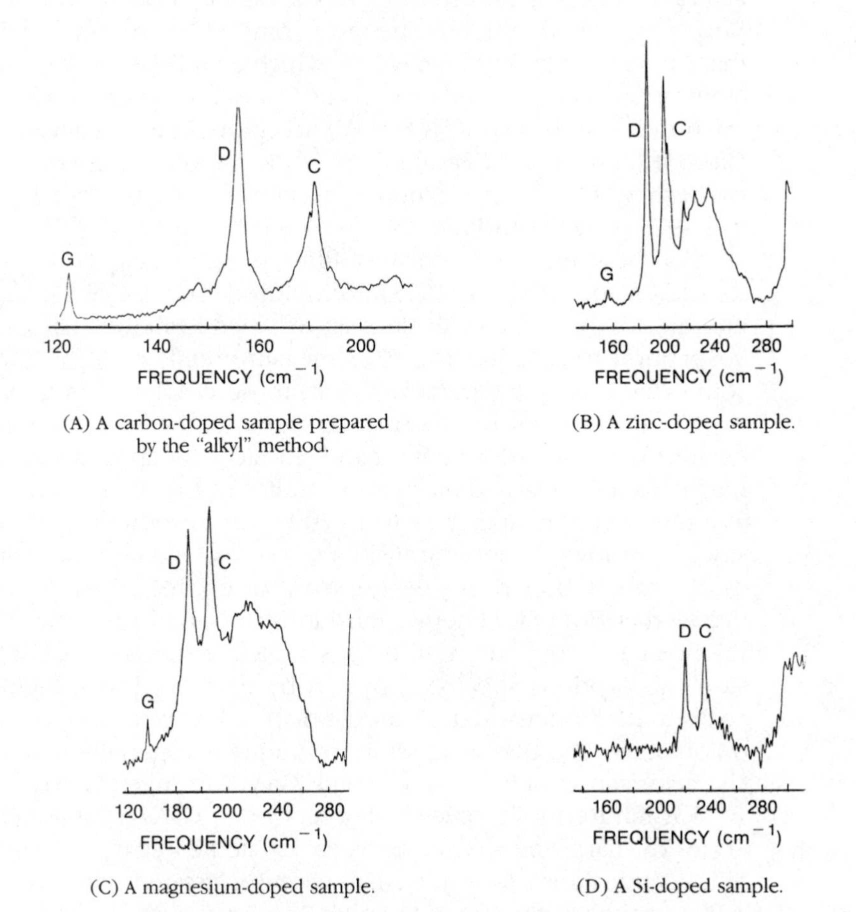

(A) A carbon-doped sample prepared by the "alkyl" method.

(B) A zinc-doped sample.

(C) A magnesium-doped sample.

(D) A Si-doped sample.

Fig. 11–32. Experimental recordings of photothermal ionization spectra obtained for four different acceptor impurities in GaAs. *(After Kirkman, Stradling, and Lin-Chung[7])*

$$1S_{3/2}(\Gamma_8^+)\text{–}2P_{3/2}(\Gamma_8^-) \equiv G \text{ line},$$
$$1S_{3/2}(\Gamma_8^+)\text{–}2P_{5/2}(\Gamma_8^-) \equiv D \text{ line},$$
$$1S_{3/2}(\Gamma_8^+)\text{–}2P_{5/2}(\Gamma_7^-) \equiv C \text{ line}.$$

In general, the D and C lines are clearly observable for each acceptor with characteristic C to D separation of about 15 cm^{-1}. The experimentally measured energies of the G, D, and C transitions and their differences ($\Delta GD, \Delta DC$) for the samples studied by Kirkman *et al.*[7] are listed in Table 11–3, which is reproduced in its entirety from Ref. 7. Table 11–3 also lists theoretical values for the ground state binding energy and for the separations ΔGD and ΔDC which were calculated by the theoretical approaches of Baldereschi and Lipari[11,12] and Lin-Chung and Henvis[249] using the valence band parameters derived from the cyclotron resonance measurements of Skolnick *et al.*[250] Note that both methods provide binding energies for the excited states which are in reasonable agreement with experiment and with each other, but there are significant differences in the ground state binding energies calculated by the two methods.

Kirkman *et al.*[7] estimated the acceptor ground state binding energies from the measured C transition energies for the various samples. It is assumed that the $2P_{5/2}(\Gamma_7^-)$ level, which is the final state for the C transition, is accurately described by the effective mass approximation, and that the

Table 11–3. Observed Energies of the *G*, *D*, and *C* lines as Listed in Table 2 of Kirkman *et al.*[7]

Sample	$1S_{3/2} - 2P_{3/2}$ (G) cm^{-1}	$1S_{3/2} - 2P_{5/2}(\Gamma_8)$ (D) cm^{-1}	$1S_{3/2} - 2P_{5/2}(\Gamma_7)$ (C) cm^{-1}	ΔGD cm^{-1}	ΔDC cm^{-1}	Deduced Central Cell Shift (cm^{-1})	Calculated Ionization Energy (meV) of $1S_{3/2}$ (Γ_8) State	Ionization Energy from Luminescence (meV) (White *et al.*[25])
Sample I	122.5 ± 0.5	156.1 ± 0.5	171.6 ± 0.5	33.6 ± 1	15.5 ± 1	5.7	26.84	26.0
Undoped LPE						17.4	27.06	
Sample II	156.3 ± 1.0	187.2 ± 1.0	202.0 ± 2.0	30.9 ± 2	14.8 ± 3	36.8	30.61	30.7
Zn doped VPE						48.5	30.83	
Sample III	not seen	186.4 ± 1.0	200.9 ± 2.0		14.5 ± 2	36.1	30.47	30.7
Zn doped VPE						47.8	30.69	
Sample IV	137.9 ± 1.0	169.9 ± 1.0	186.0 ± 1.0	32.0 ± 2	16.1 ± 2	20.3	28.63	28.4
Mg doped LPE						32.0	28.85	
Sample V	not seen	220.1 ± 1.0	235.1 ± 1.0		15.0 ± 2	65.0	34.71	34.5
Si doped LPE						76.7	34.93	
Sample VI	122.5 ± 0.5	156.1 ± 0.5	171.6 ± 0.5	33.6 ± 0.1	15.5 ± 1	5.7	26.84	26.0
Undoped LPE						17.4	27.06	
Sample VII	122.3 ± 0.4	156.0 ± 0.5	170.9 ± 0.8	33.7 ± 0.9	14.9 ± 1.3	5.4	26.75	26.0
"Alkyl"						17.1	26.97	
Manganese	815	847	863	321	16			
doped p-GaAs (Chapman and Hutchinson[25])						700	113	
Theoretical Values								
Lipari and Baldereschi[11,12]	117.9	151.7	163.4	33.8	11.7	—	25.82	—
Lin-Chung and Henvis[249]	107.3	139.4	151.3	32.2	11.8	—	24.48	—

theoretical estimates of its binding energy are relatively accurate. The binding energy of the $1S_{3/2}(\Gamma_8^+)$ ground state is then determined by adding the theoretical value for the $2P_{5/2}(\Gamma_7^-)$ excited state to the observed energy of the *C* line. Obviously, the difference between the binding energies for various acceptors are more accurate than the absolute values because of uncertainties in the calculated binding energy of the $2P_{5/2}$ excited state. For the purposely doped samples, the binding energies of the Zn, Mg, and Si acceptors determined from the FTIR spectra are in excellent agreement with the values obtained from PL which are also listed in Table 11–3. The undoped samples all contained the same residual acceptor which, on comparison with the PL results, was identified as carbon (binding energy ~26.0 meV).

Table 11–3 also lists results for the Mn acceptor reported by Chapman and Hutchinson.[251] These results are somewhat remarkable for the consistency of the separation of the lines with those of the much shallower acceptors. With a binding energy as large as 113 meV central cell effects would be expected even for the excited states. Although no PL results are listed for GaAs:Mn, the IR results are in good agreement with the PL studies discussed in Sec. 11.2.7.3.

The reader is directed to Ref. 7 for additional discussions of the theoretical approaches, weaker spectral features which are not central to acceptor identification, and a treatment of Zeeman splitting of acceptor states.

It is appropriate to present some practical examples of the application of FTIR spectroscopy to the identification of residual acceptors in an LEC GaAs substrate. Figure 11–33 shows an FTIR transmission spectrum obtained by Moore *et al.*[77] for a ~0.5 mm thick sample of bulk p-type GaAs; the detector for this spectrum was a Ge:Ga photoconductor. This spectrum, extending from 100 to 2000 cm^{-1}, clearly demonstrates the power and sensitivity of a modern FTIR spectrometer. In the spectral range below 200 cm^{-1}, sharp line absorption spectra due to the *C* and *D* lines of several shallow acceptors are evident. An expanded version of this spectral range is shown in Fig. 11–34, with some of the acceptors identified. Note that the task of identifying residual acceptor impurities when more than one species is present is complicated by the fact that the difference between the *D* and *C* lines (ΔDC) for a given acceptor is nearly equivalent to the differences between the spectral features due to different acceptor species. For this sample thickness, which is typical of LEC GaAs wafers, the minimum detectable concentration for shallow acceptors is approximately 1×10^{15} cm^{-3}.

Recently, secondary optical excitation into the EL2 absorption band (see Sec. 11.2.5.3) has been used to create a nonequilibrium population of holes in semi-insulating GaAs which neutralizes residual shallow acceptors.[252] It is then possible to observe optically induced FTIR absorption

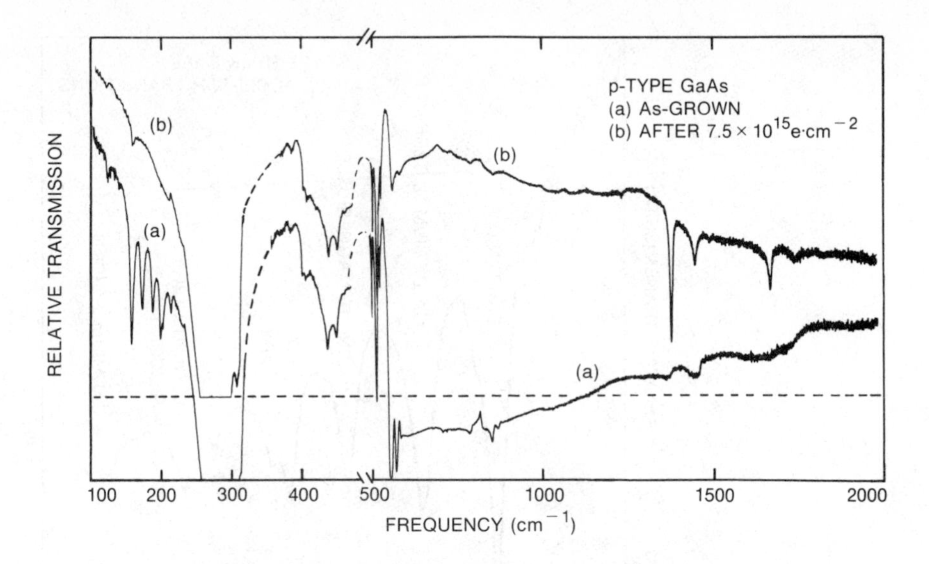

Fig. 11–33. Infrared transmission spectra of p-type GaAs sample before and after electron irradiation. Shallow acceptors are seen in the region 140 to 210 cm^{-1}, the 78-meV level is observed near 600 cm^{-1} and the absorption characteristic of the negative charge state of a double acceptor is observed near 1400 cm^{-1}. *(After Moore, Shanabrook, and Kennedy[77])*

spectra characteristic of residual shallow acceptor impurities in the normally semi-insulating GaAs. Wagner *et al.*[252] have observed such optically induced 1*S* to 2*P* transitions at carbon and zinc shallow acceptors in semi-insulating GaAs which they illuminated with 1.06-μm light from a Nd:YAG laser. The LEC GaAs samples studied had residual acceptor concentrations in the range 10^{15} to 10^{16} cm^{-3} as determined by local vibrational mode and Raman spectroscopy. These results demonstrate that under some circumstances it is possible to circumvent the usual limitation that characterization of shallow acceptors by FTIR spectroscopy is only applicable to p-type samples.

11.3.4 The Double Acceptor in GaAs

FTIR spectra are shown in Fig. 11–33 for a p-type sample of LEC GaAs which was grown from a Ga-rich melt.[77] Such crystals often exhibit an acceptor with a binding energy of 78 meV. A variety of experimental studies have led to the conclusion that the 78-meV level arises from the neutral charge state of a *double* acceptor.[69–77] Note that spectrum (*a*) in Fig. 11–33, for the as-grown p-type sample in which the Fermi level is located in the shallow single acceptor levels about 30 meV from the valence band, ex-

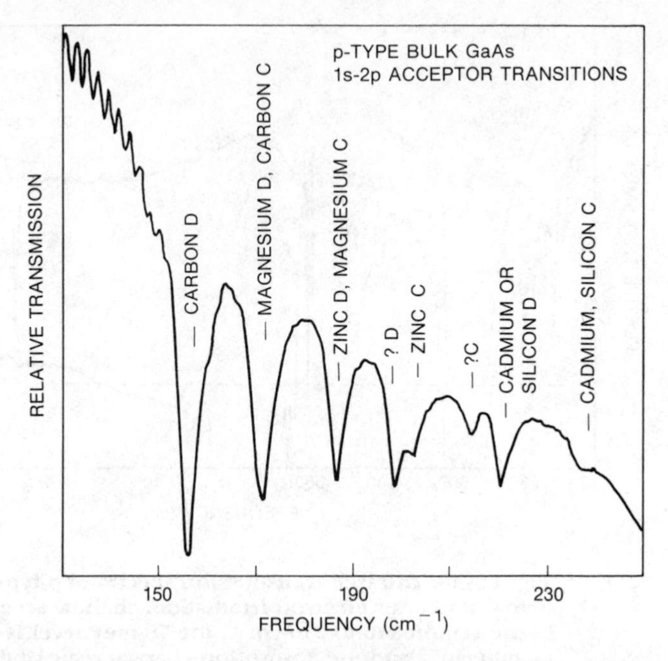

Fig. 11–34. Expanded version of the infrared transmission spectrum of Fig. 11–33 in the spectral region below 200 cm^{-1}. The sharp line absorption spectra due to the *C* and *D* transitions of several shallow acceptors are shown, as labeled.

hibits the sharp line spectra below 200 cm^{-1} characteristic of the neutral shallow acceptor impurities. The overabsorbed spectral range above these lines, between 250 and 310 cm^{-1} is the restrahlung range mentioned above. At higher energy, near 600 cm^{-1}, the intracenter absorption of the completely uncompensated 78-meV acceptor is observed. The transmission spectrum of the same sample after an electron irradiation of 7.5×10^{15} electrons cm^{-2} (2 meV) is shown in (*b*) of Fig. 11–33. This spectrum differs from spectrum (*a*) in three respects: the low-energy shallow single acceptor absorption lines are no longer observed, the strength of the absorption due to the 78-meV acceptor is significantly reduced, and new absorption lines at 1388.2 and 1457.6 cm^{-1} have appeared. As discussed in detail in Ref. 77, these lines and their 69 cm^{-1} separation are characteristic of the negative charge state of a double acceptor with binding energy of 203 meV.

It was concluded by Moore *et al.*[77] that the carrier (hole) removal effected by the electron irradiation left the shallow acceptors completely compensated and therefore unobservable in the IR spectrum, and partially compensated the 78-meV double acceptor so that its neutral state absorption strength was diminished and new absorption features due to its compensated (negatively charged) level at 203 meV became observable. The chemical identity of this double acceptor has not been established with certainty. However, the leading candidates for its identity are a boron impurity on an As site[74] or a Ga_{As} antisite defect. At this writing the Ga_{As} assignment is favored by the limited experimental evidence available.[76,77]

11.3.5 Local Vibrational Mode Spectroscopy

In this section, local vibrational mode (LVM) spectroscopy is discussed very briefly as it applies specifically to the characterization of GaAs. The general subject of LVM is treated in more detail in several excellent reviews.[8,253–255] LVM spectroscopy is applicable to substitutional impurity atoms in semiconductors having masses which are lighter than those of the major constituent atoms (host lattice). For such light impurities a localized vibrational mode occurs with characteristic frequency higher than those of the host lattice fundamental optical phonons. This displacement to higher energy is roughly proportional to $(2k/m)^{1/2}$, where k is the force constant of the impurity binding and m is the reduced mass. The impurity gives rise to optical absorption (LVM) at energies above the restrahlung bands of the lattice. The lighter the impurity atom, the higher the energy of the LVM and the more highly localized the mode. The degree of localization refers to the extent to which the kinetic energy of the mode is shared with the nearest neighbor (NN) atoms of the lattice.

The role of the impurity mass in determining the frequency of the LVM allows the identification of impurities on the basis of the observed FTIR spectra. Sensitivity to mass differences is such that isotopes differing by only 1 amu can be distinguished spectroscopically. Furthermore, high-resolution FTIR spectroscopy can detect very small shifts determined by different configuration of the isotopes of the NN atoms,[256] a crucial capability for the characterization of GaAs. Because there are two isotopes of gallium, ^{69}Ga and ^{71}Ga, with relative abundance of 60.4 percent and 39.6 percent, respectively, there are five different configurations of Ga isotopes which can surround an impurity on an As site in GaAs. The five distinct values of the reduced mass can lead to a fivefold splitting of the impurity LVM.

Figure 11–35 shows the high-resolution five-line LVM spectrum observed for carbon atoms (acceptors) located on As sites in GaAs.[257] (Calculations of the frequencies and relative intensities of the five spectral components are discussed in Ref. 8.) Such fine structure is not observed for substitutional atoms on Ga sites because of the 100 percent abundance of ^{75}As. Obviously, this fine structure provides a valuable means for the determination of the site of a substitutional impurity atom in GaAs. A second example of its usefulness is the LVM spectroscopy of B impurities in GaAs. Boron is a persistent inadvertent impurity in GaAs crystals grown in boron

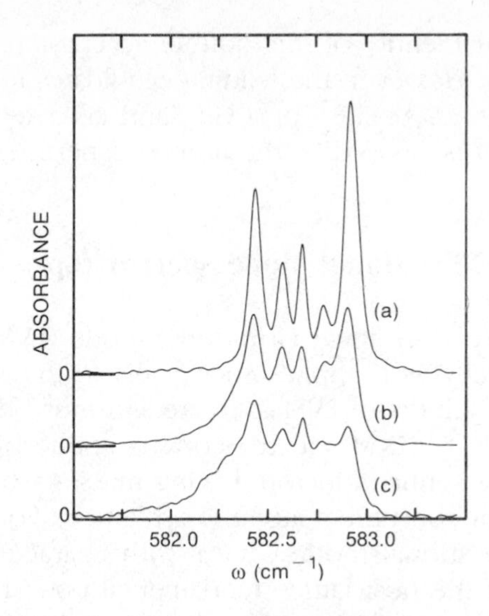

Fig. 11–35. High-resolution five-line local vibrational mode spectrum observed for carbon atoms located on As sites in GaAs. The spectra shown are after a 2-MeV electron irradiation of fluence (*a*) 2 × 10^{16} cm^{-2}, (*b*) 5 × 10^{15} cm^{-2}, and (*c*) 0 cm^{-2}. Spectrum (*a*) is also observed in unirradiated semi-insulating GaAs. ***(After Shanabrook, Moore, and Kennedy[257])***

oxide encapsulant. As a group III impurity, the B is expected to substitute primarily for Ga in GaAs. LVM spectroscopy has shown that although the B is usually located on Ga sites as expected, some B_{As} antisites have been discovered in electron irradiation studies, and the resulting LVM exhibits fivefold splitting.[258,259]

Shanabrook and coworkers[257] have shown that it is also possible to distinguish the neutral and negative charge states of the carbon acceptors on the basis of the lineshape of the LVM spectrum. The interaction between the bound hole and the LVM is apparently responsible for a broadening of the spectrum for the neutral acceptor. This ability to distinguish between the neutral and negative charge states of the acceptor means that it is possible to determine the compensation ratio of an impurity from low-temperature LVM measurements.[257]

Because of the ability of LVM to determine the site (Ga or As) of the light impurity atoms in GaAs, the technique is particularly valuable for the study of amphoteric group IV impurities such as Si which can locate on either sublattice, acting as acceptors or donors depending upon location, and can transfer from one sublattice to the other during annealing procedures.[260]

A typical benchmark for the sensitivity of LVM measurements is the minimum detectable concentration of carbon in GaAs:C, which is ~5 × 10^{14} cm^{-3} for samples of wafer thickness, assuming a data acquisition time of a few minutes.[8] Although calibration of the absolute value of the impurity concentration as determined by LVM spectroscopy is complicated by a number of factors, the relative concentrations in different samples can be determined to a high level of accuracy by direct comparison of their LVM spectra. As is the case of all of the optical measurements described in this review, LVM spectroscopy is nondestructive and, as demonstrated in the work described above, it can be used to profile the spatial distribution of impurities and, in some cases, their compensation.

LVM spectroscopy may also be applicable to native defect identification in cases where the defect is paired with an impurity or the LVM band can be associated with particular defect by some other means. An excellent example of this is the recent work of Song *et al.*[261] in which they have observed a pair of LVM bands (730 cm^{-1} and 714 cm^{-1} at 80 K) in semi-insulating GaAs which they have related to the ground state and metastable state of EL2 defects. This association is made on the basis of the similarity between the optical and thermal conditions required to interconvert the two LVM bands and those required to interconvert the ground and metastable states of EL2 defects.[261] If these measurements are confirmed and further exploited, they should represent a truly remarkable application of the LVM technique to the identification and characterization of defects in GaAs.

References

1. E. W. Williams, and H. Bebb, *Semiconductors and Semimetals, Vol. 8,* ed. R. K. Willardson and A. C. Beer, New York: Academic Press, 321 (1972).
2. P. J. Dean, and D. C. Herbert in *Excitons, Solid State Sciences 14,* ed. K. Cho, Berlin: Springer-Verlag, 55 (1979).
3. P. J. Ashen, P. J. Dean, D. T. J. Hurle, J. B. Mullin, and A. M. White, *J. Phys. Chem Solids,* 36:1041 (1975).
4. P. J. Dean, in *Progress in Solid State Chemistry, Vol. 8,* ed. J. O. McCaldin and Y. G. Somorjai, Oxford: Pergamon Press, 1 (1973).
5. A. K. Ramdas and S. Rodriguez, *Rep. Prog. Phys.,* 8:1297 (1981).
6. C. J. Armistead, P. Knowles, S. P. Najda, and R. A. Stradling, *J. Phys. C: Solid State Phys.,* 17:6415 (1984).
7. R. F. Kirkman, R. A. Stradling, and P. J. Lin-Chung, *J. Phys. C: Solid State Phys.,* 11:419 (1978).
8. W. M. Theis, in *Spectroscopic Characterization Techniques for Semiconductor Technology II, SPIE,* 524:45 (1985) and references therein.

9. H. Bebb and E.W. Williams, in *Semiconductors and Semimetals,* 8:182, ed. R.K. Willardson and A.C. Beer, New York: Academic Press (1972).
10. W. Kohn, in *Solid State Physics,* Vol. 5; ed. F. Seitz and D. Turnbull, New York: Academic Press, 257 (1957).
11. A. Baldereschi and N.O. Lipari, *Phys. Rev. B,* 8:2697 (1973).
12. A. Baldereschi and N.O. Lipari, *Phys. Rev. B,* 9:1525 (1974).
13. P.J. Dean, D.J. Robbins, and S.G. Bishop, *Solid State Commun.,* 32:379 (1979).
14. P.J. Dean, D.J. Robbins, and S.G. Bishop, *J. Phys. C: Solid State Phys.,* 12:5567 (1979).
15. A.T. Hunter and T.C. McGill, *Appl. Phys. Lett.,* 40:169 (1982).
16. R.J. Almassey, D.C. Reynolds, C.W. Litton, K.K. Bajaj, and G.L. McCoy, *Solid State Commun.,* 38:1053 (1981).
17. D.C. Reynolds, P.C. Colter, C.W. Litton, and E.B. Smith, *J. Appl. Phys.,* 55:1610 (1984).
18. S.S. Bose, B. Lee, M.H. Kim, and G.E. Stillman, *Appl. Phys. Lett.,* 51:937 (1987).
19. P.J. Dean and M.S. Skolnick, *J. Appl. Phys.,* 54:346 (1983); P.J. Dean, M.S. Skolnick, and L.L. Taylor, *J. Appl. Phys.,* 55:957 (1984).
20. G.H. Wannier, *Phys. Rev.,* 52:191 (1937); N.F. Mott, *Proc. Roy. Soc. A,* 167:384 (1938).
21. J. Frenkel, *Phys. Rev.,* 37:1276 (1931).
22. W.L. Bloss, E.S. Koteles, E.M Brody, B.J. Sowell, J.P. Salerno, and J.V. Gormley, *Solid State Commun.,* 54:103 (1985) and references therein.
23. L. Schultheis and C.W. Tu, *Phys. Rev. B,* 32:6978 (1985) and references therein.
24. J.R. Haynes, *Phys. Rev. Lett.,* 4:361 (1960).
25. A.M. White, P.J. Dean, D.J. Ashen, J.B. Mullin, M. Webb, B. Day, and P.D. Greene, *J. Phys. C: Solid State Phys.,* 6:L243 (1973).
26. R.A. Cooke, R.A. Hoult, R.F. Kirkham, and R.A. Stradling, *J. Phys. D: Appl. Phys.,* 11:945 (1978).
27. A.M. White, P.J. Dean, L.L. Taylor, R.C. Clarke, D.J. Ashen, and J.B. Mullin, *J. Phys. C: Solid State Phys.,* 5:1727 (1972).
28. J. Shah, R.C.C. Leite, and R.E. Nahory, *Phys. Rev.,* 184:811 (1969).
29. A.M. White, P.J. Dean, and B. Day, *J. Phys. C: Solid State Phys.,* 7:1400 (1974).
30. W. Ruhle and W. Klingenstein, *Phys. Rev. B,* 18:7011 (1972).
31. A.M. White, I. Hinchliffe, P.J. Dean, and P.D. Greene, *Solid State Commun.,* 10:497 (1972).

32. M. A. Gilleo, P. T. Bailey, and D. E. Hill, *Phys. Rev.,* 174:898 (1968).
33. D. Bimberg, M. Sondergeld, and E. Grobe, *Phys. Rev. B,* 4:3451 (1971).
34. J. A. Rossi, C. M. Wolfe, G. E. Stillman, and J. O. Dimmock, *Solid State Commun.,* 8:2021 (1970).
35. H. R. Fetterman, J. Waldman, C. M. Wolfe, G. E. Stillman, and C. D. Parker, *Appl. Phys. Lett.,* 21:434 (1972).
36. R. A. Stradling, L. Eaves, R. A. Hoult, N. Niura, P. E. Simmonds, and C. C. Bradley, in *Gallium Arsenide and Related Compounds, Inst. of Phys. Conf.,* 17:65 (1973).
37. G. Munschy, *Phys. Status Solidi (b),* 53:377 (1972).
38. J. J. Hopfield, *Proc. Intl. Conf. Phys. Semiconductors,* Paris 1964, Paris: Dunod, 725 (1964).
39. T. Skettrup, M. Suffczynski, and W. Gorzkowski, *Phys. Rev. B,* 4:512 (1971).
40. E. H. Bogardus and H. B. Bebb, *Phys. Rev.,* 176:993 (1968).
41. R. Ulbrich and B. Moreth, *Solid State Commun.,* 14:331 (1974).
42. T. D. Harris and M. S. Skolnick, *Defects in Semiconductors, Materials Science Forum Vols. 10–12,* ed. H. J. von Bardeleben, Switzerland: Trans Tech Publ., Ltd., 1219 (1986).
43. R. A. Faulkner, *Phys. Rev.,* 175:991 (1968).
44. P. J. Dean, *J. Luminescence,* 7:51 (1973) and A. Baldereschi, *ibid.,* p. 79.
45. D. G. Thomas and J. J. Hopfield, *Phys. Rev.,* 150:680 (1966).
46. J. C. Phillips, *Bands and Bonds,* New York: Academic Press, 54 (1973).
47. P. J. Dean, A. M. White, E. W Williams, and M. G. Astles, *Solid State Commun.,* 9:1555 (1971).
48. W. Ruhle, W. Schmid, R. Meck, N. Stath, J. U. Fischbach, I. Strottner, K. W. Benz, and M. Pilkuhn, *Phys. Rev. B,* 18:7022 (1978).
49. D. J. Wolford, J. A. Bradley, K. Fry, and J. Thompson, *17th Intl. Conf. Phys. Semiconductors,* ed. J. D. Chadi and W. A. Harrison, New York: Springer, 627 (1985).
50. M. Leroux, G. Neu, J. P. Contour, J. Massies, and C. Verie, *J. Appl. Phys.,* 59:2996 (1986).
51. R. Schwabe, W. Seifert, F. Bugge, R. Bindemann, V. F. Agekyan, and S. V. Pogarev, *Solid State Commun.,* 55:167 (1985).
52. T. N. Morgan, B. Welber, and R. N. Bhargava, *Phys. Rev.,* 166:751 (1968) and C. H. Henry, P. J. Dean, and J. D. Cuthbert, *ibid.,* p. 754.
53. H. J. Guislain, L. De Wolf, and P. J. Clauws, *Electron. Mat.,* 7:83 (1978).
54. F. Willmann, D. Bimberg, and M. Blatte, *Phys. Rev. B,* 7:2473 (1973).
55. H. J. Queisser and C. S. Fuller, *J. Appl. Phys.,* 37:4895 (1966).

56. E. F. Gross and V. I. Safarov, *Sov. Phys. Semiconductors,* 1:241 (1967).
57. E. F. Gross, V. I. Safarov, V. E. Sedov, and V. A. Maruschak, *Sov. Phys. Semiconductors,* 2:277 (1969).
58. Z. G. Wang, H. P. Gislason, and B. Monemar, *J. Appl. Phys.,* 58:230 (1985).
59. H. P. Gislason, B. Monemar, Z. G. Wang, Ch. Uihlein, and P. L Liu, *Phys. Rev. B,* 32:3723 (1985).
60. B. Monemar, H. P. Gislason, and Z. G. Wang, *Phys. Rev. B,* 31:7919 (1985).
61. H. P. Gislason, B. Monemar, M. E. Pistol, P. J. Dean, D. C. Herbert, S. Depinna, A. Kanaah, and B. C. Cavenett, *Phys. Rev. B,* 31:3774 (1985).
62. B. Monemar, H. P. Gislason, W. M. Chen, and Z. G. Wang, *Phys. Rev. B,* 33:4424 (1986).
63. H. Kunzel and K. Ploog, *Appl. Phys. Lett.,* 37:416 (1980).
64. D. C. Reynolds, K. K. Bajaj, C. W. Litton, E. B. Smith, P. W. Yu, W. T. Masselink, F. Fischer, and H. Morkoç, *Solid State Commun.,* 52:685 (1984).
65. M. S. Skolnick, T. D. Harris, C. W. Tu, C. W Brennan, and M. D. Sturge, *Appl. Phys. Lett.,* 46:427 (1985).
66. L. Eaves and P. D. Halliday, *J. Phys. C: Solid State Phys.,* 17:L705 (1984).
67. M. S. Skolnick, C. W. Tu, and T. D. Harris, *Phys. Rev. B,* 33:8468 (1986).
68. M. S. Skolnick, *Proc. 18th Intl. Conf. Phys. Semiconductors,* ed. O. Engstrom, Singapore: World Scientific, 1389 (1987).
69. S. G. Bishop, B. V. Shanabrook, and W. M Moore, *J. Appl. Phys.,* 56:1785 (1984), and references therein, and S. G. Bishop and B. V. Shanabrook, *Semi-Insulating III-V Materials,* (Kah-nee-ta), ed. D. C. Look and J. S. Blakemore, Nantwich: Shiva, 302 (1984).
70. P. W. Yu, D. E. Holmes, and R. T. Chen, *Inst. Phys. Conf.,* 63:209 (1982).
71. P. W. Yu and D. C. Reynolds, *J. Appl. Phys.,* 53:1263 (1982).
72. K. R. Elliott, D. E. Holmes, R. T. Chen, and C. G. Kirkpatrick, *Appl. Phys. Lett.,* 40:898 (1982).
73. P. W. Yu, W. C. Mitchel, M. G. Mier, and S. S. Li, and W. L. Wang, *Appl. Phys. Lett.,* 41:532 (1982).
74. L. B. Ta, H. M. Hobgood, and R. N. Thomas, *Appl. Phys. Lett.,* 41:1091 (1982).
75. K. R. Elliott, *Appl. Phys. Lett.,* 42:274 (1983).
76. W. C. Mitchel, G. J. Brown, D. W. Fischer, P. W. Yu, and J. E. Lang, *J. Appl. Phys.,* 62:2320 (1987).
77. W. J. Moore, B. V. Shanabrook, and T. A. Kennedy, *Semi-Insulating III-V Materials* (Kah-nee-ta), ed. D. C. Look and J. S. Blakemore, Nantwich: Shiva, 453 (1984).

78. D. M. Eagles, *J. Phys. Chem. Solids,* 16:76 (1960).
79. R. A. Street and W. Senske, *Phys. Rev. Lett.,* 37:1292 (1976) and *Phys. Rev. B,* 20:3267 (1979).
80. S. Nakashima, T. Hattori, and Y. Yamaguchi, *Solid State Commun.,* 25:137 (1978).
81. H. Venghaus, P. J. Dean, P. E. Simmonds, and J. C. Pfister, *Z. Phys. B,* 30:125 (1978).
82. H. Tews, H. Venghaus, and P. J. Dean, *Phys. Rev. B,* 19:5178 (1979).
83. R. J. Elliot and A. F. Gibson, *An Introduction to Solid State Physics and Its Applications,* London: Macmillan, 190 (1974).
84. J. Bourgoin and M. Lannoo, *Point Defects in Semiconductors II, Solid State Sciences 35,* Berlin: Springer, 92 (1983).
85. J. H. Schulman and C. C. Klick, in *Solid State Physics, Vol. 5,* ed. F. Seitz and D. Turnbull, New York: Academic Press, 97 (1957).
86. M. Lax, *J. Chem. Phys.,* 20:1752 (1952).
87. G. A. Allen, *Brit. J. Appl. Phys.,* 1:593 (1968).
88. W. H. Koschel, S. G. Bishop, and B. D. McCombe, *Solid State Commun.,* 19:521 (1976); *Proc. 13th Conf. Phys. Semiconductors,* ed. E. G. Fumi, Rome: Tipografia Marves, 1065 (1977).
89. H. J. Stocker and M. Schmidt, *J. Appl. Phys.,* 47:2450 (1976).
90. A. M. White, *Semi-Insulating III-V Materials* (Nottingham), ed. G. J. Rees, Orpington: Shiva, 3 (1980).
91. J. J. Hopfield, *J. Phys. Chem. Solids,* 10:110 (1959).
92. D. K. Brice, *Phys. Rev.* 188:1280 (1969).
93. T. H. Keil, *Phys. Rev. A,* 140:601 (1965).
94. Y. Toyozawa, in *Dynamical Processes in Solid State Optics,* ed. Kubo and Kamimura, New York: Benjamin, 90 (1967).
95. K. Huang, and A. Rhys, *Proc. Roy. Soc.,* (London), A204:406 (1950).
96. E. W. Williams, *Phys. Rev.,* 168:922 (1968).
97. H. Birey and J. Sites, *J. Appl. Phys.,* 51:619 (1980).
98. E. W. Williams and A. M. White, *Solid State Commun.,* 9:279 (1971).
99. C. J. Hwang, *Phys. Rev.,* 180:827 (1969).
100. C. J. Hwang, *J. Appl. Phys.,* 39:1654 (1968).
101. C. J. Hwang, *J. Appl. Phys.,* 39:4307 (1968).
102. D. Curie and J. S. Prener, in *Physics and Chemistry of II-VI Compounds,* eds. M. Aven and J. S. Prener, Amsterdam: North Holland, 455 (1967).
103. M. B. Pannish, H. J. Queisser, L. Derick, and S. Sumski, *Solid State Electron.,* 9:311 (1966).

104. J. van de Ven, W.J.A.M. Hartmann, and L.J. Giling, *J. Appl. Phys.,* 60:3735 (1986).

105. J.S. Prener and F.E. Williams, *J. Chem. Phys.,* 25:261 (1956).

106. G.M. Martin, A. Mitonneau, and A. Mircea, *Electron. Lett.,* 13:191 (1977).

107. D.E. Holmes, R.T. Chen, K.R. Elliott, and C.G. Kirkpatrick, *Appl. Phys. Lett.,* 40:46 (1982).

108. S. Makram-Ebeid, P. Langlade, and G.M. Martin, *Semi-Insulating III-V Materials* (Kah-nee-ta), ed. D.C. Look and J.S. Blakemore, Nantwich: Shiva, 184 (1984).

109. H.C. Gatos and J. Lagowski, in *Microscopic Identification of Electronic Defects in Semiconductors, MRS Symposia Proc.,* ed. N.M. Johnson, S.G. Bishop, and G.D. Watkins, Pittsburgh, 46:153 (1985).

110. E.R. Weber and P. Omling, *Festkörperprobleme,* 25:623 (1985).

111. G.M. Martin and S. Makram-Ebeid, in *Deep Centers in Semiconductors,* ed. S.T. Pantelides, New York: Gordon and Breach, 399 (1986).

112. W.J. Turner, G.D. Pettit, and N.G. Ainslie, *J. Appl. Phys.,* 34:3274 (1963).

113. P.W. Yu, *Solid State Commun.,* 32:1111 (1979).

114. P.W. Yu and D.C. Walters, *Appl. Phys. Lett.,* 41:863 (1982).

115. A. Mircea-Roussel and S. Makram-Ebeid, *Appl. Phys. Lett.,* 38:1007 (1981).

116. P. Leyral and G. Guillot, *Semi-Insulating III-V Materials* (Evian) ed. S. Makram-Ebeid and B. Tuck, Nantwich: Shiva, 166 (1982).

117. B.V. Shanabrook, P.B. Klein, E.M. Swiggard, and S.G. Bishop, *J. Appl. Phys.,* 54:336 (1983).

118. P. Leyral, G. Vincent, A. Nouailhat, and G. Guillot, *Solid State Commun.,* 42:67 (1982).

119. B.V. Shanabrook, P.B. Klein, and S.G. Bishop, *15th Intl. Conf. Phys. Semiconductors,* ed M. Averous, Amsterdam: North Holland, *Physica,* 117B and 118B:173 (1983).

120. P.W. Yu, D.E. Holmes, and R.T Chen, *Inst. Phys. Conf. Ser.,* 63:200 (1982).

121. P.W. Yu, *Solid State Commun.,* 43:953 (1982).

122. P.W. Yu, *Appl. Phys. Lett.,* 44:330 (1984).

123. P.W. Yu, *Phys. Rev. B,* 29:2283 (1984).

124. L. Samuelson, P. Omling, and H.G. Grimmeis, *Appl. Phys. Lett.,* 45:521 (1984).

125. M. Tajima, H. Tanino, and K. Ishida, *Defects in Semiconductors, Materials Science Forum Vols. 10–12,* ed. H.J. von Bardeleben, Switzerland: Trans Tech Publ., Ltd., 1265 (1986).

126. P. W. Yu, *Phys. Rev. B,* 31:8259 (1985).
127. D. Bois and G. Vincent, *J. Physique Lett.,* 38:351 (1977).
128. G. Vincent, D. Bois, and A. Chantre, *J. Appl. Phys.,* 53:3643 (1982).
129. G. M. Martin, G. Jacob, A. Goltzene, C. Schwab, and G. Poiblaud, *Inst. Phys. Conf. Ser.,* 59:281 (1981).
130. G. M. Martin, *Semi-Insulating III-V Materials,* (Nottingham), ed. G. J. Rees, Orpington: Shiva, 13 (1980).
131. G. M. Martin, *Appl. Phys. Lett.,* 39:747 (1981).
132. L. Samuelson, P. Omling, E. R. Weber, and H. G. Grimmeis, *Semi-Insulating III-V Materials,* (Kah-nee-ta), ed. D. C. Look and J. S. Blakemore, Nantwich: Shiva, 268 (1984).
133. P. Omling, L. Samuelson, and H. G. Grimmeis, *17th Intl. Conf. Phys. Semiconductors,* ed. J. D. Chadi and W. A. Harrison, New York: Springer, 751 (1985).
134. H. Tajima, *Appl. Phys. Lett.,* 46:484 (1985).
135. H. Tajima, *Defects in Semiconductors, Materials Science Forum Vols. 10–12,* ed. H. J. von Bardeleben, Switzerland: Trans Tech Publ., Ltd., 493 (1986).
136. M. Kaminska, M. Skowronski, J. Lagowski, J. M. Parsey, and H. C. Gatos, *Appl. Phys. Lett.,* 43:302 (1983).
137. M. R. Brozel, I. Grant, R. M. Ware, D. J. Stirland, and M. S. Skolnick, *J. Appl. Phys.,* 56:1109 (1984).
138. P. Dobrilla and J. S. Blakemore, *J. Appl. Phys.,* 58:208 (1985); *J. Appl. Phys.,* 61:1442 (1987).
139. M. Castagne, J. P. Fillard, and J. Bonnafe, *Solid State Commun.,* 54:653 (1985).
140. M. Tajima, *Jpn. J. Appl. Phys.,* 21:L227 (1982).
141. M. Tajima and Y. Okada, *12th Intl. Conf. on Defects in Semiconductors, Physica B,* 116:404 (1983).
142. *Defect Recognition and Image Processing in III-V Semiconductors, Materials Science Monographs, 31,* ed. J. P Fillard, Amsterdam: Elsevier (1985).
143. M. S. Skolnick, L. J. Reed, and A. D. Pitt, *Appl. Phys. Lett.,* 44:447 (1984).
144. M. S. Skolnick, D. A. O. Hope, and B. Cockayne, *Semi-Insulating III-V Materials* (Kah-nee-ta), ed. D. C. Look and J. S. Blakemore, Nantwich: Shiva, 446 (1984).
145. J. M. Baranowski, J. W. Allen, and G. L. Pearson, *Phys. Rev.,* 167:627 (1967); *Phys. Rev.,* 167:758 (1968).
146. S. G. Bishop, P. J. Dean, P. Porteous, and D. J. Robbins, *J. Phys. C: Solid State Phys.,* 13:1331 (1980).

147. U. Kaufmann and J. Schneider, *Festokörper Probleme,* 20:87 (1980).

148. S.G. Bishop, in *Deep Centers in Semiconductors,* ed. S.T. Pantelides, New York: Gordon and Breach, 541 (1986).

149. J.M. Baranowski, in *Deep Centers in Semiconductors,* ed. S.T. Pantelides, New York: Gordon and Breach, 691 (1986).

150. S.G. Bishop, in *Optical Characterization Techniques for Semiconductor Technology, Proc. Soc. Photo-Opt. Instr. Eng.,* 276:2 (1981).

151. B. Clerjaud, *J. Phys. C: Solid State Phys.,* 18:3615 (1985).

152. J.W. Allen, in *Deep Centers in Semiconductors,* ed. S.T. Pantelides, New York: Gordon and Breach, 627 (1986).

153. D.J. Robbins and P.J. Dean, *Adv. Phys.,* 27:499 (1978).

154. M.S. Skolnick, P.J. Dean, A.D. Pitt, Ch. Uihlein, H. Krath, B. Deveaud, and E.J. Foulkes, *J. Phys. C: Solid State Phys.,* 16:1967 (1983).

155. H.P. Gislason, Z.G. Wang, and B. Monemar, *J. Appl. Phys.,* 58:240 (1985).

156. T.C. Lee and W.W. Anderson, *Solid State Commun.,* 2:265 (1964).

157. W. Schairer and M. Schmidt, *Phys. Rev. B,* 10:2501 (1974).

158. M. Ilegems, R. Dingle, and L.W. Rupp, Jr., *J. Appl. Phys.,* 46:3059 (1975).

159. C.J. Ballhausen, *Introduction to Ligand Field Theory,* New York: McGraw-Hill (1962).

160. F.A. Cotton, *Chemical Applications of Group Theory,* New York: Wiley-Interscience, (1971).

161. S.G. Bishop, D.J. Robbins, and P.J. Dean, *Solid State Commun.,* 33:119 (1980).

162. M. deWit and T.L. Estle, *Phys. Rev.,* 132:195 (1963).

163. P.E.R. Nordquist, P.B. Klein, S.G. Bishop, and P.G. Siebenmann, *Gallium Arsenide and Related Compounds, 1980, Inst. Phys. Conf. Ser. No. 56* (1981).

164. G.A. Slack and B.M. O'Meara, *Phys. Rev.,* 163:335 (1967).

165. F.S. Ham and G.A. Slack, *Phys. Rev. B,* 4:777 (1971).

166. P.B. Klein, J.E. Furneaux, and R.L. Henry, *Appl. Phys. Lett.,* 42:638 (1983); *Phys. Rev. B,* 29:1947 (1984).

167. S.G. Bishop, P.B. Klein, R.L. Henry, and B.D. McCombe, *Semi-Insulating III-V Materials* (Nottingham), ed. G.J. Rees, Orpington: Shiva, 161 (1980).

168. B.V. Shanabrook, P.B. Klein, and S.G. Bishop, in *Proc. 12th Intl. Conf. on Defects in Semiconductors,* ed. C.A.J. Ammerlaan, Amsterdam: North Holland, *Physica,* 116B:444 (1982).

169. J.S. Blakemore, *Semi-Insulating III-V Materials* (Nottingham), ed. G.J. Rees, Orpington: Shiva, 29 (1980).

170. A. M. Hennel, C. D. Brandt, K. Y. Ko, and L. M. Pawlowicz, *Defects in Semiconductors, Materials Science Forum Vols. 10–12,* ed. H. J. von Bardeleben, Switzerland: Trans Tech Publ., Ltd., 645 (1986).
171. W. Ulrici, L. Eaves, K. Friedland, D. P Halliday, K. J. Nash, and M. S. Skolnick, *J. Phys. C: Solid State Phys.,* 19:L525 (1986).
172. W. Kutt, D. Bimberg, M. Maier, H. Krautle, F. Kohl, and E. Bauser, *Appl. Phys. Lett.,* 44:1078 (1984).
173. B. Clerjaud, C. Naud, B. Deveaud, B. Lambert, B. Plot, G. Bremond, C. Benjeddou, G. Guillot, and A. Nouilhat, *J. Appl. Phys.,* 58:4207 (1985).
174. W. Ulrici, K. Friedland, L. Eaves, and L. P. Halliday, *Phys. Stat. Sol. (b),* 131:719 (1985).
175. C. D. Brandt, A. M. Hennel, L. M. Pawlowicz, F. P. Dabkowski, J. Lagowski, and H. C. Gatos, *Appl. Phys. Lett.,* 47:607 (1985).
176. A. V. Vasil'ev, G. K. Ippolitova, E. M. Omel' yanovskii, and A. I. Ryskin, *Sov. Phys. Semiconductors,* 10:341 (1976).
177. A. Mircea-Roussel, G. M. Martin, and J. E. Lowther, *Solid State Commun.,* 36:171 (1980).
178. U. Kaufmann, H. Ennen, J. Schneider, R. Worner, J. Weber, and F. Kohl, *Phys. Rev. B,* 25:5598 (1982).
179. W. Ulrici, L. Eaves, K. Friedland, D. P. Halliday, and J. Kreissl, *Defects in Semiconductors, Materials Science Forum Vols. 10–12,* ed. H. J. von Bardeleben, Switzerland: Trans Tech Publ., Ltd., 639 (1986).
180. M. S. Skolnick, P. J. Dean, M. J. Kane, C. Uihlein, D. J. Robbins, W. Hayes, B. Cockayne, and W. R. MacEwan, *J. Phys. C: Solid State Phys.,* 16:L767 (1983).
181. G. Armelles, J. Barrau, and D. Thebault, *J. Phys. C: Solid State Phys.,* 17:6883 (1984).
182. G. Armelles, J. Barrau, D. Thebault, and M. Brousseau, *J. de Physique,* 45:1795 (1984).
183. M. Godlewski and A. M. Hennel, *Phys. Stat. Sol. (b),* 88:K11 (1978).
184. H. Ennen, U. Kaufmann, and J. Schneider, *Solid State Phys.,* 34:603 (1980).
185. A. M. Hennel and S. M. Uba, *J. Phys. C: Solid State Phys.,* 11:4565 (1978).
186. B. Deveaud, B. Lambert, and A. M. Hennel, *4th 'Lund' Intl. Conf. on Deep Level Impurities in Semiconductors* (Eger, Hungary) 1983, unpublished.
187. B. V. Kornilov, L. V. Marchukov, and V. K. Ergakov, *Sov. Phys. Semiconductors,* 8:141 (1974).
188. R. W. Haisty and G. R. Cronin, *Physics of Semiconductors,* ed. M. Hulin, Dunod: Paris, 1161 (1964).

189. J. M. Baranowski, M. Grynberg, and E. M. Magerramov, *Phys. Status Solidi (b),* 50:433 (1972).

190. W. J. Brown and J. S. Blakemore, *J. Appl. Phys.,* 43:2242 (1972).

191. D. Wasik and M. Baj, *Defects in Semiconductors, Materials Science Forum Vols. 10–12,* ed. H. J. von Bardeleben, Switzerland: Trans Tech Publ., Ltd., 627 (1986).

192. U. Kaufmann and J. Schneider, *Solid State Commun.,* 25:1113 (1978).

193. W. Drozdzewicz, A. M. Hennel, Z. Wasilewski, B. Clerjuad, F. Gendron, C. Porte, and R. Germer, *Phys. Rev. B,* 29:2438 (1984).

194. H. Ennen, U. Kaufmann, and J. Schneider, *Appl. Phys. Lett.,* 38:356 (1981).

195. W. Ulrici, L. Eaves, K. Friedland, L. P. Halliday, J. Kreissl, and B. Ulrici, *Defects in Semiconductors, Materials Science Forum Vols. 10–12,* ed. H. J. von Bardeleben, Switzerland: Trans Tech Publ., Ltd., 669 (1986).

196. J. S. Blakemore and S. Rahimi in *Semiconductors and Semimetals, Vol. 20,* ed. R. K. Willardson and A. C. Beer, New York: Academic Press, 233 (1984).

197. R. N. Hall and J. H. Racette, *J. Appl. Phys.,* 35:379 (1964).

198. N. Kullendorf, L. Jansson, and L. A. Ledebo, *J. Appl. Phys.,* 54:3202 (1983).

199. N. S. Averkiev, T. K. Ashirov, and A. A. Gutkin, *Sov. Phys. Semiconductors,* 17:61 (1983).

200. V. V. Batavin, V. M. Mikhaelyan, and G. V. Popova, *Sov. Phys. Semiconductors,* 6:1616 (1973).

201. E. C. Lightowlers and C. M. Penchina, *J. Phys. C: Solid State Phys.,* 11:L405 (1978).

202. E. C. Lightowlers and M. O. Henry, *Proc. Intl. Conf. Phys. Semiconductors* (Edinburgh) 1978, ed. B. L. H. Wilson, *Inst. of Phys. Conf.,* Bristol and London, 307 (1979).

203. F. Voillot, J. Barrau, M. Brousseau, and J. C. Brabant, *J. Physique Lett.,* 41:L415 (1980).

204. A. M. White, *Solid State Commun.,* 32:205 (1979).

205. G. Picoli, B. Deveaud, and D. Galland, *Semi-Insulating III-V Materials* (Nottingham), ed. G. J. Rees, Shiva:Orpington, 254 (1980); *J. Physique,* 42:133 (1981).

206. J. J. Krebs and G. H. Stauss, *Phys. Rev. B,* 16:971 (1977).

207. B. Clerjaud, A. M. Hennel, and G. Martinez, *Solid State Commun.,* 33:983 (1980).

208. L. Eaves, T. Englert, T. Instone, C. Uihlein, P. J. Williams, and H. C. Wright, *Semi-Insulating III-V Materials* (Nottingham), ed. G. J. Rees, Orpington: Shiva, 145 (1980).

209. N. Killoran, B. C. Cavenett, and W. E. Hagston, *Semi-Insulating III-V Materials* (Nottingham), ed. G. J. Rees, Orpington: Shiva, 190 (1980).
210. N. Killoran and B. C. Cavenett, *Solid State Commun.,* 43:261 (1982).
211. L. Eaves, T. Englert, and C. Uihlein, *Physics in High Magnetic Fields (Solid State Sciences),* ed. S. Chikazumi and M. Muira, Berlin: Springer, 130 (1981).
212. L. Eaves and C. Uihlein, *J. Phys. C: Solid State Phys.,* 15:6257 (1982).
213. C. Uihlein and L. Eaves, *Phys. Rev. B,* 26:4473 (1982).
214. G. Picoli, B. Deveaud, and B. Lambert, *Proc. 12th Intl. Conf. Defects in Semiconductors,* Amsterdam (1982).
215. G. Picoli, B. Deveaud, B. Lambert, and A. Chomette, *J. Physique Lett.,* 44:L85 (1983).
216. J. Barrau, M. Brousseau, S. P. Austen, and C. A. Bates, *J. Phys. C: Solid State Phys.,* 16:4581 (1983).
217. D. Thebault, J. Barrau, G. Armelles, and M. Brousseau, *J. Phys. C: Solid State Phys.,* 17:2011 (1984).
218. P. W. Yu, *Semi-Insulating III-V Materials* (Evian), ed. S. Makram-Ebeid and B. Tuck, Nantwich: Shiva, 305 (1982).
219. P. Leyral, F. Litty, S. Loualiche, and G. Guillot, *Solid State Commun.,* 38:333 (1981).
220. B. Deveaud, G. Picoli, B. Lambert, and G. Martinez, *Phys. Rev. B,* 29:5749 (1984).
221. B. Deveaud, G. Picoli, Y. Zhou, and G. Martinez, *Solid State Commun.,* 46:359 (1983).
222. L. J. Vieland, *J. Appl. Phys.,* 33:2007 (1962).
223. P. W. Yu and Y. S. Park, *J. Appl. Phys.,* 50:1097 (1979).
224. J. Hallais, A. Mircea-Roussel, J. P. Farges, and G. Poiblaud, *Inst. Phys. Conf. Ser. No. 33b,* 220 (1977).
225. R. Zucca, *Inst. Phys. Conf. Ser. No. 33b,* 228 (1977).
226. P. B. Klein, P. E. R. Nordquist, and P. G. Siebenmann, *J. Appl. Phys.,* 51:4861 (1980).
227. A. S. Jordan and G. A. Nikolakopoulou, *J. Appl. Phys.,* 55:4194 (1984).
228. B. V. Shanabrook, P. B. Klein, P. G. Siebenmann, H. B. Dietrich, R. L. Henry, and S. G. Bishop, *Semi-Insulating III-V Materials* (Evian), ed. S. Makram-Ebeid and B. Tuck, Nantwich: Shiva, 310 (1982).
229. J. T. Edmund, *J. Appl. Phys.,* 31:1428 (1960).
230. J. J. Wysocki, *J. Appl. Phys.,* 31:1686 (1960).
231. J. Barrera, *Proc. Conf. Act. Dev. for Microwaves and Int. Opt.,* Ithaca: Cornell University, 135 (1975).

232. B. Tuck, *Semi-Insulating III-V Materials* (Kah-nee-ta), ed. D.C. Look and J.S. Blakemore, Nantwich: Shiva, 2 (1984).

233. A.S. Jordan, *Semi-Insulating Materials* (Evian), ed. S. Makram-Ebeid and B. Tuck, Nantwich: Shiva, 233 (1982).

234. S. Makram-Ebeid, D. Gautard, G.M. Devillard, and G.M. Martin, *Appl. Phys. Lett.,* 40:161 (1982).

235. W.Y. Lum, H.H. Wieder, and W.H. Koshel, S.G. Bishop, and B.D. McCombe, *Appl. Phys. Lett.,* 30:1 (1977).

236. W.H. Koschel, S.G. Bishop, B.D. McCombe, W.Y. Lum, and H.H. Wieder, *Inst. Phys. Conf. Ser. No. 33a,* 98 (1977).

237. W.Y. Lum and H.H. Wieder, *Appl. Phys. Lett.,* 31:213 (1977).

238. W.Y. Lum and H.H. Wieder, *J. Appl. Phys.,* 49:6187 (1978).

239. D.C. Look, P.W. Yu, J.E. Ehret, Y.K. Yeo, and R. Kwor, *Semi-Insulating III-V Materials* (Evian), ed. S. Makram-Ebeid and B. Tuck, Nantwich: Shiva, 372 (1982).

240. G.E. Stillman, C.M. Wolfe, and J.O. Dimmock, in *Semiconductors and Semimetals, Vol. 12,* ed. R.K. Willardson and A.C. Beer, New York: Academic Press, 169 (1977).

241. E.F. Haller, in *Festkörperprobleme, Vol. 26,* ed. P. Grosse, Braunschweig: Verwieg, 203 (1986).

242. G.E. Stillman, D.M. Larsen, and C.M. Wolfe, *Phys. Rev. Lett.,* 27:989 (1971).

243. T.S. Low, G.E. Stillman, A.Y. Cho, H. Morkoç, and A.R. Calawa, *Appl. Phys. Lett.,* 40:611 (1982).

244. T.S. Low, G.E. Stillman, D.M. Collins, C.M. Wolfe, S. Tiwari, and L.F. Eastman, *Appl. Phys. Lett.,* 40:1034 (1982).

245. T.S. Low, G.E. Stillman, T. Nankanisi, T. Udagawa, and C.M. Wolfe, *Appl. Phys. Lett.,* 41:183 (1982).

246. G.E. Stillman, C.M. Wolfe, and J.O. Dimmock, *Solid State Commun.,* 7:921 (1969).

247. M. Ozeki, K. Kitihara, K. Nakai, A. Shibatomi, K. Dazai, S. Okawa, and O. Ryzuzan, *Jpn. J. Appl. Phys.,* 16:1617 (1977).

248. C.M. Wolfe, G.E. Stillman, and J.O. Dimmock, *J. Appl. Phys.,* 41:504 (1970).

249. P.J. Lin-Chung and B.W. Henvis, *Phys. Rev B,* 12:630 (1975).

250. M.S. Skolnick, A.K. Jain, R.A. Stradling, J. Leotin, J.C. Ousset, and S. Askenazy, *J. Phys. C: Solid State Phys.,* 9:2809 (1976).

251. R.A. Chapman and W.G. Hutchinson, *Phys. Rev. Lett.,* 18:443 (1967).

252. J. Wagner, H. Seelewind, and P. Koidl, *Appl. Phys. Lett.,* 49:1080 (1986).

253. W. G. Spitzer, in *Festkörperprobleme, Vol. 11,* ed. O. Madelung, London: Pergamon Press, 1 (1971).

254. R. C. Newman, *Infrared Studies of Crystal Defects,* London: Taylor and Francis (1974).

255. J. S. Barker, Jr., and A. J. Sievers, *Rev. Mod. Phys.,* 47(Suppl. No. 2): S1(1975).

256. W. M. Theis, K. K. Bajaj, C. W. Litton, and W. G. Spitzer, *Appl. Phys. Lett.,* 41:70 (1982).

257. B. V. Shanabrook, W. J. Moore, and T. A. Kennedy, *Phys. Rev. B,* 30:3563 (1984).

258. G. A. Gledhill, R. C. Newman, and J. Woodhead, *J. Phys. C: Solid State Phys.,* 17:L301 (1984).

259. W. J. Moore, B. V. Shanabrook, and T. A. Kennedy, *Solid State Commun.,* 53:957 (1985).

260. R. C. Newman, in *Microscopic Identification of Electronic Defects in Semiconductors, Mat. Res. Soc. Symp. Proc.,* ed. N. M. Johnson, S. G. Bishop, and G. D. Watkins, 46:459 (1985).

261. C. Song, W. Ge, D. Jiang, and C. Hsu, *Appl. Phys. Lett.,* 50:1666 (1987).

Chapter 12

GaAs-based Strained-Layer Superlattices

G. C. Osbourn

12.1 Introduction

Until the early 1980s, most of the studies of semiconductor heterostructures focused on structures which were grown from closely lattice matched materials.[1] The restriction to lattice-matched materials is important for structures with relatively thick layers because significant lattice mismatch causes the generation of large densities of misfit dislocations. These dislocations seriously degrade the quality of lattice mismatched heterostructures. However, closely lattice matched materials are not required in order to avoid misfit dislocations in relatively thin layered heterostructures. If the mismatched layers are kept below a critical thickness value (h_c), it is energetically favorable for the layers to entirely accommodate the lattice mismatch with coherent, elastic strains.[2–4] In this case, the lattice mismatch does not cause poor-quality heterostructures.

Subsequent work has emphasized that this aspect of thin layers can be used to grow high-quality superlattices and quantum wells from lattice mismatched semiconductors.[5–10] Superlattices are especially suitable for the use of mismatched materials since the layer thickness values of interest are typically less than the h_c values for elastic strain accommodation of mismatches on the order of a few percent. These strained-layer superlattices (SLSs) provide previously unexploited freedom in the choice of their layer materials, so that there is a great variety of potential SLS structures. The study of SLS materials is motivated by (*a*) the large number of mostly unexplored SLS materials, (*b*) the interesting new physics associated with the large built-in strains in the SLS layers, and (*c*) the possibility of new device materials and new device concepts based on the use of individual strained layers and SLSs. Strained-layer superlattices and strained quantum-well research have become very active fields in recent years. Examples of group IV, III-V, and II-VI strained-layer materials systems have been examined and a number of device structures have been fabricated and operated.

12.2 Structural Properties of Strained-Layer Systems

12.2.1 Dislocation-free Strained Layers

Strained-layer superlattices consist of thin, alternating layers of materials which are lattice mismatched in bulk form but which elastically strain to uniformly match up the lattice constants of the materials in the planes parallel to the SLS interfaces. An unstrained set of mismatched layers and the corresponding set of strained layers in an SLS are shown schematically in Fig. 12–1. The resulting value of the SLS lattice constant in the planes parallel to the SLS interfaces ($a^{\|}$) is intermediate between the unstrained lat-

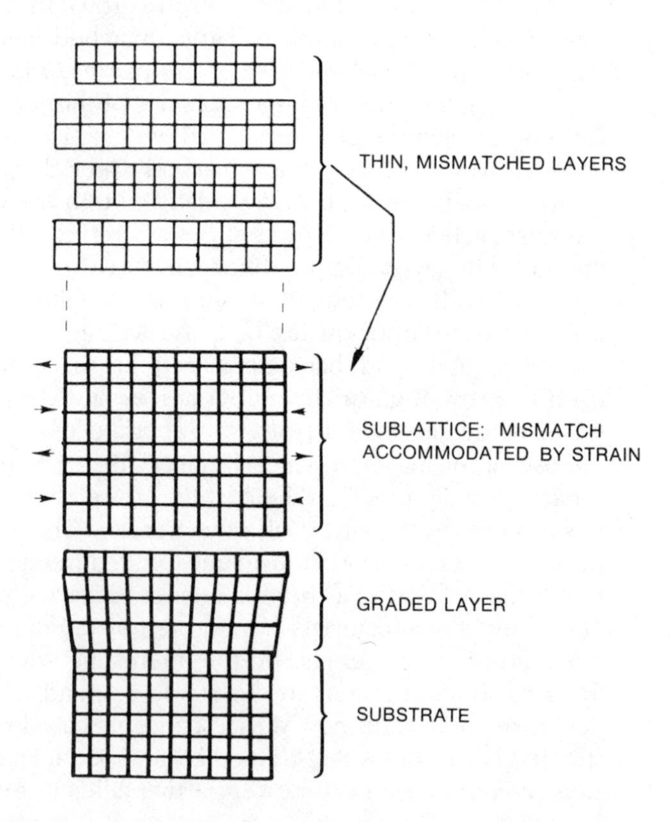

Fig. 12–1. Schematic of a strained-layer superlattice (SLS).

tice constant values of the SLS layers. The $a^{\|}$ value can be theoretically obtained by minimizing the elastic strain energy of the pair of layers, and the result is given by[4,11]

$$a^{\|} = a_1\left[1 + \frac{f}{1 + (G_1h_1/G_2h_2)}\right], \qquad (12\text{–}1A)$$

$$G_i = 2\left[C_{11}^i + C_{12} - \frac{2(C_{12}^i)^2}{C_{11}^i}\right], \qquad i = 1,2, \qquad (12\text{–}1B)$$

where a_1 is the unstrained lattice constant of layer no. 1, h_1 and h_2 are the layer thicknesses of the individual superlattice layers, f is the lattice mismatch of the unstrained layer materials, G_1 and G_2 are shear moduli of the two layers, and C_{11}^1, C_{12}^1 (C_{11}^2, C_{12}^2) are the elastic constants of material 1 (material 2), respectively. Equation 12–1 is valid for the range of bulk lattice mismatches to be encountered between the layers in high-quality SLSs (having mismatches of up to a few percent). The values of the lattice constants of the SLS layers in the direction perpendicular to the SLS interfaces ($a_{1,2}^{\perp}$) are in general different from each other and are not intermediate between the unstrained lattice constant values of the SLS layers. This is a result of the Poisson effect and causes the layers to be tetragonal rather than cubic. The a values contain a strain contribution associated with the biaxial stresses in the SLS layers, and these strain contributions ($\epsilon_{1,2}^{\perp}$) are given by

$$\epsilon_i^{\perp} = -\left(\frac{2C_{12}^i}{C_{11}^i}\right)\epsilon_i^{\|}, \qquad i = 1,2, \qquad (12\text{–}2)$$

where $\epsilon_1^{\|}$, $\epsilon_2^{\|}$ are the strain contributions to $a^{\|}$ in each of the SLS layers. The tetragonal layer distortions determined by Eqs. 12–1 and 12–2 can be resolved into purely hydrostatic and purely (100) uniaxial strain components for SLS layer growth along with ⟨100⟩ direction. This unique lattice structure leads to interesting ion channeling phenomena which are under current investigation.[12] In addition, the significant layer strains have interesting effects on the electronic band structure of these materials which are described in Secs. 12.3.2 and 12.3.3. Note that $a^{\|}$ is the lattice constant which determines the lattice matching of the SLS with other materials. It is desirable to grow an SLS on a material which has $a^{\|}$ lattice constant equal to the $a^{\|}$ of the SLS. However, $a^{\|}$ will not necessarily coincide with that of a commercially available substrate. If this is the case for the SLS of interest, it is desirable to grow an intermediate graded layer (or buffer layer) between the substrate and the SLS such that the portion of the graded (buffer) layer which is adjacent to the SLS is also lattice matched to the SLS. This type of structure is also indicated schematically in Fig. 12–1. A consequence of the presence of a mismatch in the graded buffer layers is that misfit dislocations are typically generated in these regions. Although these defects tend to propagate upward through the graded region during crystal growth, it has been observed that SLSs tend to act as barriers against dislocation propagation.[13–15] As a result, it is possible to obtain high-quality SLS material even when this material is grown on buffer layers with high defect densities. It should be noted here that buffer layers are not needed if the SLS as a whole is sufficiently thin. In this case, the $a^{\parallel}$ value of Eq. 12–1 is not valid and the entire SLS strains to match up its $a^{\parallel}$ with the lattice constant of the substrate. This approach is clearly most appropriate for structures with one or a few strained quantum-well layers.

The absence of misfit dislocations in the layers of SLSs and the presence of layer strain can be experimentally verified by a number of techniques. Transmission electron microscopy (TEM) can be used to examine the SLS layers on an atomic scale. Figure 12–2 is a TEM micrograph of a GaAs/$In_{0.2}Ga_{0.8}As$ SLS with 23-nm layer thicknesses and ~1.3-percent lattice mismatch between the layers. The strained layers of Fig. 12–2 are clearly free of misfit dislocations. Photoluminescence imaging techniques can be used to look for dark line defects (resulting from dislocations) over a much larger length scale.[15] Entire wafers can be evaluated using the luminescence approach. Layer strains (and compositions) can be obtained from an analysis of high-resolution X-ray scattering measurements.[16] Both $a^{\parallel}$ and $a^{\perp}$ components can be obtained in this fashion. Ion channeling techniques also allow determination of the layer strains.

Fig. 12–2. Cross-section transmission electron micrograph of a $In_{0.2}Ga_{0.8}As$/GaAs SLS. The layers are roughly 20 nm thick; the darker bands are the InGaAs regions.

12.2.2 Critical Layer Thicknesses

The SLS structure described in the previous section is only valid when the SLS layer thicknesses are less than the critical layer thickness, h_c. This parameter therefore is a crucially important parameter for SLS structural design. Matthews obtained a theoretical result for h_c using energy balance arguments.[4] The h_c value (approximately) is inversely proportional to the lattice mismatch in this model. Initial experimental studies were in good numerical agreement with this theoretical prediction.[4,15,17] However, a controversy now exists over both the experimental values and the theoretical basis for the h_c values. This began with the work on SiGe/Si SLSs by People and Bean[18] and has extended into the results for the InGaAs/GaAs system.[19,20] Significantly larger h_c values than expected from Matthews' work are apparently observed by some workers, and People has developed an equation based on a modified energy balance argument which produces larger h_c values.[18] Although widespread agreement on h_c values has not yet been obtained, recent work has emphasized that spuriously large h_c values can be obtained when an experimental technique does not have sufficient sensitivity to observe low densities of misfit dislocations.[21] High-sensitivity photoluminescence imaging studies have recently reconfirmed the Matthews result for the InGaAs/GaAs system.[22] In addition, recent theoretical work has provided an alternative explanation for the SiGe results based on the physics of Si dislocation generation when the thermodynamic h_c has been exceeded.[23] This model produces a second, metastable "critical thickness" which must also be exceeded (for certain growth conditions) in order for measurable densities of dislocations to be formed on the time scale of the crystal growth. The metastable "critical thickness" so obtained is in good numerical agreement with experimental values.[23]

12.2.3 Strained-Layer Superlattice Stability

The h_c values discussed in the previous section form the boundary between SLS structures which accommodate mismatch entirely by elastic deformation and those which accommodate it in part by misfit dislocations. These values also form the boundary between SLS structures which are thermodynamically stable against dislocation generation and those which are not. This point is perhaps not intuitively obvious, since large elastic layer deformations will clearly result in an increased strain energy with respect to the unstrained layer. The relevant comparison for SLS stability, however, is between the energy of the strained layer and that of a layer containing misfit dislocations. These dislocations also cost energy to generate. For thin layers, it costs less energy to strain the superlattice layers than it does to generate defects. This concept has been known from the earliest consideration of strained-layer epitaxy.[2–4,24] Nevertheless, confusion over this point occurred in the early stages of SLS experimental work.[25] Of course, all superlattice structures grown from miscible layer materials, whether lattice matched or not, are in principle unstable against interlayer diffusion.

A more relevant stability issue is whether or not the strain energy in the SLS layers causes these materials to be more easily damaged when subjected to various forms of stress or processing. Early studies of SLS degradation during CW-photopumped operation were interpreted as evidence for the "intrinsic" instability of SLS materials when subjected to high optical or electrical current densities.[25] Other studies, however, demonstrated that SLS materials, under certain conditions, can withstand ion-implantation damage and subsequent thermal annealing at the growth temperature.[26] In addition, a recent study of radiation effects in GaP/GaAsP SLSs indicate that the carrier removal rate for these structures under gamma and neutron irradiation was in fact less than for similar alloy samples.[27] Strained-layer superlattices have also been observed to tolerate repeated thermal cycling and hydrostatic pressure cycling.[28] Finally, more recent studies of SLSs operated under pulsed laser operation and CW LED operation have provided more favorable results regarding the operational lifetimes of these materials as lasers and emitters.[29–31] The stability of SLS materials against intentional damage seems plausible in light of the small strain energy per atom (10^{-4} to 10^{-3} eV for GaP/GaAsP SLSs) available compared with the large energy per atom required to remove an atom from its lattice site in the material (a few eV). The accumulated experimental evidence has shown that SLS structures can be quite robust and are suitable for use in electronic and optoelectronic devices.

12.3 GaAs-based SLS System

12.3.1 Alloy Combinations

The ternary GaAs-based strained layer systems of primary interest are GaAsP, InGaAs, GaAsSb, and combinations including AlGaAs, InAlAs and InGaP. Much of the early work on SLS material properties was carried out on the GaP/GaAsP and GaAs/InGaAs systems. On the other hand, current work has primarily shifted to the AlGaAs/InGaAs system. This latter combination extends the range of obtainable bandgap, barrier height and carrier effective mass beyond those possible from the lattice matched

GaAs/AlGaAs system. In addition, the large range of lattice mismatch values available (~7 percent) in the AlGaAs/InGaAs system allows the study of a variety of strain-induced material and device properties over a wide range of layer distortions.

12.3.2 Tailorable SLS Bandgaps and Optical Properties

All quantum-well and superlattice structures exhibit bandgaps which in general can depend on layer thicknesses through the quantum size effect. The particular gap values also depend on the conduction and valence band energies of the layer materials and the way in which the two sets of bulk energy bands line up at the layer interfaces.

For SLSs it is important to consider the additional effects of the large layer strains.[5,32] These strains produce tetragonal distortions, which are known to shift bulk energy levels and to split certain band degeneracies. The strain shifts have a direct effect on the quantum-well structure and thus on the energy levels of the SLS. Since layer strains are also related to the in-plane lattice constant of the SLS, it is possible to vary the band gap and the lattice constant independently for SLSs made out of ternary materials. This was first demonstrated experimentally for the GaP/GaAsP SLS system.[33] The comparison of theoretical and experimental results from this study is shown in Fig. 12–3. The experimental results (given by the diamonds and circles) represent a combination of photoluminescence and photocurrent measurements of the SLS bandgaps. Experimental values of the SLS lattice constant were obtained from X-ray diffraction measurements. Although only one set of SLS E_g results is given in the figure, it is in fact possible to obtain any of the E_g values between those shown for the SLS and those of the bulk alloy GaAsP. The "quarternarylike" flexibility that this property gives is especially interesting since it is not exhibited by the individual ternary materials which form the layers of the SLS. A number of new electronic and optoelectronic applications are now possible for mismatched heterojunction systems in the III-V, II-VI, and group IV families as a result of this flexibility.

The effects of layer strain on the bandgaps of a number of SLS material systems have been examined theoretically and experimentally.[5,32,34–40] These strain contributions can lead to either increases or decreases in the bandgap for different material systems. The possibility of significant strain-induced reductions in the bandgaps of small gap SLSs has led to the proposal of new III-V SLS materials based on the InAsSb system.[41] Strain-induced reduction of the bandgap has been observed experimentally in the GaP/GaAsP[33] GaSb/AlSb,[37] GaAs/InAlAs,[40] and Si/SiGe SLS systems.[39] Strain-induced increases have been observed in the GaAs/GaAsP[35] and GaAs/InGaAs[36,42] SLS systems. Layer strains clearly provide a mechanism

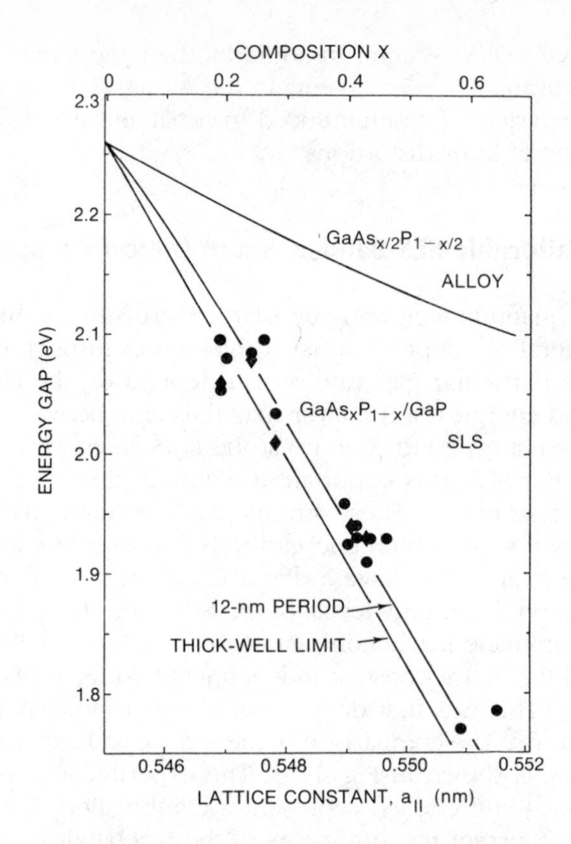

Fig. 12–3. Experimental and theoretical bandgap values for GaAsP/GaP SLSs as a function of lattice constant. The circles (diamonds) are photoluminescence (photocurrent) data, respectively. The result for the GaAsP alloy is also given. (*After Biefeld* et al.[33])

for tailoring certain optical properties of superlattice and quantum-well structures.

12.3.3 Tailorable Transport Properties

Electron transport properties in GaAs/AlGaAs heterostructures have been the subject of many investigations. More specifically, high-mobility modulation doped heterostructures have been intensively studied due to the interesting transport physics and potential field-effect–transistor applications involved. GaAs-based strained-layer structures can also exhibit improved electron mobilities due to modulation doping. For example, modulation-doped GaAs/InGaAs SLSs have exhibited enhanced electron mobilities.[43] It should be noted that electron conduction occurs in the InGaAs alloy. The physics of 2-D electron transport in the direct-gap GaAs-based strained-layer systems studied thus far is qualitatively similar to that of the GaAs/AlGaAs system.

Hole transport in the GaAs/AlGaAs system has been less extensively examined due to the relatively poor hole transport properties. Holes in bulk GaAs tend to preferentially populate the heavy hole bands due to the larger density of states, and the large masses of these bands cause the hole mobilities to be relatively small. However, breaking the cubic symmetry of the III-V material causes the two pairs of valence bands to split apart in energy. Strained-layer superlattices contain tetragonally distorted layers, and both the layered nature of the SLS and the layer strains contribute to the valence band splitting. The symmetry breaking causes the valence band dispersion in the wavevector plane parallel to the (100) SLS layer planes to be substantially modified from the bulk values.[44] For example, the $m_j = \pm 3/2$ bands have light effective mass values over an energy range which is significantly less than the zone center valence band splitting. These modifications to the valence band dispersion are illustrated schematically in Fig. 12–4 for the case of a biaxially compressed bulk III-V material. Strained-layer superlattice systems which have hole confinement in biaxially compressed layers can have preferential hole population of the 2-D light-mass portion of the valence bands. These holes would be expected to exhibit enhanced in-plane mobilities due to the lighter mass values. The ability to tailor hole masses in this fashion is of both scientific and device interest.[45] Some of the physics issues include (*a*) how small

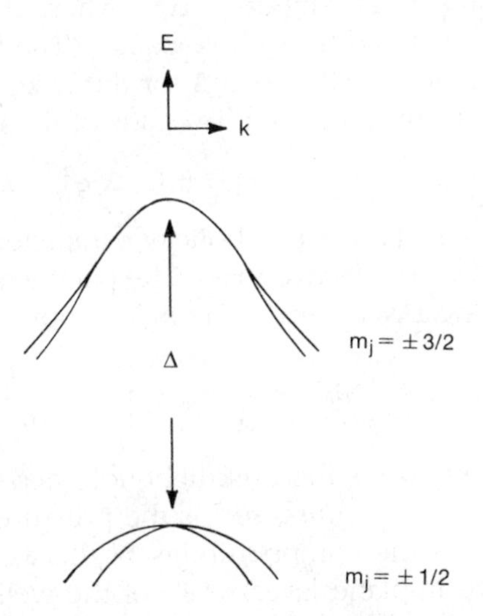

Fig. 12–4. Schematic of a valence band energy versus wavevector relation for a biaxially compressed bulk III-V semiconductor. The wavevector lies in the plane of the compression. This schematic is appropriate for relatively thin-layered (100) structures.

can the hole masses be, (*b*) how the hole masses depend on hole energy, carrier density, valence band splitting, etc., and (*c*) what high-field effects do to the hole transport?

Theoretical estimates of the values of the high stress can be obtained using $\mathbf{k} \cdot \mathbf{p}$ techniques and result in a limit to the zone center hole masses in biaxially compressed material.[44,45] The calculated value for GaAs is $\sim 0.10 m_e$. This value is to be compared with the bulk heavy hole mass value of $\sim 0.4\ m_e$. Experimental studies to directly measure the mass of the preferentially populated light hole bands have focused on the GaAs/InGaAs and the GaAs/GaAsP SLS systems.[46–49] Holes are localized in the biaxially compressed InGaAs alloy layers for the first case and in the biaxially compressed GaAs layers in the second case. Shubnikov-deHaas measurements on a variety of GaAs/InGaAs SLSs have directly demonstrated that the holes preferentially populate valence bands with small masses in the $0.11–0.17 m_e$ range. Magneto-optic studies in both material systems have confirmed the light-mass character of the near-zone center parts of the valence bands.[49]

Detailed band structure studies have been carried out to determine the behavior of the hole (cyclotron) masses with varying strain, valence splitting and hole energy for relatively thin layered SLS structures.[45,47] The calculated mass values were found to vary linearly with energy, i.e.,

$$m^*(\epsilon) \cong m_0^*[1 + 2(c/\Delta)\epsilon], \tag{12–3}$$

over an energy range $\epsilon < \Delta/2$, where Δ is the zone center splitting of the $n = 1$ light and heavy hole bands. (The $n \geq 2$ subbands occur at energies significantly greater than Δ for the cases examined.) The mass of Eq. 12–3 results from nonparabolic bands of the form

$$\epsilon[1 + (c/\Delta)\epsilon] = \hbar^2(k^{\parallel})^2/2m_0^*, \tag{12–4}$$

where c is the nonparabolicity parameter, k is the 2-D wavevector, and m_0^* is the $k = 0$ effective mass. The parameter c has a value near unity for the GaAs/InGaAs system.[47] The m_0^* is approximately

$$m_0^* \cong \frac{4}{3} m_{\text{lh}}^*(x) \left[1 + \left(\frac{1}{m_{\text{lh}}^*}\frac{dm_{\text{lh}}^*}{dP}\right)\frac{1}{K}\left|\frac{\Delta a^{\parallel}}{a^{\parallel}}\right|\right], \tag{12–5}$$

where $m_{\text{lh}}^*(x)$ is the bulk light-hole mass of the well with InAs mole fraction x, $(1/m_{\text{lh}}^*)(dm_{\text{lh}}^*/dP)$ is the pressure derivative of the bulk light-hole mass, K is the compressibility of the well layers, and $\Delta a^{\parallel}/a^{\parallel}$ is the compressive in-plane layer strain of the well layers (which approximates the hydrostatic volume change for our case of biaxial stress). Equation 12–3 can be used to directly compare hole masses with those measured by Shubnikov-deHaas measurements by setting $\epsilon = \epsilon_{\text{Fermi}}$.

The important feature to note about Eq. 12–3 is that the nonparabolicity contribution to the 2-D hole mass is inversely proportional to the relatively small and tailorable Δ value. This is in contrast to bulk light-hole and electron masses, for which the nonparabolicities scale inversely with the bandgap. As a result, the nonparabolicity contribution to the SLS hole mass can be quite significant even for Fermi levels which are a few meV below the valence band maxima. Thus, the 2-D hole masses can be intentionally varied by changing either Δ or the 2-D hole density. It should be noted that the mass expression in Eq. 12–3 in principle is valid for lattice matched systems with large valence band offsets, thin well layers, and low hole density. The Δ value in these latter systems is determined entirely by quantum size effects. In contrast, the Δ values for GaAs/InGaAs SLSs result primarily from the strain splittings and the band offsets of the GaAs/InGaAs system. Our understanding of the physics of the hole mass values is now sufficient to allow the design of specific strained quantum-well and SLS samples with tailored mass values. For example, Fig. 12–5 illustrates the carrier concentration dependence of the light hole mass value in a particular GaAs/strained InGaAs well/GaAs heterostructure. The ability to control the hole mass value over large ranges could provide a useful new tool for studying the physics of hole transport.

12.3.4 Devices

The flexibility in the choice of SLS layer materials and the tailorability of the structural, electrical, and optical properties suggest that there are many potential uses for SLS structures. A number of prototype devices have in fact been successfully fabricated in the GaAsP,[50] InGaAs,[50–52] and GaAsSb[53] SLS systems. The motivation of the early work was to demon-

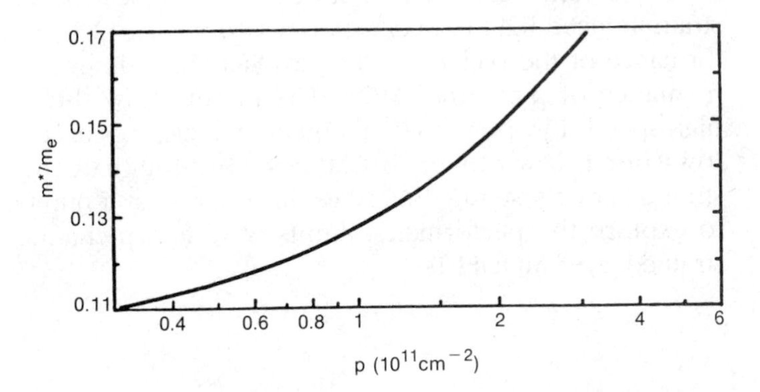

Fig. 12–5. Calculated cyclotron hole mass at the Fermi level as a function of two-dimensional hole concentration for a 10-nm strained $In_{0.15}Ga_{0.85}As$ quantum well. A constant which determines the magnitude of the nonparabolicity (parameter *c* in Eq. 12–3) has been fit to an experimental mass value at $p = 2.5 \times 10^{11}$ cm^{-2}.

strate that good device characteristics could be obtained when the active region of the device contained multiple interfaces with lattice mismatches in the percent range. Devices made in the InGaAs/AlGaAs SLS system include efficient photodetectors, infrared LEDs, pulsed and CW injection lasers and modulation-doped field-effect transistors. All of these devices contained lattice mismatches of ~1.4 percent except for the pulsed injection laser which contained ~2.5 percent.[51] The photodetector work included both ion implanted structures[54] and as-grown diode structures.[52]

Following the success of these studies, the motivation of recent SLS device research has shifted toward demonstrating specific devices which offer superior performance for a specific application. One interesting example of this new trend is the study of n-type modulation doped field-effect transistors employing strained InGaAs quantum wells.[55–58] These structures contain the two-dimensional electron gas in the strained quantum well rather than in a GaAs layer. An important advantage of this approach is that a smaller Al concentration alloy can be used for a given barrier height than is required for the GaAs/AlGaAs heterostructure system. This in turn allows the device to be free of persistent photoconductivity (PPC) problems that plague the higher Al content MODFETs. In addition, the InGaAs layer has a smaller electron effective mass than does GaAs, so that higher-speed device operation may be possible with the InGaAs-based transistors. Several research groups are now studying these novel device structures, and encouraging transconductance[55–58] and high-speed device characteristics[58] have already been reported for devices which do not exhibit PPC problems at low temperatures.

Another interesting example is the recent development of p-type modulation doped field-effect transistors which also employ strained InGaAs quantum wells.[59] These devices are designed to take advantage of the strain-induced light-hole effect discussed in Sec. 12.3.3 to improve the performance of the p-channel devices. Significant improvements in the performance of p-channel MODFETs might allow the development of a high-speed, low-power complementary-logic approach (analogous to the low-speed, low-power Si CMOS technology) in the AlGaAs/InGaAs strained-layer system. Studies are in progress at a number of laboratories to explore the performance limits of n- and p-channel AlGaAs/InGaAs strained-layer MODFETs.

References

1. See, for example, A. G. Milnes and D. L. Feucht, *Heterojunctions and Metal-Semiconductor Junctions,* New York: Academic Press (1972).

2. F. C. Frank and J. H. van der Merwe, *Proc. Roy. Soc. London,* Ser. A; 198:216 (1949).
3. J. H. van der Merwe, *J. Appl. Phys.,* 34:117 (1963).
4. J. W. Matthews and A. E. Blakeslee, *J. Cryst. Growth,* 27:118 (1974); J. W. Matthews and A. E. Blakeslee, *J. Cryst. Growth,* 29:273 (1975); J. W. Matthews and A. E. Blakeslee, *J. Cryst. Growth,* 32:265 (1976).
5. G. C. Osbourn, *J. Appl. Phys.,* 53:1586 (1982).
6. G. C. Osbourn, R. M. Biefeld, and P. L. Gourley, *Appl. Phys. Lett.,* 41:172 (1982).
7. G. C. Osbourn, *J. Vac. Sci. Technol.,* 21:469 (1982).
8. I. J. Fritz, R. M. Biefeld, and G. C. Osbourn, *Solid State Commun.,* 45:323 (1983).
9. P. L. Gourley, R. M. Biefeld, G. C. Osbourn, and I. J. Fritz, *Proc. 1982 Intl. Symp. GaAs Rel. Comp.,* Berkshire: Institute of Physics, 248 (1983).
10. I. J. Fritz, L. R. Dawson, G. C. Osbourn, P. L. Gourley, and R. M. Biefeld, *Proc. 1982 Intl. Symp. GaAs Rel. Comp.* Berkshire: Institute of Physics, 241 (1983).
11. G. C. Osbourn, *J. Vac. Sci. Technol. B,* 1:379 (1983).
12. S. T. Picraux, R. M. Biefeld, L. R. Dawson, G. C. Osbourn, and W. K. Chu, *J. Vac. Sci. Technol. B,* 1:687 (1983).
13. J. W. Matthews, A. E. Blakeslee, and S. Mader, *Thin Solid Films,* 33:253–266 (1976).
14. M. A. Tischler, T. Katsuyama, N. A. El-Masry, and S. M. Bedair, *Appl. Phys. Lett.,* 46:294–296 (1985).
15. P. L. Gourley, R. M. Biefeld, and L. R. Dawson, *Appl. Phys. Lett.,* 47:482–484 (1985).
16. V. S. Speriosu, M. A. Nicolet, S. T. Picraux, and R. M. Biefeld, *Appl. Phys. Lett.,* 45:223 (1984).
17. I. J. Fritz, S. T. Picraux, L. R. Dawson, T. J. Drummond, W. D. Laidig, and N. G. Anderson, *Appl. Phys. Lett.,* 46:967–969 (1985).
18. R. People and J. C. Bean, *Appl. Phys. Lett.,* 47:322 (1985).
19. P. J. Orders and B. I. Usher, *Appl. Phys. Lett.,* 50:980 (1987).
20. D. J. Lockwood, M. W. C. Dharma-Wardana, W. T. Moore, and R. L. S. Devine, *Appl. Phys. Lett.,* 51:361 (1987).
21. I. J. Fritz, *Appl. Phys. Lett.,* 51:1080 (1987).
22. P. L. Gourley, I. J. Fritz and L. R. Dawson, unpublished.
23. B. W. Dodson and J. Y. Tsao, *Appl. Phys. Lett.,* 51:1325 (1987).
24. J. H. van der Merwe, *CRC Crit. Rev. Solid State Mat. Sci.,* 7:209–231 (1978).

25. M.J. Ludowise, W.T. Dietze, C.R. Lewis, N.D. Camras, N. Holonyak, B.K. Fuller, and M.A. Nixon, *Appl. Phys. Lett.,* 42:487 (1983).
26. D.R. Myers, R.M. Biefeld, I.J. Fritz, S.T. Picraux, and T.E. Zipperian, *Appl. Phys. Lett.,* 44:1052–1054 (1984).
27. C.E. Barnes, G.A. Samara, R.M. Biefeld, T.E. Zipperian, and G.C. Osbourn, *Proc. 13th Intl. Conf. Defects in Semiconductors,* 471–477 (1984).
28. E.D. Jones, J.E. Schirber, I.J. Fritz, P.L. Gourley, R.M. Biefeld, L.R. Dawson, and T.J. Drummond, *Mat. Res. Soc. Symp. Proc.,* 56:241 (1986).
29. W.D. Laidig, P.J. Caldwell, Y.F. Lin, and C.K. Peng, *Appl. Phys. Lett.,* 44:653–655 (1984).
30. H. Temkin and W.T. Tsang, *J. Appl. Phys.,* 55:1413–1415 (1984).
31. T. Katsuyama, Y.J. Yang, D. Moore, N. Koram, and S. Bedair, *IEEE Trans. Electron Dev.,* ED-34:2379 (1987).
32. G.C. Osbourn, *Mat. Res. Soc. Symp. Proc.,* 25:455 (1984).
33. R.M. Biefeld, P.L. Gourley, I.J. Fritz, and G.C. Osbourn, *Appl. Phys. Lett.,* 43:759–761 (1983).
34. N.G. Anderson, W.D. Laidig, G. Lee, Y. Lo, and M. Ozturk, *Mat. Res. Soc. Symp. Proc.,* 37:223–238 (1985).
35. P.L. Gourley and R.M. Biefeld, *Appl. Phys. Lett.,* 45:749–751 (1984).
36. J.Y. Marzin and E.V.K. Rao, *Appl. Phys. Lett.,* 43:560–562 (1983).
37. P. Voison, C. Delalande, M. Voos, L.L. Chang, A. Segmuller, C.A. Chang, and L. Esaki, *Phys. Rev. B,* 30:2276–2278 (1984).
38. G. Abstreiter, H. Brugger, T. Wolf, H. Jorke, and H.J. Herzog, *Phys. Rev. Lett.,* 54:2441–2444 (1985).
39. R. People and J.C. Bean, *Appl. Phys. Lett.,* 48:538–540 (1986).
40. H. Kato, N. Iguchi, S. Chika, M. Nakayama, and N. Sano, *J. Appl. Phys.,* 59:588–592 (1986).
41. G.C. Osbourn, *J. Vac. Sci. Technol. B,* 2:176–178 (1984).
42. I.J. Fritz, L.R. Dawson, G.C. Osbourn, P.L. Gourley, and R.M. Biefeld, *Proc. 1982 Intl. Symp. GaAs Rel. Comp.,* 241–247 (1983).
43. I.J. Fritz, L.R. Dawson, and T.E. Zipperian, *Appl. Phys. Lett.,* 43:846 (1983).
44. G.L. Bir and G.E. Pikus, *Symmetry and Strain-induced Effects in Semiconductors,* New York: John Wiley & Sons (1974).
45. G.C. Osbourn, *Mat. Res. Soc. Symp. Proc.,* 37:219 (1985).
46. J.E. Schirber, I.J. Fritz, and L.R. Dawson, *Appl. Phys. Lett.,* 46:187 (1985).
47. G.C. Osbourn, J.E. Schirber, T.J. Drummond, L.R. Dawson, B.L. Doyle, and I.J. Fritz, *Appl. Phys. Lett.,* 49:732 (1986).

48. I.J. Fritz, J.E. Schirber, E.D. Jones, T.J. Drummond, and G.C. Osbourn, *Proc. 1986 Intl. Symp. GaAs Rel. Comp.*, Berkshire: Institute of Physics (1987).

49. E.D. Jones, H. Ackermann, J.E. Schirber, T.J. Drummond, L.R. Dawson, and I.J. Fritz, *Appl. Phys. Lett.*, 47:492 (1985).

50. G.C. Osbourn, P.L. Gourley, I.J. Fritz, R.M. Biefeld, L.R. Dawson, and T.E. Zipperian, *Semiconductors and Semimetals,* ed. R. Dingle, 24:459 (1987).

51. W.D. Laidig, P.J. Caldwell, Y.F. Lin, and C.K. Peng, "Strained-layer quantum-well injection laser," *Appl. Phys. Lett.*, 44:653–655 (1984).

52. T.E. Zipperian, L.R. Dawson, C.E. Barnes, J.J. Wiczer, and G.C. Osbourn, *Proc. IEEE Intl. Electron Dev. Mtg.*, 696–699 (1983).

53. J. Klem, R. Fischer, W.T. Masselink, W. Kopp, and H. Morkoç, *J. Appl. Phys.*, 55:3843–3845 (1984).

54. G.E. Bulman, D.R. Myers, J.J. Wiczer, L.R. Dawson, R.M. Biefeld, and T.E. Zipperian, *Proc. IEEE Intl. Electron Dev. Mtg.*, 719–722 (1984).

55. W.T. Masselink, A. Ketterson, J. Kelm, W. Kopp, and H. Morkoç, *Electron. Lett.*, 21:937 (1985).

56. T.E. Zipperian and T.J. Drummond, *Electron. Lett.*, 21:823 (1985).

57. J.J. Rosenberg, M. Benlamri, P.K. Kirchner, J.J. Woodall, and G.D. Petit, *IEEE Dev. Res. Conf. Tech. Dig.*, EDL-6:491 (1985).

58. T. Henderson, M.I. Aksun, C.K. Peng, H. Morkroç, PC. Chao, P.M. Smith, K.H.G. Duh, and L.F. Lester, *Proc. IEEE IEDM,* New York: IEEE, 464 (1986).

59. T.J. Drummond, T.E. Zipperian, I.J. Fritz, J.E. Schirber, and T.A. Plut, *Appl. Phys. Lett.*, 49:461 (1986).

Index

E

F

G

Q

R

S